Biosorption Processes for Heavy Metal Removal

Pinki Saini
University of Allahabad, India

A volume in the Advances in Environmental Engineering and Green Technologies (AEEGT) Book Series

IGI Global
PUBLISHER of TIMELY KNOWLEDGE

Published in the United States of America by
IGI Global
Engineering Science Reference (an imprint of IGI Global)
701 E. Chocolate Avenue
Hershey PA, USA 17033
Tel: 717-533-8845
Fax: 717-533-8661
E-mail: cust@igi-global.com
Web site: http://www.igi-global.com

Library of Congress Cataloging-in-Publication Data

CIP PENDING
Biosorption Processes for Heavy Metal Removal
2024 Engineering Science Reference

9798369316184 (hc)
9798369345139 (sf)
9798369316191 (electronic)

This book is published in the IGI Global book series Advances in Environmental Engineering and Green Technologies (AEEGT) (ISSN: 2326-9162; eISSN: 2326-9170)

British Cataloguing in Publication Data
A Cataloguing in Publication record for this book is available from the British Library.

For electronic access to this publication, please contact: eresources@igi-global.com.

Advances in Environmental Engineering and Green Technologies (AEEGT) Book Series

Sang-Bing Tsai
Zhongshan Institute, University of Electronic Science and Technology of China, China & Wuyi University, China
Ming-Lang Tseng
Lunghwa University of Science and Technology, Taiwan
Yuchi Wang
University of Electronic Science and Technology of China Zhongshan Institute, China

ISSN:2326-9162
EISSN:2326-9170

Mission

Growing awareness and an increased focus on environmental issues such as climate change, energy use, and loss of non-renewable resources have brought about a greater need for research that provides potential solutions to these problems. Research in environmental science and engineering continues to play a vital role in uncovering new opportunities for a "green" future.

The **Advances in Environmental Engineering and Green Technologies (AEEGT)** book series is a mouthpiece for research in all aspects of environmental science, earth science, and green initiatives. This series supports the ongoing research in this field through publishing books that discuss topics within environmental engineering or that deal with the interdisciplinary field of green technologies.

Coverage

- Contaminated Site Remediation
- Renewable Energy
- Alternative Power Sources
- Air Quality
- Water Supply and Treatment
- Waste Management
- Cleantech
- Sustainable Communities
- Pollution Management
- Green Transportation

IGI Global is currently accepting manuscripts for publication within this series. To submit a proposal for a volume in this series, please contact our Acquisition Editors at Acquisitions@igi-global.com or visit: http://www.igi-global.com/publish/.

The Advances in Environmental Engineering and Green Technologies (AEEGT) Book Series (ISSN 2326-9162) is published by IGI Global, 701 E. Chocolate Avenue, Hershey, PA 17033-1240, USA, www.igi-global.com. This series is composed of titles available for purchase individually; each title is edited to be contextually exclusive from any other title within the series. For pricing and ordering information please visit http://www.igi-global.com/book-series/advances-environmental-engineering-green-technologies/73679. Postmaster: Send all address changes to above address.

Titles in this Series

For a list of additional titles in this series, please visit: http://www.igi-global.com/book-series/advances-environmental-engineering-green-technologies/73679

Agriculture and Aquaculture Applications of Biosensors and Bioelectronics
Alex Khang (Global Research Institute of Technology and Engineering, USA)
Engineering Science Reference • © 2024 • 576pp • H/C (ISBN: 9798369320693) • US $265.00

Innovations in Engineering and Food Science
Shilpa Mehta (Auckland University of Technology, New Zealand) Fakhar Islam (NUR International University, Pakistan) and Ali Imran (Government College University, Pakistan)
Engineering Science Reference • © 2024 • 424pp • H/C (ISBN: 9798369308196) • US $265.00

Modeling and Monitoring Extreme Hydrometeorological Events
Carmen Maftei (Transilvania University of Brasov, Romania) Radu Muntean (Transilvania University of Braşov, Romania) and Ashok Vaseashta (International Clean Water Institute, USA)
Engineering Science Reference • © 2024 • 338pp • H/C (ISBN: 9781668487716) • US $250.00

Novel AI Applications for Advancing Earth Sciences
Sudesh Yadav (Department of Higher Education, Government PG College, Ateli, India) Satya Prakash Yadav (GL Bajaj Institute of Technology and Management, India) Pethuru Raj (Edge AI Division, Reliance Jio Platforms Ltd., Bangalore, India) Prabhakar Tiwari (Madan Mohan Malaviya University of Technology, India) and Victor Hugo C. de Albuquerque (Department of Teleinformatics Engineering (DETI), Federal University of Ceará, Brazil)
Engineering Science Reference • © 2024 • 405pp • H/C (ISBN: 9798369318508) • US $265.00

Exploring Ethical Dimensions of Environmental Sustainability and Use of AI
Hemachandran Kannan (Woxsen University, India) Raul Villamarin Rodriguez (Woxsen University, India) Zita Zoltay Paprika (Corvinus University of Budapest, Hungary) and Abejide Ade-Ibijola (University of Johannesburg, South Africa)
Engineering Science Reference • © 2024 • 426pp • H/C (ISBN: 9798369308929) • US $265.00

Perspectives on Global Biodiversity Scenarios and Environmental Services in the 21st Century
Naveen Kumar Chourasia (Government College, Bichhua, India) and Kavita Chahal (Government College, Bichhua, India)
Engineering Science Reference • © 2023 • 294pp • H/C (ISBN: 9781668490341) • US $250.00

IGI Global
PUBLISHER of TIMELY KNOWLEDGE

701 East Chocolate Avenue, Hershey, PA 17033, USA
Tel: 717-533-8845 x100 • Fax: 717-533-8661
E-Mail: cust@igi-global.com • www.igi-global.com

To my dearest son, Tanmay, whose unwavering support and intellect have been the driving force behind this academic endeavor.

With love and pride.

Pinki Saini

Table of Contents

models and methods in the evaluation of heavy metals biosorption can elucidate mechanisms, analyze experimental data, predict outcomes, and optimize processes. After a brief analysis of the literature in the field of biosorption, it has been found that the most valuable models provide information about process kinetics, maximum biosorption capacity and thermodynamics. Additionally, the preferred methods for studying the interactive effects of process variables and creating a precise mathematical model that describes the process accurately are highly sought after. Thus, the main purpose of the chapter consists in exposing the available mathematical models and methods that can be applied for the evaluation of all factors that could affect the biosorption of heavy metals. The isotherms, kinetic and thermodynamic models, as well as the statistical tools suitable for evaluating biosorption were presented, with a special emphasis on their application, advantages, as well as disadvantages.

Heavy metals are threatening to human health as they are arduous to remove from water bodies. Many technologies like photocatalysis, reverse osmosis, precipitation, and adsorption are used to get rid of those heavy metals from water. Among all these methods, adsorption is one of the most efficient methods to remove those heavy metals. In this chapter, the adsorption of heavy metals by dead biomass like algae, fungi, bacteria, and other biomass from waste will be discussed elaborately. The mechanisms of adsorption by these biomasses will be discussed, and the factors affecting their adsorption efficiencies are also included. The newest approach and emerging techniques are also considered. The future aspects and limitations are also considered.

The treatment of heavy-metal-bearing wastewaters is a critical environmental challenge, as access to clean drinking water becomes increasingly difficult due to pollution and rising water treatment costs. Traditional adsorbents derived from fossil fuels are not sustainable and can contribute to secondary pollution. To address this issue, researchers have turned to biogenic sorbents made from modern biomass as a promising alternative. These biosorbents utilize biological waste materials that would otherwise contaminate water systems, promoting a circular economy and sustainable water treatment practices. This chapter explores the potential of biogenic sorbents for water decontamination, focusing on their use in the removal of heavy metals from contaminated waters. Further, this chapter delves into the preparation methods, adsorbent types, adsorption mechanisms, and regeneration techniques employed with these biosorbents.

Biosorption technique is a growing alternative to existent conventional treatment techniques used in heavy metal remediation and associated with the production of toxic sludge. This chapter demonstrated the use of the technique and its processes, the biosorbents used, and the sorbates that they target. Evidently, many plant by-products, bacteria, fungi, and algae species are used in the process in their living or dead forms to take up a variety of heavy metals. The process is regulated by factors such as biosorbent dosage, temperature, pH, and initial metal concentration. There is need to advance research on the technique to optimize its metal removal efficacy.

Agricultural wastes have excellent potential as substitute biosorbents to reduce pollution in the environment. These include organic contaminants, pesticides, dyes, heavy metals, and other impurities from soil and water. In the chapter, the authors focused on the adsorption capacity, remediation mechanism, and possible uses of agricultural wastes as biosorbents as well as their viability and efficacy. The biosorption capabilities of a variety of agricultural wastes, including crop residues, fruit peels, rice husks, and sugarcane bagasse, has been extensively studied. The processes of biosorption generally encompass ion exchange, complexation, physical adsorption, and chemical interactions between the target pollutants and the functional groups on the biosorbents. This study emphasizes the use of agricultural waste as an efficient alternative for artificial absorbents and promotes a circular economy by addressing environmental issues. Thus, the development of effective biosorption techniques using agricultural wastes is a pressing need for achieving environmental sustainability.

Environmental pollution poses a significant threat to ecosystems and human health worldwide. Traditional methods of environmental remediation often fall short in addressing the complexities of contemporary pollutants. The emergence of genetic engineering technologies has opened new avenues for innovative solutions to environmental challenges. This chapter aims to explore the potential of genetically engineered bacteria as a cutting-edge tool for environmental remediation. By delving into the principles, applications, and concerns surrounding this technology, the chapter seeks to contribute to the understanding of its feasibility and sustainability. However, as we navigate the frontiers of this innovative approach, it is imperative to acknowledge and address the environmental concerns and conduct thorough risk assessments associated with the deployment of genetically engineered bacteria. Ecological impact, potential risks to

Preface

Heavy metals are persistent and non-degradable pollutants that can have devastating ecological impacts when released into the environment. Municipal and industrial wastewaters are particularly prone to contamination by these hazardous substances, presenting a significant challenge to water treatment technologies. Effective and affordable effluent treatment and metal recovery technologies are therefore essential for managing contaminated wastewater, not only to meet regulatory requirements but also to tap into the potential resource of valuable metals that can be recovered from the waste. Biosorption is a fundamental property of living and dead biomass that allows for the rapid binding and concentration of both organic and inorganic compounds from dilute aqueous solutions. In the context of wastewater treatment, biosorption refers to the use of biosorbents - materials derived from non-living biomass - to selectively remove target substances from contaminated water. The application of microorganisms such as bacteria, algae, yeasts, and fungi as biosorbents for heavy metal removal has gained momentum due to their high surface-to-volume ratio, widespread availability, rapid kinetics of adsorption and desorption, and low cost, making them a promising avenue for future wastewater treatment technologies.

The book *Biosorption Processes for Heavy Metal Removal* explores the importance of removing heavy metals from waste streams and offers an overview of biosorbents and biosorption technology, with a focus on the underlying features of biosorption. It compiles a collection of review articles that examine the current understanding of biosorption mechanisms and biochemistry, the performance of biosorbents, and practical considerations for biosorbent preparation and engineering. The book also covers physicochemical evaluations of biosorbents, process optimization, and factors affecting biosorption efficiency. The authors emphasize the need for a comprehensive understanding of the mechanism underlying microbial biosorption of heavy metals and related elements. Additionally, the book provides an overview of biosorption applications for the removal of nutrients, organic pollutants, and metals in wastewater treatment across various contexts.

The book serves as a reference work for scientists, environmental safety engineers, and R&D specialists interested in advancing biosorption research and developing industrial applications for heavy metal separation technologies. Operators of wastewater treatment plants, sludge treatment and disposal facilities, as well as researchers and students in the field of environmental engineering, will also find this book of interest. It is appropriate for Administrators, Policy Makers, Consultants, Industry Persons, Academicians, Scientists, Researchers, graduate and post-graduate students working and studying in the areas of environmental sciences, environment safety, microbial prospecting, biochemical engineering industrial microbiology, food biotechnology, plant biotechnology, microbial biotechnology. The book will serve as an indispensable guide for anyone looking for a comprehensive overview of the subject.

The book is organized into 13 chapters. Brief introductory information and main points of each of the chapters follows:

Chapter 1, "Heavy Metal Pollution and Biosorption," concerns over the increase in environmental pollution caused by heavy metals, posing a significant threat to public health. This issue is worsened by the continuous rise in human activities, particularly industrial processes and urbanization, leading to the uncontrolled release of pollutants into the environment. Elevated levels of heavy metals can harm organisms and plants, disrupting the normal functions of vital organs and glands. Given the ongoing increase in human activities, it is crucial to regularly assess and monitor the levels of heavy metals in the environment. The main focus of this chapter is to provide a detailed discussion on key heavy metals, the environmental pollution they cause in air, soil, and water, and the associated impacts on human health. The chapter also gives an overview of various remedies to address the toxicity effects of heavy metals.

Chapter 2, "Mechanisms of Biosorption: An Overview," primarily concentrates on various techniques of biosorption and their mechanisms employed to eliminate heavy metals from diverse aqueous sources. The process of biosorption for heavy metals is affected by factors like the condition of the biomass, the nature of biomaterials, the chemical characteristics of the metal solution, and environmental conditions such as pH and temperature. The mechanisms involved in biosorption are primarily contingent on the specific type of biosorbent utilized, highlighting the connection between the respective biomass and the sorption of metals.

Chapter 3, "Factors Affecting Biosorption Efficiency: Process Optimization and Performance Evaluation," outlines the mechanisms involved in the biosorption process such as adsorption, absorption, ion exchange, surface complexation, and precipitation. It also details the factors influencing the effectiveness of biosorption including pH, temperature, biosorbent weight, biosorbent particle size, biosorbent concentration, time and agitation, as well as the type and composition of biosorbents and the initial concentration of heavy metals. Additionally, addresses the interference of competing ions and enhancing selectivity for target pollutants are crucial considerations in optimizing biosorption efficiency. The immobilization of heavy metals also plays a significant role in affecting the overall efficiency of the biosorption process. When scaling up biosorption processes, maintaining efficiency and cost-effectiveness presents technical and engineering challenges that need to be addressed.

Chapter 4, "Heavy Metals in Foodstuffs: Presence, Bioaccumulation, Origins, Health Risk, and Remediation," deals with the presence of heavy metal contamination in food items represents a notable global threat to human health. This review delves into the existing knowledge on the origins, mechanisms of bioaccumulation, and the associated health risks posed by heavy metals in diverse food products. A comprehensive approach, merging scientific research, policy interventions, and public awareness, is essential to tackle the issue of heavy metal contamination in food. The chapter emphasizes the need for stringent regulations, the promotion of sustainable agricultural practices, and investments in innovative remediation technologies as pivotal measures to guarantee food safety and protect public health.

Chapter 5, "Mathematical Modeling of Microbial Bioremediation: A Key Step in Process Scale-Up," discusses the available mathematical models and methods applicable for evaluating all factors affecting the biosorption process. The exploration will include isotherms, kinetic and thermodynamic models, as well as relevant statistical tools suitable for assessing biosorption. A special emphasis has been placed on the practical application of these models. The efficiency of biosorption and bioaccumulation processes can be significantly enhanced when implemented under controlled conditions, with knowledge of optimal values for influencing factors and their respective degrees of influence. The judicious selection and application of appropriate models and methods in the evaluation of the process will facilitate the elucidation of process mechanisms, analysis of experimental data, prediction of responses based on operational conditions, and ultimately contribute to process optimization.

Chapter 6, "Various Biosorbents and Their Mechanisms for the Removal of Heavy Metals From Wastewater," gives an overview about bioadsorption being a process wherein non-living biological resources are utilized to immobilize and concentrate heavy metals and metalloids from aqueous solutions. Various types of microorganisms, including algae, fungi, bacteria, and yeast, serve as bioadsorbents in this process. Additionally, many low-cost biomaterials typically considered as waste, such as peat, tree bark, peanut husk, etc., can be employed for the adsorption of heavy metal ions. The chapter explains effectiveness of the adsorption process depends on numerous factors, making the optimization of these factors crucial for achieving the highest efficiencies with different biomaterials. To enhance the effectiveness further, improvements to these biomasses can be made through modifications.

Chapter 7, "Biogenic Sorption: An Intelligent and Sustainable Approach to Reclaiming and Reutilizing Heavy Metals in the Modern Era," delves into the potential of biogenic sorbents for water decontamination, with a specific focus on their application in removing heavy metals from polluted waters. Furthermore, the chapter explores preparation methods, various adsorbent types, adsorption mechanisms, and regeneration techniques associated with these biosorbents. The chapter aims to offer valuable insights into the effectiveness and applicability of biogenic sorbents in water treatment. The economic viability and versatile applications of microbial biosorbents make them a promising solution for addressing heavy metal contamination and mitigating the environmental impacts of industrial activities. The development and use of bacterial biosorbents present a promising avenue for sustainable heavy metal removal and wastewater treatment.

Chapter 8, "Biosorption Processes for Heavy Metal Removal in Aqueous Media," discusses the application of biosorption, a bioremediation technique aimed at removing heavy metals from various environmental compartments. The exploration of prospects in this field could potentially broaden the use of biosorption for addressing pollution on a larger, field-scale level. Bioremediation techniques, including biosorption, are preferred due to their high recovery efficiency, natural and safe characteristics, environmental friendliness, and cost-effectiveness compared to conventional methods. The increasing preference for biosorption in remediating heavy metals from aqueous solutions is evident, driven by its associated advantages. The chapter emphasizes the necessity to transition biosorption experiments from the laboratory to field-scale applications for real-world effectiveness. Achieving such progress involves optimizing the process through chemical treatment of biosorbents, their immobilization using suitable carriers, and genetic modifications to enhance their capability to uptake heavy metal contaminants.

Chapter 9, "Potential of Agricultural Wastes as Alternative Biosorbents: Use of Agriculture Waste as Biosorbent," extensively explores specific facets of this multifaceted investigation, providing a comprehensive understanding of the potential, mechanisms, and practical applications of agricultural waste as biosorbents. This exploration aims to contribute to the necessary paradigm shift for a more sustainable coexistence of agriculture and the environment. Each agricultural waste material has been scrutinized for its unique characteristics, adsorption capabilities, and potential roles in environmental remediation. It also investigates how these biosorbents operate in the removal of pollutants. Their effectiveness in eliminating a broad spectrum of pollutants, ranging from organic contaminants to heavy metals, necessitates a thorough examination of the underlying mechanisms. The chapter sheds light on the transformative potential of biosorbents derived from agricultural waste, bridging the fields of agriculture, waste management, and environmental sustainability.

Chapter 10, "Microbial Architects of Environmental Restoration: Exploring the Frontiers of Genetically Engineered Bacteria in Remediation Strategies," is dedicated to exploring the transformative landscape of environmental bioremediation facilitated by genetically engineered bacteria, with a specific focus on remediating heavy metal pollutants. It emphasizes the ecological and health benefits in comparison to conventional physico-chemical strategies, advocating for a synergistic approach that combines microbiological, ecological, and biochemical insights with innovative field engineering designs. The chapter sheds light on the evolving nature of genetically engineered bacteria as a tool for environmental remediation. While these bacteria offer eco-friendly alternatives, concerns related to genetic pollution and risk mitigation require careful attention, necessitating a delicate balance between environmental benefits and potential risks. The chapter specifically highlights the contribution of molecular genetics to risk mitigation, particularly in addressing environmental and public health concerns.

Chapter 11, "The Potential of Agricultural Waste Chars as Low-Cost Adsorbents for Heavy Metal Removal From Water," addresses the significant challenge of heavy metal contamination in water and explores an eco-friendly and cost-effective solution to tackle water contamination issues. The focus is on utilizing pyrolysis char derived from agricultural waste for efficiently removing heavy metals from water. It offers a comprehensive overview of using pyrolysis char as an adsorbent, emphasizing the optimal agricultural waste types for pyrolysis, detailing the process, and assessing the sustainability of this approach. Additionally, it outlines potential challenges and suggests future research directions in this promising area. The chapter further delves into sources of heavy metal contamination, associated risks, and existing mitigation technologies for treating contaminated waters.

Chapter 12, "Heavy Metal Bioaccumulation in Food Chains and Health Risks," conducts a comprehensive examination of the complex interplay among heavy metal contamination, the food chain, and the resulting health risks to human populations. The investigation involves a multifaceted exploration of the sources, pathways, and mechanisms involved in the bioaccumulation of heavy metals within the food web. The chapter places special emphasis on understanding the dynamic interactions within soil-plant-animal systems that facilitate the uptake and transfer of heavy metals across trophic levels. By amalgamating interdisciplinary insights from environmental science, toxicology, and public health, this book chapter serves as a valuable resource for researchers, policymakers, and practitioners engaged in the crucial task of addressing heavy metal contamination in the food chain.

Chapter 13, "Recovery of Heavy Metals by Biosorption and Regeneration of the Adsorbents," focusses on the regeneration or recovery of various adsorbent materials after the removal of heavy metals from water sources, which is a crucial step to reduce the toxicity of the metal-loaded adsorbent. The recovery of heavy metals and the regeneration of the adsorbent are pivotal considerations for enhancing the efficiency of both the adsorbent and the overall process. It is imperative to address these aspects because the release of metal-loaded adsorbents into the environment is not safe and should only occur after the successful recovery of metals from the spent adsorbents.

Pinki Saini
University of Allahabad, India

Acknowledgment

This reference book stands as a culmination of extensive research, collaboration, and dedication. I wish to express my sincere gratitude to those whose support has been instrumental in the realization of this project.

To my family, whose unwavering support and understanding have sustained me throughout the writing process, I express my deepest gratitude. Your encouragement has been my anchor.

Lastly, I want to thank the publishers and editorial team for their expertise and commitment to excellence. Your dedication to the dissemination of knowledge has made this book possible.

Sincerely,

Pinki Saini
University of Allahabad, India

Chapter 1
Heavy Metal Pollution and Biosorption

Himanshi Srivastava
University of Allahabad, India

Pinki Saini
https://orcid.org/0000-0001-9083-6765
University of Allahabad, India

Anchal Singh
University of Allahabad, India

Sangeeta Yadav
University of Allahabad, India

ABSTRACT

The escalation of environmental pollution by heavy metals has emerged as a global concern in recent years, posing a significant threat to public health. This phenomenon is exacerbated by the ongoing rise in anthropogenic activities, notably industrial processes and urbanization, which indiscriminately release pollutants into the environment without effective control and mitigation measures. Subsequently, these metals traverse the food chain, impacting animals and humans. Elevated levels of heavy metals can have deleterious effects on organisms and plants, disrupting the metabolic functions of vital organs and glands. Continuous assessment and monitoring of heavy metal levels in the environment are imperative due to the escalating anthropogenic activities. The primary objective of this chapter is to provide an illustrative discussion on important heavy metals, environmental pollution caused by heavy metals, encompassing air, soil, and water. Various toxic effects on human health have been considered, and corresponding remedies have been presented.

DOI: 10.4018/979-8-3693-1618-4.ch001

INTRODUCTION

Water is the essential resource required to sustain life on this planet. Water plays a major role in the civilization, growth and economy of developed and developing countries (Paul and Sinha, 2013). Rapid industrialization, increasing population, urbanization and careless utilization of natural resources is one of the most frequent reasons of water pollution and presently being the most emerging problem of each developing country (Carolin et al., 2017; Vardhan et al., 2019). This is due to large amount of industrial effluent, dyes, organic matter, radio-nuclides, pharmaceutical, pesticides, heavy metals, plastics, suspended solids, parasites and pathogens being present and disposed into the natural water bodies (Villarín and Merel, 2020). Sewage and slugs are also part of above environmental pollutants which cause danger to aquatic biota and deteriorates water quality (Sinha and Paul, 2014). Actually, the quality of water is a critical concern for living beings and it is directly associated with human welfare. Industrial waste water is source of heavy metals like cadmium (Cd), chromium (Cr), lead (Pb), arsenic (As), mercury (Hg), cobalt (Co), zinc (Zn), copper (Cu), silver (Ag), iron (Fe) and platinum (Pt). These toxic heavy metals are every day released into the water from different natural and anthropogenic sources.

Among these, some heavy metals like Cu, Fe, Cr and Zn are required as nutrients in trace amounts for different life processes in living organisms including plant, animals and microorganisms but become toxic at higher concentrations (Tchounwou et al., 2012). On the other hand, heavy metals like Pb, As, Hg, and Cd have no biological function (Dudka and Adriano, 1997; Ghannam et al., 2015). These are nonbiodegradable and accumulate in the plants, animal as well as human bodies through food chain and their higher concentration leads to adverse effects (Praveena et al., 2010; Paul and Sinha, 2015; Aze-hEngwa et al., 2019). World health Organization (WHO), Food and Agriculture Organization (FAO), and United States Environmental Protection Agency (USEPA) has reported that long term exposure of heavy metals have been associated with muscular dystrophy, multiple sclerosis, Alzheimer's disease, osteoporosis, and cancer (Fu and Xi, 2020; Gerardo et al., 2020; Srivastava et al., 2017), retarded central nervous activities, damage to the lung tissues, nasal mucous membranes and pharynx congestion, liver, kidney associated nephritis, renal tumour, extensive lesions in the kidneys, anuria, cardiovascular diseases, changes blood composition, increases blood pressure and headache are some of the heavy metals associated risks reported by several authors (Florea and Büsselberg, 2006; Jaishankar et al., 2014; Vaishaly et al., 2015). The nature of toxicity along with degree of change in organ system will depend on the heavy metal ion concentration, and route of exposure that interrupts biochemical as well as metabolic processes of body (Bobaker et al., 2019; Mokarram et al., 2020). Therefore, removal of heavy metal from water and food should be essential to prevent the problem of bioaccumulation and biomagnifications. Three main approaches to remove heavy metals are physical, chemical and biological. Physical and chemical methods involved are precipitation, sedimentation, coagulation, membrane filtration, ion exchange, reverse osmosis etc (Rawat et al., 2013; Sahmoune, 2018). These traditional techniques are effective at higher concentration of metal ions, generate excessive slugs, are expensive and ineffective at very low concentration of dissolved metals on the order of 1–100 mg/L (Tang et al., 2008). However, compared with the traditional physical and chemical techniques, biological technique has a broad application prospect in view of its high efficiency, economy, environment-friendly and ability to biosorb heavy metal even at very low concentration i.e. about 1-100 mg/L (Wang et al., 2020; Yin et al., 2018). The heavy metals significantly affect the microbial community in the environment. Studies have shown that numerous microorganisms are tolerant to heavy metals and play important roles in removal and remediation of heavy metals (Gadd 1990). In recent years, heavy metal bioremediation technology based on lactic acid bacteria has attracted extensive attentions of scientists (Wang et al., 2014). It involves biosorption as well as bioaccumulation. Many reports show the biosorption of heavy metals by bacteria and fungi (Rehman et al. 2010). Several scientists have reported heavy metal biosorption by LAB (Bhakta et al., 2012; Lin et al., 2005; Schut et al., 2011). Biosorption is the binding and concentration of heavy metals from aqueous solutions (even very dilute ones) by certain types of inactive, dead, microbial biomass. It can be defined as the selective sequestering of metal soluble species that result in the immobilization of the metals by microbial cells. It is a quick, independent and metabolically passive process responsible for the selective sequestration of heavy metal ions by dead/inactive biomaterials. It is a simple physicochemical phenomenon resembling to conventional adsorption or ion exchange method. The only difference is the nature of sorbent which in this case is the material of biological origin known as biosorbent. The objective of this review is to highlight the properties, sources of heavy metals, heavy metals pollution together with their toxicological effects on humans.

Heavy Metals

The group of metallic elements characterised by their high atomic weight and density are known as heavy metal. Heavy metals/metalloids are generally defined as having density more than 5 g cm−3 (Mishra et al., 2023). Several elements, including Cu, Co, Zn, Fe, Mn, and Mo, are biologically essential in trace amounts but become toxic at elevated concentrations. In contrast, certain toxic heavy metals/metalloids, such as arsenic (As), lead (Pb), mercury (Hg), and cadmium (Cd), are nonessential and harmful to both the environment and human health (Mishra et al., 2023, Lenntech, 2004; Battarbee et al., 1988; Garbarino et al., 1995, Hawkes, 1997).While traces of these elements naturally occur in rocks and aquifers, their concentrations have been identified as elevated in various locations, stemming from both natural and anthropogenic sources such as industrial process, mining and use of product like battery, fertilizers, paints can release them into the environment and increase the heavy metal concentration (Mhlanga et al., 2007). Heavy metals have a tendency to accumulate in ecosystem, causing risk to both the environment and human health (Figure 1). They are toxic to living organism even in a relatively small quantity, their toxicity depends on various factors such as the dose, exposure pathway and the molecular species as well as the age, gender, genetics and nutritional status of the exposed individual (Huton and Symon, 1986).

Arsenic

The principal inorganic forms of arsenic comprise trivalent arsenite and pentavalent arsenate. Organic forms include methylated metabolites such as Monomethylarsonic acid (MMA), dimethylarsinic acid (DMA), and trimethylarsine oxide. Arsenic environmental contamination arises from both natural processes like volcanic eruptions and soil erosion, as well as human activities (ATSDR, 2000). Various industrially produced arsenic-containing compounds are utilized in agriculture for products like insecticides, herbicides, fungicides, algaecides, sheep dips, wood preservatives, and dye-stuffs. Additionally, arsenic compounds have historical applications in veterinary medicine for tapeworm eradication in sheep and cattle (Tchounwou et al., 1999). In the medical field, arsenic-based drugs have been employed for over a century in the treatment of syphilis, yaws, amoebic dysentery, and trypanosomiasis (Centeno et al., 2005). Presently, arsenic-based drugs continue to be employed in the treatment of specific tropical diseases, including African sleeping sickness and amoebic dysentery, as well as in veterinary medicine for parasite diseases like filariasis in dogs and blackhead in turkeys and chickens. Several million individuals are subjected to chronic arsenic exposure globally, particularly in nations such as Bangladesh, India, Chile, Uruguay, Mexico, and Taiwan, where elevated concentrations of arsenic contaminate groundwater. Arsenic exposure can transpire through oral ingestion, inhalation, dermal contact, and, to a limited extent, the parenteral route. Airborne arsenic concentrations vary from 1 to 3 ng/m^3 in remote areas to 20 to 100 ng/m^3 in urban environments. Water typically contains arsenic at levels below 10 µg/L, although elevated concentrations may be found near natural mineral deposits or mining sites. Arsenic levels in various foods range from 20 to 140 ng/kg. Natural soil arsenic levels typically span 1 to 40 mg/kg, but pesticide application or waste disposal can result in substantially higher values (Tchounwouet al., 2004).

Figure 1. Heavy metal sources and health impacts

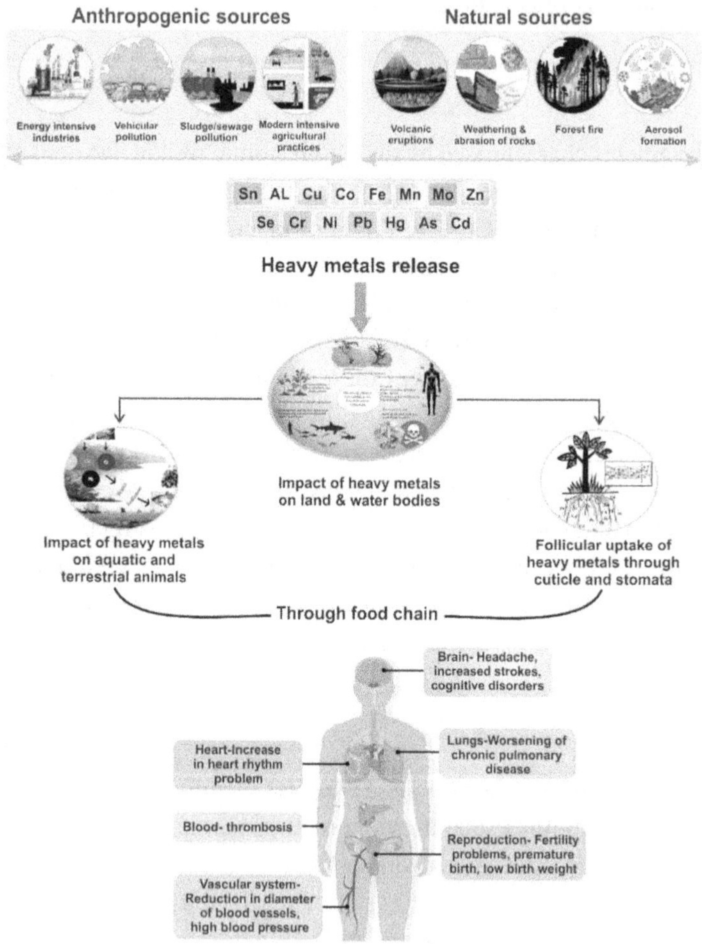

Contamination with elevated arsenic levels raises concerns due to its potential to induce various adverse health effects in humans. Numerous epidemiological studies have established a robust correlation between arsenic exposure and heightened risks of both carcinogenic and systemic health impacts. Recent reports from populous regions, including West Bengal, Bangladesh, Thailand, Inner Mongolia, Taiwan,

China, Mexico, Argentina, Chile, Finland, and Hungary, reveal populations exposed to elevated arsenic concentrations in drinking water, manifesting diverse clinico-pathological conditions such as cardiovascular and peripheral vascular diseases, developmental anomalies, neurologic and neurobehavioral disorders, diabetes, hearing loss, portal fibrosis, hematologic disorders (anemia, leukopenia, and eosinophilia), and carcinoma. Arsenic exposure adversely affects virtually all organ systems, encompassing cardiovascular, dermatologic, nervous, hepatobiliary, renal, gastrointestinal, and respiratory systems (Tchounwouet al., 2003). Studies also indicate significantly elevated standardized mortality rates for bladder, kidney, skin, and liver cancers in arsenic-polluted areas. The severity of health effects correlates with the chemical form of arsenic and is contingent upon time and dose. While evidence supporting arsenic's carcinogenicity in humans is substantial, the precise mechanisms underlying tumorigenesis remain incompletely understood. Analyzing arsenic's toxic effects is intricate due to its sensitivity to oxidation state, solubility, and various intrinsic and extrinsic factors. Studies suggest that arsenic toxicity is contingent upon exposure dose, frequency, and duration, biological species, age, gender, individual susceptibilities, as well as genetic and nutritional factors. Inorganic arsenic, particularly trivalent arsenite (AsIII), proves 2–10 times more toxic than pentavalent arsenate (AsV). AsIII can deactivate over 200 enzymes by binding to thiol or sulfhydryl groups on proteins, likely accounting for arsenic's broad-ranging effects on diverse organ systems. AsV, through phosphate substitution, interferes with numerous biochemical pathways (Hughes, 2002).

Cadmium

Cadmium, a significant environmental and occupational concern as a heavy metal, is ubiquitously present in the Earth's crust with an average concentration of approximately 0.1 mg/kg. Cadmium is emitted into the atmosphere through various natural and human activities, leading to diverse exposure routes for animals and humans. Pollution of aquatic environments with cadmium arises from absorption, industrial waste, and surface runoff, ultimately affecting both soil and sediments. Human exposure to cadmium can occur through ingestion of contaminated food, inhalation of polluted air, or consumption of water containing elevated levels of the metal. Notably, cadmium lacks attributes beneficial for plant growth and metabolic processes (Hayat et al., 2018).

Extensively utilized in various industrial applications, cadmium finds major use in alloy, pigment, and battery production. Majority of cadmium, about three fourths is used as electrode component in producing alkaline batteries. While its application in batteries has witnessed growth, its commercial usage has declined in developed nations due to environmental apprehensions. Predominant routes of cadmium exposure involve inhalation, particularly through cigarette smoke, and ingestion of contaminated food. Skin absorption is infrequent. Human exposure arises from diverse sources, including occupational engagement in primary metal industries, consumption of contaminated food, cigarette smoking, and employment in cadmium-contaminated workplaces, with smoking being a principal contributor (Unaegbu et al., 2016). Additional sources encompass emissions from industrial activities such as mining, smelting, and the manufacturing of batteries, pigments, stabilizers, and alloys. Cadmium is released into the environment through industrial processes and emissions from cadmium smelters, contributing to the contamination of sewage sludge, fertilizers, and groundwater. The metal can persist in soils and sediments for extended periods, potentially lasting several decades, and may be absorbed by plants (Engwa et al., 2015).

Cadmium is traceable in specific foods like leafy vegetables, potatoes, grains, seeds, liver, kidney, crustaceans, and molluscs. Foodstuffs rich in cadmium, such as liver, mushrooms, shellfish, mussels, cocoa powder, and dried seaweed, significantly elevate cadmium concentrations in human bodies. The circulatory system serves as a pivotal distribution route, with blood vessels considered primary organs susceptible to cadmium toxicity. Chronic inhalation exposure to cadmium particulates is commonly associated with pulmonary function alterations and chest radiographic findings indicative of emphysema. Occupational exposure to airborne cadmium particulates has been linked to diminished olfactory function. Epidemiological studies have consistently reported an association between chronic low-level cadmium exposure and reductions in bone mineral density, contributing to osteoporosis (Åkesson et al.,2006; Schutte et al., 2008).

Cadmium poses a severe threat as both a pulmonary and gastrointestinal irritant, with potential fatality upon inhalation or ingestion. Following acute ingestion, symptoms manifest rapidly, typically within 15 to 30 minutes, encompassing abdominal pain, burning sensation, nausea, vomiting, salivation, muscle cramps, vertigo, shock, loss of consciousness, and convulsions. Acute cadmium ingestion may lead to gastrointestinal tract erosion, as well as pulmonary, hepatic, or renal injury, potentially resulting in coma, contingent on the route of poisoning (Baselt, et al., 1995). Chronic exposure to cadmium exerts a depressive influence on neurotransmitter levels, including norepinephrine, serotonin, and acetylcholine. Rodent studies demonstrate that prolonged inhalation of cadmium induces pulmonary adenocarcinomas. Furthermore, cadmium has the capacity to induce prostatic proliferative lesions, including adenocarcinomas, following systemic or direct exposure (Waalkes and Rehm, 1992).

Chromium

Chromium (Cr), an element naturally present in the Earth's crust, exhibits oxidation states ranging from chromium (II) to chromium (VI) (Jacobs et al., 2005). Chromium compounds predominantly exist in the stable trivalent form [Cr(III)], found naturally in ores like ferrochromite. The hexavalent form [Cr(VI)], the second-most stable state (Patlolla et al., 2009), is prevalent due to industrial activities.

Chromium enters various environmental matrices (air, water, and soil) through diverse natural and anthropogenic sources, with industrial facilities especially those involved in metal processing, tannery operations, chromate production, stainless steel welding, ferrochrome, and chrome pigment production contributing significantly. Elevated environmental chromium concentrations are associated with releases from metallurgical, refractory, and chemical industries, primarily occurring in the hexavalent form [Cr(VI)]. This hexavalent state is a recognized toxic industrial pollutant and classified as a human carcinogen by regulatory and non-regulatory agencies. The health risk linked to chromium exposure varies based on its oxidation state, ranging from low toxicity in the metallic form to high toxicity in the hexavalent form. High levels of cadmium are present in various sources such as leafy vegetables, oilseeds, crops, organ meat, and nuts (Schaefer et al., 2020), with rice being a primary source of cadmium ingestion in China. The dismantling of Chinese e-waste exposes individuals to elevated cadmium concentrations, leading to a 60% increase in hazard quotient for those consuming rice from the affected areas (Fu et al., 2013).

Previously, Cr(VI)-containing compounds were considered predominantly anthropogenic, contrasting with the naturally ubiquitous Cr(III) in air, water, soil, and biological materials. However, recent findings reveal naturally occurring Cr(VI) in ground and surface waters, exceeding the World Health Organization's drinking water limit of 50 µg of Cr(VI) per liter. Chromium finds extensive use in industrial processes, including welding, chrome plating, dye and pigment production, leather tanning, wood preservation, and as an anticorrosive agent in cooking systems and boilers, contributing to its widespread contamination of environmental systems (Norseth, 1981; Wang et al., 2006). Non-occupational exposure stems from ingesting chromium-containing food and water, while occupational exposure primarily occurs through inhalation—the principal human exposure route. Cadmium contamination in food can occur naturally or through sewage and fertilization, entering water bodies and subsequently accumulating in fish and oysters. Approximately one-third of the total Cd(II) in the human body accumulates in the liver and kidneys. Inhalation poses a heightened risk to the lungs, the primary target organ. Breathing elevated levels of chromium (VI) can result in nasal lining irritation and ulcers. Unexpected cadmium ingestion may occur through sources like pigments and food dyes (Mead, 2010). Studies indicate that a vegetarian diet may have higher cadmium levels compared to a nonvegetarian diet, particularly in vegetables and plant products such as sunflower seeds, spinach, and potatoes, with cadmium levels exceeding 100 μg/kg (Spungen, 2019).

Annually, over 300,000 workers face exposure to chromium and chromium-containing compounds in occupational settings. In both humans and animals, trivalent chromium [Cr(III)] functions as an essential nutrient, influencing glucose, fat, and protein metabolism by enhancing insulin activity. Despite its nutritional role, occupational exposure raises significant concerns due to the heightened risk of chromium-induced diseases, particularly in workers exposed to hexavalent chromium [Cr(VI)] (Guertin, 2005).Cadmium significantly increases the risk of cardiovascular disease through smoking, as highlighted by Li et al.'s research on the impact of cadmium in tobacco (Li et al., 2019). Cadmium consumption adversely affects fertility, causes osteotoxicity, and contributes to cardiovascular diseases (Sharma et al., 2015). Moreover, iron deficiencies are associated with increased absorption of divalent metals, including cadmium, cobalt, and manganese (Meltzer et al., 2010, Gallaghar et al., 2011). This association is evident in breastfed infants with iron-poor diets, leading to elevated levels of lead, cobalt, and cadmium (Kim et al., 2014). Duc to thc scvere health risks associated with cadmium, including anaemia, insomnia, kidney damage, bladder and prostate cancers, and osteoporosis, complete avoidance is recommended (Choong et al., 2014). Chronic exposure to cadmium ions has detrimental effects.

Critical factors influencing chromium compound toxicity include oxidation state and solubility. Cr(VI) compounds, potent oxidizing agents with irritant and corrosive properties, exhibit greater systemic toxicity than Cr(III) compounds under similar conditions of quantity and solubility (De Flora et al., 1990). While the precise mechanisms of biological interaction remain uncertain, the varying toxicity is potentially linked to Cr(VI)'s ability to traverse cell membranes easily and undergo intracellular reduction to reactive intermediates. As Cr(III) is poorly absorbed, chromium toxicity predominantly stems from the Cr(VI) form, which can be absorbed through the lung, gastrointestinal tract, and, to a limited extent, intact skin. The reduction of Cr(VI) is viewed as a detoxification process when occurring away from the toxic or genotoxic target site. Conversely, if Cr(VI) reduction takes place near the cell nucleus of target organs, it may activate chromium toxicity. Extracellular reduction of Cr(VI) to Cr(III) does not readily facilitate metal transport into cells, thereby mitigating observed toxicity.

Lead

Lead, a naturally occurring bluish-gray metal found in small quantities in the Earth's crust, is subject to anthropogenic contributions, including fossil fuel combustion, mining, and manufacturing, resulting in the release of elevated concentrations. Lead is a non-biodegradable metal naturally occurring in relatively low amounts. The continuous rise in atmospheric lead levels is attributed to human activities, such as manufacturing, mining, and the combustion of fossil fuels. (Loh et al., 2016). Widely employed in various industrial, agricultural, and domestic applications, lead is utilized in the production of lead-acid batteries, ammunitions, gasoline, house paint, plumbing pipes, lead bulletsmetal products (solder and pipes), and X-ray shielding devices (Thurmer et al., 2002). Recent years have witnessed a substantial reduction in lead's industrial use, particularly in paints, ceramic products, caulking, and pipe solder (CDC, 1991).

Lead is released into the atmosphere through industrial processes and vehicle exhausts. Consequently, it can enter the soil and flow into water bodies, where plants may absorb it. Lead is a significant heavy metal with notable implications for human health. Human poisoning is commonly linked to lead-containing gasoline, mining, and battery recycling. However, unexpected sources of poisoning may include wall paints or binders used in sewer pipes (Vijayakumara et al., 2012, Manay et al., 2008). Human exposure to lead can occur through the consumption of contaminated food or drinking water (Wani et al., 2015). Lead becomes toxic to the human body when exposed to amounts surpassing the optimum levels. Children are particularly vulnerable to lead poisoning, with the severity of poisoning increasing when they come into contact with dust containing environmental lead (Loh et al., 2016). Despite being preventable, lead poisoning remains a hazardous condition, capable of affecting various organs within the body. Lead poisoning primarily affects the gastrointestinal tract and central nervous system in both children and adults (Markowitz, 2000). It can manifest as either acute or chronic. Acute exposure to lead may lead to symptoms such as headache, loss of appetite, abdominal pain, fatigue, sleeplessness, hallucinations, vertigo, renal dysfunction, hypertension, and arthritis. On the other hand, chronic exposure can result in severe outcomes, including birth defects, mental retardation, autism, psychosis, allergies, paralysis, weight loss, dyslexia, hyperactivity, muscular weakness, kidney damage, brain damage, coma, and, in extreme cases, death (Martin and Griswold, 2009). Both lead and cadmium contribute to the loss of calcium and abnormal bone metabolism, resulting in osteomalacia. Elevated levels of lead exposure can induce the movement of the plasma membrane of the blood-brain barrier into the interstitial spaces, leading to oedema (Teo et al., 1997). Additionally, lead exposure has the potential to disrupt intracellular second messenger systems and alter the functioning of the central nervous system.

Lead contamination in dust and soil poses a persistent concern, recontaminating cleaned residences and contributing to increased blood lead concentrations in children exposed to bare, contaminated soil during play. Presently, the primary source of lead poisoning in children arises from deteriorating lead paint on interior surfaces, leading to dust and chip exposure (Lanphear et al., 1998). Exposure to lead primarily occurs through inhalation of lead-contaminated dust particles or aerosols and ingestion of lead-contaminated food, water, and paints. Adults absorb 35 to 50% of lead through drinking water, with a potentially higher absorption rate in children exceeding 50%, influenced by age and physiological status. Children under six years of age face heightened risks from lead exposure, as it disrupts the development, growth, and differentiation of nerve cells, and causes bone tissue damage. Developing foetuses and children are particularly susceptible to neurotoxic effects resulting from lead exposure. Several prospective epidemiologic studies involving children under 5 years of age have indicated that even low levels of lead exposure (5–25 μg/dL in blood) can lead to the impairment of intellectual development, as evidenced by a loss of intelligence quotient points. Consequently, the Centers for Disease Control (CDC) in the United States has lowered the acceptable level of lead in children's blood from 25 to 10 μg/dL and advocated for universal screening of blood lead for all children (Brown and Morgolis, 2012).

Within the human body, lead absorption is highest in the kidney, followed by the liver and other soft tissues like the heart and brain, but the skeletal compartment represents the major body fraction. The nervous system emerges as the primary target of lead poisoning, with early symptoms including headaches, diminished attention span, irritability, memory loss, and dullness (CDC, 2001). In adults, high lead exposure has been linked to reproductive effects, such as decreased sperm count in men and spontaneous abortions in women. Acute lead exposure induces brain, kidney, and gastrointestinal damage, while chronic exposure may adversely affect blood, the central nervous system, blood pressure, kidneys, and vitamin D metabolism (USEPA, 2002; Litvak et al., 1998). In the 1980s, the National Health and Nutrition Examination Survey (NHANES) reported that an estimated 88% of children aged 1–5 years had blood lead levels (BLLs) exceeding 10 µg/dL. Subsequently, these numbers have seen a rapid decline. The American Center for Disease Control and Prevention (CDC) recently recommended a new reference value of 3.5 µg/dL based on the 98th percentile of the 2011–2014 NHANES data (Paulson and Brown, 2019). Blood lead screening programs play a crucial role in mitigating exposure risks, especially at the population level (Luo et al., 2021).

Mercury

Mercury, a heavy metal within the transition element series of the periodic table, exhibits a unique natural occurrence in three forms: elemental, inorganic, and organic, each characterized by distinct toxicity profiles. At room temperature, elemental mercury exists in liquid form with a notable vapor pressure, releasing mercury vapor into the environment. It also presents as a cation with oxidation states of +1 (mercurous) or +2 (mercuric). Methylmercury, the prevalent organic compound, results from the methylation of inorganic (mercuric) mercury by microorganisms in soil and water (Dopp et al., 2004). Mercury, an exceedingly perilous heavy metal, is present in the biosphere and has become a pervasive contaminant due to human activities, with a rising presence in the atmosphere. Upon contact with aquatic sediments, mercury transforms into the highly toxic methylmercury (Gworek et al., 2020).

As a pervasive environmental toxicant, mercury induces profound tissue alterations, leading to a spectrum of adverse health effects (Zahir et al., 2005). Human exposure to all mercury forms stems from accidents, environmental pollution, food contamination, dental care, preventive medical practices, industrial and agricultural operations, and occupational activities. Chronic, low-level exposure primarily arises from dental amalgams and fish consumption. Mercury naturally enters water through off-gassing from the Earth's crust and industrial pollution, where algae and bacteria methylate it. Organic mercury is inherently lipophilic, allowing it to easily penetrate cell membranes. Dimethylmercury, a highly toxic compound, has the capability to permeate the skin even through latex gloves. Exposure to very low doses of dimethylmercury can result in the degeneration of the central nervous system and potentially lead to fatal outcomes. Methyl mercury traverses the food chain, accumulating in fish, shellfish, and ultimately in humans (Sanfeliuet al., 2003). Elemental mercury (Hg^0) and methyl mercury (MeHg) are the most highly absorbed species. Dental amalgams, comprising over 50% elemental mercury, facilitate absorption through the lungs and oral tissues. Once in the blood, Hg^0 swiftly passes through cell membranes, including the blood-brain and placental barriers, undergoing oxidation to reactive Hg^{2+}. Methyl mercury, ingested through fish consumption, is efficiently absorbed in the gastrointestinal tract, readily crossing placental and blood-brain barriers due to its lipid solubility. Upon absorption into the human body, it enters the circulation and gives rise to various neurological issues (Rice et al., 2014). Mercury, once absorbed, exhibits a low excretion rate, with a substantial proportion accumulating in the kidneys, neurological tissues, and the liver.

All forms of mercury induce toxicity, encompassing gastrointestinal toxicity, neurotoxicity, and nephrotoxicity and their effects (Tchounwouet al., 2003). Recent research has also unveiled cardiovascular toxicity associated with mercury exposure. Levels of mercury in hair have been correlated with oxidized LDL levels in atherosclerotic lesions, acute coronary failure, and atherosclerosis (Yoshizawa et al., 2002).

Furthermore, mercury has been found to deactivate paraoxonase, an extracellular antioxidative enzyme linked to HDL dysfunction (Salonen et al., 1999). This deactivation is directly connected to the progression of atherosclerosis and an increased risk of coronary heart disease, cardiovascular disease, acute myocardial infarction, coronary heart disease, and carotid artery stenosis (Kulka, 2016). Acute exposure of the kidneys to mercury can result in acute tubular necrosis, manifesting with clinical symptoms such as acute dyspnea, altered mental status, abdominal pain, profuse salivation, tremors, vomiting, chills, and hypotension. In contrast, chronic exposure to mercury leads to injury to the epithelium and necrosis in the pars recta of the proximal tubule (Lentini et al., 2017). Symptoms of mercury-induced chronic kidney injury include tubular failure, increased urine excretion of albumin and retinol-binding protein, and a nephritic state characterized by membranous nephropathy. Brief exposure to elevated levels of metallic mercury vapors can result in symptoms such as vomiting, nausea, skin rashes, diarrhoea, lung damage, and high blood pressure. On the other hand, short-term exposure to organic mercury poisoning can lead to depression, tremors, headache, fatigue, memory problems, and hair loss. As these symptoms overlap with those of other illnesses or conditions, diagnosing mercury poisoning in such cases may be challenging (Lara-Torres et al., 2021).

Persistent exposure to mercury at chronic levels can give rise to erethism, a condition characterized by excitability, tremors in the hands, memory loss, timidity, and insomnia. Furthermore, occupational exposure to mercury, as observed by researchers, has been linked to measurable declines in performance on neurobehavioral tests assessing motor speed, visual scanning, visuomotor coordination, and verbal and visual memory. Exposure to mercury during pregnancy can have detrimental effects on the foetus, leading to potential issues such as mental retardation, cerebellar symptoms, retention of primitive reflexes, malformation, and other abnormalities (Nourmoradi et al., 2013). Recent studies have confirmed these concerns, demonstrating that pregnant women exposed to mercury through dietary intake of whale meat and fish exhibited reduced motor neuron function, memory loss, impaired speech, and compromised neural transmission in their offspring.

Other Heavy Metals

Manganese is a crucial metal for the body, but it has become a global concern with the introduction of methylcyclopentadienyl manganese tricarbonyl (MMT) as a gasoline additive, which is recognized for its toxicity. Manganese, the most abundant among the toxic heavy metals, exists in various oxidation states in nature. While manganese is essential for various physiological activities, excessive intake can lead to significant toxicity, as indicated by studies conducted by Loranger and Zayed (1995) and O'Neal and Zheng (2015). Elevated levels of manganese exposure can lead to neurotoxicity, causing manganism, a neurological disease characterized by rigidity, action tremor, a mask-like expression, gait disturbances, bradykinesia, micrographia, memory and cognitive dysfunction, and mood disorders (Young and Xiao, 2021; Srivastava and Kowshik, 2013). The symptoms of manganism closely resemble those of Parkinson's disease. However, the main distinctions lie in the insensitivity of manganism to levodopa (l-DOPA) administration and differences in the symptoms and progression of the disease (Gunther et al., 2012).

Antimony is a metalloid element that can have health impacts depending on the level and duration of exposure. Antimony, can be present in the air in nanogram amounts, originating from both natural sources such as volcanic activity and weathering, as well as human activities, leading to emissions into the atmosphere (He et al., 2019). Exposure to antimony in industrial settings, primarily through inhalation, can result in toxicity. Inhalation of antimony dust or fumes can irritate the respiratory system, leading to symptoms such as coughing, wheezing, and difficulty breathing. Prolonged exposure may result in chronic respiratory conditions. Antimony poisoning is associated with physiological deficiencies, including pancreatitis, cardiotoxicity, and respiratory issues such as pleural adhesions, chronic emphysema, chronic bronchitis, respiratory irritation, and inactive tuberculosis. Furthermore, antimony is considered carcinogenic and has adverse effects on reproductive health (Sundar and Chakravarty, 2010). Animal studies have shown that exposure to high levels of antimony can also have adverse effects on reproduction and development.

Copper is acknowledged as a crucial micronutrient for living organisms, playing a role in essential physiological functions in plants, such as the formation of chlorophyll, photosynthesis, and carbohydrate and protein metabolism. Copper is known to accumulate in the liver due to conditions like Wilson's disease. Elevated levels of copper can induce oxidative stress, making hepatic copper deposition not only pathognomonic but also pathogenic (Deering et al., 1977). Increased hepatic copper levels are also observed in cholestatic liver diseases, resulting from reduced biliary excretion of copper and not being the cause of hepatic infection (Gross et al., 1985; Yu et al., 2019). While copper deficiency can disrupt vital metabolic processes, excessive exposure can lead to toxicity (Schwartz et al., 2003).

Cobalt is widely distributed in the environment, including vegetation, soils, rocks, and water, and is commonly used in alloy production. While low levels of cobalt typically have no adverse effects, excessive discharges into the environment can pose a significant danger to humans. The impact of cobalt on the human body can be both beneficial and harmful (Domingo, 1989). Nickel is a naturally abundant element with widespread industrial applications. Emissions of nickel into the atmosphere occur from both natural and human-made sources (Li et al., 2016). The element has various detrimental effects on human health, including the development of allergies, nasal and lung cancer, as well as kidney and cardiovascular diseases (Lu et al., 2005). These health issues are attributed to the inhalation of air contaminated with nickel (Genchi et al., 2020). Thallium is present in various forms in the environment and poses a hazard to biological organisms. Its toxicity surpasses that of any other heavy metal. Monovalent thallium ions are naturally present in water and are released into the air through aqueous routes (Peter and Viraraghavan, 2005). Furthermore, industrial emissions play a significant role in elevating thallium levels in the atmosphere. Human exposure to thallium is exceptionally harmful (Kazantzis, 2000).

Sources of Heavy Metals

Heavy metals, such as lead, cadmium and mercury, comes from various sources in our environment. These heavy metals are naturally present in the Earth's crust since its formation. However, the substantial increase in the use of heavy metals has led to a significant rise in metallic substances in both terrestrial and aquatic environments (Gautam et al., 2016).Natural processes such as volcanic activity, metal corrosion, metal evaporation from soil and water, sediment re-suspension, soil erosion, and geological weathering can also contribute to an increase in heavy metal pollution (Masindi, and Muedi, 2018; Herawati et al., 2000; He et al., 2005; Walker et al., 2012).The main sources of heavy metals in the soil environment and agriculture include atmospheric deposition, livestock manure, irrigation with wastewater or polluted water (Mansour et al., 2009; Elgallal et al., 2016), metallo-pesticides or herbicides, phosphate-based fertilizers, and amendments derived from sewage sludge (Gall et al., 2015; Woldetsadik et al., 2017; El-Kady and Abdel-Wahhab, 2018).

Agricultural practices, including the use of pesticides, insecticides, fertilizers, and more, represent a secondary source of heavy metal pollution. Vegetables cultivated in greenhouses have been identified as being contaminated with heavy metals, primarily originating from anthropogenic sources (Fan et al., 2017). Advanced statistical and geospatial tools can be employed for source identification of heavy metals in indoor environments (Acosta et al., 2011; Xu et al., 2015). In China, it has been observed that greenhouse vegetables are more susceptible to cadmium (Cd) contamination compared to crops in open farmlands (Liu et al., 2011). Principal components analysis results indicated that metals such as chromium (Cr), arsenic (As), and nickel (Ni) are primarily released by weathered rocks, while metallic contaminants like mercury (Hg) and lead (Pb) are generated by industries, vehicular emissions, particulate matter (PM), and the reuse of wastewater for irrigation (Liu et al., 2011). Identifying soil contaminants and their sources is crucial due to their close association with human health (Khan et al., 2010; Rai, 2012).

Besides natural sources, conventional and emerging anthropogenic contaminants present significant human health risks through the consumption of food crops. This contamination can occur either through root transfer from soil to plant tissues or direct atmospheric deposition onto plant surfaces (Zhuang et al., 2009). The pollution caused by heavy metals is primarily due to anthropogenic activities, which include mining, smelting, foundries, and other metal-based industries. Coal-fired power plants represent a significant source of mercury (Hg) contamination in soil. Li et al. (2017) indicated that the prolonged consumption of crops such as lettuce, amaranth, water spinach, cowpea, and grains (e.g., rice) grown in Hg-contaminated soils, particularly those in proximity to thermal power plants in China, can have adverse effects on human health. Sludge generated from distilleries, as well as from the chemical, electroplating, textile, and leather industries, often contains notably high concentrations of heavy metals, including iron (Fe), copper (Cu), chromium (Cr), lead (Pb), nickel (Ni), and manganese (Mn) (Ram Chandra and Kumar, 2017). Pandey (2006) demonstrated that effluents from electroplating, containing cadmium (Cd), chromium (Cr), copper (Cu), nickel (Ni), and zinc (Zn), could have severe effects on plants, including stunted growth, necrosis and chlorosis in leaves, and even plant death.

Various processes with strong contaminant sources, including the use of wastewater for irrigation, sewage sludge as a soil amendment for food crops, and particulate matter (PM) pollution/deposition on soil and plants, present significant challenges for global food safety. Further, metal leaching from various sources such as landfills, waste dumps, excretion, livestock and chicken manure, runoffs, automobiles, and roadworks contributes to heavy metal pollution. Vehicle emissions and improper disposal of electronic waste also releases substantial heavy metal into the environment. Another study revealed that factories involved in Pb-acid battery production emitted metals bound in particulate matter (PM), which were subsequently deposited in soils and crops within the agro-ecosystem (Franca et al., 2017). The utilization of nanotechnology in various sectors, including medicine, energy, environment, and agriculture, raises environmental concerns due to the unregulated release of nanoparticles (NPs) and associated toxic metals (Xiong et al., 2017; Rai et al., 2018). Food crops, in particular, should undergo impact assessments for nano-toxicity, such as those associated with copper oxide nanoparticles (CuONPs), as the presence of NPs can have detrimental effects on both crop physiology especially reduced photosynthesis and human health (Xiong et al., 2017).

Table 1. Properties, sources, and health impacts of heavy metals

Heavy Metal	Properties	Sources	Uses	Health Impacts	References
Arsenic	Density 5.75 g/cm³, Found as three allotropic forms, Bright silvery-grey in colour, Brittle	Fish and shellfish, Meat and poultry, Dairy products, Cereals	Wood preservation, many types of glass, unique insecticide compositions, semiconductors as a doping agent, such as gallium arsenide.	Gastro-intestinal system irritation, Lung irritation, Skin changes, production of both red blood cells and white blood cells, Increased chances of cancer have been suggested, Infertility and miscarriages, Heart problems, Brain damage.	Martin et al.,2009. Kaye et al., 1986. Haynes et al.,2011.
Cadmium	Density 8.69 g/cm³, Silvery bluish tint metal, found frequently in combination with zinc, the only mineral found is Greenockite which is made up of cadmium and Sulphur	Shellfish, Mussels, Dried seaweeds, shrimps, mushrooms, liver.	Phosphate fertilizer, Pesticides, Nickel Cadmium batteries, Glassware pigmentation, Corrosion-resistant plating, Stabilizer in plastic production, nuclear reactors.	Kidneys are primarily affected causing nephrotoxicity, reproductive system failure, Calcium metabolism alterations, Bone fracture psychological disorders, Gastrointestinal disorders, Central nervous system complications, DNA impairment, Cancer, Itai-Itai disease, Osteoporosis, Renal dysfunction.	ATSDR Toxicological Profile for Cadmium; Georgia, 2012.
Chromium	Density7.15 g/cm³, Extracted as Siberian red lead, hard, shiny and steel-grey moderately active metal reacts with the majority of acids, creates a coating of chromium (III) oxide.	Fruits and vegetables, Yeasts, Meats, fish, Shell, Grains	Alloys, metal ceramics, electroplating, Manufacturing of synthetic rubies, Leather tanning, Dye paints, Chromium salts are used for colored glass	Can typically result in acute poisonings and several symptoms, such as gastrointestinal ulcers, nausea, and vomiting. vertigo, fever, diarrhea, Liver injury, toxic nephritis, and coma, death (often at 1-3g).	WHO, 2000 Barnhart, 1995 Martin, 2009 Dayan and Paine, 2001
Lead	Density 11.3 g/cm³, found in galena, dull, gray-silver metal, a mineral ore consisting of lead sulfide that can be mixed with copper, zinc, and silver.	Fruit and vegetables, Grain wine, red meat, Seafood, Soft drinks.	Hair dyes, Lead glazes for pottery, Insecticides, automobile lead-acid batteries, sheets for computer screens to protect against radiation, projectiles and ammunition, Lead crystal glass, Covering for cables, athletic gear, divers' weight belts, Container for caustic substances, Lead fittings	consequences for those High blood pressure, miscarriages, early and small deliveries, stillbirths, neurological damage, renal dysfunction, stomach ache, Damage to peripheral nerves, harm to sperm, encephalopathy, Children may experience decreased IQ, altered brain and central nervous system development, increase in antisocial behavior.	Haynes et al., 2011. ATSDR, 2007
Mercury	Density13.5336 g/cm³, It is a silvery liquid metal that is found in its native state at ambient temperature.	Seafood, Mushrooms	Utilized in barometers, thermometers, chlorine production, fillings, gold recovery, tooth fluorescent lightbulbs, photochemistry, Calomel electrodes, insecticide, rat poison, brilliant red catalyst, rectifiers, and electrical switches.	Inhalation causes eye, lung, rash, and vomiting and diarrhea, Genotoxic. Down syndrome, Birth defects, Damage to male sperm, Minamata disease, and neurological disorders, Personality changes, Paralysis, Coma and death, Learning difficulties, Speech defects, Memory loss, Tremors and muscle incoordination, Deafness, Vision issues, Teratogenic	Martin et al.,2009. Haynes et al., 2011. ATSDR, 1999. Li et al.,2017.

Continued on following page

Table 1. Continued

Heavy Metal	Properties	Sources	Uses	Health Impacts	References
Manganese	Density 7.3 g/cm^3, essential trace element, the minerals that are present are manganese dioxide, or pyrolusite, manganese carbonate, or rhodochrosite. Flexible, possesses six oxidation states.	Rice and grains, Teas, Herbs, Green beans, Oysters, Nuts, Parsley, Spinach, Bran, Soya beans, Olive oil	Due to its brittle nature, manganese is primarily employed in alloys. Manganese (IV) oxide can be used as a catalyst, rubber additive, and to remove iron impurity, induces green tint in glass, potent oxidizing agent, manganese (II) oxide is utilized in quantitative analysis, Cosmetics, Fireworks, Fertilizers, and Pesticides.	Low blood pressure Manganism, Adverse effects on the nervous system, Sluggishness, Changes in behavior, Weakness, Lethargy, Parkinson's Disease-like symptoms, Incoordination of movement, Damage to sperm, Loss of desire for sexual activity, Pneumonia	WHO,2000, Dobson et al., 2004. Haynes et al., 2011. Marcus et al., 2013. ATSDR, 2012
Copper	Density: 8.96g/cm^3, reddish-gold, found in minerals such as bornite, sometimes referred to as peacock ore, and chalcopyrite, which is a copper, iron, and sulfur-containing mineral, Simple to work with, excellent heat and electrical conductivity	Liver, Oyster, Spirulina, Shiitake mushrooms, Dark chocolate	Copper alloys to make brass and bronze, fertilizer, pipes, coins, plating, copper wires, and preservation of wood barrier lotion, Chemical analyses for Fehling's solution sugar, electronification, Water purification uses copper sulphate as an algicide. Using copper sulfate to treat agricultural mildew.	Flu-like symptoms, diarrhea, vomiting, discomfort in the eyes, light headedness, oral irritation, severe gastroenteritis. Hepatocellular degeneration, Cytotoxic, renal and liver damage. Anxiety, restlessness, agitation, and insomnia, Wilson's disease, Unintentional poisoning Death, Hepatic cirrhosis, Demyelination	Kaye et al., 1986. Haynes et al.,2011. Sharma et al.,2004. Tsafrir 2018.
Cobalt	Density: 8.86 g/cm^3, silvery-blue, found in ores known as cobaltite, which is composed of sulfur, cobalt and arsenic.	Chocolate, butter, cheese, meat.	Magnets (when combined with nickel and aluminum alloys), Within jet turbines and gas turbine generators, electroplating, used in enamels, paint, porcelain, pottery, and glass. Use of radioactive 60 cobalt Use radiation to preserve food.	Skin and respiratory problems, Congestion, oedema, decreased lung function upon inhalation, lung hemorrhage during inhalation. Prolonged inhalation causes: wheezing, asthma, respiratory irritation, lung function decline, pneumonia and fibrosis, cardiac effects, liver and kidney congestion, conjunctiva, cardiomyopathy, nausea and vomiting, diarrhea, allergic dermatitis, and liver diseases.	Kaye et al., 1986.

Impact of Heavy Metal Pollution

Heavy metal pollution poses a significant threat to environmental and human health. The release of heavy metals into the air, water, and soil from various industrial activities has led to widespread contamination. These toxic metals can accumulate in ecosystems, adversely affecting plant and animal life. In aquatic environments, heavy metals may disrupt the balance of ecosystems and harm aquatic organisms, leading to cascading effects throughout the food chain. Additionally, the absorption of heavy metals by crops grown in contaminated soil raises concerns about food safety and human exposure. Chronic exposure to heavy metals has also been linked to a range of health problems. Addressing this issue is crucial to safeguarding both environmental sustainability and human well-being.

Environment Impact

The heavy metals released into the environment can contaminate soil, water bodies and the atmosphere. This contamination disrupts ecosystem and effects the balance of nature and results in death of plant and aquatic life and reducing biodiversity (Zhao et al., 2016; Jawed et al., 2020). In developing countries, the environmental issue related to heavy metal contamination is becoming serious due to increase in anthropogenic and geologic activities. Various sources contribute to environmental pollution, including wastewater irrigation, pesticides, animal manures, fertilizers, leaded paint, mine ore waste (mine tailing), sewage sludge, petroleum distillate spillage, coal combustion residues, and waste dumpings (Chibuike and Obiora, 2014). The use of untreated sewage and wastewater has led to the contamination of agricultural lands, causing the absorption of heavy metals by crops intended for human consumption. Additionally, pollution can result from the flooding of seas and rivers, introducing sewage and contaminated water onto the land, as well as accidents involving vehicles transporting toxic chemicals. The persistent nature of heavy metals, characterized by their resistance to microbial or chemical degradation, allows them to remain in the soil for prolonged periods. Alongwith, excessive industrialisation and urbanisation trend, advancement in traffic activities has considerably contributed to the accumulation of heavy metal discharge by vehicles in environment (Zeng,2008; Chen et al., 2014). Traffic emissions can lead to heavy metal pollution in agricultural areas, potentially contaminating the crops that are growing there. Plants cultivated in these soils manifest diminished growth, performance, and yield. In agricultural domains, the absorption of heavy metals via the soil-crop system could exert a primary influence on exposure to heavy metals (Okunola et al., 2007).

Soils exhibit significant contamination attributed to the deposition of metals from both vehicular emissions and industrial activities, along with the translocation of metals deposited on road surfaces and in industrial proximity (Harrison et al., 1981). The consequential environmental, ecological, and health impacts of these heavy metals have prompted numerous studies on their presence in soils near major roads and industrial sites (Turer and Maynard, 2003). According to Okunola et al. (2007) the accumulation of these metals in plants and crops cultivated in these areas poses a considerable risk when consumed by humans and animals, particularly livestock, either directly or indirectly, potentially reaching lethal levels rapidly. The threat of heavy metals to food safety and the environment is a major concern for governments and societies globally. The escalating industrialization and urbanization in developing countries have intensified heavy metal pollution in agricultural soils (Adesuyi et al., 2015). Heavy metal pollutants, being non-biodegradable with extended half-lives in soil, raise apprehensions about their far-reaching effects on biological systems, including soil microorganisms and other biota. Frequently cited heavy metals in contaminated soils, with potential hazards, include Cd, Cr, Pb, Zn, Fe, and Cu (Alloway, 1995; Akoto et al., 2008). Metal concentrations in soil typically range from less than one to as high as 100,000 mg/kg. Heavy metals constitute a primary category of inorganic contaminants, with expansive land areas contaminated due to the use of sludge, municipal compost, pesticides, fertilizers, and emissions from municipal waste incineration, metalliferous mines, and smelting industries (Halim et al., 2002). Metal pollution originating from industrial activities, vehicular emissions, and waste disposal sites is well-documented in soil dust and agricultural soils (Addo et al., 2012). Changes in soil properties are linked to environmental alterations resulting from human activities such as mining, manufacturing, and the use of synthetic products like pesticides, paints, batteries, industrial waste, and the application of industrial or domestic sludge on land. These activities contribute to the pervasive issue of heavy metal contamination in urban and agricultural soils. Once heavy metal gets into the soil it affects the overall plant growth and harm soil microorganism results in reducing agricultural productivity (Okunola, et al., 2007). The presence and impact of heavy metals in soil ecosystems are influenced by various natural factors, such as parent material, climate, and soil processes, alongside anthropogenic activities like industry, agriculture, and transportation. Their bioavailability is shaped by physical factors like temperature, phase association, adsorption, and sequestration. Chemical factors, affecting speciation at thermodynamic equilibrium, complexion kinetics, lipid solubility, and octanol/water partition coefficients, also play a role (Hamelink et al., 1994). Biological factors, including species characteristics, trophic interactions, and biochemical/physiological adaptations, significantly contribute to the complex dynamics of heavy metal interactions in the environment (Verkleji, 1993).

Water pollution encompasses various definitions, often indicating the accumulation of substances in water to levels that adversely affect animals or humans. The toxicity of all metals increases at higher concentrations, posing a threat to water quality (Balkis, 2012). Metals are introduced into water bodies through runoff from various sources, including villages, towns, cities, and industries. These metals accumulate in water sediments and, even in trace amounts, can pose significant toxicity risks to both humans and ecosystems. The degree of toxicity varies based on factors such as the type of metal, its chemical properties, its biological role, the exposed organism, and the life stage during exposure.

Elevated levels of metals can induce harm to organisms, with certain heavy metals like mercury, plutonium, and lead being particularly toxic, lacking known vital or beneficial effects on organisms. A significant global issue, water pollution necessitates continual assessment and refinement of water resource policies at all levels, ranging from international to individual aquifers and wells. It is posited that water pollution stands as a primary cause of global deaths and diseases, contributing to over 14,000 daily fatalities (Balkis, 2012). When the heavy metal leaches into groundwater and surface water, it alters the aquatic life and contaminate drinking water sources (Mohan et al., 2018). The impact of metal toxicity extends throughout the food chain. As humans typically occupy the highest trophic level, the accumulation of heavy metals intensifies along the food chain, resulting in heightened risks for human health. Industrial and domestic wastes are commonly discharged into sewage systems, leading to elevated concentrations of heavy metals in raw sewage. In sewage treatment processes, heavy metals may not undergo sufficient degradation. Instead, they are typically removed either in the final effluent or in the produced sludge. The composition and contaminants in sewage entering water bodies depend on the sewage treatment methods employed. To address the problems associated with untreated sewage disposal into rivers and seas, strict regulations have been implemented, and advanced technologies have been developed to minimize pollutant levels released into water systems.

Air pollution is widely acknowledged as a hazardous form of environmental contamination, particularly concerning the presence of heavy metals. Pollutants can enter the atmosphere in various forms, including particles, droplets, and gaseous substances, either independently or in association with particles or droplets. Particles and droplets typically have limited travel distances, settling on the ground shortly after release, although smaller particles can travel longer distances. Gaseous particles, on the other hand, can be transported over significant distances by air masses. Both natural and human activities contribute to the release of particulate matter (PM), particularly fine particles and dust. Natural sources include sand storms, volcanic activity, soil erosion, and the weathering of rocks. Anthropogenic sources involve industrial activities, fossil fuel combustion, vehicle exhaust, smelting, and other human-related processes. Internal combustion engines and jet engines contribute to atmospheric pollution, although measures like catalytic converters and unleaded petrol have mitigated vehicle-related pollution (Masindi and Muedi, 2018). Additionally, the application of pesticides, as well as the use of refrigerators, aerosols, and radioactive materials, serves as additional sources of atmospheric pollutants (Walker et al., 2012).

The anthropogenic mobilization of heavy metals in the biosphere has emerged as a significant facet of geochemical cycling, notably observed in regions with industrial activities releasing substantial quantities of these metals into the atmosphere, soil, and plants, surpassing natural emission levels (Bilos et al., 2001). Elevated concentrations of heavy metals in the air pose a threat to vegetation, resulting in injury and diminished crop yields. The implications extend to becoming a substantial menace to plant survival in industrial zones. The presence of particulate matter can lead to severe health issues, infrastructure deterioration, acid rain formation, corrosion, eutrophication (due to particles falling into water during rainfall), and the formation of haze. Furthermore, heavy metals present in the soil can contribute to the generation of airborne particles and dust, thereby influencing air quality. Prolonged inhalation of significant amounts of these airborne particles may contribute to an increased burden of heavy metals in the human body, posing potential health risks (Gray et al., 2003).

Food Chain Contamination

Heavy metals can accumulate as they move up in the food chain, causing a threat to predators and, eventually, humans (Waqas et al., 2015). The accumulation of heavy metals in an environment is subject to numerous variables, encompassing parent material, soil properties, and human activities like industrial production, traffic, farming, and irrigation. Extensive land areas can experience heavy metal contamination originating from smelters, waste incinerators, industrial wastewater, and the application of sludge, municipal compost, pesticides, and fertilizers. Soil properties play a pivotal role in food production, and the contamination of this essential resource by heavy metals, along with their subsequent absorption and bioaccumulation in food crops, raises significant environmental and health issues, particularly in impoverished nations. The concentrations of heavy metals in the soil are influenced by factors such as soil type, plant genotype, and their interactions (Ding et al., 2013). Compared to organic manure, mineral fertilizers contain higher concentrations of heavy metals, resulting in elevated levels of heavy metal pollution in the soil (Hu et al., 2013). Metals, being nonbiodegradable with extended half-lives, cannot be broken down by biological species. Consequently, they persist in the body parts and surroundings of organisms, posing health risks (Nabulo et al., 2011). The bioaccumulation of heavy metals in vegetables presents a health hazard, as these metals have the potential to transfer from contaminated land and water into the food chain (Stasinos and Zabetakis, 2013).

Crops and seafood have the capacity to absorb heavy metals, leading to human exposure through the consumption of contaminated food. Vegetables and crops cultivated in areas with heavy metal contamination accumulate significant amounts of these toxic substances from the soil, particularly in cereal grains and various plant parts. As a result, humans are exposed to these contaminants through the consumption of both contaminated food and drinking water, resulting in a dietary intake that surpasses the maximum tolerable daily/weekly intake values (MTDI/MTWI) established by regulatory bodies. For example, research indicates that the consumption of contaminated vegetables alone may contribute to over 50% of the provisional tolerable daily intake of individual elements (Mishra et al., 2023). This highlights the substantial impact of heavy metal contamination on the human dietary intake and underscores the need for stringent measures to mitigate the risks associated with such exposure.

Various heavy metals including Cr, Zn, Cd, and Pb have been identified in edible vegetables such as Coriander sativum, *Ocimum basilicum, Beta vulgaris*, and *Raphanus sativus* in Iran. The concentration of these heavy metals, measured in mg/kg, has surpassed the maximum limits set by the World Health Organization (WHO) in some of these vegetables. The elevated levels of heavy metal make these vegetables unsafe for human consumption, posing an increased risk of chronic diseases such as cancer in humans (Manzoor et al., 2018). Human exposure to heavy metals primarily occurs through the consumption of edible vegetables, constituting approximately 90% of the overall intake (Martorell et al., 2011). The remaining 10% is attributed to skin contact and inhalation of polluted dust (Khan et al., 2014). The escalating demand for food in recent decades has elevated food safety as a significant public health concern, prompting extensive research by scientists and researchers on the health risks associated with the ingestion of heavy metals, pesticides, and food products contaminated with toxins (Jaishankar et al., 2014).

Impact on Human Health

Heavy metals impede metabolic functions through dual mechanisms: firstly, by accumulating within vital organs and glands, including but not limited to the heart, brain, kidneys, bone, and liver, thereby causing disruption to their normal physiological processes. Secondly, they exert their detrimental effects by displacing essential nutritional minerals from their native sites, consequently impeding the biological functions of these minerals (Singh et al., 2011). Prolonged exposure to toxic metals can result in significant health issues. Heavy metals like lead, Mercury and cadmium are toxic to humans, causing a range of health issues for example lead exposure is associated with cognitive impairment, developmental delay in children and damage to the nervous system. Mercury exposure can result in neurological disorder, while cadmium exposure can cause kidney damage. The health risks associated with toxic metals depend on their concentrations in specific media and the duration of exposure. Even at low levels, prolonged and chronic exposure to hazardous metals can lead to health problems (Mahalakshmi, 2012).

Heavy metals can disrupt the normal functioning of major organs in human beings, including the brain, heart, and liver. Additionally, they can decrease the activity of the central nervous system. These metals enter the body through the consumption of beverages, food, and inhaled air, accumulating in the tissues of living organisms. Among various heavy metals, chronic exposure to low doses of cancer-causing ones may induce various types of cancer (Junaid et al., 2016). Heavy metals can also lead to a delay in human growth and disturb bioregulatory systems, contributing to functional or psychological disorders such as neurodegenerative pathologies like Alzheimer's, Parkinson's diseases, and chronic fatigue syndrome. Intoxication by metals such as Pb and Hg has been associated with autoimmune disorders like rheumatoid arthritis. Respiratory issues can occur due to the inhalation of particles containing heavy metals from the air; The presence of soil-bound heavy metals can contribute to the production of airborne particles and particulate matter, consequently influencing air quality. Prolonged inhalation of substantial quantities of these airborne particles may lead to an augmented accumulation of heavy metals within the human body, thereby presenting potential health risks. (Gray et al., 2003).

Invertebrates are more sensitive to toxic metals compared to vertebrates, and since they are a crucial part of the food chain, any changes in their abundance can impact human beings as well (Xin et al., 2015). The consumption of vegetables contaminated with heavy metals poses significant health risks for humans, including gastrointestinal cancer, compromised immunological mechanisms, mental growth retardation, and malnutrition (Iyengar and Nair, 2000; Carrizales et al., 2006; Hu et al., 2013; Dickin et al., 2016). Heavy metals have the potential to accumulate in human bones or fatty tissues through dietary intake, leading to the depletion of essential nutrients and weakened immunological defenses (Turkdogan et al., 2003). Specific heavy metals, such as Al, Cd, Mn, and Pb, are also suspected to contribute to intrauterine growth retardation (Rai, 2018).Assessments of health risks related to the consumption of food crops in a developing country, conducted across 30 agro-ecological zones using health indices, indicated that the consumption of vegetables contaminated with heavy metals, especially Mn and Cu, posed greater risks to human health compared to the consumption of contaminated fruits (Shaheen et al., 2016). Another study by Obiora et al. (2016) reported heavy metal contamination, particularly with Pb and Mn, in vegetables grown near a Pb-Zn mine, which could potentially lead to Alzheimer's disease and manganism. Cui et al. (2005) highlighted the systematic health risks associated with a combination of Pb and Cd, rather than the effects of individual metals in isolation, revealing renal dysfunction in a population that ingested foods contaminated with multiple metals.

Heavy Metal Removal Approaches

Efficiently and economically addressing metal ion-polluted discharges poses challenges for researchers and water treatment experts. Various methods have been employed for treating wastewater contaminated with metallic ions, with the aim of safeguarding human health and preserving ecosystems (Wang et al., 2019). These techniques include coagulation-flocculation (Kurniawan et al., 2006; Barakat, 2011; Abbas et al., 2016), chemical precipitation (Ojedokun & Bello, 2016; Burakov et al., 2018), electrodialysis (Pedersen, 2003; Barakat, 2011; Alalwan et al., 2020), membrane filtration (Fu & Wang, 2011; Abbas et al., 2016), and ion exchange (Barakat, 2011; Huang et al., 2007; Taka et al., 2017). However, these methods come with drawbacks, including high construction, implementation, and operational costs, as well as chemical usage. They also require substantial power requirements, operate under critical conditions, and exhibit lower efficacy, especially for metal ion concentrations below 100 mg L^{-1}. Furthermore, these practices generate toxic biosolid waste, posing challenges for disposal and causing harm to ecosystems (Sud et al., 2008; Vuković et al., 2010; Marín-Rangel et al., 2012).

Specifically, the flocculation and coagulation technique for heavy metal removal faces hindrances due to high chemical usage costs. Simultaneously, electroanalysis encounters challenges such as significant power consumption, high construction and implementation costs, and operational expenses. The precipitation technique, although effective in removing sediment, presents difficulties in managing vast amounts of released sediment. Ion exchange has several drawbacks, including its nonselective nature and sensitivity to pH and concentrated heavy metal solutions (Al-Enezi et al., 2004). Moreover, the ion exchange technique, while highly efficient, contributes to environmental pollution due to the regeneration process of polymeric-made material resin (Abbas et al., 2016; Taka et al., 2017; Acharya et al., 2018). In contrast, adsorption entails the transfer of metal ions from the solution phase to a solid phase. Recently, modified biopolymers have been employed for removing heavy metals from contaminated water. Conventional methods of heavy metal removal often result in the generation of toxic sludge, and the operational costs associated with these methods are typically high. As a result, much research has focused on utilizing living organisms as an eco-friendly and sustainable solution for heavy metal removal (Barakat, 2011). Table 2 illustrates the various heavy metal approaches with their advantages and disadvantages.

Table 2. Advantages and limitations of various heavy metal removal approaches

Heavy Metal Removal Approach	Advantages	Limitations	References
Adsorption using conventional adsorbents	Minimal initial costs, simple operation under various pH conditions, strong metallic bond capabilities, high efficacy (up to 99%)	Outcomes dependent on adsorbent material, increased costs due to expensive adsorbents, no potential for adsorbent recovery, limited ion selectivity based on absorbent substance type	Perić et al., 2004; Barakat, 2011; Fu & Wang, 2011
Biosorption utilizing different bio-adsorbents	Inexpensive bio-adsorbents, superior efficacy with regenerative potential, production of minimal residue quantities, recovery of metallic ions without additional nutritional requirements	Bio-adsorbents may saturate earlier, limited progress potential for biological techniques, inability to biologically alter the oxidation state of removed metallic ions	Mishra et al., 2010; Manzoor et al., 2013; Jiménez-Cedillo et al., 2013
Chemical precipitation	Simple process with easy operational parameters, cost-effective, efficient removal of a broad range of metallic ions	Production of large residue amounts, increased running costs for sludge removal, inadequate settling, ineffective for removing metallic ions at minimum levels, reduced efficacy under low pH and high ion concentration	Khosravi & Alamdari, 2009; Abbas et al., 2016; Alalwan et al., 2020
Electrodialysis	Highly selective in metals separation, no chemical consumption, potential for attaining pure metals	Significant initial and operating expenses, substantial power requirements, membrane clogging issue, regular maintenance needed, process efficacy influenced by operational parameters like current, intensity, and pH	Barakat, 2011; Ojedokun & Bello, 2016; Taka et al., 2017
Flocculation and coagulation	Sludge with excellent settlement rate, good dewatering properties, effective removal of metallic ions and water turbidity	Residue production, high costs, large chemical consumption, additional running costs for sludge removal	Fu & Wang, 2011; Abbas et al., 2016; Taka et al., 2017
Flotation treatment	Increased generation of intensified sludge, good selectivity for targeted metal removal, high elimination effectiveness, short retention times with large overflow quantities	Substantial setup expenses, significant running and maintaining costs	Kurniawan et al., 2006; Fu & Wang, 2011; Taka et al., 2017
Ion-exchange	High elimination effectiveness, potential for achieving ppb levels of metallic ions, highly selective metallic ions elimination, efficient recovery rate, capacity for treating large volumes	Substantial initial cost, including resin cost, fewer metallic ions eliminated, resin recovery process leads to additional contamination, removal capacity varies with resin types	Ojedokun & Bello, 2016; Alalwan et al., 2020
Membrane filtration and ultrafiltration method	Highly selective process, minimal pressure needed, compact design, lower dependency on chemicals, excellent efficacy (up to 95%), minimal solid residue	Complicated process, substantial setup, running, and maintenance expenses, membrane clogging, reduced efficiency with additional metallic ions, excessive energy use	Al-Rashdi et al., 2013; Abbas et al., 2016; Taka et al., 2017
Photocatalysis	Simultaneous elimination of metallic ions and organic matter, fewer dangerous side products	Lengthy reaction periods, limited applications	Kajitvichyanukul et al., 2005; Barakat, 2011; Taka et al., 2017
Reverse osmosis	Primarily used in seawater desalination, applicable for removing various organic, mineral, and bacterial pollutants	Solid residue production, elevated pressure requirements, large operating expenses, high power consumption, ineffective for micro-organic contaminants elimination	Kurniawan et al., 2006; Taka et al., 2017

Biosorption of Heavy Metals

Biosorption is the passive absorption of toxic substances through dead or inactive biological materials or biomass, specifically through their cell wall (Singh et al., 2023). This process involves the passive uptake, exclusion, or binding of metal or metalloid species from aqueous solutions through metabolically independent or physicochemical pathways, utilizing low-cost materials derived from biological sources (Patrick, 2003; Gadd, 2009). Various active and inactive organisms have been employed as biosorbents to remove heavy metal ions from aqueous solutions. The term "biosorption" is a combination of "bio," referring to the association with biological material, and "sorption," a physicochemical process involving the attachment of one substance to another, encompassing both absorption and adsorption. Absorption involves the entry of molecules or ions into a bulk phase (liquid or solid), while adsorption is a physical process wherein molecules or ions bind to the surface of another molecule.

Biosorbents are known to be rich in functional groups, such as carboxyl, sulfate, phosphate, and amine groups, which play a crucial role in the removal of various heavy metal contaminants. This method offers a quick, economical, and eco-friendly approach with minimal sludge generation, comparable to conventional ion exchange technology, where dead microorganisms can be utilized. When sorption occurs through live biomass, it is referred to as bioaccumulation, a metabolism-dependent active process (Halttunen et al., 2007; Kinoshita et al., 2016). Singh et al. (2021) highlighted that biosorption is an effective strategy for heavy metal removal, and its efficiency can be enhanced by using different strains of probiotic bacteria. These bacteria have the potential to produce extracellular polymeric substances (EPS), and both the bacteria and extracted EPS can detoxify the adverse effects of heavy metals. Various biomasses across different genera exhibit significant biosorption capacity. This encompasses a range of living or deceased microorganisms, such as algae, bacteria, fungi, and yeast, as well as their components, including exopolysaccharides. Additionally, materials such as seaweeds, plant materials, and various forms of industrial and agricultural waste serve as natural sorbents in the biosorption process. Biosorption involves both a solid phase and a liquid phase (solvent) containing the dissolved species to be sorbed (sorbate/metal ions). This process is characterized as physicochemical, passive, rapid, reversible, and metabolic-independent. The binding properties of biological materials are primarily attributed to the functional groups present on their surfaces. Mechanisms implicated in the biosorption process comprise complexation, adsorption-complexation on surfaces and pores, ion exchange, membrane filtration, microprecipitation, condensation of heavy metal hydroxides onto the biosurface, and surface adsorption (Singh and Saini, 2020). Fig. 2 illustrates the various mechanism involved in biosorption process. Biosorption studies by different biosorbents, factors affecting biosorption, mechanisms of biosorption process, heavy metal accumulation, modelling and kinetics of biosorption process, regulations etc. has been discussed in subsequent chapters in detail.

Figure 2. Biosorption process and their mechanism (Singh et al., 2021)

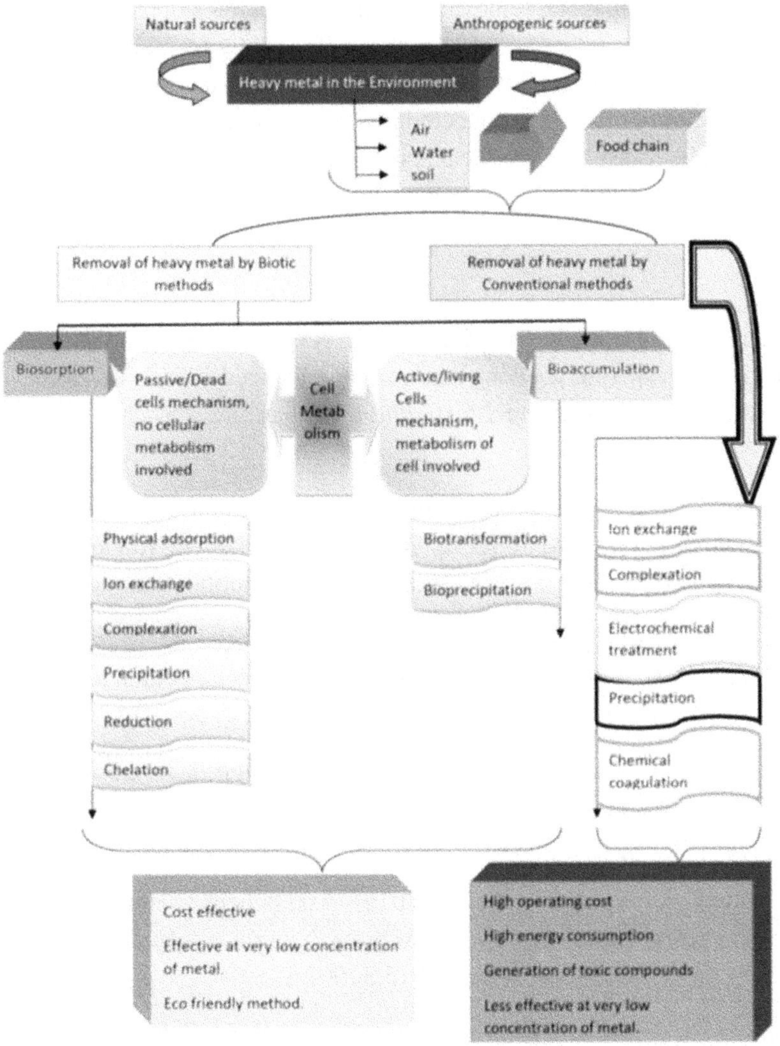

CONCLUSION

Heavy metal pollution poses a significant threat to both the environment and human health, necessitating urgent attention and comprehensive mitigation strategies. The widespread industrial activities, agricultural practices, and improper waste disposal contribute to the release of heavy metals into air, water, and soil, leading to contamination of ecosystems. Exposure to elevated levels of heavy metals, including lead, cadmium, mercury, and arsenic, has been linked to various adverse health effects. Addressing heavy metal pollution requires a multifaceted approach, involving stringent regulations, sustainable industrial practices, and effective waste management. Additionally, public awareness and education play a crucial role in minimizing exposure risks and promoting healthier lifestyles. The development and implementation of innovative technologies, like biosorption offers promising avenues for mitigating the impact of heavy metal pollution on both the environment and human health. The continual exploration of biosorption mechanisms and the optimization of biosorbent materials hold the potential for further enhancing the efficacy of this method in heavy metal removal. Overall, the findings in this field not only contribute to the development of innovative and sustainable solutions for heavy metal pollution but also underscore the importance of considering biosorption as a viable component in comprehensive environmental remediation strategies.I n light of the interconnectedness of environmental and human health, a concerted effort is needed from policymakers, industries, and the public to achieve sustainable solutions.

REFERENCES

Abbas, A., Al-Amer, A. M., Laoui, T., Al-Marri, M. J., Nasser, M. S., Khraisheh, M., & Atieh, M. A. (2016). Heavy metal removal from aqueous solution by advanced carbon nanotubes:critical review of adsorption applications. *Separation and Purification Technology*, *157*, 141–161. doi:10.1016/j.seppur.2015.11.039

Acharya, J., Kumar, U., & Rafi, P. M. (2018). Removal of Heavy Metal Ions from Wastewater by Chemically Modified Agricultural Waste Material as Potential Adsorbent-A Review. *International Journal of Current Engineering and Technology*, *8*(3). Advance online publication. doi:10.14741/ijcet/v.8.3.6

Acosta, J. A., Faz, A., Martínez-Martínez, S., Zornoza, R., Carmona, D. M., & Kabas, S. (2011). Multivariate statistical and GIS-based approach to evaluate heavy metals behavior in mine sites for future reclamation. *Journal of Geochemical Exploration*, *109*(1-3), 8–17. doi:10.1016/j.gexplo.2011.01.004

Addo, M. A., Darko, E. O., Gordon, C., Nyarko, B. J. B., Gbadago, J. K., Nyarko, E., Affum, H. A., & Botwe, B. O. (2012). Evaluation of heavy metals contamination of soil and vegetation in the vicinity of a cement factory in the Volta region, Ghana. *International Journal of Science and Technology*, *2*, 40–50.

Adesuyi, A. A., Njoku, K. L., & Akinola, M. O. (2015). Assessment of Heavy Metals Pollution in Soils and Vegetation around Selected Industries in Lagos State, Nigeria. *Journal of Geoscience and Environment Protection*, *3*(7), 11–19. doi:10.4236/gep.2015.37002

Adriano, D. C., Knox, A., Geebelen, W., Vangronsveld, J., & Mench, M. 2000. Univ of Georgia. Unpublished data.

Agency for Toxic Substances and Disease Registry. (1999). *Toxicological profile for mercury*. Academic Press.

Agency for Toxic Substances and Disease Registry. (2007a). *Toxicological profile for arsenic*. Academic Press.

Agency for Toxic Substances and Disease Registry. (2007b). *Toxicological profile for lead*. Academic Press.

Agency for Toxic Substances and Disease Registry. (2012). Toxicological profile for cadmium. Academic Press.

Agency for Toxic Substances and Disease Registry (ATSDR). (2000). *Toxicological Profile for Arsenic TP-92/09. Georgia*. Center for Disease Control.

Åkesson, A., Bjellerup, P., Lundh, T., Lidfeldt, J., Nerbrand, C., Samsioe, G., ... Järup, L. (2006). Cadmium-induced effects on bone in a population-based study of women. *Environmental Health Perspectives, 114*(6), 830–834. doi:10.1289/ehp.8763 PMID:16759980

Akoto, O., Ephraim, J. H., & Darko, G. (2008). Heavy Metals Pollution in Surface Soils in the Vicinity of Abundant Railway Servicing Workshop in Kumasi, Ghana. *International Journal of Environmental Research, 2*(4), 359-364.

Al-Enezi, G., Hamoda, M. F., & Fawzi, N. (2004). Ion exchange extraction of heavy metals from wastewater sludges. *Journal of Environmental Science and Health. Part A, Toxic/Hazardous Substances & Environmental Engineering, 39*(2), 455–464. doi:10.1081/ESE-120027536 PMID:15027828

Al-Rashdi, B. A. M., Johnson, D. J., & Hilal, N. (2013). Removal of heavy metal ions bynanofiltration. *Desalination, 315*, 2–17. doi:10.1016/j.desal.2012.05.022

Alalwan, H. A., Kadhom, M. A., & Alminshid, A. H. (2020). Removal of heavy metals fromwastewater using agricultural byproducts. *Journal of Water Supply: Research & Technology - Aqua, 69*(2), 99–112. doi:10.2166/aqua.2020.133

Alloway, B. J. (1995). Soil Processes and Behavior of Heavy Metals. In B. J. Alloway (Ed.), *Heavy Metals in Soils* (p. 368). Blackie Academic and Professional Publication. doi:10.1007/978-94-011-1344-1

Balkis, N. (Ed.). (2012). *Water Pollution*. InTech. doi:10.5772/1418

Barakat, M. A. (2011). New trends in removing heavy metals from industrial wastewater. *Arabian Journal of Chemistry, 4*(4), 361–377. doi:10.1016/j.arabjc.2010.07.019

Baselt, R. C., & Cravey, R. H. (1995). *Disposition of Toxic Drugs and Chemicals in Man* (4th ed.). Year Book Medical Publishers.

Battarbee, R. W., Anderson, N. J., Appleby, P. G., Flower, R. J., Fritz, S. C., Haworth, E. Y., ... Stevenson, A. C. (1988). *Lake acidification in the United Kingdom 1800-1986*. Academic Press.

Bhakta, J. N., Ohnishi, K., Munekage, Y., Iwasaki, K., & Wei, M. Q. (2012). Characterization of lactic acid bacteria-based probiotics as potential heavy metal sorbents. *Journal of Applied Microbiology, 112*(6), 1193–1206. doi:10.1111/j.1365-2672.2012.05284.x PMID:22404232

Bilos, C., Colomobo, J. C., Skorupka, C. N., & Rodriguez Presa, M. J. (2001). Source distribution and variability of airborne trace metals in La Plata city area, Argentina. *Environmental Pollution, 111*(1), 149–158. doi:10.1016/S0269-7491(99)00328-0 PMID:11202709

Bobaker, A. M., Alakili, I., Sarmani, S. B., Al-Ansari, N., & Yaseen, Z. M. (2019). Determination and assessment of the toxic heavy metal elements abstracted from the traditional plant cosmetics and medical remedies: Case study of Libya. *International Journal of Environmental Research and Public Health, 16*(11), 1957. doi:10.3390/ijerph16111957 PMID:31159472

Brown, M. J., & Margolis, S. (2012). Lead in drinking water and human blood lead levels in the United States. In The Morbidity and Mortality Weekly Report (MMWR). Washington, DC: Center for Disease Control and Prevention (CDC).

Burakov, A. E., Galunin, E. V., Burakova, I. V., Kucherova, A. E., Agarwal, S., Tkachev, A. G., & Gupta, V. K. (2018). Adsorption of heavy metals on conventional and nanostructured materials for wastewater treatment purposes: A review. *Ecotoxicology and Environmental Safety, 148*, 702–712. doi:10.1016/j.ecoenv.2017.11.034 PMID:29174989

Carolin, C. F., Kumar, P. S., Saravanan, A., Joshiba, G. J., & Naushad, M. (2017). Efficient techniques for the removal of toxic heavy metals from aquatic environment: A review. *Journal of Environmental Chemical Engineering, 5*(3), 2782–2799. doi:10.1016/j.jece.2017.05.029

Carrizales, L., Razo, I., Tellez-Hernandez, J. I., Torres-Nerio, R., Torres, A., Batres, L. E., Cubillas, A.-C., & Diaz-Barriga, F. (2006). Exposure to arsenic and lead of children living near a Cu-smelter in San Luis Potosi, Mexico: Importance of soil contamination for exposure of children. *Environmental Research, 101*(1), 1–10. doi:10.1016/j.envres.2005.07.010 PMID:16171795

Centers for Disease Control and Prevention (CDC). (2001). Managing Elevated Blood Lead Levels Among Young Children: Recommendations From the Advisory Committee on Childhood Lead Poisoning Prevention. CDC.

Centers for Disease Control (CDC). (1991). *Preventing Lead Poisoning in Young children: A statement by the Centers for Disease Control*. CDC.

Chen, Y., Song, R., Yang, R. F., Wu, N., & Li, J. (2014). A novel dopamine D3 receptor antagonist YQA14 inhibits methamphetamine self-administration and relapse to drug-seeking behaviour in rats. *European Journal of Pharmacology, 743*, 126–132. doi:10.1016/j.ejphar.2014.09.026 PMID:25261038

Chibuike, G. U., & Obiora, S. C. (2014). Heavy Metal Polluted Soils: Effect on Plants and Bioremediation Methods. *Applied and Environmental Soil Science, 2014*, 752708. Advance online publication. doi:10.1155/2014/752708

Choong, G., Liu, Y., & Templeton, D. M. (2014). Interplay of calcium and cadmium in mediating cadmium toxicity. *Chemico-Biological Interactions, 211*, 54–65. doi:10.1016/j.cbi.2014.01.007 PMID:24463198

Cui, Y., Zhu, Y., Zhai, R., Huang, Y., Qiu, Y., & Liang, J. (2005). Exposure to metal mixtures and human health impacts in a contaminated area in Nanning, China. *Environment International, 31*(6), 784–790. doi:10.1016/j.envint.2005.05.025 PMID:15979144

De Flora, S., Bagnasco, M., Serra, D., & Zanacchi, P. (1990). Genotoxicity of chromium compounds: A review. *Mutation Research, 238*(2), 99–172. doi:10.1016/0165-1110(90)90007-X PMID:2407950

Deering, T. B., Dickson, E. R., Fleming, C. R., Geall, M. G., McCall, J. T., & Baggenstoss, A. H. (1977). Effect of d-penicillamine on copper retention in patients with primary biliary cirrhosis. *Gastroenterology, 72*(6), 1208–1212. doi:10.1016/S0016-5085(77)80012-7 PMID:870369

Ding, C., Zhang, T., Wang, X., Zhou, F., Yang, Y., & Yin, Y. (2013). Effects of soil type and genotype on lead concentration in rootstalk vegetables and the selection of cultivars for food safety. *Journal of Environmental Management, 122*, 8–14. doi:10.1016/j.jenvman.2013.02.026 PMID:23542226

Domingo, J. L. (1989). Cobalt in the environment and its toxicological implications. *Reviews of Environmental Contamination and Toxicology, 108*, 105–132. doi:10.1007/978-1-4613-8850-0_3 PMID:2646660

Dopp, E., Hartmann, L. M., Florea, A. M., Rettenmier, A. W., & Hirner, A. V. (2004). Environmental distribution, analysis, and toxicity of organometal (loid) compounds. *Critical Reviews in Toxicology*, *34*(3), 301–333. doi:10.1080/10408440490270160 PMID:15239389

Dudka, S., & Adriano, D. C. (1997). Environmental impacts of metal ore mining and processing: A review. *Journal of Environmental Quality*, *26*(3), 590–602. doi:10.2134/jeq1997.00472425002600030003x

El-Kady, A. A., & Abdel-Wahhab, M. A. (2018). Occurrence of trace metals in foodstuffs and their health impact. *Trends in Food Science & Technology*, *75*, 36–45. doi:10.1016/j.tifs.2018.03.001

Elgallal, M., Fletcher, L., & Evans, B. (2016). Assessment of potential risks associated with chemicals in wastewater used for irrigation in arid and semiarid zones: A review. *Agricultural Water Management*, *177*, 419–431. doi:10.1016/j.agwat.2016.08.027

Engwa, A. G., Ihekwoaba, C. J., Ilo, U. S., Unaegbu, M., Ayuk, L. E., & Osuji, A. G. (2015). Determination of some soft drink constituents and contamination by some heavy metals in Nigeria. *Toxicology Reports*, *2*, 384–390. doi:10.1016/j.toxrep.2015.01.014 PMID:28962372

Engwa, G. A., Ferdinand, P. U., Nwalo, F. N., & Unachukwu, M. N. (2019). Mechanism and health effects of heavy metal toxicity in humans. *Poisoning in the modern world-new tricks for an old dog, 10*, 70-90.

Fan, Y., Li, H., Xue, Z., Zhang, Q., & Cheng, F. (2017). Accumulation characteristics and potential risk of heavy metals in soil-vegetable system under greenhouse cultivation condition in Northern China. *Ecological Engineering*, *102*, 367–373. doi:10.1016/j.ecoleng.2017.02.032

Femina Carolin, C., Senthil Kumar, P., Saravanan, A., Janet Joshiba, G., & Naushad, M. (2017). Efficient techniques for the removal of toxic heavy metals from aquatic environment: A review. *Journal of Environmental Chemical Engineering*, *5*(3), 2782–2799. doi:10.1016/j.jece.2017.05.029

Florea, A. M., & Büsselberg, D. (2006). Occurrence, use and potential toxic effects of metals and metal compounds. *Biometals*, *19*(4), 419–427. doi:10.1007/s10534-005-4451-x PMID:16841251

Franca, F. C. S. S., & Albuquerque, A. M. A. (2017). Heavy metals deposited in the culture of lettuce (Lactuca sativa L.) by the influence of vehicular traffic in Pernambuco, Brazil. *Food Chemistry*, *215*, 171–176. doi:10.1016/j.foodchem.2016.07.168 PMID:27542464

Fu, F., & Wang, Q. (2011). Removal of heavy metal ions from wastewaters: A review. *Journal of Environmental Management*, *92*(3), 407–418. doi:10.1016/j.jenvman.2010.11.011 PMID:21138785

Fu, J., Zhang, A., Wang, T., Qu, G., Shao, J., Yuan, B., Wang, Y., & Jiang, G. (2013). Influence of e-waste dismantling and its regulations: Temporal trend, spatial distribution of heavy metals in rice grains, and its potential health risk. *Environmental Science & Technology*, *47*(13), 7437–7445. doi:10.1021/es304903b PMID:23725497

Fu, Z., & Xi, S. (2020). The effects of heavy metals on human metabolism. *Toxicology Mechanisms and Methods*, *30*(3), 167–176. doi:10.1080/15376516.2019.1701594 PMID:31818169

Gadd, G. (2009). Heavy Metal Pollutants: Environmental and Biotechnological Aspects. In *Encyclopedia of Microbiology* (pp. 321–334). Elsevier. doi:10.1016/B978-012373944-5.00149-8

Gadd, G. M. (1990). Fungi and yeasts for metal binding. In H. L. Ehrlich & C. L. Brierley (Eds.), *Microbial Mineral Recovery* (pp. 249–275). McGraw-Hill.

Gall, J. E., Boyd, R. S., & Rajakaruna, N. (2015). Transfer of heavy metals through terrestrial food webs: A review. *Environmental Monitoring and Assessment, 187*(4), 201. doi:10.1007/s10661-015-4436-3 PMID:25800370

Gallagher, C. M., Chen, J. J., & Kovach, J. S. (2011). The relationship between body iron stores and blood and urine cadmium concentrations in US never-smoking, non-pregnant women aged 20–49 years. *Environmental Research, 111*(5), 702–707. doi:10.1016/j.envres.2011.03.007 PMID:21507392

Garbarino, J. R., Hayes, H., & Roth, D. (1995). Contaminants in the Mississippi River US Geological survey circular. Academic Press.

Gautam, P. K., Gautam, R. K., Chattopadhyaya, M. C., Banerjee, S., Chattopadhyaya, M. C., & Pandey, J. D. (2016). *Heavy metals in the environment: Fate, transport, toxicity and remediation technologies. Thermodynamic profiling of pollutants*. Academic Press.

Genchi, G., Carocci, A., Lauria, G., Sinicropi, M. S., & Catalano, A. (2020). Nickel: Human health and environmental toxicology. *International Journal of Environmental Research and Public Health, 17*(3), 679. doi:10.3390/ijerph17030679 PMID:31973020

Gerardo, B., Cabral Pinto, M., Nogueira, J., Pinto, P., Almeida, A., Pinto, E., Marinho-Reis, P., Diniz, L., Moreira, P. I., Simões, M. R., & Freitas, S. (2020). Associations between trace elements and cognitive decline: An exploratory 5-year follow-up study of an elderly cohort. *International Journal of Environmental Research and Public Health, 17*(17), 6051. doi:10.3390/ijerph17176051 PMID:32825289

Ghannam, H. E., El Haddad, E. S. E., & Talab, A. S. (2015). Bioaccumulation of heavy metals in tilapia fish organs. *Journal of Biodiversity and Environmental Sciences, 7*(2), 88–99.

Gray, C. W., McLaren, R. G., & Roberts, A. H. C. (2003). Atmospheric accessions of heavy metals to some New Zealand pastoral soils. *The Science of the Total Environment, 305*(1–3), 105–115. doi:10.1016/S0048-9697(02)00404-7 PMID:12670761

Gross, J. B. Jr, Ludwig, J., Wiesner, R. H., McCall, J. T., LaRusso, N. F., & Ivanova, A. (1985). Abnormalities in tests of copper metabolism in primary sclerosing cholangitis. *Gastroenterology, 89*(2), 272–278. doi:10.1016/0016-5085(85)90326-9 PMID:4007418

Guertin, J. (2005). Toxicity and health effects of chromium (all oxidation states). In J. Guertin, J. A. Jacobs, & C. P. Avakian (Eds.), *Chromium (VI) Handbook* (pp. 216–234). CRC Press.

Günther, V., Lindert, U., & Schaffner, W. (2012). The taste of heavy metals: Gene regulation by MTF-1. *Biochimica et Biophysica Acta, 1823*(9), 1416–1425. doi:10.1016/j.bbamcr.2012.01.005 PMID:22289350

Gworek, B., Dmuchowski, W., & Baczewska-Dąbrowska, A. H. (2020). Mercury in the terrestrial environment: A review. *Environmental Sciences Europe, 32*(1), 128. Advance online publication. doi:10.1186/s12302-020-00401-x

Halim, M., Conte, P., & Piccolo, A. (2002). Potential Availability of Heavy Metals to Phytoextraction from Contaminated Soils Induced by Exogenous Humid Substances. *Chemosphere*, *52*(1), 265–275. doi:10.1016/S0045-6535(03)00185-1 PMID:12729711

Halttunen, T., Salminen, S., & Tahvonen, R. (2007). Rapid Removal of Lead and Cadmium from Water by Specific Lactic Acid Bacteria. *International Journal of Food Microbiology*, *114*(1), 30–35. doi:10.1016/j.ijfoodmicro.2006.10.040 PMID:17184867

Hamelink, J. L., Landrum, P. F., Harold, B. L., & William, B. H. (Eds.). (1994). *Bioavailability: Physical, Chemical, and Biological Interactions*. CRC Press Inc.

Harsha Vardhan, K., Kumar, P. S., & Panda, R. C. (2019). A review on heavy metal pollution, toxicity and remedial measures: Current trends and future perspectives. *Journal of Molecular Liquids*, *290*, 111197. doi:10.1016/j.molliq.2019.111197

Hayat, M. T., Nauman, M., Nazir, N., Ali, S., & Bangash, N. (2018). Environmental hazards of cadmium: past, present, and future. *Cadmium Toxic. Toler. Plants Physiol. Remediat.*, 163–183. doi:10.1016/B978-0-12-814864-8.00007-3

Haynes, W. M. (2011). *CRC handbook of chemistry and physics* (W. M. Haynes, Ed.; 92nd ed.). CRC Press.

He, M., Wang, N., Long, X., Zhang, C., Ma, C., Zhong, Q., Wang, A., Wang, Y., Pervaiz, A., & Shan, J. (2019). Antimony speciation in the environment: Recent advances in understanding the biogeochemical processes and ecological effects. *Journal of Environmental Sciences (China)*, *75*, 14–39. doi:10.1016/j.jes.2018.05.023 PMID:30473279

He, Z. L., Yang, X. E., & Stoffella, P. J. (2005). Trace elements in agroecosystems and impacts on the environment. *Journal of Trace Elements in Medicine and Biology*, *19*(2-3), 125–140. doi:10.1016/j.jtemb.2005.02.010 PMID:16325528

Herawati, N., Suzuki, S., Hayashi, K., Rivai, I. F., & Koyama, H. (2000). Cadmium, copper, and zinc levels in rice and soil of Japan, Indonesia, and China by soil type. *Bulletin of Environmental Contamination and Toxicology*, *64*, 33–39. doi:10.1007/s001289910006 PMID:10606690

Hu, J., Wu, F., Wu, S., Sun, X., Lin, X., & Wong, M. H. (2013). Phytoavailability and phytovariety codetermine the bioaccumulation risk of heavy metal from soils, focusing on Cd-contaminated vegetable farms around the Pearl River Delta, China. *Ecotoxicology and Environmental Safety*, *91*, 18–24. doi:10.1016/j.ecoenv.2013.01.001 PMID:23399100

Huang, S. S., Liao, Q. L., Hua, M., Wu, X. M., Bi, K. S., Yan, C. Y., Chen, B., & Zhang, X. Y. (2007). Survey of Heavy Metal Pollution and Assessment of Agricultural Soil in Yangzhong District, Jiangsu Province, China. *Chemosphere*, *67*(11), 2148–2155. doi:10.1016/j.chemosphere.2006.12.043 PMID:17275882

Hughes, M. F. (2002). Arsenic toxicity and potential mechanisms of action. *Toxicology Letters*, *133*(1), 1–16. doi:10.1016/S0378-4274(02)00084-X PMID:12076506

Hutton, M., & Symon, C. (1986). The quantities of cadmium, lead, mercury and arsenic entering the UK environment from human activities. *The Science of the Total Environment*, *57*, 129–150. doi:10.1016/0048-9697(86)90018-5 PMID:3810138

Iyengar, V., & Nair, P. (2000). Global outlook on nutrition and the environment: Meeting the challenges of the next millennium. *The Science of the Total Environment, 249*(1-3), 331–346. doi:10.1016/S0048-9697(99)00529-X PMID:10813462

Jaishankar, M., Tseten, T., Anbalagan, N., Mathew, B. B., & Beeregowda, K. N. (2014). Toxicity, mechanism, and health effects of some heavy metals. *Interdisciplinary Toxicology*, *7*(2), 60–72. doi:10.2478/intox-2014-0009 PMID:26109881

Jawed, A., Saxena, V., & Pandey, L. M. (2020). Engineered nanomaterials and their surface functionalization for the removal of heavy metals: A review. *Journal of Water Process Engineering*, *33*, 101009. doi:10.1016/j.jwpe.2019.101009

Jiménez-Cedillo, M. J., Olguín, M. T., Fall, C., & Colin-Cruz, A. (2013). As (III) and As (V)sorption on iron-modified non-pyrolyzed and pyrolyzed biomass from Petroselinum crispum(parsley). *Journal of Environmental Management*, *117*, 242–252. doi:10.1016/j.jenvman.2012.12.023 PMID:23376307

Junaid, M., Hashmi, M. Z., Malik, R. N., & Pei, D. S. (2016). Toxicity and oxidative stress induced by chromium in workers exposed from different occupational settings around the globe: A review. *Environmental Science and Pollution Research International*, *23*(20), 20151–20167. doi:10.1007/s11356-016-7463-x PMID:27562808

Kajitvichyanukul, P., Ananpattarachai, J., & Pongpom, S. (2005). Sol-gel preparation andproperties study of TiO2 thin film for photocatalytic reduction of chromium (VI) in thephotocatalysis process. *Science and Technology of Advanced Materials*, *6*(3–4), 352–358. doi:10.1016/j.stam.2005.02.014

Kaye, G. W. C., & Laby, T. H. (1986). *Tables of physical and chemical constants* (15th ed.). John Wiley and Sons Ltd.

Kazantzis, G. (2000). Thallium in the environment and health effects. *Environmental Geochemistry and Health*, *22*(4), 275–280. doi:10.1023/A:1006791514080

Khan, S., Rehman, S., Khan, A. Z., Khan, M. A., & Shah, M. T. (2010). Soil and vegetables enrichment with heavy metals from geological sources in Gilgit, northern Pakistan. *Ecotoxicology and Environmental Safety*, *73*(7), 1820–1827. doi:10.1016/j.ecoenv.2010.08.016 PMID:20810165

Khan, S., Reid, B. J., Li, G., & Zhu, Y. G. (2014). Application of biochar to soil reduces cancer risk via rice consumption: A case study in Miaoqian village, Longyan, China. *Environment International*, *68*, 154–161. doi:10.1016/j.envint.2014.03.017 PMID:24727070

Khosravi, J., & Alamdari, A. (2009). Copper removal from oil-field brine by coprecipitation. *Journal of Hazardous Materials*, *166*(2–3), 695–700. doi:10.1016/j.jhazmat.2008.11.079 PMID:19157701

Kim, Y., & Park, S. (2014). Iron deficiency increases blood concentrations of neurotoxic metals in children. *Korean Journal of Pediatrics*, *57*(8), 345–350. doi:10.3345/kjp.2014.57.8.345 PMID:25210521

Kinoshita, H., Ohtake, F., Ariga, Y., & Kimura, K. (2016). Comparison and Characterization of Biosorption by Weissellaviridescens MYU 205 of Periodic Group 12 Metal Ions. *Animal Science Journal*, *87*(2), 271–276. doi:10.1111/asj.12425 PMID:26223952

Kulka, M. (2016). A review of paraoxonase 1 properties and diagnostic applications. *Polish Journal of Veterinary Sciences, 19*(1), 225–232. doi:10.1515/pjvs-2016-0028 PMID:27096809

Kurniawan, T. A., Chan, G. Y., Lo, W. H., & Babel, S. (2006). Physico-chemical treatmenttechniques for wastewater laden with heavy metals. *Chemical Engineering Journal, 118*(1–2), 83–98. doi:10.1016/j.cej.2006.01.015

Lanphear, B. P., Matte, T. D., Rogers, J., Clickner, R. P., Dietz, B., Bornschein, R. L., Succop, P., Mahaffey, K. R., Dixon, S., Galke, W., Rabinowitz, M., Farfel, M., Rohde, C., Schwartz, J., Ashley, P., & Jacobs, D. E. (1998). The contribution of lead-contaminated house dust and residential soil to children's blood lead levels. A pooled analysis of 12 epidemiologic studies. *Environmental Research, 79*(1), 51–68. doi:10.1006/enrs.1998.3859 PMID:9756680

Lara-Torres, S., Figueiredo, D., Paz, S., Gutiérrez, A. J., Rubio, C., González-Weller, D., Revert, C., & Hardisson, A. (2021). Determination and risk assessment of toxic metals in lipsticks from Europe and China. *Journal of Trace Elements in Medicine and Biology, 67*, 126792. doi:10.1016/j.jtemb.2021.126792 PMID:34022566

Lenntech, K. (2004). *Water treatment and air purification.* Rotter Dam Seweg.

Lentini, P., Zanoli, L., Granata, A., Signorelli, S. S., Castellino, P., & Dell'Aquila, R. (2017). Kidney and heavy metals – the role of environmental exposure. *Molecular Medicine Reports, 15*(5), 3413–3419. doi:10.3892/mmr.2017.6389 PMID:28339049

Li, H., Fagerberg, B., Sallsten, G., Borné, Y., Hedblad, B., Engström, G., Barregard, L., & Andersson, E. M. (2019). Smoking-induced risk of future cardiovascular disease is partly mediated by cadmium in tobacco: Malmö Diet and Cancer Cohort study. *Environmental Health, 18*(1), 56. doi:10.1186/s12940-019-0495-1 PMID:31200698

Li, M., Pi, H., Yang, Z., Reiter, R. J., Xu, S., Chen, X., ... Zhang, Y. (2016). Melatonin antagonizes cadmium-induced neurotoxicity by activating the transcription factor EB-dependent autophagy–lysosome machinery in mouse neuroblastoma cells. *Journal of Pineal Research, 61*(3), 353–369. doi:10.1111/jpi.12353 PMID:27396692

Li, R., Wu, H., Ding, J., Gan, L., & Li, Y. (2017). Mercury pollution in vegetables, grains and soils from areas surrounding coal-fired power plants. *Scientific Reports, 7*(1), 46545. doi:10.1038/srep46545 PMID:28484233

Lin, C., Lu, W., & Wu, Y. (2005). Agricultural soils irrigated with acidic mine water: Acidity, heavy metals, and crop contamination. *Soil Research (Collingwood, Vic.), 43*(7), 819–826. doi:10.1071/SR04148

Litvak, P., Slavkovich, V., Liu, X., Popovac, D., Preteni, E., Capuni-Paracka, S., ... Graziano, J. (1998). Hyperproduction of erythropoietin in nonanemic lead-exposed children. *Environmental Health Perspectives, 106*(6), 361–364. doi:10.1289/ehp.98106361 PMID:9618353

Liu, P., Zhao, H., Wang, L., Liu, Z., Wei, J., Wang, Y., Jiang, L., Dong, L., & Zhang, Y. (2011). Analysis of heavy metal sources for vegetable soils from Shandong Province, China. *Agricultural Sciences in China, 10*(1), 109–119. doi:10.1016/S1671-2927(11)60313-1

Loh, N., Loh, H.-P., Wang, L.K., & Wang, M.-H.-S. (2016). Health effects and control of toxic lead in the environment. *Natural Resources and Control Processes*, 233–284. doi:10.1007/978-3-319-26800-2_5

Loranger, S., & Zayed, J. (1995). Environmental and occupational exposure to manganese: A multimedia assessment. *International Archives of Occupational and Environmental Health*, *67*(2), 101–110. doi:10.1007/BF00572233 PMID:7545647

Lu, H., Shi, X., Costa, M., & Huang, C. (2005). Carcinogenic effect of nickel compounds. *Molecular and Cellular Biochemistry*, *279*(1-2), 45–67. doi:10.1007/s11010-005-8215-2 PMID:16283514

Luo, X., Huo, X., Zhang, Y., Cheng, Z., Chen, S., & Xu, X. (2021). Increased intestinal permeability with elevated peripheral blood endotoxin and inflammatory indices for e-waste lead exposure in children. *Chemosphere*, *279*, 130862. doi:10.1016/j.chemosphere.2021.130862 PMID:34134434

Mahalakshmi, M. (2012). Characteristic levels of heavy metals in canned tuna fish. *Journal of Toxicology and Environmental Health Sciences*, *4*(2), 43–45. doi:10.5897/JTEHS11.079

Mañay, N., Cousillas, A. Z., Alvarez, C., & Heller, T. (2008). Lead contamination in Uruguay: The "La Teja" neighborhood case. *Reviews of Environmental Contamination and Toxicology*, *195*, 93–115. PMID:18418955

Mansour, S. A., Belal, M. H., Abou-Arab, A. A. K., & Gad, M. F. (2009). Monitoring of pesticides and heavy metals in cucumber fruits produced from different farming systems. *Chemosphere*, *75*(5), 601–609. doi:10.1016/j.chemosphere.2009.01.058 PMID:19237184

Manzoor, J., Sharma, M., & Wani, K. A. (2018). Heavy metals in vegetables and their impact on the nutrient quality of vegetables: A review. *Journal of Plant Nutrition*, *41*(13), 1744–1763. doi:10.1080/01904167.2018.1462382

Marín-Rangel, V. M., Cortés-Martínez, R., Villanueva, R. A., Garnica-Romo, M. G., & Martínez-Flores, H. E. (2012). As (V) biosorption in an aqueous solution using chemically treated lemon (Citrus aurantifolia Swingle) residues. *Journal of Food Science*, *77*(1), T10–T14. doi:10.1111/j.1750-3841.2011.02466.x PMID:22122309

Martin, S., & Griswold, W. (2009). Human health effects of heavy metals. *Environmental Science and Technology Briefs for Citizens*, *15*, 1–6.

Martorell, I., Perelló, G., Martí-cid, R., Llobet, J.M., Castell, V., Domingo, J.L. (2011). *Human exposure to arsenic, cadmium, mercury, and lead from foods in Catalonia, Spain: Temporal Trend*. Academic Press.

Masindi, V., & Muedi, K. L. (2018). Environmental contamination by heavy metals. In *Heavy Metals*. InTech. doi:10.5772/intechopen.76082

Mead, M. N. (2010). Cadmium confusion: Do consumers need protection? *Environmental Health Perspectives*, *118*(12), 528–534. doi:10.1289/ehp.118-a528 PMID:21123140

Meltzer, H. M., Brantsaeter, A. L., Borch-Iohnsen, B., Ellingsen, D. G., Alexander, J., Thomassen, Y., Stigum, H., & Ydersbond, T. A. (2010). Low iron stores are related to higher blood concentrations of manganese, cobalt, and cadmium in non-smoking, Norwegian women in the HUNT 2 study. *Environmental Research*, *110*(5), 497–504. doi:10.1016/j.envres.2010.03.006 PMID:20381026

Mhlanga, S. D., Mamba, B. B., Krause, R. W., & Malefetse, T. J. (2007). Removal of organic contaminants from water using nanosponge cyclodextrin polyurethanes. Journal of Chemical Technology & Biotechnology: International Research in Process. *Environmental & Clean Technology*, *82*(4), 382–388.

Mishra, S., Singh, G., Gupta, A., & Tiwari, R. K. (2023). Heavy Metal/Metalloid Contamination: Their Sources in Environment and Accumulation in Food Chain. In R. P. Singh, P. Singh, & A. Srivastava (Eds.), *Heavy Metal Toxicity: Environmental Concerns, Remediation and Opportunities*. Springer. doi:10.1007/978-981-99-0397-9_2

Mishra, V., Balomajumder, C., & Agarwal, V. K. (2010). Zn (II) ion biosorption onto the surfaceof eucalyptus leaf biomass: Isotherm, kinetic, and mechanistic modeling. *Clean (Weinheim)*, *38*(11), 1062–1073. doi:10.1002/clen.201000030

Mohan, V. (2018). Across India, high levels of toxins in groundwater. *The Times of India*.

Mokarram, M., Saber, A., & Sheykhi, V. (2020). Effects of heavy metal contamination on river water quality due to release of industrial effluents. *Journal of Cleaner Production*, *277*, 123380. doi:10.1016/j.jclepro.2020.123380

Moore, T. A., Black, A., Centeno, J. A., Harding, J. S., & Trumm, D. A. (Eds.). (2005). *Metal Contaminants in New Zealand* (pp. 195–219). Resolutionz Press.

Nabulo, G., Black, C. R., & Young, S. D. (2011). Trace metal uptake by tropical vegetables grown on soil amended with urban sewage sludge. *Environmental Pollution*, *159*(2), 368–376. doi:10.1016/j.envpol.2010.11.007 PMID:21129831

Norseth, T. (1981). The carcinogenicity of chromium. *Environmental Health Perspectives*, *40*, 121–130. doi:10.1289/ehp.8140121 PMID:7023928

Nourmoradi, H., Foroghi, M., Farhadkhani, M., & Vahid Dastjerdi, M. (2013). Assessment of lead and cadmium levels in frequently used cosmetic products in Iran. *Journal of Environmental and Public Health*, *2013*, 962727. doi:10.1155/2013/962727 PMID:24174937

O'Neal, S. L., & Zheng, W. (2015). Manganese toxicity upon overexposure: A decade in review. *Current Environmental Health Reports*, *2*(3), 315–328. doi:10.1007/s40572-015-0056-x PMID:26231508

Obiora, S. C., Chukwu, A., & Davies, T. C. (2016). Heavy metals and health risk assessment of arable soils and food crops around Pb–Zn mining localities in Enyigba, southeastern Nigeria. *Journal of African Earth Sciences*, *116*, 182–189. doi:10.1016/j.jafrearsci.2015.12.025

Ojedokun, A. T., & Bello, O. S. (2016). Sequestering heavy metals from wastewater using cowdung. *Water Resources and Industry*, *13*, 7–13. doi:10.1016/j.wri.2016.02.002

Okunola, O. J., Uzairu, A., & Ndukwe, G. (2007). Levels of trace metals in soil and vegetation along major and minor roads in metropolitan city of Kaduna, Nigeria. Academic Press.

Patrick, L. (2003). Toxic Metals and Antioxidants: Part II. The Role of Antioxidants in Arsenic and Cadmium Toxicity. *Alternative Medicine Review*, *8*(2). PMID:12777158

Paul, D., & Sinha, S. N. (2013). Bacteria showing phosphate solubilizing efficiency in river sediment. *Electronic Journal of Biosciences*, *1*(1), 1–5.

Pedersen, A. J. (2003). Characterization and electrolytic treatment of wood combustion fly ash for the removal of cadmium. *Biomass and Bioenergy*, *25*(4), 447–458. doi:10.1016/S0961-9534(03)00051-5

Perić, J., Trgo, M., & Vukojević Medvidović, N. (2004). Removal of zinc, copper and lead by naturalzeolite - a comparison of adsorption isotherms. *Water Research*, *38*(7), 1893–1899. doi:10.1016/j.watres.2003.12.035 PMID:15026244

Peter, A. L. J., & Viraraghavan, T. (2005). Thallium: A review of public health and environmental concerns. *Environment International*, *31*(4), 493–501. doi:10.1016/j.envint.2004.09.003 PMID:15788190

Rai, P. K. (2012). An eco-sustainable green approach for heavy metals management: Two case studies of developing industrial region. *Environmental Monitoring and Assessment*, *184*(1), 421–448. doi:10.1007/s10661-011-1978-x PMID:21465134

Rai, P. K. (2018). Heavy metal phyto-technologies from Ramsar wetland plants: Green approach. *Chemistry and Ecology*, *34*(8), 786–796. doi:10.1080/02757540.2018.1501476

Rawat, M., Ramanathan, A. L., & Kuriakose, T. (2013). *Characterisation of municipal solid waste compost (MSWC) from selected Indian cities—a case study for its sustainable utilisation*. Academic Press.

Rice, K. M., Walker, E. M., Wu, M., Gillette, C., & Blough, E. R. (2014). Environmental mercury and its toxic effects. *Journal of Preventive Medicine and Public Health*, *47*(2), 74–83. doi:10.3961/jpmph.2014.47.2.74 PMID:24744824

Sahmoune, M. N. (2018). Performance of *Streptomyces rimosus* biomass in biosorption of heavy metals from aqueous solutions. *Microchemical Journal*, *141*, 87–95. doi:10.1016/j.microc.2018.05.009

Salonen, J. T., Malin, R., Tuomainen, T. P., Nyyssönen, K., Lakka, T. A., Lehtimäki, T., & Flather, M. D. (1999). Polymorphism in high density lipoprotein paraoxonase gene and risk of acute myocardial infarction in men: Prospective nested case-control study. *British Medical Journal*, *319*(7208), 487–488. doi:10.1136/bmj.319.7208.487 PMID:10454401

Sanfeliu, C., Sebastia, J., Cristofol, R., & Rodriquez-Farre, E. (2003). Neurotoxicity of organomercurial compounds. *Neurotoxicity Research*, *5*(4), 283–305. doi:10.1007/BF03033386 PMID:12835120

Schaefer, H. R., Dennis, S., & Fitzpatrick, S. (2020). Cadmium: Mitigation strategies to reduce dietary exposure. *Journal of Food Science*, *85*(2), 260–267. doi:10.1111/1750-3841.14997 PMID:31957884

Schut, S., Zauner, S., Hampel, G., König, H., & Claus, H. (2011). Biosorption of copper by wine-relevant lactobacilli. *International Journal of Food Microbiology*, *145*(1), 126–131. doi:10.1016/j.ijfoodmicro.2010.11.039 PMID:21195499

Schutte, R., Nawrot, T. S., Richart, T., Thijs, L., Vanderschueren, D., Kuznetsova, T., Van Hecke, E., Roels, H. A., & Staessen, J. A. (2008). Bone resorption and environmental exposure to cadmium in women: A population study. *Environmental Health Perspectives*, *116*(6), 777–783. doi:10.1289/ehp.11167 PMID:18560534

Schwartz, G. G., Il'yasova, D., & Ivanova, A. (2003). Urinary cadmium, impaired fasting glucose, and diabetes in the NHANES III. *Diabetes Care, 26*(2), 468–470. doi:10.2337/diacare.26.2.468 PMID:12547882

Shaheen, N., Irfan, N., Khan, I. N., Islam, S., Islam, M., & Ahmed, M. (2016). Presence of heavy metals in fruits and vegetables: Health risk implications in Bangladesh. *Chemosphere, 152*, 431–438. doi:10.1016/j.chemosphere.2016.02.060 PMID:27003365

Sharma, H., Rawal, N., & Mathew, B. B. (2015). The characteristics, toxicity and effects of cadmium. *International Journal of Nanotechnology and Nanoscience, 3*, 1–9.

Singh, A., Ahmed, M., Saini, P., & Kaur, D. (2023). *Biosorption: Heavy Metals Removal through Biofilms and Microbial Polysaccharides. Food and Industrial Bioprocessing*. New India Publishing Agency.

Singh, A., & Saini, P. (2020). Biosorption: A Novel Biotechnological Application for Removal of Hazardous Pollutants. In *Innovations in Food Technology: Current Perspectives and Future Goals*. Springer. doi:10.1007/978-981-15-6121-4_24

Singh, A., Saini, P., Srivastava, U., Ahmed, M., & Iqbal, U. (2021). Cadmium Toxicology and its Prevention through Lactic Acid Bacteria. *Journal of Food and Agriculture Research, 1*(2), 221–248.

Singh, A., Saini, P., Srivastava, U., & Kushwaha, R. (2021). Biosorption of Cadmium and Lead through Lactobacillus reuteri and Lactobacillus plantarum and its Exopolysaccharides: A Comparative Study. *International Journal of Current Research and Review, 13*(15), 30–35. doi:10.52403/ijrr.20210506

Singh, R., Gautam, N., Mishra, A., & Gupta, R. (2011). Heavy metals and living systems: An overview. *Indian Journal of Pharmacology, 43*(3), 246–253. doi:10.4103/0253-7613.81505 PMID:21713085

Sinha, S. N., & Paul, D. (2012). Detoxification of heavy metals by biosurfactants. *Bulletin of Environmental and Scientific Research, 1*(3-4), 1–3.

Sinha, S. N., & Paul, D. (2014). Heavy metal tolerance and accumulation by bacterial strains isolated from wastewater. *Journal of Chemical, Biological and Physical Sciences, 4*, 812–817.

Sinha, S. N., Paul, D., & Biswas, K. (2016). Effects of Moringa oleifera Lam. and Azadirachta indica A. Juss. leaf extract in treatment of tannery effluent. *Our Nature, 14*(1), 47–53. doi:10.3126/on.v14i1.16440

Spungen, J. H. (2019). Children's exposures to lead and cadmium: FDA total diet study 2014–16. *Food Additives & Contaminants. Part A, Chemistry, Analysis, Control, Exposure & Risk Assessment, 36*(6), 893–903. doi:10.1080/19440049.2019.1595170 PMID:30985263

Srivastava, P., & Kowshik, M. (2013). Mechanisms of metal resistance and homeostasis in haloarchaea. *Archaea (Vancouver, B.C.), 2013*, 732864. doi:10.1155/2013/732864 PMID:23533331

Srivastava, V., Sarkar, A., Singh, S., Singh, P., De Araujo, A. S., & Singh, R. P. (2017). Agroecological responses of heavy metal pollution with special emphasis on soil health and plant performances. *Frontiers in Environmental Science, 5*, 64. doi:10.3389/fenvs.2017.00064

Stasinos, S., & Zabetakis, I. (2013). The uptake of nickel and chromium from irrigation water by potatoes, carrots, and onions. *Ecotoxicology and Environmental Safety, 91*, 122–128. doi:10.1016/j.ecoenv.2013.01.023 PMID:23433838

Sud, D., Mahajan, G., & Kaur, M. P. (2008). Agricultural waste material as potential adsorbent for sequestering heavy metal ions from aqueous solutions - a review. *Bioresource Technology, 99*(14), 6017–6027. doi:10.1016/j.biortech.2007.11.064 PMID:18280151

Sundar, S., & Chakravarty, J. (2010). Antimony toxicity. *International Journal of Environmental Research and Public Health, 7*(12), 4267–4277. doi:10.3390/ijerph7124267 PMID:21318007

Taka, A. L., & Pillay, K. (2017). Nanosponge cyclodextrin polyurethanes andtheir modification with nanomaterials for the removal of pollutants from wastewater: A review. *Carbohydrate Polymers, 159*, 94–107. doi:10.1016/j.carbpol.2016.12.027 PMID:28038758

Tchounwou, P. B., Ayensu, W. K., Ninashvilli, N., & Sutton, D. (2003). Environmental exposures to mercury and its toxicopathologic implications for public health. *Environmental Toxicology, 18*(3), 149–175. doi:10.1002/tox.10116 PMID:12740802

Tchounwou, P. B., Centeno, J. A., & Patlolla, A. K. (2004). Arsenic toxicity, mutagenesis and carcinogenesis – a health risk assessment and management approach. *Molecular and Cellular Biochemistry, 255*(1/2), 47–55. doi:10.1023/B:MCBI.0000007260.32981.b9 PMID:14971645

Tchounwou, P. B., Patlolla, A. K., & Centeno, J. A. (2003). Carcinogenic and systemic health effects associated with arsenic exposure-a critical review. *Toxicologic Pathology, 31*(6), 575–588. doi:10.1080/714044691 PMID:14585726

Tchounwou, P. B., Wilson, B., & Ishaque, A. (1999). Important considerations in the development of public health advisories for arsenic and arsenic-containing compounds in drinking water. *Reviews on Environmental Health, 14*(4), 211–229. doi:10.1515/REVEH.1999.14.4.211 PMID:10746734

Tchounwou, P. B., Yedjou, C. G., Patlolla, A. K., & Sutton, D. J. (2012). Heavy metal toxicity and the environment. *Molecular, clinical and environmental toxicology: volume 3: environmental toxicology*, 133-164.

Teo, J., Goh, K., Ahuja, A., Ng, H., & Poon, W. (1997). Intracranial vascular calcifications, glioblastoma multiforme, and lead poisoning. *AJNR. American Journal of Neuroradiology, 18*, 576–579. PMID:9090426

Thurmer, K., Williams, E., & Reutt-Robey, J. (2002). Autocatalytic oxidation of lead crystallite surfaces. *Science, 297*(5589), 2033–2035. doi:10.1126/science.297.5589.2033 PMID:12242437

Turer, D. G., & Maynard, B. J. (2003). Heavy metal contamination in highway soils. Comparison of Corpus Christi, Texas and Cincinnati, Ohio shows organic matter is key to mobility. *Clean Technologies and Environmental Policy, 4*(4), 235–245. doi:10.1007/s10098-002-0159-6

Turkdogan, M. K., Fevzi, K., Kazim, K., Ilyas, T., & Ismail, U. (2003). Heavy metals in soil, vegetables, and fruits in the endemic upper gastrointestinal cancer region of Turkey. *Environmental Toxicology and Pharmacology, 13*(3), 175–179. doi:10.1016/S1382-6689(02)00156-4 PMID:21782652

Unaegbu, M., Engwa, G. A., Abaa, Q. D., Aliozo, S. O., Ayuk, E. L., Osuji, G. A., & Onwurah, E. I. (2016). Heavy metal, nutrient, and antioxidant status of selected fruit samples sold in Enugu, Nigeria. *International Journal of Food Contamination, 3*(7), 1–8. doi:10.1186/s40550-016-0031-9

United States Environmental Protection Agency. (2002). *Lead Compounds*. Technology Transfer Network- Air Toxics Website. Available: https://www.epa.gov/cgi-bin/epaprintonly.cgi

Vaishaly, A., Blessy, M. B., & Krishnamurthy, N. (2015). Health effects caused by metal contaminated ground water. *International Journal of Advances in Scientific Research, 1*(2), 60–64. doi:10.7439/ijasr.v1i2.1798

Verkleji, J. A. S. (1993). The effects of heavy metals stress on higher plants and their use as biomonitors. In B. Markert (Ed.), *Plants as Bioindicators: Indicators of Heavy Metals in the Terrestrial Environment* (pp. 415–424). VCH.

Vijayakumar, S., Sasikala, M., & Ramesh, R. (2012). Lead poisoning—An overview. *International Journal of Phramcology and Toxicology, 2*, 70–82.

Villarín, M. C., & Merel, S. (2020). Paradigm shifts and current challenges in wastewater management. *Journal of Hazardous Materials, 390*, 122139. doi:10.1016/j.jhazmat.2020.122139 PMID:32007860

Vuković, G. D., Marinković, A. D., Čolić, M., Ristić, M. D., Aleksić, R., Perić-Grujić, A. A., & Uskoković, P. S. (2010). Removal of cadmium from aqueous solutions by oxidized and ethylenediamine-functionalized multi-walled carbon nanotubes. *Chemical Engineering Journal, 157*(1), 238–248. doi:10.1016/j.cej.2009.11.026

Waalkes, M. P., & Rehm, S. (1992). Carcinogenicity of oral cadmium in the male Wistar (WF/NCr) rat: Effect of chronic dietary zinc deficiency*1. *Fundamental and Applied Toxicology, 19*(4), 512–520. doi:10.1016/0272-0590(92)90089-Z PMID:1426709

Walker, C. H., Sibly, R. M., Hopkin, S. P., & D.B.P., T. (2012). *Principles of Ecotoxicology* (4th ed.). CRC Press.

Wang, H., Mi, X., Li, Y., & Zhan, S. (2020). 3D graphene-based macrostructures for water treatment. *Advanced Materials, 32*(3), 1806843. doi:10.1002/adma.201806843 PMID:31074916

Wang, X. F., Xing, M. L., Shen, Y., Zhu, X., & Xu, L. H. (2006). Oral administration of Cr (VI) induced oxidative stress, DNA damage and apoptotic cell death in mice. *Toxicology, 228*(1), 16–23. doi:10.1016/j.tox.2006.08.005 PMID:16979809

Wang, Y., Pan, C., Chu, W., Vipin, A., Sun, L. (2019). Environmental remediation applications of carbon nanotubes and graphene oxide: Adsorption and catalysis. *Nanomaterials, 9*(439).

Wani, A. L., Ara, A., & Usmani, J. A. (2015). Lead toxicity: A review. *Interdisciplinary Toxicology, 8*(2), 55–64. doi:10.1515/intox-2015-0009 PMID:27486361

Waqas, M., Li, G., Khan, S., Shamshad, I., Reid, B. J., Qamar, Z., & Chao, C. (2015). Application of sewage sludge and sewage sludge biochar to reduce polycyclic aromatic hydrocarbons (PAH) and potentially toxic elements (PTE) accumulation in tomato. *Environmental Science and Pollution Research International, 22*(16), 12114–12123. doi:10.1007/s11356-015-4432-8 PMID:25877899

Woldetsadik, D., Drechsel, P., Keraita, B., Itanna, F., & Gebrekidan, H. (2017). Heavy metal accumulation and health risk assessment in wastewater-irrigated urban vegetable farming sites of Addis Ababa, Ethiopia. *International Journal of Food Contamination, 4*(1), 9. doi:10.1186/s40550-017-0053-y

World Health Organization. (2000). *Air quality guidelines for Europe* (2nd ed.). World Health Organization. https://www.who.int/en/newsroom/fact-sheets/detail/lead-poisoning-and-health

Xin, Z., Wenchao, Z., Zhenguang, Y., Yiguo, H., Zhengtao, L., Xianliang, Y., Xiaonan, W., Tingting, L., & Liming, Z. (2015). Species sensitivity analysis of heavy metals to freshwater organisms. *Ecotoxicology (London, England), 24*(7–8), 1621–1631. doi:10.1007/s10646-015-1500-2 PMID:26104218

Xiong, T., Dumat, C., Dappe, V., Vezin, H., Schreck, E., Shahid, M., Pierart, A., & Sobanska, S. (2017). Copper oxide nanoparticle foliar uptake, phytotoxicity, and consequences for sustainable urban agriculture. *Environmental Science & Technology, 51*(9), 5242–5251. doi:10.1021/acs.est.6b05546 PMID:28383257

Xu, L., Lu, A., Wang, J., Ma, Z., Pan, L., Feng, X., & Luan, Y. (2015). Accumulation status, sources and phytoavailability of metals in greenhouse vegetable production systems in Beijing, China. *Ecotoxicology and Environmental Safety, 122*, 214–220. doi:10.1016/j.ecoenv.2015.07.025 PMID:26264401

Yoshizawa, K., Rimm, E. B., Morris, J. S., Spate, V. L., Hsieh, C. C., Spiegelman, D., Stampfer, M. J., & Willett, W. C. (2002). Mercury and the risk of coronary heart disease in men. *The New England Journal of Medicine, 347*(22), 1755–1760. doi:10.1056/NEJMoa021437 PMID:12456851

Young, T. R., & Xiao, Z. (2021). Principles and practice of determining metal–protein affinities. *The Biochemical Journal, 478*(5), 1085–1116. doi:10.1042/BCJ20200838 PMID:33710331

Yu, L., Liou, I. W., Biggins, S. W., Yeh, M., Jalikis, F., Chan, L.-N., & Burkhead, J. (2019). Copper deficiency in liver diseases: A case series and pathophysiological considerations. *Hepatology Communications, 3*(8), 1159–1165. doi:10.1002/hep4.1393 PMID:31388635

Zahir, A., Rizwi, S. J., Haq, S. K., & Khan, R. H. (2005). Low dose mercury toxicity and human health. *Environmental Toxicology and Pharmacology, 20*(2), 351–360. doi:10.1016/j.etap.2005.03.007 PMID:21783611

Zhao, M., Xu, Y., Zhang, C., Rong, H., & Zeng, G. (2016). New trends in removing heavy metals from wastewater. *Applied Microbiology and Biotechnology, 100*(15), 6509–6518. doi:10.1007/s00253-016-7646-x PMID:27318819

Zheng, N., Wang, Q. C., Liang, Z. Z., & Zheng, D. M. (2008). Characterization of heavy metal concentrations in the sediments of three freshwater rivers in Huludao City, Northeast China. *Environmental Pollution, 154*(1), 135–142. doi:10.1016/j.envpol.2008.01.001 PMID:18280624

Chapter 2
Mechanisms of Biosorption:
An Overview

Rakesh Dutta
Pandit Deendayal Upadhyaya Adarsha Mahavidyalaya, Behali, India

Jayashri Dutta
Gauhati University, India

ABSTRACT

Biosorption is the method to remove heavy metals from the ecosystem, and it has replaced the use of other physical and chemical methods that are used for removal of heavy metals. The main cause of heavy metal pollution is industrialization. The industrial waste that contains heavy metal leads to pollution in aquatic life systems. It is also poisonous to humans and other animals. The major highlight of this chapter will be the mechanism of various biosorption processes with the help of different biosorbents such as several micro-organisms, non-living biomass, and some herbal plants. It is eco-friendlier as well as cost effective. Nowadays, vegetable waste and fruit peels are also used as biosorbants for the removal of heavy metals.

1. INTRODUCTION

Nowadays the rapid increase in modernization of society and development of different industries leads to the pollution of heavy metal in environment. The increase in heavy metal pollution increases danger to the human being, animals, plants and life under water i.e. aquatic life. It disturbs almost all normal ecosystems in the nature. It gives rise to major illness to human. Major portion of the heavy metal pollution involves with water pollution due to industrial and agricultural use. Increase in heavy metal pollution also leads to the metal poisoning in human and living creatures. However, the use of heavy metal is essential in various industries, but pollution because of them is one of the serious issue because heavy metals are non-degradable in nature and it is accumulated in nature year after year (Perumal et al., 2021; Bilal et al., 2018).

DOI: 10.4018/979-8-3693-1618-4.ch002

The heavy metal can be classified as essential and non-essential heavy metal based on their requirement. Some heavy metal like Copper, Nickel and Zinc are classified as essential heavy metal which is required by some plants and animals in little amount. However, its excessive level in body is dangerous (Mustapha et al., 2015). The other remaining metals are non-essential heavy metal such as Cadmium, Mercury, Arsenic and Lead. The heavy metal can be removed by various physical and chemical method such as ultra-filtration, chemical precipitation, sludge reduction, chemical oxidation-reduction, ion exchange, reverse osmosis, evaporation etc. However, these methods are neither very productive nor it is cost effective and become impossible to use based on the pollutant concentration, volume and characteristics of aqueous solution (Ahn et al., 2009; Patron-Prado et al., 2010). So, these methods can be replaced by some biological method such as biosorption and bioaccumulation. These biological methods are very much productive in terms of their efficiency, cost effectiveness, very promising and fast process. The biosorption is the process of adsorption of heavy metals on the surface of different biomass. In biosorption, generally dead biomass is used because it has many advantages over the living one. So, it is called as passive method (Swiatek et al., 2014). The process of biosorption is better method over many other methods which are used for removal of pollutants such as heavy metals from aqueous solution specially when dealing with low concentrations. However, in bioaccumulation, living biomass is used for the removal of heavy metal. So, it is called active method.

In these chapter, we will mainly focus on different methods of biosorption and its mechanism to remove heavy metals from different sources of aqueous solution. The mechanism of heavy metal biosorption process is influenced by the status of biomass, types of biomaterials, chemical properties of metal solution and environmental conditions such as pH and temperature (Das et al., 2008). Mechanisms of biosorption mainly depends on the types of biosorbent used. Bioaccumulation is the metabolism dependent process also can be called as extracellular accumulation or cell surface adsorption. Whereas Intracellular metabolism is dependent on cell metabolism where transport of metal takes place across the cell membrane (Ahalya et al., 2003; Ahluwalia and Goy, 2007; Javanbakht et al., 2014). Biosorption is the metabolism independent process that is faster and reversible in comparison to metabolism dependent bioaccumulation (Tiwari et al., 2022).

2. GENERAL ASPECT

Most of the heavy metals occur naturally in the environment or on the earth surface and earth crust. The heavy metal can be defined as metals with density higher than water or high atomic weight, or atomic number (Brathwaite and Rabone, 1995; Helmenstine, 2018). It can be introduced into the environment through natural sources, which involve volcanic emissions, forest fires etc. or it can be introduced through anthropogenic sources, which involves mining, smelting, painting industries, leather tanning, battery manufacturing, petroleum refining, metal manufacturing, paints, coating industries, tanneries, printing, pesticides, agricultural runoff, etc. Typically, the presence of heavy metals is of low concentration in agricultural soils, water of rivers and lakes, but not on earth. Heavy metals are extracted from the earth's crustal ores and anthropogenic activities contribute to their migration to areas where they are naturally found in low concentrations. Due to their non-degradable nature, they are considered to be very harmful to living organisms. Heavy metals are classified in two categories, one is essential and another is non-essential heavy metals (Shamim, 2018).

Table 1. Types and examples of heavy metal

Types of Heavy Metal	Essential Heavy Metals	Non-Essential Heavy Metals
Examples of Heavy Metal	Copper, Zinc, Chromium (III) and Nickel	Lead, Mercury, Cadmium, Arsenic and Chromium (VI)

The types of heavy metal and its examples were listed in Table 1. The essential heavy metals are required in less amount by the body. The essential heavy metal is responsible for majority of the biological functions like oxidation and reduction processes, various transportation process etc., but its high concentration in the body is very much toxic to human health (Shamim, 2018; Iyengar, 1998). Essential heavy metals are also required by plants in small quantities. Deficiency of essential heavy metals in the body may cause many deficiency diseases or syndromes. However, non-essential heavy metals, as their name suggest, have no role in our body. They only induce toxic effects to the living being even when are present in the environment in very low concentrations (Koller and Saleh, 2018). Both the essential and non-essential heavy metal can cause the death depending upon the amount of ingestion by the person. But the non-essential metals are more toxic than the others. Some non-essential heavy metals like mercury may cause serious health problems that worsen over time and eventually lead to death or some heavy metal like arsenic may cause cancer. So, these heavy metals are categorized as carcinogenic (Ahalya et al., 2003).

3. METHODS FOR REMOVAL OF HEAVY METALS:

In our country, rapid growth of population and high standard of living leads to increased use of heavy metals in our everyday life. As a result, it increases the introduction of heavy metals as pollutants in the environment (Koelmel et al., 2016). All heavy metals are toxic, non-biodegradable in nature and have the potential to accumulate along food chain. They remain in the environment due to their persistence nature and it enter the food chain of living organisms and result many adverse health effects to the living being like damage to many vital organs that provide a route for deafness, blindness, infertility, kidney problems, respiratory problems, and even cause cancer which leads to death of the person. The toxic nature of these heavy metals may also disturb our normal body function like circulatory system, gastrointestinal, respiratory and also nervous systems.

Table 2. Different types of methods used for removal of heavy metal

Types of Methods	Name of Methods
Physical method	Ion-exchange, Reverse osmosis, Ultrafiltration and Electrodialysis
Chemical method	Chemical precipitation and Electrochemical treatment
Biological method	Biosorption and Bioaccumulation

Different types of methods used for removal of heavy metal are shown in Table 2. The above-mentioned methods are capable of removing heavy metals pollutants from the environment. However, the physical and chemical methods have some disadvantages like they are labour-exhaustive, lack of selectivity in the removal process, the requirement of energy in the treatment process, high cost, needed high amount of reagent, production of toxic by-product and other toxic materials which are also harmful for the environment (Shamim, 2018; Ahalya et al., 2003; Kanamarlapudi et al., 2018). When the concentration of the pollutant is very low, then these methods are unable to remove the heavy metal efficiently (Kanamarlapudi et al., 2018). Based on all these disadvantages, there is a need for a better method that is efficient, eco-friendly and inexpensive in removing heavy metal pollutants from solution having very low concentration also. In this regard, biological method is the most favorable method that gives the desired outcome.

3.1. Biological Methods

Biological method appears as an environment friendly substitute for the physical and chemical methods. It has many advantages over these methods. In these biological methods, biomass like bacteria, fungi, algae, agricultural product, vegetables & fruits, industrial waste, and herbal plant parts are used as bio sorbent for removal of heavy metal pollutant from the environment. For the biosorption of heavy metal these bio sorbents have some functional groups on their outer layer which bind to the metal ions and pulled out the metal ions from the contaminated solution. The regeneration of the bio sorbents is only possible after desorption of metal ions from the active site of bio sorbents, and these regenerated bio sorbents can be used again for the removal of heavy metal, which makes this method more economical in comparison to the other heavy metal removal methods. For large scale removal of heavy metal pollutants, it's often necessary to modify the bio sorbents both chemically and physically for better performance (Kanamarlapudi et al., 2018). In Figure 1, we have shown the kind of bio sorbent material with reference to their cost effectiveness.

Figure 1. Representative scheme showing different kind of biomass based on their cost effectiveness

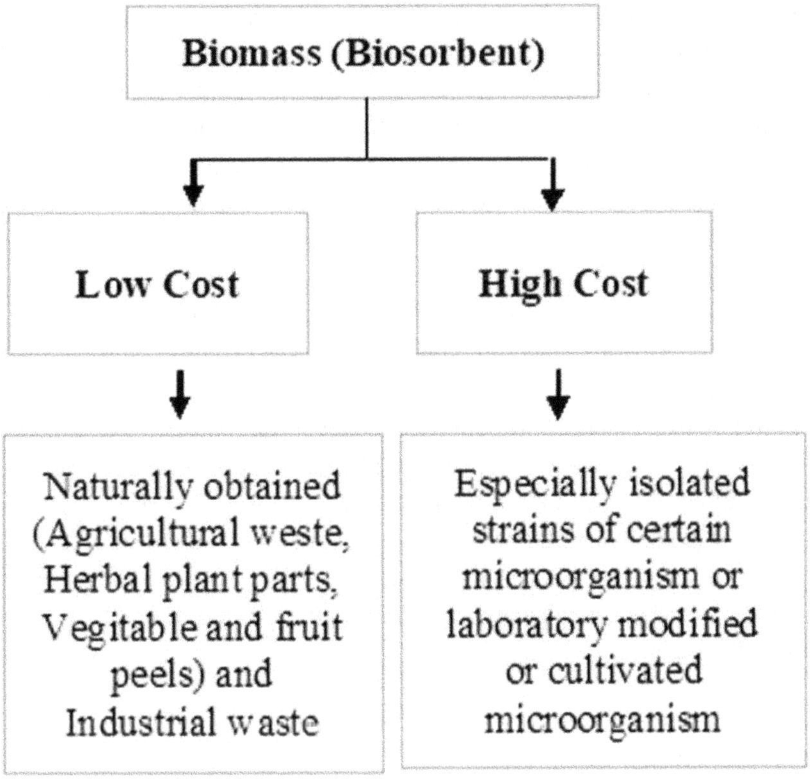

3.2. Biosorption

Biosorption is a natural treatment process which is cheap, fast and environment-friendly heavy metal removal technology (Bilal et al., 2018; Michalak et al., 2013). According to Bohumil Volesky (2007) "Biosorption has been defined as the property of certain biomolecules (or types of biomass) to bind and concentrate selected ions or other molecules from aqueous solutions. As opposed to a much more complex phenomenon of bioaccumulation based on active metabolic transport, biosorption by dead biomass (or by some molecules and/or their active groups) is passive and based mainly on the "affinity" between the bio-sorbent and sorbate." This method has many advantages over the conventional physical and chemical methods. It includes less expensive, regeneration of biological sorbent material, high removal efficiency, complete removal of metal ions, no nutrient requirement etc. In addition to the above advantages, it is highly selective for specific heavy metals removal. This method also minimized the accumulated sludge or production of other toxic materials. As a whole this biosorption process has effectiveness over a broad range of environmental conditions like pH value of solution, temperature, metallic ion concentration, *etc.* (Bilal et al., 2018; Shamim, 2018; Ahalya et al., 2003; Michalak et al., 2013).

4. CONVENTIONAL CANDIDATES AS BIO SORBENT MATERIALS

For the removal of heavy metals biologically, a variety of biomass is available in nature (Michalak et al., 2013). A large variety of bio sorbent are present like algae (microalgae, seaweeds, macroalgae), fungi, bacteria, yeasts which are mainly used because of their great affinity and good accomplishment towards heavy metal removal (Bilal et al., 2018; Michalak et al., 2013). Ideal bio sorbent used should possess desired featured like, easy to regenerate /reusability, non-toxic in nature, highly available, high affinity towards metal-binding, large scale production, no or minimal use of hazardous chemicals (Bilal et al., 2018). In some case agricultural wastes and waste from industrial operations are also used as bio sorbent candidate in heavy metal removal.

For example, solid residues obtained from the olive mill, available waste mycelia from fermentation process (Pagnanelli et al., 2002), activated mud or slurry released from sewage treatment plants (Hammaini et al., 2003), biosolids (Norton et al., 2004), submerged aquatic plants are used for the process (Keskinan et al., 2003). A particular bio sorbent is specific for the removal of a particular heavy metals, while some bio sorbent materials have a wide range of binding, collecting and removing most of the heavy metals with peculiar activity. Some bio sorbents used for removal of heavy metal are particular isolated species of some particular microorganisms and some bio sorbent undergoes some process to improve their biosorption characteristic (Ahalya et al., 2003). A very low expensive source of biomass can also act as a good bio sorbent which can be easily obtained from sea or ocean like ocean seaweeds or algae (Bilal et al., 2018).

5. MECHANISM OF BIOSORPTION

As a very promising, cost-effective, highly efficient and fast process, implementation and development, biosorption is better method over many other methods which are used for removal of pollutants such as heavy metals from aqueous solution specially when dealing with low concentrations. Mechanism of heavy metal biosorption process is influenced by the status of biomass, types of biomaterials, chemical properties of metal solution and environmental conditions such as pH and temperature (Das et al., 2008). Mechanisms of biosorption mainly depends on the types of biosorbent used. To understand the mechanism involving in metal binding, it is necessary to know the functional groups involved which are different for cell membranes of different biomass. Some of the functional groups are shown by spectroscopic studies that have potential of metal binding are hydroxyl, phosphate, amino, imine, carboxyl, sulfhydryl, amide, sulphate, ketone, phosphate, phosphodiester (Javanbakht et al., 2014).

There are two types of sorption based on the cell metabolism viz., bioaccumulation which is metabolism dependent and biosorption which is metabolism independent. Bioaccumulation is an intracellular process that takes longer time and requires energy sources also called active process. This process can be inhibited by the low temperature, lack of nutrients and metal toxicity. Biosorption is faster and reversible in comparison to bioaccumulation and also called passive process. This is an energy independent process and can be performed by both living and dead biomass. In this process, ions bound to the surface of the biosorbents (Timkova et al., 2018). In Figure 2, we have shown all the classification of mechanism regarding the biosorption process.

Figure 2. Classification of mechanisms of biosorption and bioaccumulation

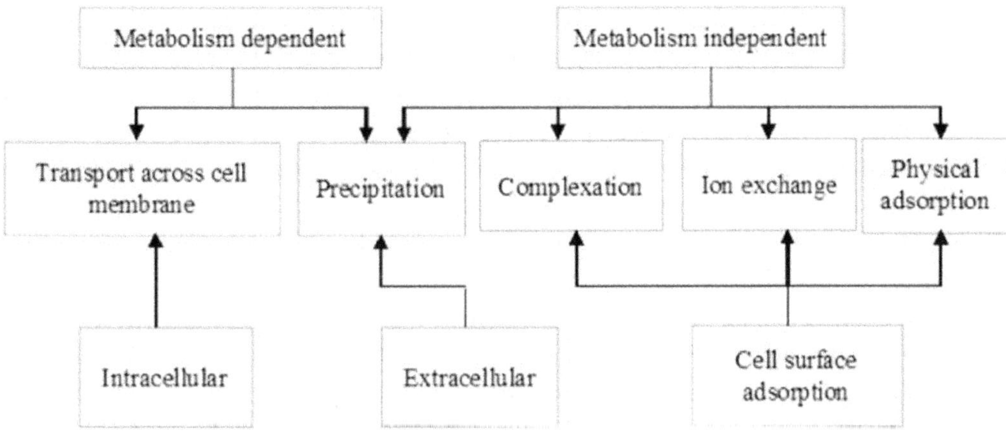

5.1. Transport Across Cell Membrane

This mechanism takes place in two stages: metabolism independent where metal ions bind to the binding sites on the surface of the cell wall of the microorganisms and metabolism dependent intracellular uptake where metal ions transport through the cell membrane into the cell. The mechanism is similar as the cells take essential ions (Redha, 2020). The removal of essential ion copper by *Aspergillus niger* is achieved by this metabolic process and that promotes internal adsorption in the cell. The bioaccumulation takes place inside the cell to accumulate metals which is very slow metabolism dependent process (Hamza et al., 2010; Javanbakht et al., 2014).

5.2. Precipitation

At the beginning, it was believed that sorption-precipitation process accumulates the metals inside the cell at higher levels of pH (Venkata et al., 2009). Precipitation takes place in both metabolism dependent and metabolism independent biosorption where insoluble metal is precipitated. Living microorganisms when encounter the environment holding foreign metal ion, their active defence system starts working and precipitate the metal ions (Redha, 2020). In the metabolism independent process, chemical reaction takes place between the functional groups present in the cell wall of microorganisms and the metal ions Kanamarlapudi et al., 2018). The process can lead to high capacity of biosorption which may hinder desorption. The metabolism independent sorption by non-living biomass extensively precipitate actinides like uranium and thorium (Gadd and White, 1992). Precipitation may take place both in the cell walls of the microbes and in the solution (Ahalya et al., 2003). The precipitation and ion exchange mechanisms are confirmed for biosorption of iron (III) by green tomato musk where iron (III) hydroxide is precipitated at pH 6. The amine and carboxyl groups present in the watermelon rind precipitate copper, lead and zinc. The ion exchange process also responsible for this biosorption process (Liu et al., 2012).

5.3. Physical Adsorption

Physical adsorption is a mechanism of biosorption where electrostatic attraction such as Van der walls force works on the surface of the biosorbent material and it is a reversible process (Raji et al., 2023). The process is sometimes affected by pH of the solution and surface area of the biosorbents (Redha, 2020). The hypothesis given by Kuyucak and Volesky, 1988 suggest that the dead biomass of yeast, algae and fungi biosorbed zinc, cobalt, copper, uranium and cadmium by electrostatic attraction between the cell wall of the microbes and the metal ions. The biosorption of chromium (III) ion takes place through physical adsorption by wheat straw and grass that is suggested based on knowledge of ionic radius of the metal ion and maximum biosorption capacity of the biomass (Chojnacka, 2006). Physical adsorption has been found as responsible mechanism for chromium (VI) biosorption by fungi *Aspergillus niger* and *Ganoderma lucidum* and biosorption of copper by *Chlorella vulgaris* and bacteria *Zoogloea ramigera* (Ahalya et al., 2003). Physical adsorption takes place in the biosorption of lead by hami melon peels at higher pH with the aid of functional group carboxylate and hydroxylate whereas at low pH, biosorption is attained through ion exchange process (Redha, 2020).

5.4. Complexation

In the process of complexation, a complex is formed through electrostatic attraction forming covalent bonds between the metal ions and the functional groups present on the cell wall of microorganisms (Redha, 2020). Chelation is an advanced form of complexation where active groups can form coordinate bond with metal ions form more than one position at the same time to form a highly stable metal complex. Mononuclear metal complexes are formed when one metal in the centre is bounded by functional groups and polynuclear complexes are formed when two or more metals are bounded by functional groups (Bilal et al., 2021). The organic acids such as citric acid, oxalic acid, lactic acid, gluconic acid etc. formed by microorganisms may chelate the metals forming metallo-organic compounds. The microbial polysaccharides and polymers contain carboxyl groups that may biosorbed heavy metals. For the accumulation of cadmium, zinc, mercury, copper by the microorganism *Pseudomonas syringae*, complexation is found as the only mechanism (Raji et al., 2023). In the determination of the most significant factor in complexation (affinity of the ligands and metals), hard and soft acids and bases (HSAB) theory is followed. The HSAB theory states that elements are categorised into hard and soft acids specially in case of metals with smaller ionic radii and hard and soft bases in case of non-metals with larger ionic radii. Hard acids have strong binding affinity to hard bases and soft acids have strong binding affinity to soft bases. For example, the soft acid Pb favour binding covalently with biosorbents which contains soft bases such as nitrogen or sulphur (Ayers, 2005). Energy dispersive electron microscope, scanning electron microscope (SEM) confirmed the complexation mechanism of chromium by *Chlorella miniate* reacting with the phosphonate, amine and carboxyl ligands (Han et al., 2006).

5.5. Ion Exchange

The process of ion exchange takes place as exchange of ions between metals and the aqueous solution. The polysaccharides present in the cell walls of microorganisms are exchanged with bivalent metal ions (Raji et al, 2023). The bacteria and fungi have polysaccharides in their cell walls which may attribute to the ion exchange mechanism to exchange the metal ions. Copper ion is biosorbed by *Gandoderma lucidium* through this process of ion exchange (Redha, 2020). The metal ions of zinc, cadmium, cobalt and copper are exchanged by the marine algae which have alginates as a salt of potassium, sodium, magnesium and calcium (Ahalya et al., 2003). A study by Kanamarlapudi et al., 2018 reported that rice straw which is a metabolism independent biosorbent can exchange cadmium ion with potassium, sodium, calcium and magnesium. They also reported that a cyanobacteria Spirulina have hydroxyl, carboxyl and phosphate functional groups that contribute to the occurrence of ion exchange process to biosorb copper (II), cadmium (II) and chromium (III). The ion exchange mechanism takes place in biosorption of lead by hami melon peels at low pH with the aid of carboxylate and hydroxylate functional groups (Redha, 2020).

In some cases, more than one biosorption mechanism that mentioned above take place concurrently.

For example, in the sorption of chromium (III) and copper (II), by soybean in contribution with carboxyl and hydroxyl group, multiple mechanisms viz., precipitation, ion exchange and chelation take place simultaneously (Witek Krowiak and Kumar Reddy, 2013). Ion exchange, complexation and precipitation together works in the sorption of iron and manganese by tomato husk in presence of carboxylic and phenolic groups (Garcia Mendieta et al., 2012). In Table 3, we have shown different metal ion which undergoes sorption in presence of different biomass along with their functional groups and the mechanism which is responsible for biosorption process. From the above table we can see that, different mechanism is responsible for the same functional group present in different biomass.

5.6. Oxidation and Reduction

Oxidation and reduction process can be possible only by the prokaryotes which can oxidize metal such as Cr^{6+}, Mn^{2+}, Fe^{2+}, Co^{2+}, Cu, AsO^{2}, SeO_3^{2} and reduce the metals such as Mn^{4+}, Fe^{3+}, Co^{3+}, AsO_4^{2-}, SeO_3^{2-} while gaining the energy from those reactions (Perpetuo et al., 2014). The solubility of the metal ions increases when reduction takes place. For example, when Mn^{4+} and Fe^{3+} reduce to Mn^{2+} and Fe^{2+} respectively, their solubility increases. Microorganisms can reduce the heavy metals by lowering their oxidation state to a less bioactive metallic element by proficiently immobilizing them (Valls and Lorenzo, 2002; Gadd, 2004).

Table 3. The mechanisms associated with the respective biomass and metal sorption

Heavy Metal	Biomass	Functional Group	Mechanism	References
Pb, Cd, Ni, Zn	*Aspergillus niger, Mucor rouxii*	Carboxylate, phosphate, amine	Ion exchange	Yan and Viraraghavan, 2008
Fe (III)	Bengal gram husk	Hydroxyl, carboxyl, amide	Complexation	Ahalya et al., 2006
Cd, Co, Cu, Ni, Zn	*Aspergillus niger*	Carboxyl	Ion exchange	Akthar et al., 1996
Cu, Pb	*Fusarium equiseti* KR706303 *Penicilium citrium* KR706304	Carbonyl, hydroxyl, amide	Complexation	Akinkunmi et al., 2016
Pb (II), Zn (II)	*Oceanobacillus profundus*	Carboxyl, sulphate, phosphate	Chelation, Complexation	Mwandira et al., 2020
Cd (II)	*Pseudomonus plecoglossicia*	-NO, -OH, -CH, -CONH	Electrostatic interaction, Ion exchange	Guo et al., 2012
Cd (II)	Rice straw	C=C, carboxyl, hydroxyl	Chelation, ion exchange	Ding et al., 2012
Cu (II), Pb(II)	Barley straw	Carboxylic, hydroxyl	Complexation, Ion exchange	Pehlivan et al., 2012
Cr (VI)	Inactivated fungal biomass	Carboxyl, sulfhydryl, amides, phosphate	Ion exchange, Physical adsorption, Electrostatic interaction, Complexation	Ramrakhiani et al., 2011
Cu (II)	*Pichia pastoris* X33	Amino, hydroxyl, carbonyl, C=O, -CH, C-N	Physical adsorption	Chen et al., 2021
Sb (III)	*Cyanobacteria mycrocystis*	Carboxyl, hydroxyl, amino group	Complexation	Sun et al., 2011
Cd (II), Cr (III), Cu (II)	*Spirulina* sp.	Phosphate, carboxyl, hydroxyl	Ion exchange, Physical adsorption	Chojnacka et al., 2005
Cr (IV)	Date pits/tea waste	$-NH_2$, -COOH, $-CO-CH_2$	Complexation	Albadarin et al., 2013
Cd (II), Cr (III), Cu (II)	*Cyanobacteria Spirulina*	Hydroxyl, carboxyl, phosphate	Ion exchange	Kanamarlapudi et al., 2018
Fe, Mn	Tomato husk	Carboxylic, phenolic	Ion exchange, Complexation, Precipitation	Garcia Mendieta et al., 2012

6. FACTORS AFFECTING BIOSORPTION

The process of biosorption is mainly depends on different environmental conditions such as pH, temperature, metal ion concentration in the solution, the dose of bio sorbent applied, bio sorbent competing for extraction of metal ions, functional group present in the sorbent and time of contact with metal ion (Bilal et al., 2018; Ahalya et al., 2003). The factors affecting biosorption can be classified into two categories which are physical and chemical properties. The chemical factors affecting biosorption include molecular weight, ionic radius of metal ions, the oxidation state of metal ions and structural tropology of microbial cell surface, *etc*. In other way the physical factors include process characteristic like pH range, temperature range, sorbate concentration applied, biological material used, contact time of sorbent with the metal ions, dose of sorbent applied, *etc*. Michalak et al., 2013). Among all the above factors, the pH factor is the most significant and deciding factor in case of sorbent metal binding for extraction of heavy metal (Bilal et al., 2018; Ahalya et al., 2003; Michalak et al., 2013). The pH of the targeted solution is responsible for the redox reactions that took place during binding, the chemistry of pollutant solutions, precipitation, hydrolysis, and competition of metallic ions and affinity towards sorbent molecule (Ahalya et al., 2003; Michalak et al., 2013). The value of optimum pH varies for every biological material or bio sorbent molecule. The medium is responsible for the protonation and deprotonation of functional group, which also alter the capacity of biosorption by the sorbent molecule (Bilal et al., 2018).

Another major factor that is responsible for the biosorption process is the temperature. An increase or decrease in temperature changes the thermodynamic function, which cause variation in biosorption mechanism. In case of increase in temperature, the rate of biosorption of lead metal by algae as the sorbent was found to be increase. However, in the similar condition it causes decrease in the rate of biosorption of iron, manganese and zinc metal. Sometimes very high temperature or increase in temperature denature the structure of biological sorbent material (Bilal et al., 2018). In the temperature range of 20-35 °C, it shows no change in the biosorption mechanism (Ahalya et al., 2003).

The contact time also alter the biosorption process. As the contact time increases up to optimum time, the rate of biosorption increases. But after some time of biosorption, the rate of sorption attained equilibrium state because of the saturation of all binding sites. After this equilibrium position, the binding of heavy metal to the sorbent material stop. For every biomass, the duration of contact time is also varying (Bilal et al., 2018).

In addition to the above condition, the presence of other metal ions also affect the removal of a particular metal ion by occupying the binding site. For example, biosorption of uranium metal ion by *Rhizopus arrhizus* is reduced by the presence of other metal ions such as iron and zinc ions (Ahalya et al., 2003; Tsezos and Volesky, 1982).

7. FUTURE ASPECTS

With increasing urbanization and industrialization, heavy metals along with other pollutants increases which creates major health concern for the people mainly the urban people. Therefore, biosorption using living and non-living biomass will be a better option to expel the pollutants from the environment. The utilization of naturally derived and waste biomass as biosorbent materials forms the foundation of an innovative and economically efficient technology. This technology holds the greatest potential for widespread implementation in the purification of industrial wastewater contaminated with heavy metals (Das et al., 2008). In near future, the removal of heavy metal from the contaminated water will be purely based on the bio sorbent materials. In this regard, more research work should be carried out to increase the use of agricultural waste, fruits and vegetable and herbal plant parts as biosorbent of heavy materials for the biosorption heavy metal from contaminated water.

8. CONCLUSION

The used of bio sorbent materials for the purpose of biosorption process is a very cost-effective and eco-friendly method than the other physical and chemical method for the extraction of potentially hazardous heavy metals from the environment. This process of removal technique of heavy metal pollutants still favourable for the solution having very low concentration of pollutants. Different researchers have studies different biosorption process and their individual mechanism of metal uptake. During this process of recovery of the bio sorbents for repeated use, intense care should be taken, so that the extracted heavy metal along with the bio sorbent molecule should not be exposed again to the environment. In case of removing heavy metal from treating wastewater, the desired bio sorbent material should be used which have an affinity for different metals or multiple bio sorbents material should be used which have different binding specificity for different metals at the same time. It will reduce time taken for the removal process and steps in the biosorption method. As we known that the removal of a specific metal can be hindered by the presence of other similar heavy metal ions. In this regard, precautions must be taken during the removal process. While during the treatment of wastewater using biosorption, safe and eco-friendly bio sorbent materials like vegetables waste, fruits, herbal plant parts or agricultural waste should be used in place of other chemical bio sorbents. The used of microorganisms as biomass for biosorption should be replaced by the agricultural waste, which will have reduced the generation of further waste to the environment. For more development in the field of biosorption, further research investigations and studies should be continued in the direction of utilization of low-cost bio sorbents materials with a maximum binding affinity for heavy metal removal. Further steps must be taken to enhance the use of biosorption on a large scale in industries to increase the impact of biosorption in environment.

REFERENCES

Ahalya, N., Kanamadi, R. D., & Ramachandra, T. V. (2006). Biosorption of iron(III) from aqueous solutions using the husk of Cicer arientinum. *Indian Journal of Chemical Technology*, *13*(2), 122–127.

Ahalya, N., Ramachandra, T. V., & Kanamadi, R. D. (2003). Bio sorption of heavy metals. *Research Journal of Chemistry and Environment*, *7*(4), 71–78.

Ahluwalia, S. S., & Goyal, D. (2007). Microbial and plant derived biomass for removal of heavy metals from wastewater. *Bioresource Technology*, *98*(12), 2243–2257. doi:10.1016/j.biortech.2005.12.006 PMID:16427277

Ahn, C. K., Park, D., Woo, S. H., & Park, J. M. (2009). Removal of cationic heavy metal from aqueous solution by activated carbon impregnated with anionic surfactants. *Journal of Hazardous Materials*, *164*(2-3), 1130–1136. doi:10.1016/j.jhazmat.2008.09.036 PMID:19022570

Akthar, M. N., Sastry, K. S., & Mohan, P. M. (1996). Mechanism of metal ion biosorption by fungal biomass. *Biometals*, *9*(1), 21–28. doi:10.1007/BF00188086

Albadarin, A. B., Mangwandi, C., Walker, G. M., Allen, S. J., Ahmad, M. N., & Kharaisheh, M. (2013). Influence of solution chemistry on Cr (VI) reduction and complexation onto date pits/tea waste biomaterials. *Journal of Environmental Management*, *114*, 190–201. doi:10.1016/j.jenvman.2012.09.017 PMID:23134975

Ayers, P. W. (2005). An elementary deviation of the hard/soft-acid/base principle. *The Journal of Chemical Physics*, *122*(14), 141102. Advance online publication. doi:10.1063/1.1897374 PMID:15847508

Bilal, M., Hafiz, M. N., Tahir, R., Ali, R., Faran, N., & Juan Eduardo, S. H. (2018). Bio sorption interplay between marine algae and potentially toxic elements-a review. *Marine Drugs, 16*(2), 1–16. doi:10.3390/md16020065 PMID:29463058

Bilal, M., Ihsanullah, I., Younas, M., & Shah, M. U. H. (2021). Recent advances in applications of low-cost adsorbents for the removal of heavy metals from water: A critical review. *Separation and Purification Technology, 278*, 119510. doi:10.1016/j.seppur.2021.119510

Brathwaite, R. L., & Rabone, S. D. C. (1995). Heavy metals sulphide deposits and geochemical surveys for heavy metals in New Zealand. *Journal of the Royal Society of New Zealand, 15*(4), 363–370. doi:10.1080/03036758.1985.10421713

Chen, X., Tian, Z., Cheng, H., Xu, G., & Zhou, H. (2021). Assorption process and mechanism of heavy metal ions by di ff erent components of cells, using yeast (Pichia pastoris) and Cu^{2+} as biosorption models. *RSC Advances, 28*(28), 17080–17091. doi:10.1039/D0RA09744F PMID:35479686

Chojnacka, K. (2006). Biosorption of Cr(III) by wheat straw and grass: A systematic characteristics of new biosorbents. *Polish Journal of Environmental Studies, 15*(6), 845–852.

Chojnacka, K., Chojnacka, A., & Gorecka, H. (2005). Biosorption of Cr(III), Cd(II) and Cu(II) ions by blue-green algae *Spirulina Sp. Chemosphere, 59*, 75–84. doi:10.1016/j.chemosphere.2004.10.005 PMID:15698647

Das, N., Vimala, R., & Karthika, P. (2008). Biosorption of heavy metals - An overview. *Indian Journal of Biotechnology, 7*(2), 159–169.

Ding, Y., Jing, D., Gong, H., Zhou, I., & Yang, X. (2012). Biosorption of aquatic Cd(II) by unmodified rice straw. *Bioresource Technology, 114*, 20–25. doi:10.1016/j.biortech.2012.01.110 PMID:22445266

Gadd, G. M. (2004). Micobial Influence on Metal Mobilility and Application for Bioremediation. *Geoderma, 122*(2-4), 109–119. doi:10.1016/j.geoderma.2004.01.002

Gadd, G. M., & White, C. (1992). Removal of thorium from stimulated acid process streams by fungal biomass: Potential for thorium desorption and reuse of biomass and desorbent. *Journal of Chemical Technology and Biotechnology, 55*(1), 39–44. doi:10.1002/jctb.280550107

Garcia Mendieta, A., Olguin, M. T., & Solache Rios, M. (2012). Biosorption properties of green tomato husk for iron, manganese and iron-manganese from aquoes solution. *Desalination, 284*, 167-174.

Guo, J., Zheng, X. D., Chen, Q. B., Zhang, L., & Xu, X. P. (2012). Biosorption of Cd(II) from aqueous solution by *Pseudomonas plecoglossicida. Current Microbiology, 65*(4), 350–355. doi:10.1007/s00284-012-0164-x PMID:22706778

Hammaini, A., Gonzalez, F., Ballester, A., Blazquez, M. L., & Munoz, J. A. (2003). Simultaneous uptake of metals by activated sludge. *Minerals Engineering, 16*(8), 723–729. doi:10.1016/S0892-6875(03)00166-3

Hamza, S. M., Ahmed, H. F., & Mohammmad, E. A. M. (2010). Optimization of cadmium, zinc and copper biosorption in an aquoes solution by *Saccharomyces cerevisiae. The Journal of American Science, 6*, 597–604.

Han, X., Wong, Y. S., & Tam, N. F. Y. (2006). surface complexation mechanism and modelling in Cr(III) biosorption by a microalgal isolate, *Chloerlla miniata. Journal of Colloid and Interface Science*, *303*(2), 365–371. doi:10.1016/j.jcis.2006.08.028 PMID:16962604

Helmenstine, A. M. (2018). *Heavy metal definition and list*. https://www.thoughtco.com

Iyengar, G. V. (1998). Re-evaluation of the trace element content in Reference man. *Radiation Physics and Chemistry*, *51*(4-6), 545–560. doi:10.1016/S0969-806X(97)00202-8

Javanbakht, V., Alavi, S. A., & Zilouei, H. (2014). Mechanisms of heavy metal removal using microorganisms as biosorbent. *Water Science and Technology*, *69*(9), 1775–1787. doi:10.2166/wst.2013.718 PMID:24804650

Kanamarlapudi, R. K., Chintalpudi, V. K., & Muddada, S. (2018). Application of bio sorption for removal of heavy metals from waste water. In Tech Open, 70-116.

Kanamarlapudi, S. L. R. K., Chintalpudi, V. K., & Muddaada, S. (2018). Application of Biosorption for Removal of Heavy Metals from Waste Water. School of Environmental Sciences, 69-116.

Keskinan, O., Goksu, M. Z. L., Yuceer, A., Basibuyuk, M., & Forster, C. (2003). Heavy metal adsorption characteristics of a submerged aqueous plant (Myriophyllum spicatum). *Process Biochemistry (Barking, London, England)*, *39*(2), 179–183. doi:10.1016/S0032-9592(03)00045-1

Koelmel, J., Prasad, M. N. V., Velvizhi, G., Butti, S. K., & Venkata, M. S. (2016). Metalliferous waste in India and knowledge explosion in Metal Recovery Techniques and processes for the prevention of pollution. *Environ. Mat. Was.*, *15*, 339–390. doi:10.1016/B978-0-12-803837-6.00015-9

Koller, M., & Saleh, H. M. (2018). Introductory chapter: introducing heavy metals. In Tech Open, 1-11.

Kuyucak, N., & Volesky, B. (1988). Biosorbents for recovery of metals from industrial solutions. *Biotechnology Letters*, *10*(2), 137–142. doi:10.1007/BF01024641

Liu, C., Ngo, H. H., & Guo, W. (2012). Watermelon rind: Agro waste or superior biosorbent. *Applied Biochemistry and Biotechnology*, *167*(6), 1699–1715. doi:10.1007/s12010-011-9521-7 PMID:22222432

Michalak, I., Chojnacka, K., & Krowiak, A. W. (2013). State of the art for the bio sorption process-A review. *Applied Biochemistry and Biotechnology*, *170*(6), 1389–16. doi:10.1007/s12010-013-0269-0 PMID:23666641

Mustapha, M. U., & Halimmon, N. (2015). Microorganisms and bio sorption of heavy metals in the environment: A review paper. *Journal of Microbial & Biochemical Technology*, *7*(5), 253–256. doi:10.4172/1948-5948.1000219

Norton, L., Kangaratnam, B., & McKenzie, T. (2004). Bio sorption of zinc from aqueous solutions using bio solids. *Advances in Environmental Research*, *8*(3-4), 629–635. doi:10.1016/S1093-0191(03)00035-2

Pagnanelli, F., Toro, L., & Veglio, F. (2002). Oilve mill solid residues as heavy metal sorbent material: A preliminary study. *Waste Management (New York, N.Y.)*, *22*(8), 901–907. doi:10.1016/S0956-053X(02)00086-7 PMID:12423052

Patrón-Prado, M., Acosta-Vargas, B., Serviere-Zaragoza, E., & Méndez-Rodríguez, L. C. (2010). Copper and cadmium biosorption by dried seaweed sargassum sinicola in saline wastewater. *Water, Air, and Soil Pollution, 210*(1-4), 197–202. doi:10.1007/s11270-009-0241-3

Perpetuo, E. A., Souza, C. B. & Nascimento, C. A. O. (2014). Engineering Bacteria for Bioremediation. *Progress in Molecular and Environmental Engineering - From Analysis and Modelling to Technology Applications,* 605-632.

Perumal, K., Antony, J., & Muthuramalingam, S. (2021). Heavy metal pollutants and their spatial distribution in surface sediments from Thondi coast, Palk Bay, South India. *Environmental Sciences Europe, 33*(2), 63–83. doi:10.1186/s12302-021-00501-2

Raji, Z., Karim, A., Karam, A., & Khalloufi, S. (2023). Adsorption of Heavy Metals: Mechanisms, Kinetics, and Applications of Various Adsorbents in Wastewater Remediation -. *Review*, 775–805.

Ramrakhiani, L., Majumder, R., & Khowala, S. (2011). Removal of hexavalent chromium by heat inactivated fungal biomass of *Termitomyces clypeatus. Chemical Engineering Journal, 171*(3), 1060–1068. doi:10.1016/j.cej.2011.05.002

Redha, A. A., & Redha, A. A. (2020). Removal of heavy metals from aqueous media by biosorption Removal of heavy metals from aqueous media by biosorption. *Arab Journal of Basic and Applied Sciences, 27*(1), 183–193. doi:10.1080/25765299.2020.1756177

Shamim, S. (2018). Biosorption of heavy metals. In Tech Open, 21-49.

Sun, F., Wu, F., Liao, H., & Xing, B. (2011). Biosorption of antimony(V) by freshwater cyanobacter *Mycrocystis* biomass. *Chemical Engineering Journal, 171*, 1082–1090. doi:10.1016/j.cej.2011.05.004

Swiatek, M. Z., & Krzywonos, M. (2014). Potentials of biosorption and bioaccumulation processes for heavy metal removal. *Polish Journal of Environmental Studies, 23*(2), 551–561.

Timková, I., Sedláková-Kaduková, J., & Pristaš, P. (2018). Biosorption and Bioaccumulation Abilities of Actinomycetes/Streptomycetes Isolated from Metal Contaminated Sites. *Separations, 5*(4), 54. doi:10.3390/separations5040054

Tiwari, S., Bhardwaj, U., & Dhingra, N. (2022). Towards Further Understanding the Mechanism of Biosorption Through Various Biosorbents. *International Journal of Pharmaceutical Sciences and Research, 13*(4), 1441–1455. doi:10.13040/IJPSR.0975-8232.13(4).1441-55

Tsezos, M., & Volesky, B. (1982). The mechanism of Uranium Biosorption by Rhizopus arrhizus. *Biotechnology and Bioengineering, 24*(2), 385–401. doi:10.1002/bit.260240211 PMID:18546310

Umas, B., & Roslan, H. A. (2016). Mechanism of Biosorption of Pb (II) and Cu (II) ions using Dead Biomass of Fusarium equiseti strain UMAS and Penicillium citrinum. *Journal of Biochemistry. Microbiology, 4*(2), 1–6.

Valls, M., & Lorenzo, V. (2022). Exploiting the Genetic and Biochemical Capacities of Bacteria for the Remediation of Heavy Metal Pollution. *FEMS Microbiology Reviews, 26*, 327–338. doi:10.1016/S0168-6445(02)00114-6 PMID:12413663

Venkata, A., Kumar, A., Darwish, N. A., & Hilal, N. A. (2009). Study of various parameters in the biosorption of heavy metals on activated sludge. *World Applied Sciences Journal*, *5*, 32–40.

Volesky, B. (2007). Biosorption and Me. *Water Research*, *41*(18), 4017–4029. doi:10.1016/j.watres.2007.05.062 PMID:17632204

Witek-Krowiak, A., & Harikishore Kumar Reddy, D. (2013). Removal of microelemental Cr(III) and Cu(II) by using soybean meal waste - Unusual isotherms and insights of binding mechanisms. *Bioresource Technology*, *127*, 350–357. doi:10.1016/j.biortech.2012.09.072 PMID:23138058

Yan, G., & Viraraghavan, T. (2008). Mechanism of Biosorption of Heavy Metals by *Mucor rouxii*. *Engineering in Life Sciences*, *4*(4), 363–371. doi:10.1002/elsc.200820237

Yilmaz, A. B., Mustafa, K. S., Deniz, Y., & Cemal, T. (2010). Metals (major, essential to non-essential) composition of different tissues of three demersal fish species from Iskenderun Bay. *Food Chemistry*, *123*(2), 410–415. doi:10.1016/j.foodchem.2010.04.057

Chapter 3
Factors Affecting Biosorption Efficiency:
Process Optimization and Performance Evaluation

Sudha Subburaj
https://orcid.org/0000-0001-7584-4820
College of Engineering Guindy, Anna University, Chennai, India

A. Lakshmi Kanthan Bharathi
https://orcid.org/0000-0002-3037-4164
College of Engineering Guindy, Anna University, Chennai, India

ABSTRACT

Biosorption, an eco-friendly, cost-effective method for removing heavy metal pollutants from the environment, gained great attention. This utilizes biological sources such as microbial and agricultural biomass for the removal of heavy metals and other hazardous materials from industrial effluents and other wastes that directly come into the environment. The efficiency is influenced by factors such as temperature, pH, biomass concentration, initial ion concentration, contact time, agitation rate, type of biosorbent, competing ions. The source of biosorbents ranges from microbial sources to plant biomass following different mechanisms. In this chapter, the influencing factors are examined widely, and it spotlights optimization by modifying the parameters. The outcomes hold implications for designing an effluent removal process with the right process factors for maximum efficiency. The findings provided give valuable insight to the researchers, industrial experts, students, policy makers, and academicians who are seeking sustainable solutions to reduce heavy metal pollution.

DOI: 10.4018/979-8-3693-1618-4.ch003

1. INTRODUCTION

The industrialization era changed the world from an agro-based economy to an industry-based economy that utilizes large quantities of mineral and human resources. This significantly increased the use of resources, leading to the generation and accumulation of industrial waste in the environment. The industrial waste more commonly pollutes water, air, and land. Apart from the common pollutants from the industries, the heavy metal pollutants which include arsenic, mercury, lead, phosphates, nitrates, sulfur, cadmium, nickel, copper, zinc, cobalt, magnesium, chromium, selenium, and other metalloids (Sedlakova-Kadukova et al., 2019) leave an irreversible impact on the ecosystem. The discharged heavy metal pollutants into the ecosystems are difficult to assimilate and remove, leading to major environmental modification in biomagnification. Heavy metal contamination is noticeable in the environment due to sources like metal processing industries, mining industries, dying industries, agricultural processes, and waste generated from medical applications. Environmental poisoning by heavy metals has gone over the allowed level, threatening all forms of life. According to the Comprehensive Environmental Response Compensation and Liability Act (CERCLA) of the United States of America, the maximum allowable concentration of certain heavy metals in water is 0.01 mg/L for Ar, 0.05 mg/L for Cd, 0.01 mg/L for Cr, 0.015 mg/L for Pb, 0.002 mg/L for Hg, and 0.05 mg/L for Ag, respectively. On the other hand, according to the Indian standards for heavy metals, the standard for soil is 3–6 mg/kg for Cd, 135–270 mg/kg for Cu, 75–150 mg/kg for Ni, 250–500 mg/kg for Pb, and 300–600 mg/kg for Zn, respectively (Ayangbenro et al., 2019). Heavy metals are carcinogenic and cause permanent damage to the skin, kidneys, blood, and irregularity in blood pressure, and retardation in cognitive skills of children. The heavy metals have prolonged effects on the environment since their half-life is comparatively longer and removing them is difficult.

1.1 Heavy Metal Removal

The removal of heavy metals from the contaminated sources is termed as heavy metal removal. The heavy metals are removed by different means of methods that can be classified into conventional methods and modern methods. The conventional methods employed for the removal of heavy metals are ion floatation (Hoseinian et al., 2020), filtration, and ion-exchange process (Bashir et al., 2019), evaporation, chemical precipitation process (Wang et al. 2019), electrochemical precipitation, catalysis (Gupta et al., 2021), Chemical coagulation/flocculation (Zou et al., 2021), reverse osmosis (Sunil et al., 2018), membrane separation, chemical oxidation, adsorption, and reduction (Shahrokhi-Shahraki et al., 2021). The segregation of heavy metal through the above conventional methods is not efficient in terms of cost, substrate efficiency, recovery, need for advanced equipment, maintenance of equipment, and application for low-concentration solutions.

On considering the downsides of the above-mentioned existing methods, the biosorption process stands out in the positive aspects of recycling and recovery of heavy metals. The first study on bio-sorption was reported in 1951 and due to its captivating properties and environment friendliness, it has gained more interest from the researchers and industries. In 1970, the awareness of environmental pollution and wastewater treatment led to the search for new techniques that landed on biosorption technique due to the advantages over the conventional sorption techniques such as cost-effectiveness, ability to apply for different materials, less or no sludge formation, and high adsorption performance.

1.2 Biosorption

Figure 1. Schematic diagram of the biosorption process

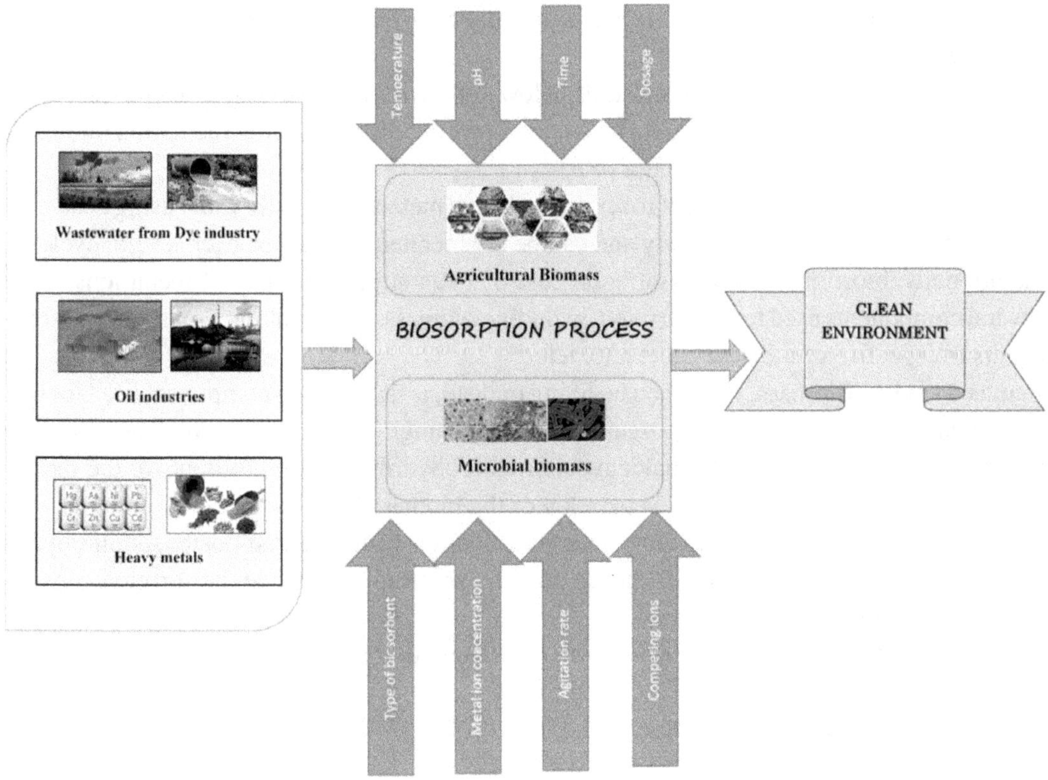

Biosorption is defined as the process of removing pollutants using biologically active substances (Volesky, 1994). Biosorption is the physicochemical and biological process developed for the removal of desired heavy metals from the wastewater stream and any deposited source achieved by various biological substances with different sizes namely bacteria, fungi, and algae as biosorbents (Kaushik, 2009). The microorganisms have been adapted over the past millions of years owing to their ability to tolerate harsh conditions (Hrynkiewicz and Baum, 2014).

The biosorption process involves two routes which include adsorption and absorption by the activities of biological agents that can be of living or dead mass or the products obtained from them. The biosorption process that involves active material is called active biosorption or bioaccumulation whereas the biosorption process involving the dead biomass is called passive biosorption. The availability and generation of microbial biomass are comparatively cost-effective and found copious in the environment. Figure 1 shows the schematic diagram of the biosorption process and the influencing factors.

This chapter deals with the definition of biosorption, different biosorbents, and their application in removing heavy metals. The main objectives of this chapter are to divulge factors such as pH, competing ions, temperature, initial metal ion concentration, contact time, biosorbent concentration, and agitation that influence the biosorption process. In addition to that, biosorption optimization and performance evaluations, limitations and future scope of biosorption are discussed.

1.3 Advantages of Biosorption

Biosorption offers several advantages over conventional methods of heavy metal removal.

a) It is a cost-effective solution, as it can utilize waste or by-products as biosorbents which is easily available and eco-friendly.
b) When the concentration of pollutants falls below 100 mg/L, their neutralization process cannot be achieved by conventional removal methods, in these circumstances, bioremediation through living and non-living biomass plays a major role.
c) Biosorption is also highly efficient in removing heavy metals, especially when using dead biomass.
d) It has a high metal binding capacity and can be regenerated and reused over several cycles.
e) Additionally, biosorption is a versatile process that can attract a variety of metal ions. There are several components used for biosorption, including biomass, cross-linking agents, and pretreatment agents irrespective of the heavy metal toxicity.
f) Reduced sludge: Biomass, which can be living or dead, is the primary component used for biosorption. It can be derived from microorganisms such as fungi, bacteria, or microalgae. Cross-linking agents, such as polysulfone, formaldehyde, or chlorohydrin, are used to improve the mechanical strength, chemical resistance, and separation of the biomass from the effluent.
g) Pretreatment agents, such as alkaline detergent, sodium hydrogen carbonate, sodium hydroxide, sodium carbonate, or supercritical carbon dioxide, are used to enhance the sorption capacity and rate of metal uptake in the biomass.
h) The biosorbents have a high surface-to-volume ratio with the heavy metal sludge.

1.4 Mechanism of Biosorption

Although biosorption has several advantages, there are many essential objectives to be satisfied for microorganisms to remove heavy metals from the environment as effectively as possible. The biosorption mechanism is a complex process that is influenced by numerous parameters and involves independent complex mechanisms. The efficiency of the biosorption process is improved by optimizing the valuable parameters and removing the inhibiting parameters. The mechanism of biosorption is generally based on the physico-chemical interaction between metal ions and functional groups on the surface of the sorbents. The functional groups such as amide, amine, carbonyl, carboxyl, hydroxyl, imine, imidazole, sulfonate, sulfhydryl, thioether, phenolic, phosphate, and phosphodiester groups; nevertheless, the presence of some functional groups limits the biosorption of contaminants (Park et al., 2010). Biosorption mechanisms include chemisorptions, complexation, reduction, chelation, adsorption-complexation on surface and pores, ion exchange, precipitation, heavy metal hydroxide condensation, and surface adsorption (Priya et al. 2022). Figure 2 shows mechanisms involved in the biosorption of the microbial biomass.

Figure 2. Various mechanisms involved in biosorption

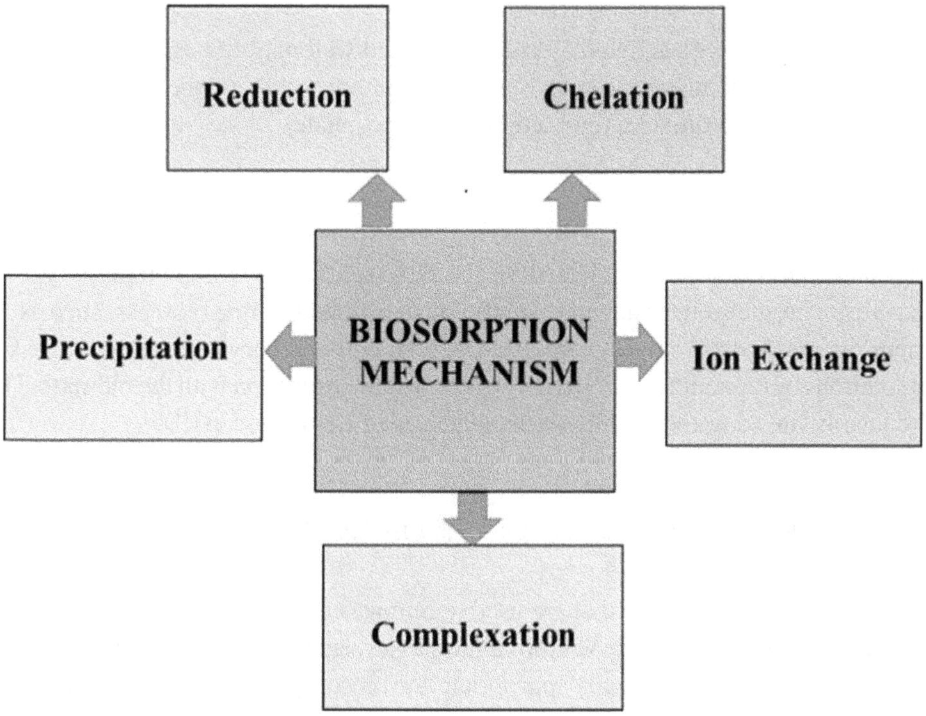

a) Ion Exchange

In ion exchange, an ion from a solution, electrolyte, or molten salt is swapped for an equally charged ion bonded to an insoluble and immobile solid substance. This is a reversible stoichiometric chemical process. Three-dimensional inorganic matrices, synthetic organic resins, and new-generation hybrid materials are frequently utilized as ion-exchange material matrices.

b) Chelation

Multiple functional groups with donor electron pairs can be concurrently donated to a metal atom by organic molecules. This may lead to the metal atom forming a ring structure and this process is referred to as chelation.

c) Precipitation

An initial method for precipitating metals is sorption precipitation, which occurs in cells and cell walls with higher pH values. Precipitation may be independent or dependent on cellular metabolism. In the previous case, an active microbial defense process is typically associated with the extraction of metal from the solution.

d) Complexation

The interaction of the metal with active molecules in the cell wall may end up in complex structures on the cell surface. Heavy metal ions have been observed to interact with a wide range of groups, such as carboxy, hydroxy-carboxy, hydroxide, thiol, amino, and phosphate.

e) Reduction

A large body of research has been published on the biosorption of various biomass types to remove harmful hexavalent chromium from aqueous solutions. Inactivated fungal biomass, such as Aspergillus niger, Rhizopus oryzae, Saccharomyces cerevisiae, and Penicillium chrysogenum, remove Cr(VI) from water-based solutions by reduction to Cr(III) as they come into interaction with the biomass. This removal process is frequently linked to the simultaneous reduction of Cr(VI) to Cr(III).

2. BIOSORBENT MATERIALS AND TYPES OF BIOSORBENTS

Biosorbents are biological substances that are used to remove contaminants either actively or passively from a contaminated solution or water. Microbial biomasses, agricultural waste, and industrial byproducts are examples of biological materials appropriate for biosorption processes. These materials have been hailed as promising substrates for metal removal by biosorption because of their high efficiency, low cost, and great abundance.

2.1 Agricultural Biosorbents

Agricultural biosorbents are sorbents based on agricultural wastes or materials from agriculture. According to current research, the agricultural and food industry wastes – such as hazelnut and coconut shells, rice husk (RH), pecan nut shells, jackfruit peels, powder pomegranate and corn peels, potato peel, clove leaves, sugarcane bagasse, water chestnut shells, and peels of various fruits and vegetables - are used as potential biosorbents (Jimenez-Cedilloa et al.,2013; Kamsonlian et al., 2012; Shafique et al., 2012 Aminuddin et al., 2023; Kusuma et al., 2023). The use of agricultural materials has increased due to the expensive cost, chemical environment, and creation of sludge from the usage of inorganic adsorbents.

Aminuddin et al., (2023) used rice husk and bio charcoal rice husk for the removal of engine oil, and the results showed that the rice husk has 2.7g/g of sorption capacity. Wastewater from industries contains many toxic materials that cause pollution to the freshwater bodies. Sabri Can Karadeniz et al., (2023), used Lolium perenne seeds as a low-cost adsorbent for the removal of Safranin T dye from the aqueous medium. M. Samimi and E. Mansou., (2023), analyzed the adsorption efficiency of Falcaria vulgaris biomass for the adsorption of Co(II) from the aquatic ecosystem at a biosorbent-to-metal ratio of 10:1, pH 4.88, and contact period of 70 minutes with the extraction efficiency of 80.941%. Markovi̇́c et. al., (2023) used raw eggshell in the biosorption of copper ions considering the influence of process varying parameters like initial mass of sorbent, pH value of the solution, the concentration of Cu^{2+} ions, and stirring rate. Kusuma et al., (2023) investigated the biosorption capacity of waste clove leaves treated with sodium hydroxide for the removal of methyl blue from wastewater. This method is used in dye-based industries, is cost-effective and the biosorbent is recyclable. Pomegranate and corn peels were found to have a remarkable adsorption capacity to remove heavy metal ions from industrial wastewater and were employed to remove Cr, Cd, and Pb from wastewater and the resulting recovery efficiency was 97.4%, 91.15%, and 89.4% respectively (Ati et. al., 2021). Table 1 shows plant-based (Agricultural) biosorbents applied for the removal of respective adsorbates.

Table 1. Agricultural biosorbents applied for the biosorption process

Reference	Biosorbent	Adsorbate
Aminuddin, I. H et al., (2023)	Rice Husk	Engine Oil
Sabri Can Karadeniz et al., (2023)	Lolium perenne seeds	Safranin T dye
Samimi and Mansouri; (2023)	Falcaria vulgaris	Co(II) u
Markovi'c et. al., (2023)	Raw eggshell	Copper ions
Kusuma et al., (2023)	clove leaves	Methylene blue
Ati et al., (2021)	Powder Pomegranate And Corn Peels	Chromium, Cadmium, Lead
Cheraghi et al. (2016)	Tea waste	Arsenic
Hadadi et al., (2023)	Potato Peel (Solanum tuberosum)	Methylene Blue
Shafique et al.(2012)	Chir pine leaves	Arsenic
Sumathi and Alagumuthu (2014)	Activated moringa oleifera leaves carbon	Arsenic
M.Samimi et al., (2023)	Eucalyptus Globulus fruits	Cadmium

2.2 Microbial Biosorbents

Microorganisms involved in the biosorption process are termed microbial biosorbents. Algae, fungi, and bacteria are involved in biosorption processes as microbial biosorbents. The microorganisms have excellent assimilating mechanisms through respiration, fermentation, and co-metabolism that help in increased efficiency of the biosorption rate of heavy metals. The micro-organisms can easily be modified through genetic modification to suit extreme experimental conditions. The metabolites and by-products produced by the microorganisms help in the efficient uptake of heavy metals at different optimum process conditions (Ramírez Calderon et al., 2020). The biosorption of heavy metals by microbial biomass is influenced by elements such as the type of metal to be removed, microorganism type, pH, contact duration, initial metal concentration, temperature, biosorbent dose, and condition (Varia et al., 2014; Abu Hasan et al., 2016; Khadivinia et al., 2014; Munoz et al., 2016).

The biosorption utilizes biomass of different origins and different natures. The nature of biomass can be living, or dead biomass that can be obtained from different sources namely crops and residue, industrial residue, sewage, Forest waste and residue, and municipal solid waste (Verma and Kuila, 2019). The living cells of Cyanobacterial biomass have activated reaction centers at the bacterial cell wall that absorb and adsorb the heavy metal cation Attraction of the heavy metal cations on the cell surface leads to the transfer of heavy metals into the cell cytoplasm and assimilation occurs. Whereas in dead bacterial biomass, the heavy metals directly enter the substratum of the cell (Kalita, N., & Baruah, P. P. et al. 2023).

Table 2. Microbe-based biosorbents used for various heavy metals

Ref.	Biosorbent	Heavy Metal
Tchounwou et al., 2012	Saccharomyces cerevisiae	Ni, Pb
Abioye, O. P. et al 2018	B. subtilis	
Carreira, A. R et al. 2023	C. minutissima	Zn, Cd, Cu
Safari and Ahmady-Asbchin (2018)	Cladophora crispata	Cu (III)
Tattibayeva et al. (2022)	Chlorella vulgaris ZBS1	Cr(VI)
Smjecanin et al. (2022)	Spirogyra & Vaucheria	U(VI)
Nagababu et al. (2022)	Gracilaria Rhodophyta biochar	Cr(VI)
Almomani and Bhosale (2021)	Spirulina platensis	Cu(II)
Saad et al. (2021)	Chitosan – Ulva lactuca	Cd(II)

(Abioye et al., 2018) conducted an assessment on the bioremediation of Pb, Cr, and Cd from effluent. They utilized different species such as Bacillus subtilis, Aspergillus niger, B. megaterium, and Penicillium to carry out the process. Among these species, Bacillus megaterium showed the highest success in reducing the total Pb concentration from 2.125 mg/L to 0.029 mg/L, followed by Bacillus subtilis. Aspergillus niger demonstrated remarkable capability in lowering Cd concentration from 0.41 mg/L to 0.029 mg/L after 20 days. The greatest reduction in Cr concentration (from 1.375 mg/L to 0.075 mg/L) and the Penicillium species (from 1.38 mg/L to 0.13 mg/L), as well as Bacillus subtilis, were achieved (Abioye et al., 2018; Fathollahi et al. 2021). For the removal of Cr^{6+}, Cu, and Ni, zeolite mobilization was found to be highly effective, with removal rates of 99.9%, 98.3%, and 90.2% respectively, as reported by Kim and Ni. The bacteria consortia were able to remove approximately 75-85% of the metals within 2 hours of contact time (Kim et al., 2015). Table 2 shows microbial biosorbents for the removal of specific heavy metal sources.

3. BIOSORPTION EFFICIENCY

This literature review explores biosorption as an environmentally friendly method for removing heavy metals from aqueous media. Different types of biomass, including algae, fungi, bacteria, and plants, can be used for biosorption, with various mechanisms such as physical adsorption, ion exchange, complexation, precipitation, and transport across cells. The review also considers the potential of biosorption as a competitor to traditional industrial methods (Ali Redha, A. 2020). The biosorption efficiency of living biomass is higher compared to dead biomass due to their metabolic rate and assimilation rate (Priya et al., 2022).

The efficiency of the biosorption process is affected by various factors such as pH, temperature, the weight of the biosorbent, the particle size of the biosorbent, the concentration of biosorbent, time and agitation of biosorbents, type, and composition of biosorbents, initial concentration of heavy metals (Priyadarshanee & Das, 2021). In addition to the above factors, immobilization of heavy metals also plays a major role in affecting the efficiency of the biosorption process.

The effectiveness of biosorbents in removing chromium was determined by comparing the initial and final concentrations of total chromium before and after biosorption. To assess the removal efficiency (E) (Eq.1) and biosorption capacity (q_t) (Eq. 2), the following expressions were utilized (Deshmukh et al., 2017).

Percentage of removal efficiency, $E = (C_i - C_f) \times 100 / C_i$ (1)

Biosorption capacity, $q_t = (C_o - C_e) \times V / M$ (2)

where q_t represents the uptake of total chromium in mg/g of biosorbent, V denotes the volume of the metal solution in liters that comes into contact with the biosorbent, $C_{o,}$ and C_e indicate the initial and residual concentrations of the metal in the solution in mg/L, and M represents the mass of the biosorbent in grams.

4. KEY FACTORS INFLUENCING BIOSORPTION EFFICIENCY

The biosorbents have a greater affinity to the adsorbate, as it undergoes mechanisms like adsorption, absorption, ion exchange, surface complexion, and precipitation during the process of biosorption. Although the biosorbents have a greater affinity to a wide range of metals, dye sorbents, biomass, and environmental conditions the metal uptake efficiency and effectiveness of biosorption are influenced by various key factors. The factors that mostly influence the process of biosorption are discussed in this chapter and the key factors are shown in the schematic Figure 3. Table 3 provides an overview of the biosorption process for respective influencing factors. The efficiency of biosorption differs based on different factors that are specific to different biomass in the sorption processes.

Figure 3. Factors influencing biosorption efficiency

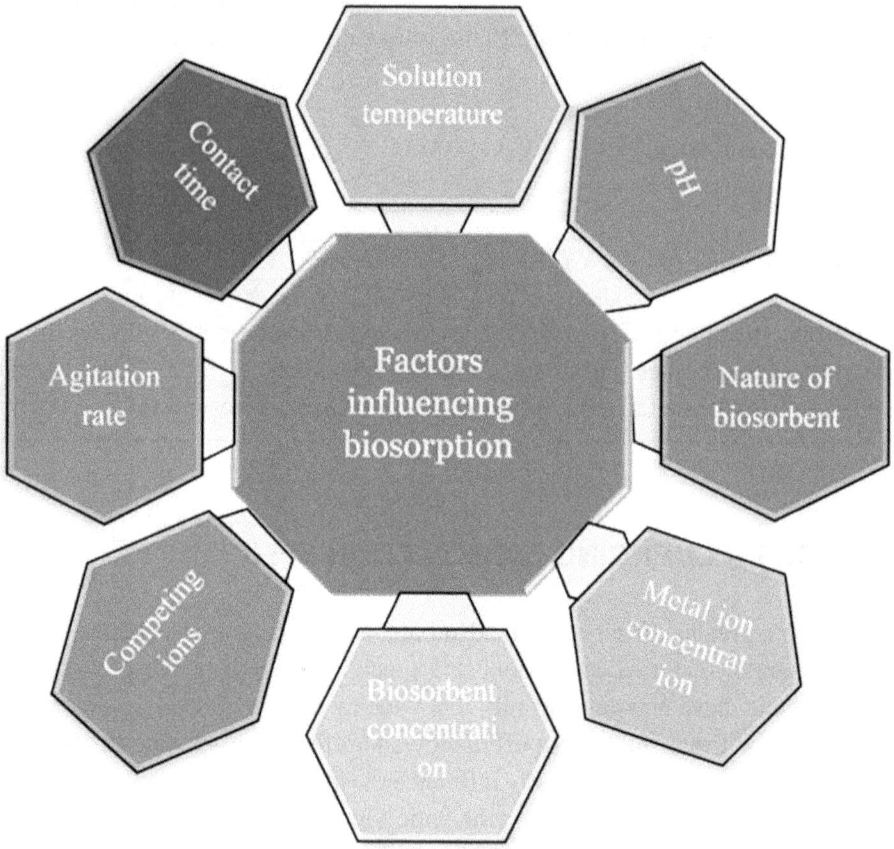

Table 3. Overview of various factors influencing biosorption

Biosorbent	Heavy Metal	Nature of Biosorbent	Temperature (°C)	pH/ Acidity	Contact Time (min)	Biomass Dosage	Metal Ion Concentration	Ref
Chlorella vulgaris ZBS1	Cr(VI)	Microbial based	24.85	2	120	0.125 g/L;	10 mg/L	Tattibayeva et al. (2022)
Spirogyra & Vaucheria	U(VI)	Microbial based	25	2	40	100	50 mg	Smjecanin et al. (2022)
Gracilaria Rhodophyta biochar	Cr(VI)	Microbial based	30	-	90	0.3g/100ml	60 mg	Nagababu et al. (2022)
Spirulina platensis	Cu(II)	Microbial based	25 ± 0.5	7	120	25 ± 0.1	50 mg	Almomani and Bhosale (2021)
Chitosan - Ulva lactuca	Cd(II)	Microbial based	25	5.5	120	0.1	-	Saad et al. (2021)
B. subtilis	Cu (III)	Microbial based	25 to 35	-	-	-	-	(Tchounwou et al., 2012).
Ç. minutissima	Zn, Cd, Cu	Microbial based	10 - 28	-	-	-	-	(Tchounwou et al., 2012).
Cladophora crispata	Cu (III)	Microbial based	15 - 25	-	-	-	-	(Tchounwou et al., 2012).
Soybean and its waste	Pb(II) Cd (II) Zn (II)	Agricultural based	5; 25; 50	1.0 - 6.6	5 -180	4.0 - 40.4 (g/L)	11.66–416.45 mg/L 9.22–230.54mg/L 13.08–209.25 mg/L	Fertu, D. I et al. 2022
Mango leaf	Phosphorus removal	Agricultural based	25	4 -8	60	1,2 g		Dey, S et al. 2023
modified clove leaves	Cu(II)	Agricultural based	30 - 48	6	30	-	-	H.S. Kusuma et al., 2023
Eucalyptus globulus fruit	Pb(II) Cd (II) Zn (II)	Agricultural based	–	6	80	9g	1g	M.Samimi., 2024
Falcaria vulgaris	Cr(VI)	Agricultural based	–	5.5	70	-	-	M. Samimi And E. Mansouri., 2023
Biomass from waste Polyglutamic acid	Cr(VI)	Industrial based	20-35	7	60	2 g/L	200 mg/L	Zhang et al., 2020
Gelidium amansii	Pb^{2+}	Microbial based	45	4.5	180	1g/L	200mg/L	El-Naggar et al., 2018
Pomegranate and corn peels	Cr, Cd, Pb	Agricultural based	25-45	5	60	1 g	500 mg/dm	Estabraq Mohammed Ati eta l., (2021)
Potato peels	Cd(II)	Agricultural based	-	6	1	0.1 g/L	-	Marwa El-Azazy et al., (2019)
Potato Peel (Solanum tuberosum)	Methylene Blue	Agricultural based	25	7	-	1g	1 g/L	Hadadi et al., (2023)
Mesorhizobium amorphae	Cu(II)	Microbial based	28	5	30	-	0.5g/L	O.A.Mohamad et al., (2012)
Raw eggshell	Cu^{2+}	Agricultural based	60	5	60	1g	500 mg/dm3	M.Marckovic et al., (2023)

4.1 Initial Metal Concentration

The concentration of metal ions at the beginning of the biosorption process holds greater significance (Shen, L et al., 2018; Iqbal et al., 2022). The relationship between metal concentration and metal sorption involves the possibility of electrostatic interactions between functional groups. Biosorption efficiency decreases as the initial metal concentration of metal ions increases. In other words, the ratio of metal to biosorbent is increased when the availability of binding sites decreases, resulting in a proportional reduction in the amount of metal that can be absorbed by the biosorbent (Castro et al., 2017).

Initially, the binding sites of the biosorbents are free, which helps in increased absorption of biosorbents. As the concentration of metal ions increases, most of the binding sites become filled, rendering them inaccessible for further binding. Consequently, this obstructs the sorption process (Singh, 2020). At low initial metal concentrations, the binding sites on the biosorbents are unsaturated, resulting in a higher percentage of metal adsorption. This is due to the low ratio between the initial amount of metal ions and the sorption sites available. As a result, there is an increase in biosorption yields and a higher probability of effectively removing the metal. According to Sen et al. (2017), a combination of Leptolyngbya subtilis and Limnococcus limneticus cyanobacterial consortia was most effective in removing Cr (VI) from a 15 mg/L initial metal concentration. However, their efficiency in removing Cr (VI) decreased when the initial metal concentration was 10 mg/L, 20 mg/L, or 25 mg/L. On the other hand, Anabaena flocculosa and Spirulina platensis showed an increase in Hg^{2+} removal as the concentration of Hg^{2+} increased from 10 to 50 mg/L. In S. platensis, the biosorption increased from 131.3 mg/g to 227.0 mg/g when the Hg^{2+} concentration was increased from 10-50 mg/L. A similar trend was observed in A. flocculosa, with biosorption increasing from 96.2 mg/g to 262.5 mg/g as the Hg^{2+} concentration increased. Al-Homaidan et al. (2015) found that S. platensis exhibited a gradual increase in Cu adsorption, from 47.65% to 90.61%, as the concentration increased from 10 to 100 mg/L. However, it was noted that beyond 150 mg/L, the binding sites became saturated. According to Safari and Ahmady-Asbchin (2018), the range of metal concentrations tested for biosorption was 25 to 225 mg/L. The study revealed that the highest level of biosorption occurred at an ideal metal concentration of 175 mg/L. Mohamad, O. A et al., 2012 investigated the sorption of copper with an initial concentration of 100 mg L^{-1} of Cu (II), the maximum rate of removal of Cu (II) by the M. amorphae was 37, 49, 53, 64, and 64%, respectively from the varied doses indicating the removal efficiency decreased by increasing the initial concentration. In some cases, an increase in the concentration of biosorbents increases the removal percentage which is due to the availability of the active sites on the surface of the adsorbent, resulting in more metal uptake. Figure 7(e) shows the Cr(VI) metal removal or adsorption percentage increases with the increase in metal ion concentration up to the optimum level and reduces after the optimum (Zhang et al., 2020). Therefore, the biosorption capacity increases with higher initial metal concentrations until the maximum capacity is reached in the case of fixed biomass but decreases efficiency.

4.2 pH of the Solution

In biosorption processes, the pH of the solution is very essential in determining the efficiency. Metal biosorption is highly pH dependent which is found repeatedly in almost all bio-sorption systems including bacteria, fungi, and algae (Oyewole et al., 2018). The influence of pH on the biosorption of heavy metals is both significant and intricate, impacting both the metal ions and the biosorbent itself. The optimum pH for biosorption varies depending on the metal ions and the biomass used. The process can be operated under a wide range of pH, but extreme pH conditions can affect both the metal sorption capacity and the viability of living cells. The surface of the adsorbent becomes positively charged during high pH and negatively charged during low pH which affects the sorption efficiency. It influences metal precipitation, speciation, and the availability of metal ions for biosorption.

The ideal pH for biosorption can be determined by using the isoelectric point (pHIEP) or point of zero charge (pHpzc). pHIEP is the pH at which the net surface charge on the biomass equals zero. The above pH values help in determining the optimal condition for biosorption (Salam, 2015). Metal precipitation and speciation affect the availability of metal ions for biosorption. High solution pH (above pHpzc) deprotonates the functional groups on the biomass surface, resulting in a net negative surface charge. This increases the biosorption of positively charged metal ions. In contrast, low solution pH (below pHpzc) protonated the functional groups, resulting in a net positive surface charge. This increases the biosorption of metal complexes existing as anions at that particular pH. The net surface charge of the biomass (binding sites, also called the functional group) is also controlled by pH, which in turn affects the biosorption process. This charge affects the binding affinity of the biomass for metal ions. Further metal ions can be categorized into three classes based on their behavior at different pH ranges. Usually, the meta lions are categorized as class I, class II, and class II (Carreira et al., 2023). Class I metals (pH = 4 - 7) namely Al(III), Be(III), Cu(II), Cr(III), Co(II), Fe(III), Ni(II), Pb(II), Zn(II), and $UO^{22+}$are strongly bound at near neutral pH but are not bound or can be desorbed at low pH. Class II metals (pH < 4) namely $PtCl_4^{2-}$, SeO_4^{2-}, and CrO_4^{2-} behave oppositely, binding strongly at low pH but weakly at high pH. Class III metals such as Ag(I), Au(III) Hg(II) are pH-independent and are bound most strongly among all and their binding is pH-independent. This categorization helps narrow down the search for the optimum pH for metal biosorption. Metal ion categorization helps in understanding the pH-dependent behavior of different metals during biosorption.

The biosorption efficiency increases with higher pH and reduces with lower pH. The bio-sorption efficiency increased with the rise in the pH level of the solution and was found to be approximately 10.8 mg. g^{-1} at pH = 2, and nearly doubled at pH = 5 with 21.62 mg. g^{-1} during adsorption of Cu^{+2} ions by using egg shells as bio-sorbent which was due to the presence of calcium carbonate in eggshells and similarly the solution with lower pH has a reduced ability for bio-sorption (Markovic et al., 2023). The optimal pH for maximum cadmium biosorption by biomass from Eucalyptus globulus fruit is 6. Eucalyptus globulus fruit biomass has a higher affinity for cadmium ions at acidic pH due to the carboxyl, hydroxyl, and amino functional groups, which interact with the metal ions by electrostatic attraction and ion exchange (Samim, 2024). For biosorption of Co(II) with Falcaria vulgaris biomass, the optimal pH value was 5.5 whereas the removal or bio-sorption efficiency decreases when the pH is lower than 5.5 due to competition of H+ and Co(II) ions. Conversely, the bio-sorption efficiency decreases when the pH value is higher than 5.5 as the Co(II) precipitates as hydroxide (Samimi and Mansou., 2023). At a fixed dye concentration of 0.02 mb. L^{-1} of methylene blue using modified clove leaves as bio-sorbent, the pH of methylene blue was maintained at 6 (Kusuma et al.2023). Metal biosorption typically involves complex processes such as ion exchange, chelation, and physical adsorption, as well as the trapping of ions in the capillaries and gaps of a biosorbent's cell structure network (Dursun, 2006). Figure 4 shows the effect of change in pH change of Zn^{+2} and Fe^{+3} removal efficiency by using dried algal powder and the activated carbon of U. lactuca and C. officinalis as biosorbents (Joshi et al., 2023). Figure 4 & Figure 7(c) show that biosorption efficiency increases with increasing pH up to the optimum pH level and reduces above the optimum (Ameen et al., 2021). The results of various research support the idea that pH plays a crucial role in biological processes, influencing both growth rates and the capacity for adsorption in substances (Deng et al., 2007).

Figure 4. Effect of pH change on Zn^{+2} and Fe^{+3} removal efficiency by dried algal powder and the activated carbon of U. lactuca and C. officinalis as biosorbents (Reused with permission from Ameen et al., 2021)

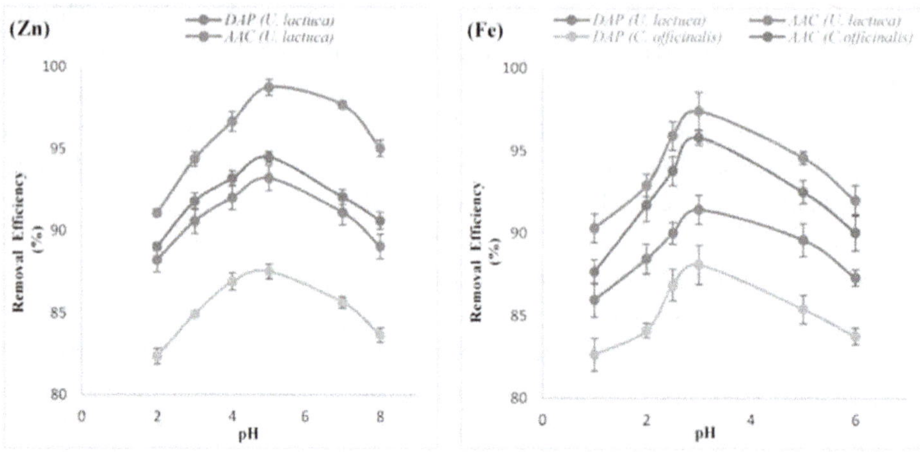

4.3 Solution Temperature

Temperature can have varying effects on biosorption efficiency. In some cases, metal uptake is independent of temperature, while in others, increases in temperature can decrease biosorption capacity. Low temperatures may promote metal sorption, while high temperatures can lead to a decrease in capacity, indicating an exothermic process. The temperature can affect the net surface charge of the biomass, metal speciation, and the rate of metal uptake or selectivity of the metal-bound onto the binding sites. The effect of temperature on biosorption capacity can differ depending on the variety of biosorbents and the type of metal being adsorbed. In certain instances, the bio-sorption capability may be improved by raising the temperature. This is because greater temperatures have the potential to reduce the viscosity of the solution, which increases the pace at which the heavy metal ions diffuse through the bio-adsorbent layers encouraging more removal of heavy metals.

In other words, the temperature parameter influences the biosorption efficiency in different ways. Concerning the active binding sites on the surface of the biosorbents, the rise in temperature increases the availability of active binding sites which in turn increases the biosorption efficiency. At lower temperatures, physical adsorption occurs due to van der Waals forces. This form of adsorption is characterized by low-energy adsorption and the molecules that are adsorbed are not bound to a specific location on the surface (Kilic et al. 2014). The decrease in the biosorption of metal ions as temperature increases could be due to the impairment of active binding sites in the biomass or a greater tendency to release metal ions from the interface into the solution (Meena et al. 2005; Sari, 2008). The optimal range of temperatures for bio-sorption efficiency is 20–35°C; while extreme temperatures, such as 50°C, boost bio-sorption occasionally, they can also permanently harm the microbial cells, which lowers metal adsorption (Oyewole et al., 2018). There is a threshold, beyond which the biosorption capability drops, even if greater temperatures can sometimes improve biosorption as shown in Figure 7(b)

Few researchers noted that temperature did not affect metal absorption. For example, Puranik and Paknikar, 1999 found that citrobacter strain as a biosorbent shows no significant variation in biosorption of lead, cadmium, and zinc between 4 and 55°C. Also, Khani et al.,2023 found that temperature has no significant effect during the biosorption of strontium ions by Chlorella vulgaris. At a temperature of 28°C the Cu^{2+} removal has increased greatly at 5 pH during the removal of Copper (II) with Mesorhizobium amorphae Strain (Mohamad et al.,2012). Similarly, from 37 to 70°C, B. sphaericus shows no variations in uptake of copper, nickel, or chromium (AI Daghistani et al., 2012). While using rice husk as an adsorbent for the removal of engine oil the efficiency was high when the temperature was raised to 160°C (Aminuddin et. al., 2023). Overall, the relationship between solution temperature and biosorption mechanism is complex and depends on the specific system and conditions.

4.4. Contact Time

Figure 5. Effect of contact duration on the biosorption capacity of Pb(II), Cd(II), and Zn(II) ions by soybean waste (Reused with permission from Fertu et al., 2022)

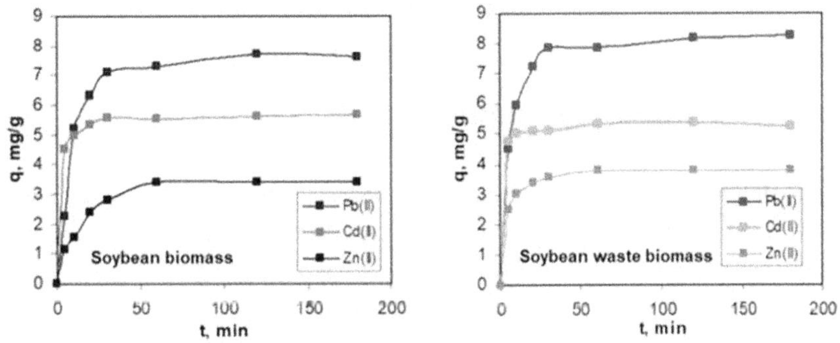

Contact time is an important factor in the biosorption process as it determines the time required for the equilibrium or isotherm studies. The duration of contact between the biosorbent and the metal ions affects the biosorption process. The efficiency or capacity of biosorption increases with contact time until equilibrium is reached. At time $t = 0$, the availability of binding sites on biosorbents to the heavy metals is high in numbers. As time progresses the heavy metals occupying the binding sites results in reduced efficiency of biosorption. Rapid sorption is desirable as it reduces the size of the biosorption column and makes the process more cost-effective. The actual contact time required for equilibrium can vary widely, ranging from 10 to 2880 minutes. Some studies have observed rapid kinetics at the initial contacts, with a significant percentage of metal adsorbed within the first few minutes. Longer contact times generally result in higher metal uptake (Gracia et al., 2016; Kalita and Joshi, 2017; Podder and Majumder, 2018).

A study by Zinicovscaia et al. (2018) demonstrated that Spirulina platensis can effectively remove 97% of Re (VII) from industrial effluents and batch solutions within just 20 minutes of contact time. Sivakami et al. (2015) found that the uptake of cadmium, lead, and mercury by Oscillatoria limosa increases as the contact time prolongs. Their study also suggests that the absorption of metals varies depending on their concentration and the duration of contact. On the other hand, Sun et al. (2011) discovered that the absorption rate of antimony (V) by Microcystis sp. increases proportionally with time. This indicates that the contact time required can differ between species and metals. Cain et al. (2008) conducted an experiment on Anabaena flocculosa and Spirulina platensis to observe how they absorb Hg^{2+}. They discovered that initially, there is a rapid uptake of Hg^{2+} which then decreases over time. Both strains follow a biphasic pattern, with the first 20-30 minutes of contact removing 60 mg/g and 30 mg/g respectively. Over the next 14 hours, A. flocculosa removes 131 mg/g of Hg^{2+} while Spirulina platensis removes 96 mg/g in 24 hours. About 90% of Cu^{2+} can be effectively eliminated within 90 minutes using Spirulina platensis. Daniela Ionela Fertu et al (2022) analyzed the effect of contact duration between soybean waste on the biosorption capacity of Pb(II), Cd(II), and Zn(II) ions from aqueous solutions as shown in Figure 5 & Figure 7(a) and, the efficiency of biosorption remains equilibrium after the optimum contact time. The biosorption process becomes more expensive when the contact time is too lengthy, and the process's overall efficiency is drastically reduced when the parameter's value is too low than the optimum value. Safari and Ahmady-Asbchin (2018) confirmed this finding and offered a detailed explanation of how Fischerella ambigua ISC67 can efficiently remove Zn (II) from water. They also emphasized the significance of factors like pH, temperature, initial metal concentration, and duration of contact. The biosorption capability of biosorbents made from groundnut husks showed that 1-hour contact time had the maximum Cr adsorption of 94.41% and adsorption capacity of 0.58 mg/g; the quick initial uptake may have been caused by the total chromium building up quickly on the surface of biosorbents (Kebru Tegegn et al.,2022).

4.5. Biosorbent Concentration/Dosage

The concentration of biomass is an important factor in biosorption efficiency. Increasing biomass concentration generally enhances removal efficiency, but excessive concentrations above the equilibrium sorption capacity can lead to reductions in biosorption capacity. This is because a higher biomass concentration leads to an increase in the surface area and the number of active binding sites available for metal uptake. As a result, the biosorbent can effectively bind and remove more metal ions from the solution. Excessive biomass concentration can cause biomass aggregation, which decreases the effective specific surface area and active binding sites. This aggregation can lead to a decrease in the removal efficiency of heavy metals. The total amount of biosorbent used in most published studies ranges from 0.5 and 6.0 g/L.

The biosorbent dose significantly affected the amount of biosorption, with the maximum biosorption capacity recorded at 0.5 g/L at various initial copper concentrations of 22, 75, 53, 63, 71, 68, 89, 90, 110, and 93 mg/g, respectively. However, the decreased dosages of biosorbent frequently resulted in increased absorption and decreased elimination efficacy (Mohamad et al.,2012). Fathollahi, Khasteganan, Coupe et al. (2021) assessed the effect of low, medium, and high bacterial biomass dosage for removing seven heavy metals such as Cd (86), Cu (66), Ni (81), Pb (107), Mn (24), Cr (54) and Zn (42). Where biomass dosage of 1 to 5 g/L exhibited larger effective sites and biosorption effectiveness than a low biomass dosage. This is due to the increased number of active binding sites on the biosorbent's surface. Raising the biosorbent dosage reduces interaction between metal ions for adsorption to functional groups, especially when multiple metal ions are present. Zhang et al., (2020) investigated Cr(VI) removal with 2 g/L Polyglutamic acid as biosorbent and concluded that the removal rate increases with an increase in dosage but decreases the adsorption efficiency as depicted in Figure 7(d).

Figure 6. Influence of agitation speed on biosorption capacity and efficiency of Cu(II)
(Reused with permission from Mohamad et al., 2012)

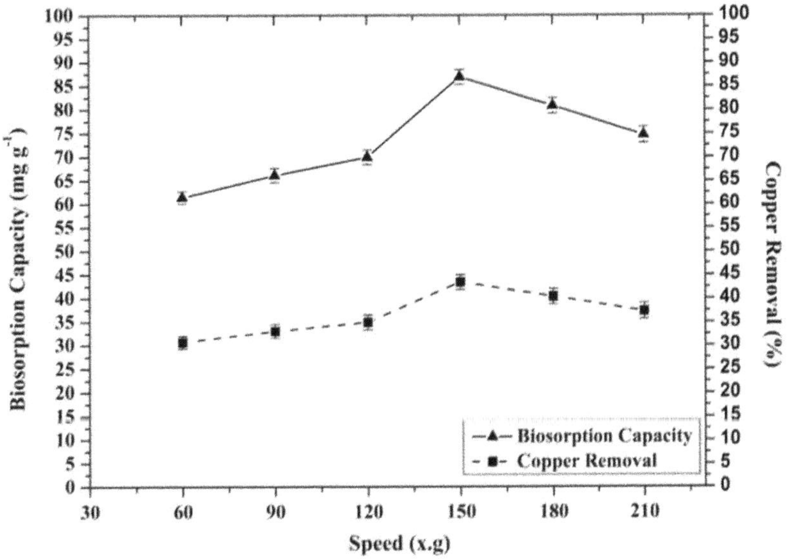

4.6. Agitation Rate/Shaking Speed

Agitation speed affects biosorption efficiency. Increasing the agitation speed enhances the mass transfer of metal ions from the bulk fluid to the binding sites on the biosorbent. This allows for more contact between the metal ions in the solution and the active binding sites, thereby increasing the removal efficiency of the metal ions. On the other hand, excessively high agitation speeds can make the suspension non-homogeneous, resulting in reductions in removal efficiency. Therefore, moderate agitation speeds are advisable for increased removal efficiencies during the biosorption of metals. At 200 rpm shaking speed or agitation rate, 30 min of contact time, and 6 pH with a fixed quantity of bio-sorbent say 1 g/L the efficiency of Methylene blue is 93% with modified clove as biosorbent (Kusuma et al., 2023). Fadel et al. (2014), investigated the impact of varying agitation speeds (100, 150, 200, and 250 rpm) for 30 minutes on the Mn^{+2} biosorption with 4.8 mg Mn^{2+}/l and 0.1 g of live biomass S. cerevisiae. Where accumulation begins at 150 rpm and found the optimum agitation speed for heavy metal settling in biosorption media is 150 rpm for chromium biosorption. Similarly, biosorption of Copper (II) using Non-Living Mesorhizobium amorphae at 150 g shows improvement in biosorption capacity from 61.4 mg/g to 87.0 mg/g. However, decreased removal of Cu (II) was noticed for the increasing agitation speed above 150 g as shown in Figure 6. (Mohamad et al., 2012). Kebru et al., (2022) examined the impact of agitation speed on biosorption efficiency at 80, 120, 160, and 200 rpm for the removal of chromium using groundnut husk as sorbent, and the maximum level of chromium of 90.02% was removed at 160 rpm.

Figure 7. Influence of agitation speed on biosorption capacity and efficiency of Cu(II) (Reused with permission from Mohamad et al., 2012)

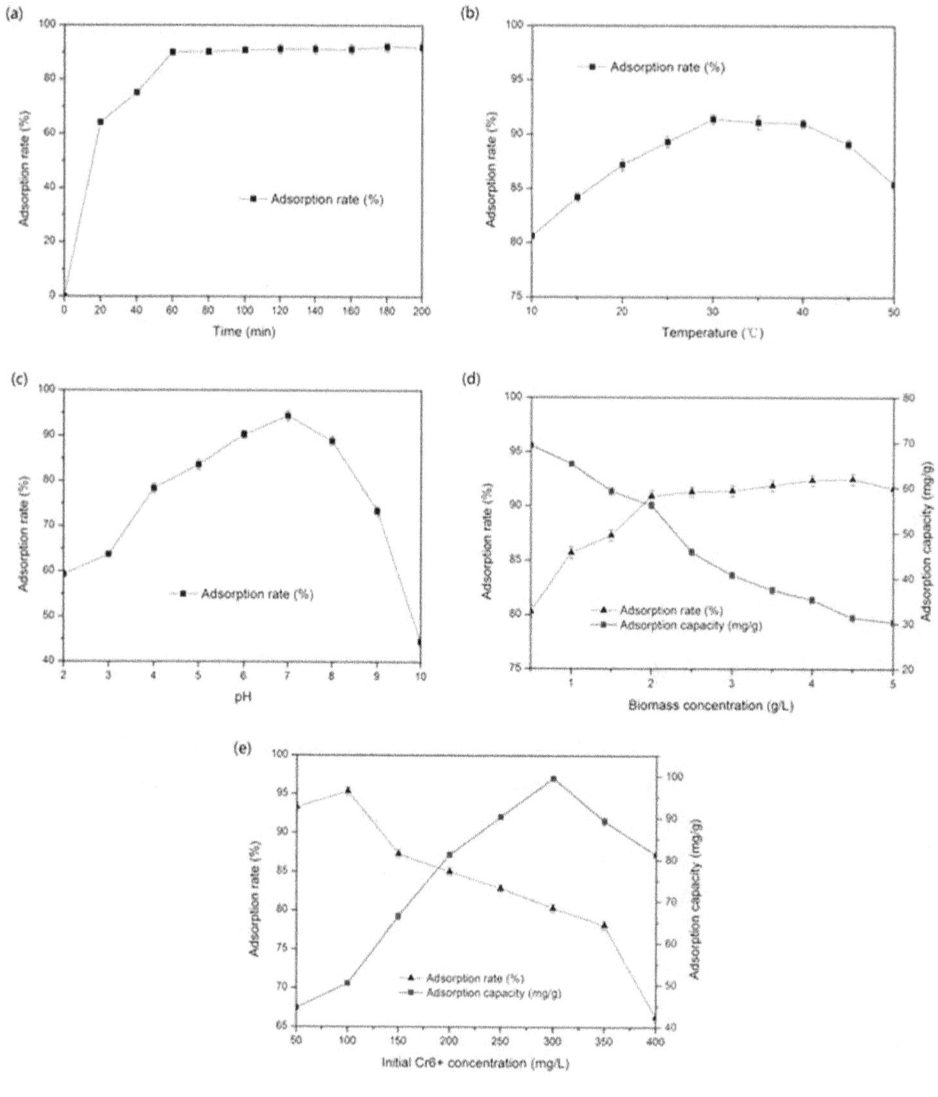

4.7. Competing Ions

Competing ions in biosorption mechanisms refer to the presence of multiple metallic ions in industrial effluents, which can make the treatment process challenging. The effluents from industries have more than one metal ion in it, henceforth it becomes very hard for the treatment using biosorbents. More precisely, if the coexisting ions incline the same sites of the metal that is intended to be removed, they compete with it for the same binding sites. This competition can lower the specificity of the biosorbent. Still, if the coexisting ions have a preference for different binding sites, it may not affect the biosorption process significantly. The presence of competing ions can be predicted using Pearson's metals classification, which helps understand the chemical coordination characteristics of elements in multiple metallic ionic solutions. Sulaymon et al. (2013) investigated the competitive properties of Pb (II), Cd (II), Cu (II), and

As (III) during the process of biosorption by employing a mixture of constituting 92% of Oscillatoria princeps, 2% of Oscillatoria subrevis, and 1% of Oscillatoria formosa in Cyanophyta and 3% of Spirogyra aequinoctialis in Chlorophyta and other species. For single metal systems, the biosorbent affinity was the highest for Cu, followed by Cd and Pb. However, in binary, ternary, and quaternary systems, these constants were drastically reduced. This happens because the metals under investigation preferred the same binding locations, indicating competing binding behaviors.

Cd (II) is a soft metal whereas As (III) is a hard metal and Pb (II) as well as Cu (II) are transition metals. Both hard and soft metals can bind to biomass materials. The result might be the case for as long the ions are present at the optimal adsorption pH in their pure ionic forms (Pereira et al., 2011; Sulaymon et al., 2013) which was because at low pH (optimal 4), it was thermodynamically viable for $As2O_3$ t_o dissolve in arsenous acid and this kind of dissolution is expressed as Eq.3 and the arsenous acid breaks down at a pH of 9.2 (Nordstrom et al., 2014) as expressed in Eq. 4.

$$As3O3(_cr)_+H2O\ (l) \leftrightarrow H_2AsO3\ (aq)\ _\Delta G\ (-ve_) \quad (3)$$

$$H3AsO3(aq) \leftrightarrow H2AsO3-\ (_aq)+H+(_aq) \quad (4)$$

T$_h$ese fa$_c$tors collectively influence the net surface charge of the biomass, metal speciation, and the rate of metal uptake, ultimately affecting the performance of the biosorption process.

5. IMMOBILIZATION OF BIOSORBENT FOR ENHANCED EFFICIENCY

Immobilization is the process of making cells, biomass, catalysts, cells enzymes, etc. fixed under a separate phase that does not comprise the bulk phase (Robinson, 1998). Zhang et al., (2020), used waste from the polyglutamic acid industry for biosorption of Cr(VI) by immobilization with sodium alginate and found the optimal adsorption factors: pH 7, 35 °C temperature, 200 mg/L of initial Cr(VI), 2 g/L of biomass after 60 minutes. The biosorption efficiency of Cr(VI) under immobilized circumstances of Polyglutamic acid was 96.3±0.4%. The desorption rate was 94.4±0.8% when 1 mol/L HCl was applied to the waste biomass (Zhang et al., 2020). Wu et al., (2004), used immobilized activated sludge into Ca-alginate beads for biosorption of three heavy metals copper (II), zinc (II), and chromium (II) from an aqueous solution. A biphasic adsorption pattern was found in the study and beads reached a sorption equilibrium in roughly 20 hours, and the sorbed heavy metal had no alteration over time and also, and the temperature influence was not noticed under a test range of 10 to 40 °C, which shows the immobilization is independent of temperature. El-Naggar et al. (2018), optimized by immobilization of Gelidium amansii biomass for Pb2+ removal from an aqueous med^{iu}m as shown in Figure 8, which resulted in 100% removal of Pb2+ with Gelidium amansii bioso^{rb}ent following the factors as follows: 200 mg/L tial Pb2+ concentration, 45°C tempera^{tu}re, 4.5 pH, and 1g/L Gelidium amansii as biosorbent. Figure 9 shows the adsorbent efficiency of the Gelidium amansii biosorbent in Pb2+ removal.

Figure 8. Immobilized Gelidium amansii biosorbent on sodium alginate beads: (A) funnel with sodium alginate beads, (B) sodium alginate beads immobilized with Gelidium amansii biosorbent (El-Naggar et al., 2018)

Figure 9. Adsorbent efficiency of the Gelidium amansii biosorbent in Pb^{2+} removal (El-Naggar et al., 2018)

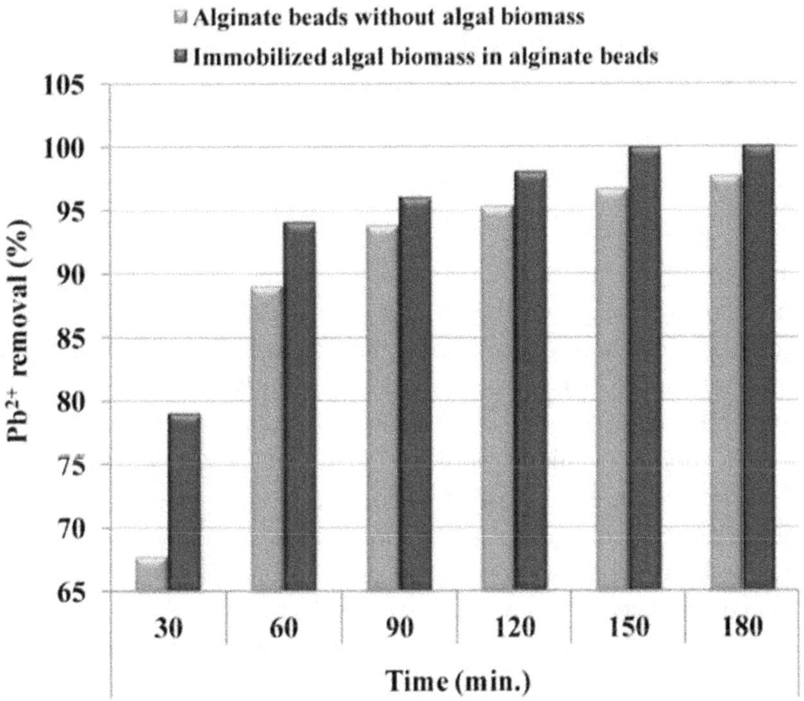

6. LIMITATION OF BIOSORPTION

The biosorption process an effective and environmentally friendly process for removing pollutants from wastewater, faces certain limitations and challenges:

a) Specificity and selectivi***ty: Biosorbents may exhibit*** specificity and selectivity for certain pollutants, making them less effective for a broad range of contaminants.
b) Biosorbent regeneration: ***Regenerating biosorbent***s for multiple cycles can be challenging and may involve harsh conditions, impacting the viability and performance of the biosorbent.
c) Competing ions: The pre***sence of compet***ing ions in the wastewater can reduce the biosorption efficiency of the desired pollutant.
d) Scale-up challenges: B***ench-scale studies m***ay not accurately represent the performance of biosorption at larger scales.
e) Kinetic limitations: ***It has slower kineti***cs compared to other treatment methods.
f) Biosorbent stability***: The stability and d***urability of some biosorbents may be compromised under certain conditions and have a shorter lifetime.
g) Mass transfer limit***ations: Inadequate mass tr***ansfer of pollutants to biosorbent surfaces can limit the efficiency of the process.
h) Cost of biosorben***ts: Certain biosorben***ts especially those modified or engineered, can be expensive to produce.
i) Biosorbent availa***bility: Some biosorbents*** may have limited availability, especially those derived from specific organisms or agricultural by-products.

j) Environmental im***pact: Disposal of spe***nt biosorbents or regeneration by-products may pose environmental challenges.
k) Lack of standar***dization: Lack of standa***rdization protocols for biosorption studies can hinder comparability between different research findings.

7. FUTURE STUDY

The role of microbial activity and the structural attributes of biosorbents contribute significantly to the biosorption capacity, further emphasizing the need for a holistic approach in environmental remediation studies. Industrial effluents include a variety of contaminants, including those of attraction, in contrast to laboratory solutions (Park et al. 2010). Although the key factors like metal content, initial concentration, temperature, and pH of industrial wastewater are known to have a significant impact on the process of removing heavy metals, very few studies have been observed on biosorbent selectivity (Li et al., 2022).

Investigating the simultaneous elimination of several coexisting contaminants is therefore necessary. Reducing the expenses related to biomass growing while enhancing the process's environmental advantages is another goal to be accomplished (Mahari et al., 2022; Ciani et al., 2023). On the other hand, biosafety issues and contamination risks need to be considered for engineering microbial biosorbents like cyanobacteria (Cui et al., 2021). The development of all-purpose biosorbents that are capable of eliminating a wide range of contaminants is needed. Using combination biosorbents, which are made up of many types of biomass, is one such possible alternative. "Combination" biosorbents might be a more practical method of designing biosorbent systems, even though they tend to make the characterization of these biosorption systems even more difficult.

It is important to carefully consider the possible dangers and advantages of reusing biomass polluted with heavy metals into value-added commodities, as well as ecological and financial benefits. The potential uses of these substances in several sectors of utilization, such as nanotechnology, are hampered by several safety issues. Nanoparticles may readily pass through biological barriers and enter any organ in a living being because of their tiny size. Numerous harmful consequences, such as immune system modification, oxidative stress, carcinogenesis, and DNA damage, have been linked to their biological accumulation (Liu et al., 2022; Das et al., 2023). Metal nanoparticles' size, shape, surface charge, and aggregation state dictate a number of their toxicological characteristics (Fernandes et al., 2022). If the synthesized nanoparticles are combined with wastewater-based heavy metal removal, the risk evaluation of these elements should be considered more precisely to avoid damage and dangers (Solano et al., 2021). Identification of the physicochemical characteristics, hazardous nature, and risk assessment and management procedures of the process must be considered.

In addition, addressing the interference of competing ions and enhancing selectivity for target pollutants. In the area of scaling up biosorption processes, maintaining efficiency and cost-effectiveness poses technical and engineering challenges. Improving the kinetics and stability of the biosorbents to meet the demands of real-time wastewater treatment and prolonged usage of the sorbent. There is a need for establishing universal benchmarks and protocols for accurate assessment and comparison of biosorption processes. Caution must be taken during the disposal of biosorbents by employing environmentally friendly methods, and addressing potential ecological remediation.

8. CONCLUSION

Heavy metal pollution is considered one of the most concerning environmental pollution. The biosorption process is an efficient method to remove heavy metals from the ecosystem. The biosorption process engages a wide range of substrates that range from microbial biomass to plant biomass. The mechanism of the biosorption process is complex and differs for different biomass. The biosorption process is influenced by various factors. The biosorption process is influenced significantly by a few of the influencing factors namely solution temperature, pH of the solution, contact time, type of biosorbent, initial concentration of metal ions, and biosorbent dosage. The optimization of biosorption efficiency requires a nuanced understanding of these factors, recognizing their dynamic relationships and impacts on the overall remediation performance. As the world seeks innovative and eco-conscious solutions for water and soil remediation, the insights provided in this study serve as a foundation for future research endeavors. Incorporating biosorption in the removal of heavy metals from industrial pollutants reduces the cost of effluent treatment and increases the efficient heavy metal removal in less time. As there is a wide range of biosorbents, the versatility of biosorbents and each biosorbents having unique properties make it versatile for a wide range of heavy metal removal. By advancing our understanding of the factors influencing biosorption, researchers and environmental practitioners can tailor strategies to address specific challenges, fostering the development of sustainable practices in the realm of environmental science and engineering. The study on factors influencing biosorption elucidates a complex interplay of variables crucial for the success of environmentally sustainable remediation strategies. Through a comprehensive analysis of pH, temperature, concentration gradients, and the inherent characteristics of biosorbents and contaminants, this research highlights the intricate nature of biosorption processes.

REFERENCES

Abioye, O. P., Oyewole, O. A., Oyeleke, S. B., Adeyemi, M. O., & Orukotan, A. A. (2018). *Biosorption of lead, chromium, and cadmium in tannery effluent using indigenous microorganisms*. Academic Press.

Abu Hasan, H., Sheikh Abdullah, S. R., Tan Kofli, N., & Yeoh, S. J. (2016). Interaction of environmental factors on simultaneous biosorption of lead and manganese ions by locally isolated Bacillus cereus. *Journal of Industrial and Engineering Chemistry, 37*, 295–305. Advance online publication. doi:10.1016/j.jiec.2016.03.038

Al-Daghistani, H. (2012). Bio-remediation of Cu, Ni, and Cr from rotogravure wastewater using immobilized, dead, and live biomass of indigenous thermophilic bacillus species. *The Internet Journal of Microbiology, 10*(1).

Al-Homaidan, A. A., Alabdullatif, J. A., Al-Hazzani, A. A., Al-Ghanayem, A. A., & Alabbad, A. F. (2015). Adsorptive removal of cadmium ions by Spirulina platensis dry biomass. *Saudi Journal of Biological Sciences, 22*(6), 795–800. doi:10.1016/j.sjbs.2015.06.010 PMID:26587009

Ali Redha, A. (2020). Removal of heavy metals from aqueous media by biosorption. *Arab Journal of Basic and Applied Sciences, 27*(1), 183–193. doi:10.1080/25765299.2020.1756177

Almomani, F., & Bhosale, R. R. (2021). Bio-sorption of toxic metals from industrial wastewater by algae strains Spirulina platensis and Chlorella vulgaris: Application of isotherm, kinetic models and process optimization. *The Science of the Total Environment*, *755*, 142654. doi:10.1016/j.scitotenv.2020.142654 PMID:33082037

Ameen, M. M., Moustafa, A. A., Mofeed, J., Hasnaoui, M., Olanrewaju, O. S., Lazzaro, U., & Guerriero, G. (2021). Factors Affecting Efficiency of Biosorption of Fe (III) and Zn (II) by Ulva lactuca and Corallina officinalis and Their Activated Carbons. *Water (Basel)*, *13*(23), 3421. doi:10.3390/w13233421

Aminuddin, I. H., Taufik, S. H., Puasa, N. A., Radziff, S. B. M., Zamree, N. D., Shaharudddin, N. A., Che Abdullah, C. A., Rahman, M. F., Azmi, A. A., & Ahmad, S. A. (2023). Biosorption of Engine Oil Using Rice Husk in a Filtration System. *Sustainability (Basel)*, *15*(19), 14599. doi:10.3390/su151914599

Ati, E. M., Abbas, R., Latif, A. S., & Ajmi, R. N. (2021). Factors Affecting Biosorption of Heavy Metals By Powder Pomegranate And Corn Peels. *NVEO-Natural Volatiles & Essential Oils Journal*.

Ayangbenro, A. S., Babalola, O. O., & Aremu, O. S. (2019). Bioflocculant production and heavy metal sorption by metal resistant bacterial isolates from gold mining soil. *Chemosphere*, *231*, 113–120. doi:10.1016/j.chemosphere.2019.05.092 PMID:31128345

Bashir, L. A., Malik, L. A., Ahad, S., Manzoor, T., Bhat, M. A., Dar, G. N., & Pandith, A. H. (2019). Malik, S. Ahad, T. Manzoor, M.A. Bhat, G.N. Dar, A.H. Pandith, Removal of heavy metal ions from aqueous system by ion-exchange and biosorption methods. *Environmental Chemistry Letters*, *17*(2), 729–754. doi:10.1007/s10311-018-00828-y

Cain, A., Vannela, R., & Woo, L. K. (2008). Cyanobacteria as a biosorbent for mercuric ion. *Bioresource Technology*, *99*(14), 6578–6586. doi:10.1016/j.biortech.2007.11.034 PMID:18158240

Carreira, A. R., Passos, H., & Coutinho, J. A. (2023). Metal biosorption onto non-living algae: A critical review on metal recovery from wastewater. *Green Chemistry*, *25*(15), 5775–5788. doi:10.1039/D3GC01993D

Castro, L., Blazquez, M. L., González, F., Muñoz, J. A., & Ballester, A. (2017). ´ F. Gonz´ alez, J.A. Munoz, ˜ A. Ballester, Biosorption of Zn (II) from industrial effluents using sugar beet pulp and F. vesiculosus: From laboratory tests to a pilot approach. *The Science of the Total Environment*, *598*, 856–866. doi:10.1016/j.scitotenv.2017.04.138 PMID:28458203

Cheraghi, M., Sobhanardakani, S., Lorestani, B., & Zandipak, R. (2016). Tea Wastes Efficiency on Removal of Cd (II) From Aqueous Solutions. *Archives of Hygiene Sciences*, *5*(3).

Ciani, M., & Adessi, A. (2023). Cyanoremediation and phyconanotechnology: Cyanobacteria for metal biosorption toward a circular economy. *Frontiers in Microbiology*, *14*, 1166612. doi:10.3389/fmicb.2023.1166612 PMID:37323915

Cui, J., Xie, Y., Sun, T., Chen, L., & Zhang, W. (2021). Deciphering and engineering photosynthetic cyanobacteria for heavy metal bioremediation. *The Science of the Total Environment*, *761*, 144111. doi:10.1016/j.scitotenv.2020.144111 PMID:33352345

Das, S., Sultana, K. W., Ndhlala, A. R., Mondal, M., & Chandra, I. (2023). Heavy Metal Pollution in the Environment and Its Impact on Health: Exploring Green Technology for Remediation. *Environmental Health Insights*, *17*. Advance online publication. doi:10.1177/11786302231201259 PMID:37808962

Deng, L., Zhu, X., Wang, X., Su, Y., & Su, H. (2007). Biosorption of copper (II) from aqueous solutions by green alga Cladophora fascicularis. *Biodegradation*, *18*(4), 393–402. doi:10.1007/s10532-006-9074-6 PMID:17091350

Deshmukh, P. D., Khadse, G. K., Shinde, V. M., & Labhasetwar, P. (2017). Cadmium removal from aqueous solutions using dried banana peels as an adsorbent: Kinetics and equilibrium modeling. *Journal of Bioremediation & Biodegradation*, *8*(3), 3. doi:10.4172/2155-6199.1000395

Dey, S., Veerendra, G. T. N., Manoj, A. P., & Babu, P. A. (2023). Performance of plant leaf biosorbents for biosorption of phosphorus from synthetic water. *Cleaner Materials*, *8*, 100191. doi:10.1016/j.clema.2023.100191

Dursun, A. Y. (2006). A comparative study on determination of the equilibrium, kinetic and thermodynamic parameters of biosorption of copper (II) and lead (II) ions onto pretreated Aspergillus niger. *Biochemical Engineering Journal*, *28*(2), 187–195. doi:10.1016/j.bej.2005.11.003

El-Azazy, M., El-Shafie, A. S., Issa, A. A., Al-Sulaiti, M., Al-Yafie, J., Shomar, B., & Al-Saad, K. (2019). Potato peels as an adsorbent for heavy metals from aqueous solutions: Eco-structuring of a green adsorbent operating Plackett–Burman design. *Journal of Chemistry*, *2019*, 2019. doi:10.1155/2019/4926240

El-Naggar, N. E. A., Hamouda, R. A., Mousa, I. E., Abdel-Hamid, M. S., & Rabei, N. H. (2018). Biosorption optimization, characterization, immobilization and application of Gelidium amansii biomass for complete Pb2+ removal from aqueous solutions. *Scientific Reports*, *8*(1), 13456. doi:10.1038/s41598-018-31660-7 PMID:30194341

Fathollahi, A., Khasteganan, N., Coupe, S. J., & Newman, A. P. (2021). A meta-analysis of metal biosorption by suspended bacteria from three phyla. *Chemosphere*, *268*, 129290. doi:10.1016/j.chemosphere.2020.129290 PMID:33383280

Fernandes, M., Padrão, J., Ribeiro, A. I., Fernandes, R. D. V., Melro, L., Nicolau, T., Mehravani, B., Alves, C., Rodrigues, R., & Zille, A. (2022). Polysaccharides and metal nanoparticles for functional textiles: A review. *Nanomaterials (Basel, Switzerland)*, *12*(6), 6. doi:10.3390/nano12061006 PMID:35335819

Fertu, D. I., Bulgariu, L., & Gavrilescu, M. (2022). Modeling and optimization of heavy metals biosorption by low-cost sorbents using response surface methodology. *Processes (Basel, Switzerland)*, *10*(3), 523. doi:10.3390/pr10030523

García, R., Campos, J., Cruz, J. A., Calderon, M. E., Raynal, M. E., & Buitron, G. (2016). Biosorption OF CD, CR, MN, and PB from aqueous solutions BY Bacillus SP strains isolated from industrial waste-activated sludge. Tip, 19, 5-14. doi:10.1016/j.recqb.2016.02.001

Gupta, A., Mishra, A., Singh, R., & Thakur, I. S. (2021). Fabrication of calcite based biocomposites for catalytic removal of heavy metals from electroplating industrial effluent. *Environmental Technology & Innovation*, *21*, 101278. doi:10.1016/j.eti.2020.101278

Hadadi, A., Imessaoudene, A., Bollinger, J. C., Cheikh, S., Manseri, A., & Mouni, L. (2023). Dual Valorization of Potato Peel (Solanum tuberosum) as a Versatile and Sustainable Agricultural Waste in Both Bioflocculation of Eriochrome Black T and Biosorption of Methylene Blue. *Journal of Polymers and the Environment*, *31*(7), 1–16. doi:10.1007/s10924-023-02780-7

Hoseinian, F. S., Rezai, B., Kowsari, E., & Safari, M. (2020). A hybrid neural network/genetic algorithm to predict Zn(II) removal by ion flotation. *Separation Science and Technology*, *55*(6), 1197–1206. doi: 10.1080/01496395.2019.1582543

Hrynkiewicz, K., & Baum, C. (2014). Application of microorganisms in bioremediation of environment from heavy metals. *Environmental Deterioration and Human Health: Natural and Anthropogenic Determinants*, 215-227.

Iqbal, J., Javed, A., & Baig, M. A. (2022). Heavy metals removal from dumpsite leachate by algae and cyanobacteria. *Bioremediation Journal*, *26*(1), 31–40. doi:10.1080/10889868.2021.1884530

Jimenez-Cedilloa, M. J., Olguin, M. T., Fall, C., & Colin-Cruz, A. (2013). As(III) and As(V)sorption on iron-modified non-pyrolyzed and pyrolyzed biomass from Petroselinum crispum(parsley). *Journal of Environmental Management*, *117*, 242–252. doi:10.1016/j.jenvman.2012.12.023 PMID:23376307

Joshi, H. K., Vishwakarma, M. C., Kumar, R., Sharma, H., Bhandari, N. S., & Joshi, S. K. (2023). The biosorption of Zn2+ by various biomasses from wastewater: A review. *Journal of Water Process Engineering*, *56*, 104389. doi:10.1016/j.jwpe.2023.104389

Kalita, D., & Joshi, S. R. (2017). Study on bioremediation of Lead by exopolysaccharide producing metallophilic bacterium isolated from extreme habitat. *Biotechnology Reports (Amsterdam, Netherlands)*, *16*, 48–57. Advance online publication. doi:10.1016/j.btre.2017.11.003 PMID:29167759

Kalita, N., & Baruah, P. P. (2023). (Cyanobacteria as a potent platform for heavy metals biosorption: Uptake, responses, and removal mechanisms. *Journal of Hazardous Materials Advances*, *11*, 100349. doi:10.1016/j.hazadv.2023.100349

Kamsonlian, S., Suresh, S., Ramanaiah, V., Majumder, C., Chand, S., & Kumar, A. (2012). Bio-sorptive behavior of mango leaf powder and rice husk for arsenic (III) from aqueous solutions. *International Journal of Environmental Science and Technology*, *9*(3), 565–578. doi:10.1007/s13762-012-0054-6

Karadeniz, S. C., Isik, B., Ugraskan, V., & Cakar, F. (2023). Agricultural Lolium perenne seeds as a low-cost biosorbent for Safranine T adsorption from wastewater: Isotherm, kinetic, and thermodynamic studies. *Physics and Chemistry of the Earth Parts A/B/C*, *129*, 103338. doi:10.1016/j.pce.2022.103338

Kaushik, P., & Malik, A. (2009). Fungal dye decolorization: Recent advances and future potential. Environmental. *International Journal (Toronto, Ont.)*, *35*, 127–141. PMID:18617266

Kebru, Yusuf, Sasikumar, & Gorfu. (2022). Biosorbent Efficacy of Groundnut Husk for the Elimination of Chromium from the Effluent of Mojo Tannery Industry, Ethiopia. *International Journal of Biomaterials*. doi:10.1155/2022/9997348

Khadivinia, E., Sharafi, H., Hadi, F., Zahiri, H. S., Modiri, S., Tohidi, A., Mousavi, A., Salmanian, A. H., & Noghabi, K. A. (2014). Cadmium biosorption by a glyphosate-degrading bacterium, a novel biosorbent isolated from pesticide-contaminated agricultural soils. *Journal of Industrial and Engineering Chemistry*, *20*(6), 4304–4310. doi:10.1016/j.jiec.2014.01.037

Khani, M. H., & Khamseh, A. G. (2023). Statistical analysis, equilibrium, and dynamic study on the biosorption of strontium ions on Chlorella vulgaris. *Journal of Radioanalytical and Nuclear Chemistry*, *332*(8), 3325–3334. doi:10.1007/s10967-023-09026-9

Kilic, Z., Atakol, O., Aras, S., Cansaran-Duman, D., & Emregul, E. (2014). Biosorption properties of zinc (II) from aqueous solutions by Pseudevernia furfuracea (L.) Zopf. *Journal of the Air & Waste Management Association*, *64*(10), 1112–1121. doi:10.1080/10962247.2014.926299

Kim, I. H., Choi, J. H., Joo, J. O., Kim, Y. K., Choi, J. W., & Oh, B. K. (2015). Development of a microbe-zeolite carrier for the effective elimination of heavy metals from seawater. *Journal of Microbiology and Biotechnology*, *25*(9), 1542–1546. doi:10.4014/jmb.1504.04067 PMID:26032363

Kusuma, H. S., Aigbe, U. O., Ukhurebor, K. E., Onyancha, R. B., Okundaye, B., Simbi, I., Ama, O. M., Darmokoesoemo, H., Widyaningrum, B. A., Osibote, O. A., & Balogun, V. A. (2023). Biosorption of Methylene blue using clove leaves waste modified with sodium hydroxide. *Results in Chemistry*, *5*, 100778. doi:10.1016/j.rechem.2023.100778

Li, S., Dai, M., Wu, Y., Fu, H., Hou, X., Peng, C., & Luo, H. (2022). Resource utilization of electroplating wastewater: Obstacles and solutions. *Environmental Science. Water Research & Technology*, *8*(3), 484–509. doi:10.1039/D1EW00712B

Marković, M., Gorgievski, M., Štrbac, N., Grekulović, V., Božinović, K., Zdravković, M., & Vuković, M. (2023, January 20). (shen). Raw Eggshell as an Adsorbent for Copper Ions Biosorption—Equilibrium, Kinetic, Thermodynamic and Process Optimization Studies. *Metals*, *13*(2), 206. doi:10.3390/met13020206

Meena, A. K., Mishra, G. K., Rai, P. K., Rajagopal, C., & Nagar, P. N. (2005). Removal of heavy metal ions from aqueous solutions using carbon aerogel as an adsorbent. *Journal of Hazardous Materials*, *122*(1–2), 161–170. doi:10.1016/j.jhazmat.2005.03.024 PMID:15878798

Mohamad, O. A., Hao, X., Xie, P., Hatab, S., Lin, Y., & Wei, G. (2012). Biosorption of copper (II) from aqueous solution using non-living Mesorhizobium amorphae strain CCNWGS0123. *Microbes and Environments*, *27*(3), 234–241. doi:10.1264/jsme2.ME11331 PMID:22353770

Munoz, A.J., Espínola, F., & Ruiz, E. (2018). Removal of heavy metals by Klebsiella sp.3S1. Kinetics, equilibrium, and interaction mechanisms of Zn(II) biosorption. *J. Chem. Technol. Biotechnol.*, *93*, 1370-1380. . doi:10.1002/jctb.5503

Nagababu, A., Reddy, D. S., & Mohan, G. V. K. (2022). Toxic chrome removal from industrial effluents using marine algae: Modeling and optimization. *Journal of Industrial and Engineering Chemistry*, *114*, 377–390. doi:10.1016/j.jiec.2022.07.027

Nordstrom, D. K., Majzlan, J., & Königsberger, E. (2014). Thermodynamic properties for arsenic minerals and aqueous species. *Reviews in Mineralogy and Geochemistry*, *79*(1), 217–255. doi:10.2138/rmg.2014.79.4

Oyewole, O. A., Adamu, B. B., Oladoja, E. O., Balogun, A. N., Okunlola, B. M., Odiniya, E. E. (2018). A review on heavy metals biosorption in the environment. *Braz. J. Biol. Sci., 5*(10), 225-236. doi:10.21472/bjbs.051003

Park, D., Yun, Y. S., & Park, J. M. (2010). The past, present, and future trends of biosorption. *Biotechnology and Bioprocess Engineering; BBE, 15*(1), 86–102. doi:10.1007/s12257-009-0199-4

Pereira, S., Micheletti, E., Zille, A., Santos, A., Moradas-Ferreira, P., Tamagnini, P., & De Philippis, R. (2011). Using extracellular polymeric substances (EPS) producing cyanobacteria for the bioremediation of heavy metals: Do cations compete for the EPS functional groups and also accumulate inside the cell. *Microbiology, 157*(2), 451–458. doi:10.1099/mic.0.041038-0 PMID:20966085

Podder, M. S., & Majumder, C. B. (2018). Biological detoxification of As(III) and As(V) using immobilized bacterial cells in fixed-bed bio-column reactor: Prediction of kinetic parameters. *Groundwater for Sustainable Development, 6*, 14–42. Advance online publication. doi:10.1016/j.gsd.2017.10.004

Priya, A. K., Gnanasekaran, L., Dutta, K., Rajendran, S., Balakrishnan, D., & Soto-Moscoso, M. (2022). Biosorption of heavy metals by microorganisms: Evaluation of different underlying mechanisms. *Chemosphere, 307*, 135957. doi:10.1016/j.chemosphere.2022.135957 PMID:35985378

Priyadarshanee, M., & Das, S. (2021). Biosorption and removal of toxic heavy metals by metal tolerating bacteria for bioremediation of metal contamination: A comprehensive review. *Journal of Environmental Chemical Engineering, 9*(1), 104686. doi:10.1016/j.jece.2020.104686

Puranik, P. R., & Paknikar, K. M. (1999). Biosorption of lead, Cadmium, and Zinc by Citrobacter strain MCM B-181: Characterization studies. *Biotechnology Progress, 15*(2), 228–238. doi:10.1021/bp990002r PMID:10194398

Ramírez Calderon, ́. O. A., Abdeldayem, O. M., Pugazhendhi, A., & Rene, E. R. (2020). Current updates and perspectives of biosorption technology: An alternative for the removal of heavy metals from wastewater. *Current Pollution Reports, 6*(1), 8–27. doi:10.1007/s40726-020-00135-7

Robinson, P. K. (1998). Immobilized algal technology for wastewater treatment purposes. In N. F. Y. Tam & Y. S. Wong (Eds.), *Wastewater treatment with algae*. Springer-Verlag. doi:10.1007/978-3-662-10863-5_1

Saad, E. M., Elshaarawy, R. F., Mahmoud, S. A., & El-Moselhy, K. M. (2021). New Ulva lactuca algae-based chitosan bio-composites for bioremediation of Cd(II) ions. *J. Bioresour. Bioprod., 6*(3), 223–242. doi:10.1016/j.jobab.2021.04.002

Safari, M., & Ahmady-Asbchin, S. (2018). Biosorption of zinc from aqueous solution by cyanobacterium Fischerella ambigua ISC67: Optimization, kinetic, isotherm and thermodynamic studies. *Water Science and Technology, 78*(7), 1525–1534. doi:10.2166/wst.2018.437 PMID:30427793

Salam, K. A. (2019). Towards sustainable development of microalgal biosorption for treating effluents containing heavy metals. *Biofuel Research Journal, 6*(2), 948–961. doi:10.18331/BRJ2019.6.2.2

Samimi, M. (2024). Efficient biosorption of cadmium by Eucalyptus globulus fruit biomass using process parameters optimization. *Global Journal of Environmental Science and Management, 10*(1), 27–38.

Samimi, M., & Mansouri, E. (2023). Efficiency evaluation of Falcaria vulgaris biomass in Co (II) uptake from aquatic environments: Characteristics, kinetics and optimization of operational variables. *International Journal of Phytoremediation*, 1–11. PMID:37622683

Sari, M., & Tuzen, M. (2008). Tuzen, Biosorption of Pb (II) and Cd (II) from aqueous solution using green alga (Ulva lactuca) biomass. *Journal of Hazardous Materials*, *152*(1), 302–308. doi:10.1016/j.jhazmat.2007.06.097 PMID:17689186

Sedlakova-Kadukova, J., Kopcakova, A., Gresakova, L., Godany, A., & Pristas, P. (2019). Bioaccumulation and biosorption of zinc by a novel Streptomyces K11 strain isolated from a highly alkaline aluminum brown mud disposal site. *Ecotoxicology and Environmental Safety*, *167*, 204–211. doi:10.1016/j.ecoenv.2018.09.123 PMID:30340085

Sen, S., Dutta, S., Guhathakurata, S., Chakrabarty, J., Nandi, S., & Dutta, A. (2017). Removal of Cr (VI) using a cyanobacterial consortium and assessment of biofuel production. *Int. Biodeterior. Biodegrad.*, *119*, 211–224. https://doi.org/.ibiod.2016.10.050 doi:10.1016/j

Shafique, U., Ijaz, A., Salman, M., Zaman, W., Jamil, N., Rehman, R., & Javaid, A. (2012). Removal of arsenic from water using pine leaves. *Journal of the Taiwan Institute of Chemical Engineers*, *43*(2), 256–263. doi:10.1016/j.jtice.2011.10.006

Shahrokhi-Shahraki, R., Benally, C., El-Din, M. G., & Park, J. (2021). High-efficiency removal of heavy metals using tire-derived activated carbon vs commercial activated carbon: Insights into the adsorption mechanisms. *Chemosphere*, *264*, 128455. doi:10.1016/j.chemosphere.2020.128455 PMID:33032208

Shen, L., Li, Z., Wang, J., Liu, A., Li, Z., Yu, R., & Zeng, W. (2018). Characterization of extracellular polysaccharide/protein contents during the adsorption of Cd (II) by Synechocystis sp. PCC6803. *Environmental Science and Pollution Research International*, *25*(21), 20713–20722. doi:10.1007/s11356-018-2163-3 PMID:29754298

Singh, S. (2020). Biosorption of heavy metals by cyanobacteria: potential of live and dead cells in bioremediation. *Microb. Bioremed. Biodegrad.*, 409–423. doi:10.1007/978-981-15-1812-6_15

Smjecanin, N., Buzo, D., Masic, E., Nuhanovic, M., Sulejmanovic, J., Azhar, O., & Sher, F. (2022). Algae based green biocomposites for uranium removal from wastewater: Kinetic, equilibrium and thermodynamic studies. *Materials Chemistry and Physics*, *283*, 125998. doi:10.1016/j.matchemphys.2022.125998

Solano, R., Patiño-Ruiz, D., Tejeda-Benitez, L., & Herrera, A. (2021). Metal- and metal/oxide-based engineered nanoparticles and nanostructures: A review on the applications, nanotoxicological effects, and risk control strategies. *Environmental Science and Pollution Research International*, *28*(14), 16962–16981. doi:10.1007/s11356-021-12996-6 PMID:33638785

Sulaymon, A. H., Mohammed, A. A., & Al-Musawi, T. J. (2013). Competitive biosorption of lead, cadmium, copper, and arsenic ions using algae. *Environmental Science and Pollution Research International*, *20*(5), 3011–3023. doi:10.1007/s11356-012-1208-2 PMID:23054774

Sumathi, T., & Alagumuthu, G. (2014). Adsorption studies for arsenic removal using activated Moringa oleifera. *International Journal of Chemical Engineering*.

Sun, T., Qing, G., Su, B., & Jiang, L. (2011). Functional biointerface materials inspired from nature. *Chemical Society Reviews*, *40*(5), 2909–2921. doi:10.1039/c0cs00124d PMID:21347500

Sunil, K., Karunakaran, G., Yadav, S., Padaki, M., Zadorozhnyy, V., & Pai, R. K. (2018). Al- Ti2O6 a mixed metal oxide based composite membrane: A unique membrane for removal of heavy metals. *Chemical Engineering Journal*, *348*, 678–684. doi:10.1016/j.cej.2018.05.017

Tattibayeva, Z., Tazhibayeva, S., Kujawski, W., Zayadan, B., & Musabekov, K. (2022). Peculiarities of adsorption of Cr (VI) ions on the surface of Chlorella vulgaris ZBS1 algae cells. *Heliyon*, *8*(9), 10468. doi:10.1016/j.heliyon.2022.e10468 PMID:36105478

Tchounwou, P. B., Yedjou, C. G., Patlolla, A. K., & Sutton, D. J. (2012). Heavy Metals Toxicity and the Environment. *EXS*, *101*, 133–164. doi:10.1007/978-3-7643-8340-4_6 PMID:22945569

Varia, J., Zegeye, A., Roy, S., Yahaya, S., & Bull, S. (2014). Shewanella putrefaciens for the remediation of Au3þ, Co2þ and Fe3þ metal ions from aqueous systems. *Biochemical Engineering Journal*, *85*, 101–109. doi:10.1016/j.bej.2014.02.002

Verma, S., & Kuila, A. (2019). Bioremediation of heavy metals by microbial process. *Environmental Technology & Innovation*, *14*, 100369. doi:10.1016/j.eti.2019.100369

Volesky, B. (1994). Advances in biosorption of metals: Selection of biomass types. *FEMS Microbiology Reviews*, *14*(4), 291–302. doi:10.1111/j.1574-6976.1994.tb00102.x PMID:7917417

Wang, J. Yu., Chen, X., Du, D., Wu, R., Qu, G., Guo, X., Jia, H., & Wang, T. (2019). Non-thermal plasma oxidation of Cu(II)-EDTA and simultaneous Cu(II) elimination by chemical precipitation. *Journal of Environmental Management*, *248*, 109237. doi:10.1016/j.jenvman.2019.07.008 PMID:31310932

Wu, H. S., Zhang, A. Q., & Wang, L. S. (2004). Immobilization study of biosorption of heavy metal ions onto activated sludge. *Journal of Environmental Sciences (China)*, *16*(4), 640–645. PMID:15495972

Zhang, C., Ren, H. X., Zhong, C. Q., & Wu, D. (2020). Biosorption of Cr (VI) by immobilized waste biomass from polyglutamic acid production. *Scientific Reports*, *10*(1), 3705. doi:10.1038/s41598-020-60729-5 PMID:32111901

Zinicovscaia, I., Yushin, N., Shvetsova, M., & Frontasyeva, M. (2018). Zinc removal from model solution and wastewater by Arthrospira (Spirulina) Platensis biomass. *International Journal of Phytoremediation*, *20*(9), 901–908. doi:10.1080/15226514.2018.1448358 PMID:29873533

Zou, Z., Gu, Y., Yang, W., Liu, M., Han, J., & Zhao, S. (2021). A modified coagulation-ultrafiltration process for silver nanoparticles removal and membrane fouling mitigation: The role of laminarin. *International Journal of Biological Macromolecules*, *172*, 241–249. doi:10.1016/j.ijbiomac.2021.01.034 PMID:33454322

Chapter 4
Heavy Metals in Foodstuffs: Presence, Bioaccumulation, Origins, Health Risk, and Remediation

Hari Shankar Biswas
https://orcid.org/0000-0003-1883-5244
Surendranath College, India

Sandeep Poddar
https://orcid.org/0000-0001-9771-877X
Lincoln University College, Malaysia

ABSTRACT

Heavy metal contamination in foodstuffs poses a significant threat to human health worldwide. This chapter depicts sources, bioaccumulation mechanisms, and associated health risks of heavy metals in various food items. Metals such as lead, cadmium, mercury, and arsenic find their way into the food chain through natural processes, industrial activities, and agricultural practices. Once ingested, these metals tend to accumulate in human tissues, leading to severe health consequences. This chapter also explores the various remediation strategies employed to mitigate heavy metal contamination in food, including agricultural practices, phytoremediation, bioremediation, and technological interventions. The issue of heavy metal contamination in foodstuffs requires a multidisciplinary approach that combines scientific research, policy interventions, and public awareness. Implementing stringent regulations, promoting sustainable agricultural practices, and investing in innovative remediation technologies are crucial steps toward ensuring food safety and safeguarding public health.

INTRODUCTION

At this stage, food safety faces a significant challenge with the universal threat of heavy metal contamination in foodstuffs. The infiltration of metals such as lead, cadmium, mercury, and arsenic poses a dire risk to global human health, permeating diverse food items through natural processes, industrial activities, and agricultural practices (Landrigan, 1982; Uddin *et al.*, 2021). Once ingested, these heavy metals accumu-

DOI: 10.4018/979-8-3693-1618-4.ch004

late within human tissues, leading to severe health disorders such as neurological complications, organ damage, and increased cancer risks. Vulnerable populations, especially children and pregnant women, face heightened intensity due to their distinctive physiological susceptibilities (Ali *et al.*, 2020; Daurov, D., Zhambakin, K., & Shamekova, 2023). This chapter delves into the intricate tapestry of heavy metal contamination, exploring its origins, bioaccumulation pathways, and the profound health risks it poses to humanity. Moreover, the chapter navigates through a diverse array of remediation strategies meticulously crafted to counteract the pervasive influence of heavy metal contamination in our food supply. From time-honored agricultural practices to cutting-edge technological interventions, the chapter scrutinizes the efficacy of methods such as organic farming and advanced irrigation management in stemming the entry of metals into the food chain. Additionally, the exploration extends to nature's own remedies, as certain plant species emerge as remarkable allies in the pursuit of detoxification. Phytoremediation, a process harnessing the unique capabilities of these plants, is unveiled as a pivotal tool in cleansing our environment of heavy metal pollutants. The narrative does not stop there; it ventures into the realm of innovative technologies, where nanotechnology and bioremediation techniques converge to promise efficient and sustainable solutions for the removal of heavy metals from contaminated environments. As this chapter unfolds, it becomes evident that combating the issue of heavy metal contamination in foodstuffs necessitates a holistic and interdisciplinary approach (Ali *et al.*, 2020; Daurov, D., Zhambakin, K., & Shamekova, 2023). The fusion of rigorous scientific inquiry, targeted policy interventions, and widespread public awareness forms the bedrock of our collective efforts. It is through the implementation of stringent regulations, the promotion of sustainable agricultural practices, and strategic investments in transformative remediation technologies that we can pave the way toward a world where food safety is not just an ideal but a tangible reality, safeguarding the health and well-being of present and future generations.

The problem of heavy metal contamination in foodstuffs has become a pressing global concern, posing a significant threat to human health. Heavy metals like lead, cadmium, mercury, and arsenic find their way into our daily sustenance through natural processes, industrial activities, and agricultural practices. Once ingested, these metals accumulate within human tissues, leading to severe health disorders, including neurological complications, organ damage, and increased cancer risks (Balali-Mood *et al.*, 2021; Rai *et al.*, 2019). Vulnerable populations, especially children and pregnant women, are disproportionately affected due to their unique physiological susceptibilities. The pervasive infiltration of heavy metals into our food chain demands a comprehensive understanding of its origins and pathways. It is imperative to explore the intricate mechanisms that underpin the toxic journey of these metals from their sources to the human body. Additionally, identifying effective remediation strategies is crucial to counteracting this contamination and ensuring food safety for all. This study holds paramount significance in the realm of public health and environmental sustainability (Ali, Khan & Ilahi, 2019). Addressing heavy metal contamination in foodstuffs is critical for several reasons. Heavy metal exposure leads to a wide range of health issues, including debilitating neurological disorders, organ damage, and cancer. By understanding the sources and pathways of contamination, we can develop targeted interventions to mitigate these health risks and improve overall well-being. Children and pregnant women are particularly vulnerable to the adverse effects of heavy metals. Investigating ways to protect these demographics is not only a moral imperative but also essential for the development of future generations. Heavy metal contamination doesn't only affect human health; it also has detrimental effects on the environment. Studying the bioaccumulation pathways helps in understanding its impact on ecosystems and biodiversity, paving the way for sustainable environmental management (Chheang *et al.*, 2021). This research delves into a diverse array of remediation strategies, from traditional agricultural practices to cutting-edge technologies like nanotechnology and phytoremediation. These solutions not only offer hope for cleansing contaminated environments but also contribute to the evolution of sustainable practices in agriculture and industry. By shedding light on the complexities of heavy metal contamination, this study can inform policymaking. Implementing stringent regulations, promoting sustainable agricultural practices, and raising public awareness are crucial steps toward mitigating this issue.

Sources of Heavy Metal Contamination

Heavy metals can enter the food chain through natural processes, including geological weathering and volcanic activities. Certain minerals and rocks naturally contain traces of heavy metals, which can leach into soil and water, eventually making their way into foodstuffs (Tchounwou *et al.*, 2012). Industries such as mining, metal smelting, and manufacturing release substantial amounts of heavy metals into the environment. Effluents and emissions from factories often contain lead, cadmium, mercury, and arsenic, contaminating nearby soil and water sources. Agricultural activities contribute significantly to heavy metal contamination (Alengebawy *et al.*, 2021). The use of chemical fertilizers, pesticides, and sewage sludge can introduce metals like cadmium and lead into the soil. Improper disposal of industrial waste, such as agricultural fertilizer, amplifies this issue. Mining operations directly expose heavy metals present in ores. Smelting processes further concentrate these metals, releasing them into the air and nearby water bodies. This contamination affects not only agricultural lands but also aquatic ecosystems. Heavy metals can be transported over long distances through the air. Pollutants emitted from industries and vehicles settle on soil and water surfaces, leading to contamination (Briffa, Sinagra & Blundell, 2020). This atmospheric deposition is a widespread source of heavy metal pollution in various regions. Improper disposal of electronic waste, batteries, and other products containing heavy metals results in leaching. This leachate can infiltrate soil and groundwater, contaminating nearby agricultural fields and posing a risk to crops and livestock. Municipal and industrial wastewater often contains heavy metals from domestic and commercial sources. When untreated or inadequately treated, this wastewater contaminates water bodies used for irrigation, affecting the safety of crops and, subsequently, the food chain. This multidisciplinary analysis emphasizes the urgent need to address heavy metal contamination at its various sources, including natural occurrences and human activities (Othman *et al.*, 2021). Effective strategies and policies are essential to prevent further contamination and mitigate existing risks to human health.

Table 1. Permissible limit, sources, and side effects of some heavy metal ions

Heavy Metal Ion	Permissible Limit in Water (ppm or mg/L	Permissible Limit in Foodstuffs (ppm or mg/kg)	Sources	Side Effects
Lead (Pb^{2+})	0.01 (WHO)	2 (FDA for certain foods)	smelting, lead-containing pipes, Paint, gasoline, battery, pigments, arm industry, printing of books	Behavioral changes, kidney damage and nervous system, anemia, miscarriage of women, infertility in males, carcinogenic effects
Mercury (Hg^{2+})	0.001 (WHO)	0.5 (FAO/WHO Codex Alimentarius)	Mercury salt contamination comes from Volcanic eruptions	Damage to the nervous system, kidney, and cardiovascular system
Cadmium (Cd^{2+})	0.003 (WHO)	0.1(Codex Alimentarius Commission)	Fossil fuel, agricultural activity, cigarette smoke, paint, gasoline, smelting, battery, pigments, printing, etc.	Cancers, cardiovascular problems, respiratory issues, damage to kidney and nervous
Copper (Cu^{2+})	2 (WHO)	10 (FAO/WHO Codex Alimentarius)	Rock weathering, animal manure, fossil fuel burning, bubble bursting, aerosols	Dizziness, stomach pain, lack of muscular coordination
Arsenic (As^{3+})	0.01 (WHO)	0.1 (FAO/WHO Codex Alimentarius)	Wood preservatives, alloying, semiconductors, sewage sludge, glassware, pesticides	nervous system damage, lung and bladder cancer, diabetes, heart problems
Chromium (Cr^{3+})	0.05 (WHO)	0.5 (FAO/WHO Codex Alimentarius)	Steel industries, construction materials, glassware, pesticides	Mutagenic effects, cancers, neurological disorders, reproductive system issues
Zinc (Zn^{2+})	5 (WHO	150 (FAO/WHO Codex Alimentarius)	weathering of rocks and minerals in the Earth's crust, corrosion of zinc-coated pipes, meat, poultry, fish, and dairy, etc.	cause stomach cramps, nausea, vomiting, and diarrhea, suppress the immune system, nerve damage, resulting in numbness
Nickel (Ni^{2+})	0.1 (WHO)		Paint, galvanizing plants, porcelain enameling, nickel-cadmium batteries, plating	Respiratory problems, lung cancer, nasal cancer, gastrointestinal problems, heart diseases, epigenetic effects

Bioaccumulation of Heavy Metals

Heavy metals, a group of highly toxic elements including cadmium, lead, and mercury, find their way into the food chain primarily through plants. These contaminants infiltrate plants by seeping into the soil they grow in. The process starts with the plant's root systems, which act as sponges, absorbing water and nutrients from the surrounding soil. Unfortunately, if the soil is contaminated with heavy metals, these pollutants are soaked up alongside the essential nutrients. Once absorbed, heavy metals embark on a perilous journey within the plant (Tchounwou *et al.*, 2012). They are transported through the plant's vascular system, accumulating in various parts such as leaves, stems, and fruits. It's important to note that different plant species have varying capacities to accumulate these toxic substances. Some plants might store more heavy metals in their leaves, while others prefer their fruits as the storage site. This discrepancy in accumulation patterns makes it all the more challenging to predict and control the spread of these harmful elements within plant life (Angulo-Bejarano, Puente-Rivera & Cruz-Ortega, 2021). Aquatic ecosystems also fall prey to heavy metal contamination. In these environments, phytoplankton, microscopic plant-like organisms, form the foundation of the food chain. These tiny creatures absorb heavy metals directly from the water sources they inhabit. Consequently, heavy metals bioaccumulate as they move up the trophic ladder. Fish and shellfish, vital sources of sustenance for many human populations, have become carriers of heavy metals. These aquatic creatures accumulate these toxic substances either by ingesting contaminated organisms or by directly absorbing metals from the water and sediment around them. As a result, consumers who feast on these contaminated fish and shellfish unwittingly ingest heavy metals, which can lead to severe health complications (Gheorghe *et al.*, 2017). Terrestrial ecosystems are not exempt from this insidious cycle. Heavy metals traverse through these environments, impacting both herbivores and carnivores. Herbivorous animals, in their quest for sustenance, consume contaminated plants. Over time, these animals accumulate heavy metals in their tissues. This accumulation doesn't just stop there; it progresses up the food chain, affecting carnivores, including humans. Humans are at risk when they consume contaminated herbivores or other carnivores. Consequently, heavy metals accumulate in the human body, wreaking havoc on health (Tchounwou *et al.*, 2012; Jaishankar *et al.*, 2014).

Understanding these intricate pathways of bioaccumulation is paramount in assessing the risks associated with heavy metal contamination in various ecosystems. The implications of this contamination extend far beyond the immediate environment; it jeopardizes the safety of our food supply and the overall health of ecosystems. Consequently, effective management and remediation strategies are not just a luxury but a necessity to mitigate these risks. Implementing these strategies requires a multi-faceted approach. First and foremost, there must be a concerted effort to monitor and regulate industrial activities that are often the source of heavy metal pollution. Strict regulations can minimize the discharge of these substances into the environment, curbing their entry into the food chain. Additionally, educating farmers about responsible agricultural practices is crucial. Using organic farming methods and reducing the reliance on chemical fertilizers and pesticides can prevent heavy metals from entering the soil in the first place. Soil testing and remediation efforts can help identify contaminated areas and facilitate the cleanup process (Priya *et al.*, 2023; Sánchez-Castro *et al.*, 2023). Furthermore, it is essential to invest in wastewater treatment facilities, ensuring that industrial and municipal effluents are adequately treated before being released into natural water bodies. Efficient wastewater treatment can significantly reduce the concentration of heavy metals in aquatic environments, safeguarding aquatic life and the humans who depend on it. Moreover, public awareness campaigns play a vital role. Educating communities about the risks associated with heavy metal contamination and providing them with information on how to minimize exposure can lead to healthier dietary choices and lifestyle changes. These campaigns can empower individuals to make informed decisions about their food consumption, reducing the demand for contaminated products. In tandem with awareness campaigns, scientific research must continue to uncover innovative solutions (Obaideen *et al.*, 2022). Bioremediation, a process that uses living organisms to detoxify polluted environments, holds promise for addressing heavy metal contamination. Certain plants, bacteria, and fungi have the remarkable ability to absorb and accumulate heavy metals. Harnessing the power of these natural entities can aid in cleaning up contaminated sites, mitigating the risks associated with bioaccumulation (Kapahi & Sachdeva, 2019). Therefore, heavy metal contamination significantly threatens ecosystems and human health. By understanding the pathways of bioaccumulation and implementing comprehensive management and remediation strategies, we can safeguard our food supply and protect the delicate balance of nature. It is not just a matter of regulation or scientific innovation; it is a collective responsibility. Governments, industries, scientists, and communities must collaborate to tackle this pressing issue, ensuring a healthier, safer future for all.

Studies on the Bioaccumulation of Specific Heavy Metals in Foodstuff

Numerous studies have been conducted to investigate the bioaccumulation of specific heavy metals in various food items, shedding light on the extent of contamination and potential health risks associated with consumption. Understanding the specific pathways of bioaccumulation in different foodstuffs is crucial for developing targeted strategies to mitigate contamination. The following summaries highlight key findings from selected studies:

Research conducted by Huang *et al.* (2020) investigated the bioaccumulation of cadmium in leafy vegetables commonly consumed by humans. The study revealed that certain crops, such as spinach and lettuce, exhibited a higher tendency to accumulate cadmium. Factors such as soil pH and organic matter content were identified as significant contributors to the variability in cadmium uptake by these vegetables. A study conducted by Rai *et al.* (2019) focused on the bioaccumulation of lead in cereals and grains, staple foods for many populations. The research found that lead contamination in these crops was influenced by both soil properties and agricultural practices. Additionally, the study emphasized the importance of monitoring lead levels in irrigated water used for crop cultivation. Research led by Gelais and Costa-Pierce (2016) examined the bioaccumulation of mercury in various fish species and seafood. The study identified predatory fish, such as tuna and swordfish, as having elevated mercury levels due to biomagnification in the marine food chain. The findings underscored the need for informed choices in seafood consumption, especially for vulnerable populations such as pregnant women. Investigations by Rokonuzzaman *et al.* (2022) focused on the bioaccumulation of arsenic in rice, a staple food for a significant portion of the global population. The study revealed that the arsenic content in rice was influenced by both natural geological factors and anthropogenic inputs, such as the use of arsenic-containing pesticides. Implementing arsenic mitigation strategies in rice cultivation was highlighted as crucial for reducing human exposure. Rai *et al.* (2019) conducted a study exploring the bioaccumulation of nickel in edible plants. The research found that certain crops, including beans and carrots, showed a propensity to accumulate nickel from contaminated soils. Soil amendments and crop rotation practices were identified as potential mitigation strategies to reduce nickel uptake in edible plants. These studies collectively emphasize the need for targeted interventions based on the specific characteristics of heavy metal bioaccumulation in different food items. By incorporating these findings into regulatory frameworks and agricultural practices, it is possible to develop more effective strategies to safeguard the food supply and human health from the threats posed by heavy metal contamination.

Health Risks Associated With Heavy Metal Exposure

Bioaccumulation is the process by which substances, usually toxic chemicals or heavy metals, accumulate in living organisms over time. This accumulation occurs through various mechanisms, such as ingestion, absorption, and inhalation. Once inside an organism, these substances may be stored in tissues and organs, posing potential health risks (Fig. 1). Heavy metal exposure poses significant risks to human health, leading to a range of acute and chronic health effects. Understanding these effects and the various routes of exposure is crucial to preventing and mitigating the impact on individuals and communities. Heavy metal poisoning, resulting from exposure to high concentrations of metals like lead, mercury, and arsenic, can cause the rapid onset of symptoms such as nausea, vomiting, and abdominal pain. Severe cases can lead to organ failure and death if not treated promptly. Ingesting food or water contaminated with heavy metals can cause gastrointestinal disturbances, including diarrhea and cramps. Immediate medical attention is required to prevent dehydration and further complications (Witkowska et al., 2021). Chronic exposure to lead, mercury, and manganese can lead to neurological disorders, especially in children. These disorders include developmental delays, learning disabilities, and behavioral issues due to the damaging effects of heavy metals on the central nervous system. Certain heavy metals, such as cadmium, chromium, and arsenic, are known carcinogens. Prolonged exposure to these metals increases the risk of developing various cancers, including lung, bladder, liver, and skin cancers (Tchounwou *et al.*, 2012; Mitra *et al.*, 2022).

Heavy metal exposure can lead to reproductive problems, including infertility and miscarriages. Additionally, exposure during pregnancy can result in developmental disorders in the fetus, affecting physical and mental health later in life. Studies have linked heavy metal exposure, particularly to lead and cadmium, with an increased risk of cardiovascular diseases. These metals can damage the cardiovascular system, leading to hypertension, heart attacks, and other heart-related conditions. Contaminated food and water sources are significant routes of heavy metal exposure. Industrial discharges, agricultural runoff, and improper waste disposal contaminate water bodies and crops, leading to ingestion by humans. Inhalation of heavy metal particles or fumes occurs in occupational settings such as mining, welding, and manufacturing. Workers in these industries are at risk of respiratory issues, lung infections, and systemic toxicity due to inhalation exposure (Alengebawy *et al.*, 2021). Contact with contaminated soil, dust, or products can lead to the dermal absorption of heavy metals. Skin exposure can cause irritation, rashes, and, in severe cases, systemic toxicity as the metals are absorbed through the skin barrier. Understanding these health risks and the pathways through which heavy metals enter the human body is essential to developing effective strategies to minimize exposure and protect public health. Regulatory measures, workplace safety protocols, and public awareness campaigns are crucial in preventing the adverse health effects associated with heavy metal exposure.

Figure 1. Bioaccumulation occurs through various mechanisms

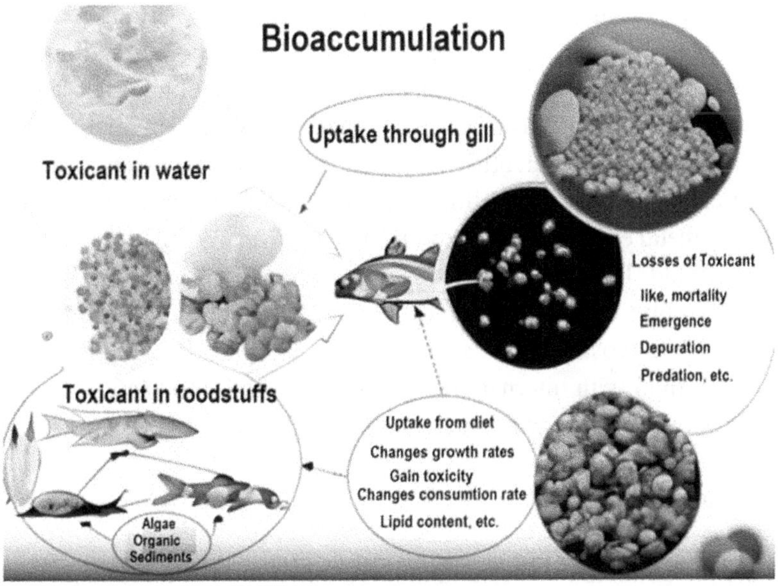

REGULATORY FRAMEWORK AND GUIDELINES FOR MITIGATING HEAVY METAL CONTAMINATION IN FOODSTUFFS

Environmental Protection Laws

Most countries have stringent environmental protection laws that dictate the permissible levels of heavy metals in soil, water, and air. These laws often set standards for remediation efforts and require compliance with specific guidelines. In the United States, for instance, the Comprehensive Environmental Response, Compensation, and Liability Act (CERCLA) and the Resource Conservation and Recovery Act (RCRA) provide the legal framework for addressing hazardous waste sites and regulating the disposal of hazardous waste. Some regions have site-specific regulations that address unique environmental challenges related to heavy metal contamination. These regulations may consider factors such as the local ecology, hydrogeology, and population density. Remediation projects often require permits from environmental regulatory agencies. These permits outline the specific activities that are allowed, as well as the standards and requirements that must be met. Compliance with these permits is crucial to avoid legal consequences. Regulatory frameworks typically include provisions for continuous monitoring of the remediation process and the affected environment. Regular reporting to regulatory authorities ensures transparency and accountability.

International Regulations

Various international organizations, such as the World Health Organization (WHO) and the United Nations Environment Programme (UNEP), provide guidelines and standards for acceptable levels of heavy metals in the environment. The Codex Alimentarius, established jointly by the Food and Agriculture Organization (FAO) and the World Health Organization (WHO), plays a central role in setting international standards for food safety. It formulates guidelines on permissible levels of heavy metals in different food categories, ensuring consistency in standards across nations. Codex standards act as a reference point for countries, aiding in the development of their national regulations to address heavy metal contamination in food. WHO provides comprehensive guidelines that specify the acceptable levels of specific heavy metals in various food items. These guidelines are meticulously researched and regularly updated to reflect the latest scientific findings. WHO's guidelines prioritize protecting vulnerable groups such as pregnant women and children, acknowledging their heightened susceptibility to the adverse effects of heavy metal exposure. FAO contributes significantly to global efforts to combat heavy metal contamination in food. Its regulations focus on sustainable agricultural practices that minimize the entry of heavy metals into the food chain. This includes promoting responsible soil management, ensuring water quality, and advocating safe agricultural techniques. By emphasizing prevention at the source, FAO regulations align with international goals of ensuring food safety. Compliance with these standards is often mandatory for countries participating in global trade agreements.

National Regulations

Several countries have implemented robust regulations to effectively control heavy metal contamination in food. These regulations often include regular monitoring of food products, stringent testing protocols, and strict enforcement mechanisms. Case studies from countries like Japan, the United States, and Sweden demonstrate successful strategies for managing heavy metal contamination. For instance, Japan's Food Safety Basic Act establishes strict standards and regular inspections, ensuring low levels of heavy metals in food products, especially seafood.

Despite the presence of regulations, implementing them at the national level poses challenges. Limited resources, inadequate infrastructure, and complex supply chains are common hurdles. Some countries address these challenges through public-private partnerships, leveraging technology for efficient monitoring, and investing in public awareness campaigns. Additionally, international cooperation and knowledge sharing between nations experiencing similar challenges play a pivotal role in overcoming implementation obstacles. Therefore, the establishment and effective implementation of regulatory frameworks at both the international and national levels are crucial to mitigating heavy metal contamination in food. International organizations like Codex Alimentarius, WHO, and FAO provide essential guidelines, while individual nations develop tailored regulations and innovative solutions to address specific challenges. The synergy between global standards and localized approaches is key to ensuring food safety and safeguarding public health against the detrimental effects of heavy metal exposure (Nkwunonwo, Odika & Onyia, 2020).

Evaluation of Existing Remediation Technologies

Common methods include chemical precipitation, ion exchange, and electrokinetic remediation. These methods often involve the addition of chemicals to the contaminated site to facilitate the removal or immobilization of heavy metals. Flaws may include high costs, the potential generation of secondary waste, and the need for extensive infrastructure. This involves using plants to extract, immobilize, or accumulate heavy metals. Although environmentally friendly, the process can be slow and may not be suitable for all contaminants or soil types. This method involves the addition of amendments to the soil to reduce the mobility and bioavailability of heavy metals. Challenges may include the need for ongoing maintenance and the potential for the leaching of stabilizing agents. Utilizes microorganisms to either transform or immobilize heavy metals. This method is often more sustainable, cost-effective, and has fewer negative environmental impacts compared to some physicochemical methods. Potential flaws may include the need for optimization, longer treatment times, and variations in effectiveness depending on the site-specific conditions. To justify the sustainability aspect of bioremediation techniques, the authors can highlight factors such as reduced energy consumption, minimal use of chemicals, the potential for in-situ treatment, and the ability of microbial populations to adapt to changing environmental conditions. Additionally, emphasizing the long-term effectiveness and the potential for synergies with natural ecosystem processes can strengthen the argument for the sustainability of bioremediation in comparison to other technologies.

Mitigating Heavy Metal Contamination in Food: Strategies for a Sustainable Future

Heavy metal contamination in food poses significant threats to public health, necessitating comprehensive mitigation strategies. This chapter explores diverse approaches, emphasizing the importance of an interdisciplinary effort to combat this pressing issue. Phytoremediation, a sustainable and eco-friendly method, utilizes plants to remove heavy metals from soil (Aghili & Golzary, 2023). The chapter delves deeply into this technique, spotlighting plant species with remarkable metal-accumulating abilities. By evaluating its effectiveness and limitations, the chapter emphasizes the potential of phytoremediation for cleansing contaminated agricultural lands. Bioremediation, employing microorganisms to degrade pollutants, offers promise in heavy metal contamination mitigation. The chapter explores recent advancements, highlighting their application in removing heavy metals from diverse environments. By understanding its mechanisms and challenges, the chapter assesses the viability of bioremediation on a larger scale. Promoting sustainable agriculture is pivotal (Sarker *et al.*, 2023).

The chapter emphasizes organic farming, proper waste management, and the judicious use of agricultural inputs. By advocating for responsible practices, the objective is to minimize environmental contamination, consequently reducing the presence of heavy metals in food. Contaminated water sources contribute significantly to heavy metal exposure. The chapter evaluates innovative water treatment technologies, highlighting their role in ensuring a clean water supply. By emphasizing the importance of proper irrigation management, the objective is to reduce the entry of heavy metals into the food chain. Public awareness is a potent tool in combating heavy metal contamination. The chapter stresses the importance of accurate information dissemination. By fostering a sense of responsibility among consumers, individuals can make informed choices, reducing their exposure to contaminated foods. Education becomes a key component in this process, empowering communities to demand safer food practices. Corporate entities play a pivotal role. The chapter explores corporate social responsibility initiatives within the food industry. By advocating for ethical sourcing, responsible waste disposal, and eco-friendly production processes, corporations can significantly mitigate their environmental impact. Collaboration between industries and local communities can lead to sustainable solutions. The chapter explores the integration of nanotechnology in remediation efforts. Nanomaterials show promise for efficiently removing heavy metals from contaminated environments. By evaluating current advancements and challenges, the chapter provides insights into the future prospects of this innovative approach. Additionally, the chapter explores emerging technologies, paving the way for novel, effective solutions. By harnessing the potential of nature, promoting sustainable practices, ensuring clean water supply, fostering public awareness, encouraging corporate responsibility, and embracing emerging technologies, societies can collectively mitigate the risks associated with heavy metal exposure (Silva, 2023). This comprehensive approach, grounded in scientific research and societal collaboration, is paramount to creating a safer, healthier future for all.

Mitigating Heavy Metal Contamination With Case Studies

One of the most tragic cases of heavy metal contamination is the Minamata Disease in Japan. In the mid-20th century, the Chisso Corporation released methylmercury into Minamata Bay, contaminating seafood (Harada, 1995). The mercury accumulated in humans and led to severe neurological disorders. This incident underscored the devastating impact heavy metals can have on communities reliant on seafood-based diets, emphasizing the urgent need for strict industrial regulations and thorough monitoring of water bodies. Another poignant case is the Itai-itai Disease, also in Japan, caused by cadmium pollution. This disease primarily affects women, causing excruciating pain and weakening bones due to cadmium-contaminated rice and water (Nishijo *et al.*, 2017). It highlighted the significance of understanding vulnerable demographics and the necessity of comprehensive agricultural reforms to prevent metal contamination in staple foods. Shifting our focus to the USA, the Flint Water Crisis is a stark example of heavy metal contamination in drinking water. Lead-contaminated water supplies in Flint, Michigan, exposed thousands to lead poisoning, especially children. This crisis emphasizes the importance of stringent regulations and constant vigilance in maintaining water quality standards, ensuring that heavy metals do not seep into the daily essentials of life.

Recent Studies on Heavy Metal Pollution in Food

In recent years, a growing body of literature has shed light on the ongoing concerns surrounding heavy metal (HM) pollution in food, emphasizing the need for continuous monitoring, comprehensive regulations, and proactive measures to safeguard public health. This section provides an overview of key findings from recent studies, addressing various aspects of heavy metal contamination in food sources.

Mercury Contamination in Seafood

Recent studies on mercury contamination in seafood have further emphasized the persistent threat to communities reliant on fish and shellfish as dietary staples. Advances in analytical techniques have allowed researchers to assess mercury levels in different seafood species, providing valuable insights into the extent of contamination. Understanding the bioaccumulation patterns of methylmercury in aquatic organisms is crucial for predicting human exposure risks. Additionally, research has explored potential mitigation strategies, such as promoting sustainable fishing practices and developing technologies for real-time monitoring of mercury levels in seafood.

Cadmium Pollution in Staple Foods

The Itai-Itai Disease remains a poignant reminder of the devastating impact of cadmium pollution on staple foods. Recent studies have focused on assessing cadmium levels in crops, particularly rice, and water sources. Advances in agricultural practices and soil management techniques have been explored to reduce cadmium uptake by plants. Additionally, efforts to identify alternative low-cadmium cultivars and the implementation of precision farming technologies have been investigated to minimize the risk of cadmium contamination in food crops.

Lead Contamination in Drinking Water

The Flint Water Crisis has prompted an increased focus on lead contamination in drinking water and its potential health implications. Recent studies have delved into the assessment of lead levels in water distribution systems, exploring novel materials for pipes and the development of efficient water treatment methods to mitigate lead leaching. The impact of lead exposure on vulnerable populations, especially children, has been a significant area of research, leading to advocacy for stricter regulations and improved infrastructure maintenance to prevent lead contamination in public water supplies.

Emerging Contaminants and Regulatory Challenges

In addition to traditional heavy metals, recent literature has addressed the presence of emerging contaminants in food sources. Studies have investigated the occurrence of metal nanoparticles, metalloids, and other novel contaminants in various food products. This evolving landscape poses challenges for regulatory frameworks, necessitating continuous updates and adaptability to emerging threats. Researchers have explored the integration of advanced technologies, such as machine learning and remote sensing, to enhance monitoring capabilities and predict potential hotspots of heavy metal contamination in food-producing regions.

Recommendations for Future Research and Policy

Based on the insights gleaned from recent studies, it is evident that heavy metal pollution in food remains a dynamic and multifaceted challenge. To address this issue effectively, future research should prioritize interdisciplinary collaborations, incorporating fields such as environmental science, agriculture, and public health. Furthermore, policymakers should consider the evolving nature of contaminants and enact flexible regulations that can adapt to emerging threats. Continuous monitoring, public awareness campaigns, and community engagement are crucial components of a comprehensive strategy to mitigate the impact of heavy metal contamination on human health. These studies underscore the ongoing relevance of heavy metal pollution in food, emphasizing the need for evidence-based strategies to protect communities from the devastating consequences witnessed in historical cases such as Minamata and Itai-itai. The lessons learned from these incidents should guide researchers, policymakers, and industry stakeholders in shaping a sustainable and secure future for global food systems.

CONCLUSION

In conclusion, this chapter has delved into the intricate web of heavy metal contamination in the food chain, elucidating the sources, bioaccumulation mechanisms, and associated health risks. It has highlighted the vulnerabilities of specific population groups, emphasizing the heightened risks faced by pregnant women and children. By scrutinizing various remediation strategies, both traditional and innovative, the chapter has evaluated their effectiveness in curbing heavy metal entry into our food. Phytoremediation, with its focus on specific plant species and bioremediation techniques, has shown promise, and the integration of nanotechnology presents an innovative frontier in this battle. The implications for policy and practice are profound. Stringent regulations are imperative, ensuring monitoring and control across the entire food supply chain. The chapter's multidisciplinary approach underscores the necessity for collaborative efforts between scientists, policymakers, and the public. Policymakers must utilize evidence-based recommendations to bolster existing frameworks, enhance food safety standards and ultimately safeguard public health. Integrating sustainable agricultural practices into policy-making is paramount, emphasizing organic farming, responsible waste management, and judicious use of agricultural inputs. Looking forward, further research is essential. A deeper exploration of phytoremediation, focusing on identifying and optimizing plant species for different types of heavy metals, can significantly advance our remediation efforts. Additionally, understanding the long-term effects of heavy metal exposure on human health, especially in vulnerable populations, requires extensive study. The integration of nanotechnology in remediation processes demands in-depth research to address challenges and harness its full potential. There is a need to assess the socio-economic impact of heavy metal contamination, especially in developing regions where agricultural practices might amplify exposure risks. Additionally, exploring public perception and awareness regarding heavy metal contamination can inform educational campaigns effectively.

Furthermore, combating heavy metal contamination demands concerted efforts. This chapter underscores the urgency for action, urging policymakers, researchers, and the public to collaborate actively. By implementing stringent regulations, embracing innovative technologies, and promoting sustainable practices, we can mitigate heavy metal risks, ensuring a safer, healthier future for generations to come. Public awareness and education are our strongest allies in this endeavor, empowering individuals to make informed choices and catalyzing the shift toward a contamination-free food chain.

ACKNOWLEDGMENT

The authors thanks the UGC Minor Research Project, Reference No.F.PSW-140/15-16 for funding. The authors also thank Dr. Indranil Kar, Principal, Surendranath College, Dr. Suchandra Chatterjee, IQAC Coordinator, Surendranath College, Dr. Lalita Das, HoD, Department of Chemistry, Surendranath College, Dr. Harisadhan Ghosh, Dr. Apurba Biswas, Dr. Noor Salam, Mrs. Sneyee Goswami, and Mr. Sourav Mishra, Department of Chemistry, Surendranath College, for their suggestions and technical help during the review process.

REFERENCES

Aghili, S., & Golzary, A. (2023). Greening the earth, healing the soil: A comprehensive life cycle assessment of phytoremediation for heavy metal contamination. *Environmental Technology & Innovation, 103241*, 103241. Advance online publication. doi:10.1016/j.eti.2023.103241

Alengebawy, A., Abdelkhalek, S. T., Qureshi, S. R., & Wang, M. Q. (2021). Heavy metals and pesticides toxicity in agricultural soil and plants: Ecological risks and human health implications. *Toxics*, *9*(3), 42. doi:10.3390/toxics9030042 PMID:33668829

Alengebawy, A., Abdelkhalek, S. T., Qureshi, S. R., & Wang, M. Q. (2021). Heavy metals and pesticides toxicity in agricultural soil and plants: Ecological risks and human health implications. *Toxics*, *9*(3), 42. doi:10.3390/toxics9030042 PMID:33668829

Ali, H., Khan, E., & Ilahi, I. (2019). Environmental chemistry and ecotoxicology of hazardous heavy metals: Environmental persistence, toxicity, and bioaccumulation. *Journal of Chemistry*, *2019*, 1–14. Advance online publication. doi:10.1155/2019/6730305

Ali, S., Abbas, Z., Rizwan, M., Zaheer, I. E., Yavaş, İ., Ünay, A., Abdel-DAIM, M., Bin-Jumah, M., Hasanuzzaman, M., & Kalderis, D. (1927). Application of floating aquatic plants in phytoremediation of heavy metals polluted water: A review. *Sustainability (Basel)*, *12*(5), 1927. Advance online publication. doi:10.3390/su12051927

Angulo-Bejarano, P. I., Puente-Rivera, J., & Cruz-Ortega, R. (2021). Metal and metalloid toxicity in plants: An overview on molecular aspects. *Plants*, *10*(4), 635. doi:10.3390/plants10040635 PMID:33801570

Balali-Mood, M., Naseri, K., Tahergorabi, Z., Khazdair, M. R., & Sadeghi, M. (2021). Toxic mechanisms of five heavy metals: Mercury, lead, chromium, cadmium, and arsenic. *Frontiers in Pharmacology*, *227*, 643972. Advance online publication. doi:10.3389/fphar.2021.643972 PMID:33927623

Briffa, J., Sinagra, E., & Blundell, R. (2020). Heavy metal pollution in the environment and their toxicological effects on humans. *Heliyon, 6*(9), e04691. Advance online publication. doi:10.1016/j.heliyon.2020.e04691 PMID:32964150

Chheang, L., Thongkon, N., Sriwiriyarat, T., & Thanasupsin, S. P. (2021). Heavy metal contamination and human health implications in the Chan Thnal reservoir, Cambodia. *Sustainability (Basel), 13*(24), 13538. doi:10.3390/su132413538

Daurov, D., Zhambakin, K., & Shamekova, M. (2023). Phytoremediation as a way to clean technogenically polluted areas of Kazakhstan. *Brazilian Journal of Biology, 83*, e271684. doi:10.1590/1519-6984.271684 PMID:37222372

Gelais, A. T. S., & Costa-Pierce, B. A. (2016). Mercury concentrations in Northwest Atlantic winter-caught, male spiny dogfish (*Squalus acanthias*): A geographic mercury comparison and risk-reward framework for human consumption. *Marine Pollution Bulletin, 102*(1), 199–205. doi:10.1016/j.marpolbul.2015.12.009 PMID:26707980

Gheorghe, S., Stoica, C., Vasile, G. G., Nita-Lazar, M., Stanescu, E., & Lucaciu, I. E. (2017). Metals toxic effects in aquatic ecosystems: modulators of water quality. *Water Quality, 87*, 59-89. doi:10.5772/65744

Harada, M. (1995). Minamata disease: Methylmercury poisoning in Japan caused by environmental pollution. *Critical Reviews in Toxicology, 25*(1), 1–24. doi:10.3109/10408449509089885 PMID:7734058

Huang, L., Wang, Q., Zhou, Q., Ma, L., Wu, Y., Liu, Q., Wang, S., & Feng, Y. (2020). Cadmium uptake from soil and transport by leafy vegetables: A meta-analysis. *Environmental Pollution, 264*, 114677. doi:10.1016/j.envpol.2020.114677 PMID:32388299

Jaishankar, M., Tseten, T., Anbalagan, N., Mathew, B. B., & Beeregowda, K. N. (2014). Toxicity, mechanism and health effects of some heavy metals. *Interdisciplinary Toxicology, 7*(2), 60–72. doi:10.2478/intox-2014-0009 PMID:26109881

Kapahi, M., & Sachdeva, S. (2019). Bioremediation options for heavy metal pollution. *Journal of Health & Pollution, 9*(24), 191203. doi:10.5696/2156-9614-9.24.191203 PMID:31893164

Landrigan, P. J. (1982). Occupational and community exposures to toxic metals: Lead, cadmium, mercury and arsenic. *The Western Journal of Medicine, 137*(6), 531. PMID:7164433

Mitra, S., Chakraborty, A. J., Tareq, A. M., Emran, T. B., Nainu, F., Khusro, A., Idris, A. M., Khandaker, M. U., Osman, H., Alhumaydhi, F. A., & Simal-Gandara, J. (2022). Impact of heavy metals on the environment and human health: Novel therapeutic insights to counter the toxicity. *Journal of King Saud University. Science, 34*(3), 101865. doi:10.1016/j.jksus.2022.101865

Nishijo, M., Nakagawa, H., Suwazono, Y., Nogawa, K., & Kido, T. (2017). Causes of death in patients with Itai-itai disease suffering from severe chronic cadmium poisoning: A nested case–control analysis of a follow-up study in Japan. *BMJ Open, 7*(7), e015694. doi:10.1136/bmjopen-2016-015694 PMID:28710217

Nkwunonwo, U. C., Odika, P. O., & Onyia, N. I. (2020). A review of the health implications of heavy metals in food chain in Nigeria. *TheScientificWorldJournal, 2020*, 1–11. Advance online publication. doi:10.1155/2020/6594109 PMID:32351345

Obaideen, K., Shehata, N., Sayed, E. T., Abdelkareem, M. A., Mahmoud, M. S., & Olabi, A. G. (2022). The role of wastewater treatment in achieving sustainable development goals (SDGs) and sustainability guideline. *Energy Nexus*, *7*, 100112. doi:10.1016/j.nexus.2022.100112

Othman, Y. A., Al-Assaf, A., Tadros, M. J., & Albalawneh, A. (2021). Heavy metals and microbes accumulation in soil and food crops irrigated with wastewater and the potential human health risk: A metadata analysis. *Water (Basel)*, *13*(23), 3405. doi:10.3390/w13233405

Priya, A. K., Muruganandam, M., Ali, S. S., & Kornaros, M. (2023). Clean-Up of Heavy Metals from Contaminated Soil by Phytoremediation: A Multidisciplinary and Eco-Friendly Approach. *Toxics*, *11*(5), 422. doi:10.3390/toxics11050422 PMID:37235237

Rai, P. K., Lee, S. S., Zhang, M., Tsang, Y. F., & Kim, K. H. (2019). Heavy metals in food crops: Health risks, fate, mechanisms, and management. *Environment International*, *125*, 365–385. doi:10.1016/j.envint.2019.01.067 PMID:30743144

Rokonuzzaman, M. D., Li, W. C., Wu, C., & Ye, Z. H. (2022). Human health impact due to arsenic contaminated rice and vegetables consumption in naturally arsenic endemic regions. *Environmental Pollution*, *308*, 119712. doi:10.1016/j.envpol.2022.119712 PMID:35798190

Sánchez-Castro, I., Molina, L., Prieto-Fernández, M. Á., & Segura, A. (2023). Past, present and future trends in the remediation of heavy-metal contaminated soil-Remediation techniques applied in real soil-contamination events. *Heliyon*, *9*(6), e16692. Advance online publication. doi:10.1016/j.heliyon.2023.e16692 PMID:37484356

Sarker, A., Al Masud, M. A., Deepo, D. M., Das, K., Nandi, R., Ansary, M. W. R., Islam, A. R. M. T., & Islam, T. (2023). Biological and green remediation of heavy metal contaminated water and soils: A state-of-the-art review. *Chemosphere*, *138861*. Advance online publication. doi:10.1016/j.chemosphere.2023.138861 PMID:37150456

Silva, J. A. (2023). Wastewater treatment and reuse for sustainable water resources management: A systematic literature review. *Sustainability (Basel)*, *15*(14), 10940. doi:10.3390/su151410940

Tchounwou, P. B., Yedjou, C. G., Patlolla, A. K., & Sutton, D. J. (2012). Heavy metal toxicity and the environment. *Molecular, Clinical and Environmental Toxicology: volume 3: environmental toxicology*, 133-164. doi:10.1007/978-3-7643-8340-4_6

Uddin, M. M., Zakeel, M. C. M., Zavahir, J. S., Marikar, F. M., & Jahan, I. (2021). Heavy metal accumulation in rice and aquatic plants used as human food: A general review. *Toxics*, *9*(12), 360. doi:10.3390/toxics9120360 PMID:34941794

Witkowska, D., Słowik, J., & Chilicka, K. (2021). Heavy metals and human health: Possible exposure pathways and the competition for protein binding sites. *Molecules (Basel, Switzerland)*, *26*(19), 6060. doi:10.3390/molecules26196060 PMID:34641604

Chapter 5
Mathematical Modeling of Microbial Bioremediation: A Key Step in Process Scale-Up

Mihaela Rosca
https://orcid.org/0000-0002-1405-0848
Ion Ionescu de la Brad Iasi University of Life Sciences, Romania & Gheorghe Asachi Technical University of Iasi, Romania

Petronela Cozma
https://orcid.org/0000-0002-9237-3423
Gheorghe Asachi Technical University of Iasi, Romania

Maria Gavrilescu
Gheorghe Asachi Technical University of Iasi, Romania & Academy of Romanian Scientists, Bucharest, Romania

ABSTRACT

Biosorption is a method that holds great promise for environmental bioremediation. However, the process mechanism is not yet fully comprehended as it relies heavily on the characteristics of the biosorbent, properties of the pollutants, and operational conditions. The knowledge and understanding of sorption kinetics and mechanisms are critical for the design and scaling up of biosorption. Applying the proper models and methods in the evaluation of heavy metals biosorption can elucidate mechanisms, analyze experimental data, predict outcomes, and optimize processes. After a brief analysis of the literature in the field of biosorption, it has been found that the most valuable models provide information about process kinetics, maximum biosorption capacity and thermodynamics. Additionally, the preferred methods for studying the interactive effects of process variables and creating a precise mathematical model that describes the process accurately are highly sought after. Thus, the main purpose of the chapter consists in exposing the available mathematical models and methods that can be applied for the evaluation of all factors that could affect the biosorption of heavy metals. The isotherms, kinetic and thermodynamic models, as well as the statistical tools suitable for evaluating biosorption were presented, with a special emphasis on their application, advantages, as well as disadvantages.

DOI: 10.4018/979-8-3693-1618-4.ch005

INTRODUCTION

Bioremediation, a powerful approach in environmental science, leverages the capabilities of microorganisms to mitigate the impact of pollutants on ecosystems. Notably, biosorption and bioaccumulation find preeminent application in the remediation of heavy metals, demonstrating their effectiveness in addressing the challenges posed by the presence of these toxic elements in natural ecosystems (Filote et al., 2022; Gavrilescu, 2004; Rosca et al., 2015). Heavy metals are persistent inorganic pollutants, the concentrations of which often exceed the limits allowed in legislative regulations for water, air, and soil. Research conducted at both laboratory and large scales has demonstrated that biological methods exhibit a high potential for removing heavy metals from wastewaters compared to conventional, physico-chemical methods (Gavrilescu, 2004; Rosca et al., 2019). This symbiotic relationship between microbes and pollutant removal processes underscores the significance of bioremediation in fostering environmental sustainability.

Over the last two decades, numerous studies have focused on biosorption and bioaccumulation processes, and advances in this field have increased interests in these techniques for addressing pollution issues. Most notably, microbial-driven mechanisms play a key role in the removal of pollutants from the environment. Bioaccumulation in microorganisms involves the gradual buildup of heavy metals from their surroundings. Microorganisms, for instance bacteria and fungi, absorb these metals from the water or other environments they inhabit. Over time, the metals accumulate within the living microorganisms' cells or on their surfaces, contributing to the concentration of heavy metals within the microbial biomass (Filote et al., 2021; Limcharoensuk et al., 2015). Bioaccumulation, unlike biosorption, is characterized by a more extended duration, often requiring long periods for the gradual buildup of heavy metals by living microorganisms. The efficiency of bioaccumulation is notably influenced by specific environmental conditions that play a crucial role in shaping the rate and extent of metal accumulation within microbial biomass (Rosca et al., 2015; Sedlakova-Kadukova et al., 2019). Biosorption is a process involving the ability of various biomasses, especially non-living microorganisms like bacteria, fungi, and yeasts, to retain metal ions. This occurs due to the presence of specific sites on the surface of these microorganisms known as active sites or functional groups. These active sites have a chemical affinity for metal ions, meaning they attract and hold onto them. When metal ions come into contact with the microorganisms, these active sites interact with the metal ions, forming a bond (e.g. by electrostatic attraction, ion-exchange processes). This bonding mechanism allows the microorganisms to capture and hold the metal ions, effectively removing them from the surrounding water or solution (Ciobanu et al., 2023; Gavrilescu, 2004; Tinega et al., 2023; Uysal et al., 2022). As non-living microbial biomass is a cost-effective source with a high pollutant retention capacity, biosorption is considered an environmentally friendly and economical option with high efficiency in removing heavy metals from wastewater (Beni & Esmaeili, 2020; Filote et al., 2021; Volesky, 2007).

The study of this process is essential for understanding how to optimize biosorption for the efficient removal of metals in various environmental applications. Moreover, scaling-up of bioremediation processes from the laboratory to an industrial scale remains a challenging task due to the complex interactions between pollutants and biosorbents and the diversity of factors that could affect the process (Rosca et al., 2021). The transfer of data from the laboratory to pilot and industrial scales is typically slow, and optimal conditions may change due to the influence of factors related to bioreactor construction and operation. Therefore, in the exploitation of biosorption, it is essential to study the influence of process-related factors (such as pH, biosorbent dose, initial concentration of the metal in solution, temperature at which the process takes place, and the contact time between the biosorbent and the metal ions) (Bulgariu et al., 2015; Gavrilescu, 2004), as well as the factors related to the construction of the bioreactors in which the process is applied (e.g., air flow, mixing rate, bed height, liquid flow, etc.) (Cozma et al., 2021; Filote et al., 2022; Rosca et al., 2015).

The design and scaling-up of bioremediation processes strongly depend on knowledge and understanding of the kinetics and mechanisms of pollutant sorption. According to studies conducted, (bio) sorption mechanisms can be analyzed by: (i) modeling sorption equilibrium results; (ii) characterizing the adsorbent material used, before and after the sorption process; (iii) using molecular dynamics; and (iv) applying density functional theory (Figure 1) (Rosca et al., 2021).

Figure 1. Framework for modeling of heavy metals biosorption performed at laboratory scale (adapted upon Rosca et al., 2021)

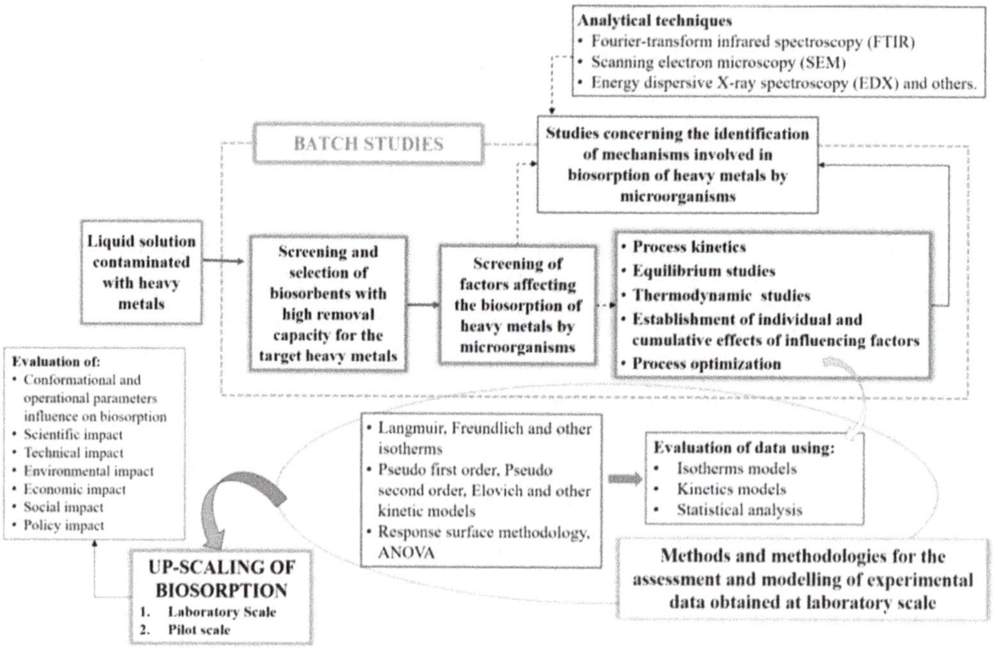

The analysis of published studies about biosorption highlighted that the most useful models are those that provide information about process kinetics, maximum biosorption capacity, and thermodynamics (Hlihor et al., 2023; Roşca et al., 2023; Tafakori et al., 2017). Critical methods include those that can be applied to investigate the interactive effects of process variables and to build a mathematical model that accurately describes the process. The information provided by applying kinetics models, isotherms, and calculating thermodynamic parameters can enhance understanding of individual interactions in a biological system, the mechanisms by which these interactions occur, and the effects that changing operational conditions may have on these processes (Bulgariu et al., 2015). Among the most used isotherms in the evaluation of biosorption are Langmuir, Freundlich, Hill, Dubinin-Radushkevitch, Redlich-Peterson, and Sips. Also, Lagergren, Ho, Weber, and Morris, and Elovich are the most applied kinetics models. To evaluate the influence exerted by these factors and to determine their values to achieve the desired performances, a high number of experiments are conventionally required. To reduce the number of experiments, along with costs, materials, and time savings, experimental data can be modelled using different statistical analysis methods (Aydar, 2018).

Nonlinear regression is a statistical technique by which the relationship between a single dependent variable (criterion) and one or more independent variables can be determined. Depending on the purpose, this relationship can be used either for predicting experimental conditions based on a small number of experimental data or for refining and evaluating a large number of data (Palmer & O'Connell, 2009). Design of experiments (DOE) is a statistical tool that can be applied to assess the statistical influence of the factors and to optimize the process. Response surface methodology (RSM), as part of DOE, has been frequently employed to obtain a valid result with a minimum of resources, effort, and time (Biswas et al., 2017). This mathematical tool has been successfully used in numerous studies and various fields for designing experiments and for setting optimal values of process variables. RSM is useful for mapping a response surface over a given region of interest, optimizing the response, and/or selecting operating conditions to achieve the required response. In addition, RSM allows the application of ANOVA to investigate the nature of influencing factors and to identify the factors that significantly affect the response. For process refinement and evaluation, experiments can be planned using various statistical models (e.g., full factorial design (FFD), Plackett-Burman design, Box-Behnken model), and the results can be examined using analysis of variance (ANOVA), which provides a set of criteria based on which the mathematical regression model that best describes the process can be established (de Oliveira et al., 2019; Rajewski & Dobrzyńska-Inger, 2021).

The efficiency of biosorption and bioaccumulation processes can be considerably improved if they are applied under controlled conditions and the optimal values of the influencing factors and their degree of influence are known. Therefore, for the exploitation of metal ion bioaccumulation processes by microorganisms, it is important to analyse the possibilities of improving the results by changing only the values of the parameters influencing the process. This involves the identification of the admissible region of space in which the variation of influencing factors does not affect the maximum performance of the process (El Mourabit, 2016; Lazli & Hellal, 2020). Therefore, choosing and applying the right models and methods in the evaluation of the process will facilitate the elucidation of the process mechanisms, the experimental data analysis, the prediction of the answers according to the operational conditions, along process optimization. In this manner, critical insights will be obtained for scaling-up the process from the laboratory to pre-industrial and industrial scale (Rosca et al., 2021).

Within this framework, the main purpose of the present chapter is to reveal the available mathematical models and methods that can be applied for the evaluation of all factors that could affect the biosorption process. The isotherms, kinetic and thermodynamic models, as well as the statistical tools suitable for evaluating biosorption will be presented, with a special emphasis on their application, advantages, as well as disadvantages.

FACTORS WITH A SIGNIFICANT INFLUENCE ON BIOSORPTION

The biosorption of metal ions by microorganisms is a multifaceted process influenced by various factors, each playing a key role in shaping the efficiency and dynamics of this environmentally significant phenomenon. This section provides an exploration of these factors, revealing their general impact and interdependencies in the context of metal biosorption processes.

Studies on the biosorption of metal ions by microorganisms have revealed that **pH** is one of the most important factors affecting both the metal biosorption capacity of microorganisms and the speciation and solubility of metals in the solution. The functional groups on the biomass surface undergo protonation or deprotonation due to the increasing acidity or alkalinity of the aqueous medium. Protonation of functional groups on the surface of the biomass releases protons, leading to a decrease in pH in the aqueous medium and an excess of protons that compete with metal ions, resulting in lower biosorption efficiency. In acidic aqueous media, heavy metals (HMs) undergo ionization, becoming more soluble and easier to be adsorbed. Conversely, a low acid-base pH can induce HMs precipitation, further decreasing their solubility and uptake by the biosorbent (Abdi & Kazemi, 2015; Bulgariu et al., 2015).

Temperature can have both positive and negative effects on biosorption. Increasing the temperature enhances the kinetic energy of metal ions, facilitating their attachment to the biosorbent surface, but it can also cause structural damage to the biosorbent. Thus, temperature variations influence the biosorption capacity of metal ions by microorganisms, with the optimal temperature, according to most studies, falling within the range of 20-35°C (Rosca et al., 2021; Shamim, 2018).

Biosorption involves the retention of metal ions on the surface of the cell wall, primarily due to the interaction between metal ions and functional groups present on the cell wall. Consequently, the **biosorbent dose** significantly influences the biosorption process, as increasing the biosorbent dose leads to an increase in the number of sites available for binding of metal ions (Abdi & Kazemi, 2015; Hlihor et al., 2015; Shamim, 2018).

The initial concentration of metal ions in the solution is another factor that can significantly affect heavy metal biosorption. As the initial concentration increases, the amount of metal ions biosorbed per unit weight of biosorbent increases until it reaches a saturation value. However, the efficiency of the biosorption process decreases with increasing concentration, as at low concentrations, the ratio of the number of available binding sites to the amount of metal ions is higher, decreasing with an increase in concentration (Hlihor et al., 2017; Shamim, 2018).

The time required to reach maximum biosorption efficiency depends on the type of biosorbent, as well as the metal ion species. Typically, the highest amount of metal is retained in the first minutes to hours, as all active sites are free and available for the biosorption of metal ions (Fertu et al., 2022; Shamim, 2018).

The particle size of the biosorbent is another critical factor to be considered, as the contact surface between the biosorbent and the metal ions in the solution can significantly influence processes. Reducing the size of the biosorbent increases the contact surface area between the biosorbent and metal ions, favoring the retention of metal ions (Abdi & Kazemi, 2015; Shamim, 2018).

In the batch system, increasing the stirring speed improves the removal rate of metals by biosorption, minimizing the resistance of the biosorbent to mass transfer. However, higher speed can damage the physical structure of the biosorbent, leading to a decrease in process efficiency (Abdi & Kazemi, 2015).

BIOSORPTION ASSESSMENT THROUGH KINETICS MODELS

The characterization of heavy metal biosorption from a kinetic standpoint is one of the most critical steps for process scale-up. Kinetic studies provide vital information about the mechanisms involved in metal ion biosorption, mass transfer, and/or chemical interactions (Bulgariu et al., 2015; Pagnanelli, 2011). The kinetics of biosorption can be described using various models based on equilibrium studies. These models are employed to correlate experimental data obtained under specific conditions, including initial metal ion concentration, biosorbent dose, pH, temperature, and various values of contact time between the biosorbent and metal ions in an aqueous solution (Bulgariu et al., 2015). In general, the biosorption kinetics are characterized by an initial phase in which the amount of metal in solution decreases rapidly, followed by a subsequent phase where the decrease becomes slower and less significant. This trend can be attributed to the abundance of active sites on the biosorbent surface at the outset, which are gradually occupied, resulting in decreased efficiency of metal ion biosorption in the second phase (Pagnanelli, 2011). Therefore, the adsorption/biosorption can be considered to occur in several steps, as follows (Pagnanelli, 2011):

1. The transfer of the dissolved substance from the solution bulk to the liquid film adjacent to the adsorbent particles;
2. The transport of the dissolved substance from the liquid film around the adsorbent particles to the particle surface (external diffusion);
3. The transfer of the dissolved substance from the particle surface to its interior, involving bond formation (intraparticle diffusion);
4. The interaction of the solute with active bonds.

In the last decades, several mathematical models have been proposed to describe the kinetics of adsorption processes, which can generally be classified as (Wang & Guo, 2020a):

I. Adsorption reaction models and empirical models (e.g. Lagergren, Ho, Elovich models, Ritchie equation, pseudo n model);
II. Diffusion models:
   - external diffusion models, for example Frusawa and Smith, Boyd, Mathews and Weber, phenomenological external mass transfer model;
   - internal diffusion models, such as: Boyd, Weber-Morris, phenomenological internal mass transfer model;
   - pore volume and surface diffusion model;
III. Models for adsorption onto active sites (e.g. Langmuir kinetic model, phenomenological active site adsorption model).

Analysis of published studies on metal ion biosorption showed that the most widely used models for describing the kinetic of biosorptions are Lagergren (pseudo I order), Ho (pseudo II order), Weber-Morris and Elovich, whose equations are presented in Table 1. Lagergren and Ho models are among the most applied in kinetic studies, as they quantify the metal uptake capacity of the biosorbent (Bulgariu et al., 2015).

The Lagergren model describes the adsorption process as irreversible and its kinetic equation is based on the following five assumptions (Largitte & Pasquier, 2016):

1. Adsorption is carried out only by specific sites and does not involve interaction between adsorbed ions;
2. Adsorption energy does not depend on the degree of surface coverage;
3. The maximum adsorption corresponds to the saturation of a single layer on the biosorbent surface;
4. The adsorbate concentration is considered to be constant;

5. Uptake of metal ions by the biosorbent is described by a first order equation.

The Ho model is based on the same assumptions as the Lagergren model, except that the uptake of metal ions by the biosorbent is described by a second-order equation (Largitte & Pasquier, 2016). The Elovich model predicts the diffusion, activation and deactivation energy of the bulk and surface of biosorbent assuming that the solute sorption rate decreases exponentially as the amount of sorbed solute increases (Kajjumba et al., 2019). The intraparticle diffusion model proposed by Weber and Morris is based on the assumption that the internal diffusion of the adsorbate is the slowest step and controls the adsorption rate, with instantaneous sorption. Therefore, the kinetics of the biosorption process depend on the pore size of the adsorbent material and the biosorbent preparation steps will significantly influence the process efficiency.

Table 1. The equations of the most widely applied models for describing the biosorption kinetic

Kinetic Model	*Equation*	*Parameters of the Model*
Lagergren (pseudo first order)	$\frac{dq_t}{dt} = k_1(q_e - q_t)$	q_e and q_t – equilibrium adsorption capacity and adsorption capacity at time t, (mg/g) k_1 – pseudo-first-order rate constant (min^{-1})
Ho (pseudo second order)	$q_t = \frac{q_e^2 k_2 t}{1 + q_e k_2 t}$	k_2 – pseudo-second-order rate constant of adsorption (g/mg min); q_e and q_t – equilibrium adsorption capacity and adsorption capacity at time t, (mg/g).
Weber-Morris	$q_t = kt^{1/2} + C_{te}$	k – the rate constant of the intraparticle transport (mg/g min$^{1/2}$) C_{te} – adsorbate concentration in solution at equilibrium (mg/L)
Elovich	$\frac{dq_t}{dt} = \alpha e^{-\beta q_t}$	q_t – adsorption capacity at time t, (mg/g); α – initial adsorption rate constant of the model (mg/g min); β – constant related to the extent of surface coverage and activation energy for chemisorption (mg/g min)

The kinetic models described above (Table 1) have been mostly applied in modelling the experimental data obtained from the metal ions biosorption by different types of microorganisms, as shown in Table 2. Espinosa-Sánchez et al. (2019) and Roşca et al. (2023) used four kinetic models, including pseudo-first order, pseudo-second order, Elovich and intraparticle diffusion, to analyze data obtained from Cr(VI) biosorption by *Rhizopus* sp., *Rhodotorula* sp., and *Bacillus megaterium*. It was found in both studies that the Elovich model is the most suitable for describing the biosorption of Cr(VI). This suggests that the rate of Cr(VI) biosorption decreases exponentially as the quantity of Cr(VI) biosorbed increases (Roşca et al., 2023) and the adsorption kinetics involve chemisorption, as reported by Turner (1975). Hlihor et al. (2023) established that Cr(VI) biosorption kinetics by *Trichoderma viride* is best described by pseudo-first order model, with physical adsorption as the dominant mechanism. However, in most studies the biosorption kinetic of metal ions such as Cd(II), Zn(II), Co(II), Cu(II), Pb(II) and Ni(II) is described by the pseudo second order model (Hajahmadi et al., 2015; Hu et al., 2020; Roşca et al., 2018; Sedlakova-Kadukova et al., 2019). According to Kajjumba et al. (2019), pseudo second kinetic model assumes that the solute adsorption rate is proportional to the available sites on the adsorbent surface, the interactions between biosorbent and solute being predominantly chemical, with ion exchange and complexation as the main mechanisms of biosorption. This model describes the kinetics of most biosorption systems, in which the adsorbent material has functional groups with different but close affinities on its surface. These groups interact directly with metal ions in the aqueous solution (Hubbe et al., 2019). In general, the accuracy of kinetic models applied in the evaluation of biosorption was determined using the coefficient of determination (R^2). The kinetic model with the best correlation is the one with the highest R^2 value (closer to 100% or 1, depending on the tool used) (Ebelegi et al., 2020; Ncibi, 2008).

Table 2. Kinetic models applied for screening of heavy metals biosorption

Biosorbent	Metal Ion	Experimental Conditions	Kinetic Models Applied	Kinetic Model Fitted	References
Streptomyces K11	Zn(II)	pH= 6.5, initial metal concentration = 1 mM, temperature = 10 - 50 °C, biosorbent dosage = 1 g/L	- pseudo-first order - pseudo-second order - Elovich model - Weber-Moris	Pseudo-second order, R^2 > 0.9995 chemisorption is the dominant mechanism	(Sedlakova-Kadukova et al., 2019)
Rhodotorula sp.	Cd(II)	pH = 6, initial metal concentration = 30 and 100 mg/L, temperature = 25 °C, biosorbent dosage = 5 g/L	- pseudo-first order - pseudo-second order	Pseudo-second order, R^2> 0.99 chemisorption is the dominant mechanism	(Roşca et al., 2018)
Bacillus megaterium		pH = 4, initial metal concentration = 50 and 100 mg/L, temperature = 25 °C, biosorbent dosage =3 g/L			
Saccharomyces cerevisiae	Cr(VI)	pH = 2, initial metal concentration = 100 mg/L, temperature = 25 °C, biosorbent dosage = 5 g/L	- pseudo-first order - pseudo-second order	Pseudo-second order, R^2 = 0.9754 chemisorption is the dominant mechanism	(Hlihor et al., 2013)
Trichoderma viride	Cd(II)	pH = 6, initial metal concentration = 25, 50 and 100 mg/L, temperature = 25, 40 and 50 °C, biosorbent dosage = 8 g/L	- pseudo-first order - pseudo-second order - Elovich model	Pseudo-second order, R^2 = 0.9265 – 0.9996 chemisorption is the dominant mechanism	(Hlihor et al., 2015)
Arthrobacter viscosus	Pb(II)	pH = 4, initial metal concentration = 25, 50 and 100 mg/L, temperature = 26 °C, biosorbent dosage = 2 g/L	- pseudo-first order - pseudo-second order - Elovich model	Pseudo-second order, R^2 = 0.9999 chemisorption is the dominant mechanism	(Hlihor et al., 2017)
Escherichia coli	Cd(II)	pH 6.5, initial metal concentration = 20 mg/L, temperature = 30°C, biosorbent dosage = 1 g /L	- pseudo-first order - pseudo-second order	Pseudo-second order, R^2 = 0.99 chemisorption is the dominant mechanism	(Tafakori et al., 2017)
Bacillus badius AK	Cd(II)	pH 7, initial metal concentration = 100 mg/L, temperature = 40°C, biosorbent dosage = 2 g /L	- pseudo-first order - pseudo-second order	Pseudo-second order, R^2 = 0.99 chemisorption is the dominant mechanism	(Vishan et al., 2019)
Pseudomonas aeruginosa PAO1 *Pseudomonas fluorescens* BME	Cu(II)	pH = 5.5, initial metal concentrations = 50 mg/L, temperature = 285, 290 and 295 K, biosorbent dosage = 1 g/L	- pseudo-second order	pseudo-second order, R^2 > 0.99 chemisorption is the dominant mechanism	(Konig-Peter et al., 2019)
Aspergillus niger	Zn(II) Co(II) Cd(II)	pH = 5, initial metal concentrations = 10 - 300 mg/L, temperature = 25°C, biosorbent dosage = 2 g/L	- pseudo-first order - pseudo-second order - Intraparticle diffusion	Pseudo-second order, R^2 > 0.998 chemisorption is the dominant mechanism	(Hajahmadi et al., 2015)
Rhizopus sp.	Cr(VI)	pH = 2, initial metal concentrations = 50 mg/L, Temperature 25±2°C, biosorbent dosage = 5 g/L,	- pseudo-first order - pseudo-second order - Elovich model - Intraparticle diffusion	Pseudo-second order if is applied the linear form, R^2> 0.9975 and Elovich if is applied the nonlinear form, R^2> 0.9953 Both models suggest that chemisorption is the dominant mechanism.	(Espinoza-Sánchez et al., 2019)
Tsukamurella paurometabola A155, *Pseudomonas aeruginosa* B237, and *Cupriavidus taiwanensis* E324	Cd(II)	pH=6, initial metal concentration = 25 mg/L, temperature = 30 °C, biosorbent dosage = 1 g/L,	- pseudo-first order - pseudo-second order	Pseudo-second order, R^2 = 1 chemisorption is the dominant mechanism	(Limcharoensuk et al., 2015)
	Zn(II)			Pseudo-first order, R^2 = 0.952 physical adsorption is the dominant mechanism	

Continued on following page

Table 2. Continued

Biosorbent	Metal Ion	Experimental Conditions	Kinetic Models Applied	Kinetic Model Fitted	References
Corynebacterium glutamicum JB-000	Cd(II)	pH = 5 and 6, initial metal concentration = 300 mg/L, temperature = 25°C, biosorbent dosage = 1.67 g/L	- pseudo-first order - pseudo-second order	Pseudo-second order, $R^2 = 0.991$ chemisorption is the dominant mechanism	(Kim et al., 2015)
Bacillus subtilis JB-008				Pseudo-second order, $R^2 = 0.971$ chemisorption is the dominant mechanism	
Bacillus subtilis JB-016				Pseudo-first order, $R^2 = 0.939$ physical adsorption is the dominant mechanism	
Bacillus catenulatus JB-022				Pseudo-first order, $R^2 = 0.986$ physical adsorption is the dominant mechanism	
Bacillus laterosporus, MTCC 1628	Cd(II)	pH = 7, initial metal concentration 50 mg/L, temperature = 30°C, biosorbent dosage = 4 g/L	- pseudo-first order - pseudo-second order	Pseudo-second order, $R^2 =$ 0.9999 chemisorption is the dominant mechanism	(Kulkarni et al., 2014)
	Ni(II)			Pseudo-second order, $R^2 =$ 0.9989 chemisorption is the dominant mechanism	
Rhodotorula sp.	Cr(VI)	Initial metal ion concentration = 100 mg/L, pH = 1, Temperature = 25°C, biosorbent dosage = 8 g/L	- pseudo-first order - pseudo-second order - Elovich model - intraparticle diffusion	Elovich model, $R^2 = 0.9887$ chemisorption is the dominant mechanism	(Roşca et al., 2023)
Bacillus megaterium				Elovich model, $R^2 = 0.9868$ chemisorption is the dominant mechanism	
Trichoderma viride	Cr(VI)	pH = 2, Initial metal ion concentration = 25 - 200 mg/L, Temperature = 25 - 50°C, biosorbent dosage = 3 - 10 g/L	- pseudo-first order - pseudo-second order	Pseudo-first order, $R^2 > 0.9$ physical adsorption is the dominant mechanism	(Hlihor et al., 2023)
Rhodococcus sp. HX-2	Pb(II)	pH = 5, initial metal ion concentration = 200 mg/L, temperature = 20 °C, biosorbent dosage = 0.5 g/L	- pseudo-first order - pseudo-second order - intraparticle diffusion	Pseudo-second order, $R^2 =$ 0.9970 chemisorption is the dominant mechanism	(Hu et al., 2020)
Bacillus xiamenensis PbRPSD202	Pb(II)	pH = 6, initial metal ion concentration = 100 - 200 mg/L, temperature = 35 °C, biosorbent dosage = 1 g/L	- pseudo-first order - pseudo-second order - intraparticle diffusion	Pseudo-second order, $R^2 > 0.984$ chemisorption is the dominant mechanism	(Mohapatra et al., 2019)

Although these are the most widely used kinetic models to describe the adsorption process, they are associated with several shortcomings. For example, Lagergren and Ho models do not have specific physical meanings which means that by their use, the mass transfer mechanisms cannot be investigated and the physical meanings of empirical kinetic models should be established. Differential kinetic models, such as phenomenological models of external/internal mass transfer and adsorption in active sites, although have specific physical meanings, their solving is more complicated, with few studies investigating mass transfer processes through these models (Wang & Guo, 2020a). In some cases, kinetic models have been applied or solved inappropriately. According to Wang & Guo (2020a), the Frusawa and Smith kinetic models can only be used to model the adsorption process with linear isotherm, but it was applied to evaluate the adsorption process of which the fitting isotherm was Langmuir. Obradovic (2020) also opined that in the literature are a significant number of cases of improper use of kinetic models, with erroneous expressions and incorrect interpretation of the results. By applying Lagergren, Ho and Elovich models it is assumed that the overall rate of the metal ions adsorption is governed by the binding rate, while in case of Weber-Morris and Boyd models is assumed that the mass transport rate governs the overall rate of the process. However, the mentioned models can only account for one of these two assumptions, and to model the complete adsorption process, more complex models need to be applied.

BIOSORPTION ASSESSMENT THROUGH ISOTHERMS MODELS

The utilization of adsorption isotherms in the assessment of biosorption is a fundamental step, providing essential insights into the biosorbent and its capacity to retain heavy metal ions (Bulgariu et al., 2015; Rangabhashiyam et al., 2014; Saadi et al., 2015). An adsorption isotherm is defined as the graphical representation of the relationship between the quantities of sorbate (heavy metal) adsorbed per unit of adsorbent and the sorbate that remains in the solution at equilibrium. The plot shows how the adsorbable solute is distributed between the solid and liquid phases at various concentrations. Equilibrium is achieved when the solute concentration remains constant in the solution (Desta, 2013). Isotherm curves are evaluated by varying the initial concentrations of solute under stable environmental conditions (constant temperature, pH, and biosorbent dosage). These studies yield valuable information for determining the nature of the interaction between the adsorbent and metal ions, optimizing the adsorption mechanism, and designing efficient adsorption systems. Generally, metal uptake increases with an increase in the initial metal ion concentration, reaching saturation at higher concentrations (Barros et al., 2013; Bulgariu et al., 2015). The graphical representation of isotherms can take various forms, including linear, favorable, strongly favorable, irreversible, and unfavorable (Barros et al., 2013).

Different models of adsorption isotherms can be derived by assuming a thermodynamic equilibrium relationship between the amount of molecules adsorbed by a unit mass of adsorbent and the amount of molecules remaining in the liquid phase. The analytical forms of the equations of adsorption isotherms depend on the biosorbent surface, as some researchers consider that adsorption can occur on a single layer or several layers (multilayer).

Isotherm models are considered as being complex models due to the structural and energetic heterogeneity of adsorbent surfaces (Bulgariu et al., 2015; Rangabhashiyam et al., 2014; Saadi et al., 2015). Most adsorption isotherms can be applied in both solid-gas and liquid-solid equilibrium systems. The adsorption isotherms used in solid-liquid systems were initially developed for the adsorption of gases on the surface of a sorbent but can also be used for the adsorption of heavy metal ions on the surface of biological material (Gavrilescu, 2004). Isotherms derived from solid-gas systems provide essential information about the properties of adsorbents, such as pore volume, pore size, energy distribution, specific surface area, and adsorption capacity (Rangabhashiyam et al., 2014; Saadi et al., 2015).

The mathematical relationships of isotherms are characterized by a limited number of adjustable parameters, providing a good description of the experimental behavior over a wide range of operating conditions. For adsorption processes in the solid-liquid system, several models of isotherms can be applied, each having two or more parameters available for the analysis of experimental data (Figure 2). The parameters of each model often provide information about the adsorption mechanism, surface properties, and sorbent affinity (Rangabhashiyam et al., 2014; Saadi et al., 2015).

Figure 2. Isotherms models that can be used for the evaluation of new types of biosorbents

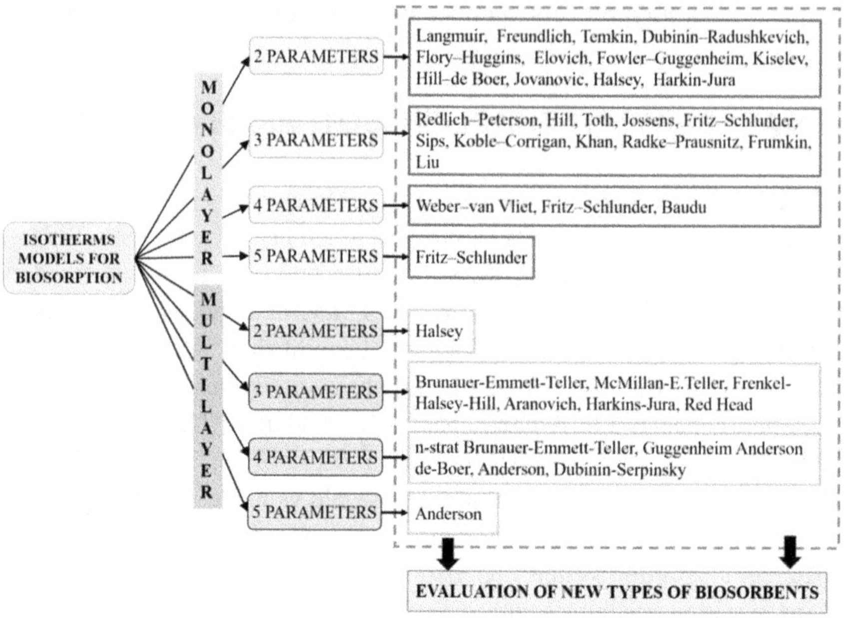

Among the most valuable isotherms applied in the evaluation of biosorption are included: Langmuir, Freundlich, Temkin, Dubinin-Radushkevitch, Redlich-Peterson, Jovanovich, Hill and Sips. These isotherm models are usually chosen due to the information that each isotherm model provides and their usefulness in describing the biosorption. In general, the linear regression method is the most widely applied in isotherm parameter calculations due to its simplicity, but it has been shown in several studies that the linearization of models causes increasing errors between experimental and predicted data. Therefore, the use of the nonlinear method is more advantageous because it preserves the same error structure when comparing the models that best fit the experimental data (Roşca et al., 2020; Wang & Guo, 2020a). The nonlinear method can provide consistent and reliable estimations for the model parameters (Wang & Guo, 2020b). Thus, Table 3 shows the nonlinear forms of the most used isotherm models.

Table 3. Nonlinear forms of the most used isotherms models for modelling heavy metals biosorption

Isotherm	*Equation*	*Parameters*	*References*
Langmuir	$R_L = \frac{1}{b \cdot C_0 + 1}$	q_{max} - the maximum amount of metal ions adsorbed under experimental conditions (mg/g) b - Langmuir constant (L/mg) C_e - equilibrium metal ion concentration in solution (mg/L) R_L – separation factor C_0 – initial metal ions concentration (mg/L)	(Begum et al., 2015)
Freundlich	$q = k \cdot C_e^{1/n}$	k – Freundlich constants reflecting the adsorption capacity (mg/g) $1/n$ – Freundlich constants reflecting the adsorption intensity, $0 < 1/n < 1$ C_e – equilibrium metal ions concentration in solution (mg/L)	(Rangabhashiyam et al., 2014; Saadi et al., 2015)
Temkin	$q_e = \frac{RT}{B} \ln K_T C_e$	K_T - Temkin isotherm equilibrium binding constant (L/mg) B - Temkin isotherm constant R - the universal gas constant (8.314 J/mol K) T - temperature (K) C_e – equilibrium metal ions concentration in solution (mg/L)	(Rangabhashiyam et al., 2014; Saadi et al., 2015).
Dubinin-Radushkevich	$q_e = Q_s \exp(-B_D \varepsilon^2)$ $\varepsilon = RT \ln\left(1 + \frac{1}{C_e}\right)$	B_D - Dubinin–Radushkevich constant (mol^2/kJ^2) ε - Polanyi potential Q_s - the maximum amount of metal ions adsorbed (mg/g); R - the universal gas constant (8.314 J/mol K) T - temperature (K) C_e - equilibrium metal ions concentration in solution (mg/L) E - the mean free energy (kJ/mol) E < 8 kJ/mol - physical adsorption 8 < E < 16 kJ/mol - ion exchange E > 16 kJ/mol - particle diffusion	(Rangabhashiyam et al., 2014; Saadi et al., 2015)
Redlich-Peterson	$q_e = \frac{A_R C_e}{1 + B_R C_e^{m_R}}$	A_R - Redlich-Peterson isotherm constant (L/g); B_R – model constant $(mg/L)^{-mR}$ m_R – model exponent, $0 < m_R < 1$	(Ayawei et al., 2017)
Sips	$q_e = \frac{q_m a_s C_e^{Bs}}{1 + a_s C_e^{Bs}}$	q_m - the maximum amount of metal ions adsorbed (mg/g); a_s - Sips equilibrium constant $(L/mg)^{Bs}$; B_s - model exponent, $0 < B_s < 1$ C_e - equilibrium metal ions concentration in solution (mg/L)	(Rangabhashiyam et al., 2014; Saadi et al., 2015)
Jovanovich	$q_e = q_m(1 - e^{K_J C_e})$	K_J –Jovanovich constant (L/mg) C_e – equilibrium metal ions concentration in solution(mg/L) q_m - the maximum amount of metal ions adsorbed (mg/g)	(Ayawei et al., 2017; Misra, 1973)
Hill	$q_e = \frac{q_m C_e^{n_H}}{K_D + C_e^{n_H}}$	q_m - the maximum amount of metal ions adsorbed (mg/g); n_H – Hill cooperativity coefficient of the binding interaction. K_D - Hill constant C_e - equilibrium metal ions concentration in solution (mg/L)	(Ayawei et al., 2017; Torkia et al., 2014)

The Langmuir model provides information on the retention capacity and the behaviour of the sorption equilibrium and is based on the following hypotheses (Begum et al., 2015):

- Each active site interacts with only one adsorbent molecule,
- The adsorbent molecules are adsorbed on well-located sites,
- There is no interaction between adjacent adsorbed molecules, and
- Adsorption sites are all energy equivalent.

This model also assumes that the biosorbent surface is homogeneous in terms of the active sorption sites distribution and the biosorption of metal ions takes place until a monolayer covering the biosorbent surface is formed (Roşca et al., 2023). The Freundlich model describes multilayer adsorption and considers that it takes place on heterogeneous surfaces, for which energy is uniformly distributed. During the adsorption, the stronger binding sites are occupied first, along with others until the adsorption energy decreases. This model considers the adsorption to be a reversible process (Rangabhashiyam et al., 2014; Saadi et al., 2015). Furthermore, this model can determine if the biosorption is favourable under the given experimental conditions (Ayawei et al., 2017). The Temkin model assumes that adsorption is multi-layered and characterized by a uniform distribution of binding energies. The heat of all molecules in the layer decreases linearly, not logarithmically, with coverage due to adsorbent-adsorbate interactions (Gebreslassie, 2020). The Dubinin-Radushkevich model is often used to differentiate between physical and chemical adsorption, as it is temperature dependent and provides information on the porosity characteristic of the biosorbent and the apparent free energy of adsorption (Saadi et al., 2015). The Redlich-Peterson model is a combination of the Langmuir and Freundlich isotherms, being described by an empirical equation that incorporates three parameters. This model considers that the adsorption mechanism is a mixture and the adsorption is not ideally monolayer (Ayawei et al., 2017). The Jovanovich model is based on the assumptions of the Langmuir model, except that this model also considers the possibility of mechanical contacts between adsorbent and adsorbant (Ayawei et al., 2017; Misra, 1973). The Hill model describes the binding of different species to homogeneous substrates. This model assumes that adsorption is a cooperative phenomenon, and the sorbent can bind to an active site on the biosorbent surface but could affect other binding sites on the biosorbent surface. Important information about the mechanism of biosorption also provides the Hill cooperativity coefficient of the binding interaction (n_H) which can have higher or lower values than the unit, making it possible to identify the mode of binding of a sorbate molecule with the active binding sites of the biosorbent. Therefore, if $n_H > 1$ then several molecules are adsorbed on a single binding site and the adsorption is called multi-molecular adsorption, and if $n_H < 1$ then only a fraction of a molecule is adsorbed on a single binding site (Torkia et al., 2014). The Sips model, which is a combination between Langmuir and Freundlich models, is used only for the adsorption of a single species of heavy metal and considers that the adsorption systems are heterogeneous and the adsorbants do not interact with each other (Rangabhashiyam et al., 2014; Saadi et al., 2015).

According to the data presented in Table 4 the Langmuir and Freundlich isotherms are the most used models, while generally, the Langmuir model fits better with the experimental data obtained from the biosorption of metal ions by microorganisms. However, some studies found that the biosorption of metal ions by various microorganisms was best described by Temkin (Hajahmadi et al., 2015), Hill (Roşca et al., 2020) or Freundlich (Jiang et al., 2017; Konig-Peter et al., 2019; Roşca et al., 2023; Tafakori et al., 2017).

Table 4. Isotherms applied for assessment of heavy metals biosorption

Biosorbent	Metal Ion	Experimental Conditions	Isotherms Applied	Form of Isotherm	Equilibrium Model Followed	References
Rhodotorula sp.	Cd(II)	Initial metal concentration between 25 and 288 mg/L, biosorbent dosage = 5 g/L, pH = 6, agitation speed = 150 rpm, temperature = 25°C, equilibrium time = 48 h.	Langmuir Freundlich Jovanovic Hill	Linear	Langmuir, R^2 = 0.997 q_{max} = 10.53 mg/g	(Roşca et al., 2018; Roşca et al., 2020)
				Non-linear	Langmuir, R^2 = 0.934 q_{max} = 10.94 mg/g	
Bacillus megaterium		Initial metal concentration between 25 and 288 mg/L, biosorbent dosage = 3 g/L, pH = 4, agitation speed = 150 rpm, temperature = 25°C, equilibrium time = 2 h.		Linear	Langmuir, R^2 = 0.999 q_{max} = 15.41 mg/g	
				Non-linear	Hill, R^2 = 0.997 q_{max} = 15.59 mg/g	
Arthrobacter viscosus	Pb(II)	Initial metal concentration between 25 and 500 mg/L, biomass dosage = 2 g/L, pH = 4, temperature = 40 °C, agitation speed = 150 rpm, equilibrium time = 24 h.	Langmuir, Freundlich Temkin Dubinin-Radushkevich Dual mode	Linear	Langmuir, R^2 = 0.9658 q_{max}= 126.58 mg/g	(Hlihor et al., 2017)
Escherichia coli	Cd(II)	Initial metal concentration between 10 and 110 mg/L, biosorbent dosage = 1 g/L, pH 6.5, temperature = 30°C, agitation speed = 100 rpm, equilibrium time = 1 h.	Langmuir Freundlich	Linear	Freundlich, R^2 = 0.98 1/n = 0.32	(Tafakori et al., 2017)
Bacillus badius AK	Cd(II)	Initial metal concentration between 50 mg/L and 200 mg/L, biosorbent dosage = 2 g/L, pH = 7, temperature = 40 °C, agitation speed = 150 rpm, equilibrium time = 30 min.	Langmuir Freundlich	Linear	Langmuir, R^2 = 0.945 q_{max}= 131.58 mg/g	(Vishan et al., 2019)
Pseudomonas aeruginosa PAO1	Cu(II)	Initial metal concentration between 5 mg/L and 250 mg/L, pH = 5.5, biosorbent dosage = 1 g/L, temperature 285, 290 and 295 K, agitation speed = 150 rpm, equilibrium time = 24 h.	Langmuir Freundlich	Non-linear	Freundlich, R^2 = 0.98, n=2.23	(Konig-Peter et al., 2019)
Pseudomonas fluorescens BME					Langmuir, R^2 = 0.98, q_{max}= 51 mg/g	
Aspergillus niger	Zn(II) Cd(II) Co(II)	Initial metal concentration between 10 mg/L and 300 mg/L, biosorbent dosage = 2 g/L, temperature = 25±2°C, pH = 5, agitation speed = 200 rpm, equilibrium time = 2 h.	Langmuir, Freundlich Temkin Sips	Non-linear	Temkin, R^2 > 0.9819	(Hajahmadi et al., 2015)
Rhizopus sp.	Cr(VI)	Initial metal concentration between 12.5 mg/ L and 300 mg /L, pH = 2, biosorbent dosage = 5 g/L, temperature = 10, 25 and 45 °C, agitation speed = 150 rpm, equilibrium time = 120 min.	Langmuir Freundlich	Linear	Langmuir, R^2>0.841 q_{max} = 8.0589 – 9.9507 mg/g	(Espinoza-Sánchez et al., 2019)
Pseudomonas aeruginosa B237, *Tsukamurella paurometabola* A155 and *Cupriavidus taiwanensis* E324	Cd(II) Zn(II)	Initial metal concentration between 5 and 100 mg/L, pH = 6, biomass dosage =1 g/L, temperature = 30 °C, agitation speed = 200 rpm, equilibrium time = 12 h.	Langmuir Freundlich	Linear	Langmuir, R^2 > 0.974 q_{max} =16.48 mg/g (Cd) q_{max} =17.67 mg/g (Zn)	(Limcharoensuk et al., 2015)
Pseudoalteromonas sp. SCSE709-6	Cd(II)	Initial metal concentration between 0.1 mmol/L and 10 mmol/L, biosorbent dosage = 16 g/L, temperature = 25°C, agitation speed = 200 rpm, equilibrium time = 8 h	Langmuir Freundlich	Non-linear	Langmuir, R^2=0.977 q_{max} = 1.38 mmol/g	(Jiang et al., 2017)
	Pb(II)				Freundlich R^2 = 0.981 n = 3.08	
	Zn(II)				Freundlich R^2 = 0.990 n = 3.45	

Continued on following page

Table 4. Continued

Biosorbent	Metal Ion	Experimental Conditions	Isotherms Applied	Form of Isotherm	Equilibrium Model Followed	References
Bacillus laterosporus, MTCC 1628	Cd(II)	Initial metal concentration between 10 and 50 mg/L, biomass dosage = 4 g/L, agitation speed = 150 rpm, pH=7, temperature = 20-40°C, equilibrium time = 180 min	Langmuir Freundlich	Linear	Langmuir, $R^2 >$ 0.9991, q_{max} = 84.03-86.20 mg/g	(Kulkarni et al., 2014)
	Ni(II)				Langmuir, $R^2 >$ 0.9991, q_{max} = 44.05-46.72 mg/g	
Stenotrophomonas maltophilia	Pb(II)	Initial metal concentration between 15.9 and 325.3 mg/L, biomass dosage = 1 g/L, agitation speed = 200 rpm, temperature = 30°C, pH = 5.5, equilibrium time = 30 min	Langmuir Freundlich	Linear	Langmuir, R^2 = 0.996, q_{max} = 133.3 mg/g	(Wierzba, 2015)
	Zn(II)	Initial metal concentration between 11.2 and 172.3 mg/L, biomass dosage = 1 g/L, agitation speed = 200 rpm, temperature = 30°C, pH = 5.5, equilibrium time = 30 min			Langmuir, R^2 = 0.990, q_{max} = 47.8 mg/g	
	Ni(II)	Initial metal concentration between 14.3 and 245.8 mg/L, biomass dosage = 1 g/L, Agitation speed = 200 rpm, temperature = 30°C, pH = 5.5, equilibrium time = 30 min			Langmuir, R^2 = 0.990, q_{max} = 54.3 mg/g	
Bacillus subtilis	Pb(II)	Initial metal concentration between 15.9 and 325.3 mg/L, biomass dosage = 1 g/L, agitation speed = 200 rpm, temperature = 30°C, pH = 5.5, equilibrium time = 30 min			Langmuir, R^2 = 0.998, q_{max} = 166.7 mg/g	
	Zn(II)	Initial metal concentration between 11.2 and 172.3 mg/L, biomass dosage = 1 g/L, agitation speed = 200 rpm, temperature = 30°C, pH = 5.5, equilibrium time = 30 min			Langmuir, R^2 = 0.990, q_{max} = 49.7 mg/g	
	Ni(II)	Initial metal concentration between 14.3 and 245.8 mg/L, biomass dosage = 1 g/L, agitation speed = 200 rpm, temperature = 30°C, pH = 5.5, equilibrium time = 30 min			Langmuir, R^2 = 0.990, q_{max} = 57.8 mg/g	
Rhodotorula sp.	Cr(VI)	Initial metal concentration between 25.29 and 402.52 mg/L, biomass dosage = 8 g/L, agitation speed = 150 rpm, temperature = 25°C, pH = 1, equilibrium time = 48 h	Freundlich Langmuir, Redlich-Peterson Jovanovic Hill Dual Mode	Non-linear	Freundlich R^2 = 0.9382 n = 3.734	(Roşca et al., 2023)
Bacillus megaterium					Freundlich R^2 = 0.9432 n = 4.817	
Trichoderma viride	Cr(VI)	Initial metal concentration between 25 and 200 mg/L, biomass dosage = 5 g/L, agitation speed = 150 rpm, temperature = 25, 40 and 50°C, pH = 2, contact time = 2.5 days	Langmuir, Freundlich, Temkin and Dubininin-Radushkevich		Langmuir, $R^2 >$ 0.8135 q_{max} = 7.98 - 23.08 mg/g	(Hlihor et al., 2023)

Continued on following page

THERMODYNAMIC MODELLING OF BIOSORPTION

Thermodynamic studies are paramount to predict whether adsorption is spontaneous or not. These provide information on the appropriate temperature range for sorption and the nature of the sorbent and sorbate at equilibrium (Batool et al., 2018). The biosorption thermodynamics is directly linked to the kinetic energy of the metal ions, so the diffusion process can be explained through thermodynamic modeling. An increase or decrease in temperature should cause a change in the concentration of metal ions removed or adsorbed by the biosorbent biomass (Hlihor et al., 2010; Liu & Liu, 2008). Thermodynamic parameters useful in understanding the biosorption mechanism can be calculated using the following equations (Eqs. 1 and 2):

$$\Delta G^0 = -RT \ln K_d \tag{1}$$

$$\ln K_d = \frac{\Delta S^o}{R} - \frac{\Delta H^0}{RT} \tag{2}$$

where ΔG^0, ΔH^0, $\Delta S^{0,}$ and T are Gibbs free energy, enthalpy, entropy, and temperature in Kelvin respectively, and R is the gas constant. The values of enthalpy (ΔH^0) and entropy (ΔS^0) are calculated from the slope and intercept of the $\ln K_d$ vs. $1/T$ plot.

The distribution coefficient (K_d) can be calculated taking into account the value of the initial metal ion concentration in solution (C_i) and the final metal ion concentration in solution (C_e) according to Eq. (3).

$$K_d = \frac{C_i - C_e}{C_e} \frac{V}{m} \tag{3}$$

where V is the volume of solution(L) and m is the weight of biosorbent (g).

The Gibbs free energy change (ΔG^0) can be calculated using Eq (4):

$$\Delta G^0 = \Delta H^0 - T\Delta S^0 \tag{4}$$

The negative value of ΔG^0 at different temperatures indicates the spontaneous nature of the biosorption, while the negative value of ΔH^0 suggests that the biosorption reaction is exothermic and the negative value of ΔS^0 denotes that the process is driven by enthalpy (Chakraborty et al., 2012). When the change in free energy (ΔG^0) varies between -20 and 0 kJ/mol, adsorption is classified as physical adsorption, while chemical adsorption occur when the free energy values vary between -80 and -400 kJ/mol (Ali & Alrafai, 2016). Liu et al. (2021) found that Pb(II) biosorption by *Enterobacter* sp. is a spontaneous and endothermic process and the physical adsorption is the mechanism by which Pb(II) ions are removed. The values of ΔG^0, ΔH^0 and ΔS^0 were -4.37kJ/mol, +1491 kJ/mol and respectively, +6.159 kJ/mol•K. Chromium (VI) biosorption by *Aspergillus niger* is spontaneous and endothermic in nature, being favourable at a higher temperature (60 °C)(Mondal et al., 2017). Vishan et al. (2019) asserted that Cd(II) biosorption by *Bacillus badius* AK was spontaneous and endothermic and ion removal is achieved by physical adsorption.

EXPERIMENTAL DESIGN METHODS FOR SCREENING, MODELLING AND OPTIMIZATION OF BIOSORPTION

The removal of heavy metal ions from aqueous solutions by biosorption is influenced by several factors that have not only individual but also cumulative effects on the process's effectiveness. Statistical models based on regression analysis are useful for evaluating the effects of factors and optimizing the process (Amini et al., 2008). Regression analysis is a statistical technique that allows us to establish the relationship between a single dependent variable (response) and one or more independent variables (factors that influence it). Depending on the desired purpose, regression analysis can be used either to predict experimental conditions based on a small number of experimental data or to refine and evaluate a large number of data (Palmer & O'Connell, 2009). In statistics, various experimental designs can be used, such as (Boshagh & Rostami, 2020; Zohrehvandi & Bashiri, 2014):

1. Single Factor Design for comparison;
2. Two Level Factorial Designs (full or fractional factorial design), Taguchi Orthogonal Matrix and Plackett-Burman Design for screening of the independent variables;
3. Central Composite Design and Box-Behnken Design to identify and optimize the transfer functions;

Depending on the considered statistical model, the relationship between the response and the independent variables can be described by a linear (Eq. 5) or quadratic (Eq. 6) mathematical equation (Meier, 2014; Montgomery, 2013).

$$Y = b_0 + \sum_{i=1}^{j} b_i X_i + \sum_{i=1}^{k-1} \sum_{j=i+1}^{k} b_{ij} X_i X_j + \varepsilon \tag{5}$$

$$Y = b_0 + \sum_{i=1}^{j} b_i X_i + \sum_{i=1}^{j} b_{ii} X_i^2 + \sum_{i=1}^{k-1} \sum_{j=i+1}^{k} b_{ij} X_i X_j + \varepsilon \tag{6}$$

where: Y = response variable, X_i = influencing factors (independent variables), b_0 = constant, b_i = coefficients for linear effects, b_{ii} = the coefficients of quadratic effect, b_{ij} = the coefficient of interaction effect, ε = error observed in answer, k = the number of variables.

Design of experiments (DOE) is a systematic and efficient statistical technique, that based on planning, conducting, analyzing and interpreting controlled tests, the relationship between several input and output variables is studied. Often, DOE has been used for the selection of variables with significant influence and process optimization (Lanati et al., 2015; Lee & Gilmore, 2006). As Jankovic et al. (2021) state, DOE "*includes a series of applied statistics tools that can be used to systematically classify and quantify the cause-and-effect relations between variables and outputs in the studied process or phenomenon, which may result (if that is the aim) in finding the settings and conditions under which the process becomes optimized*". The main tools of DOE are (Lanati et al., 2015):

- *Factorial Analysis* that can be used to identify the most important factors in the experimental phase based on a small number of experimental results.

- *Response Surface Methodology (RSM)* is used in the statistical modelling of experimental data and process optimization.
- *Mixed Design* is used when the factors are components of a mixture, and the result depends on their relative proportion.

The use of DOE in the evaluation of independent variables and process optimization involves several steps, as is shown in Figure 3. Each DOE starts with the selection of the process and the recognition of the investigation problem, while the last step includes the analysis of the data using statistical tools (ANOVA and associated statistical methods) and the interpretation of the results. By following the steps of DOE, we can gain a better understanding of the process behavior and optimize it (Jankovic et al., 2021).

Figure 3. Design of experiments (DOE) steps

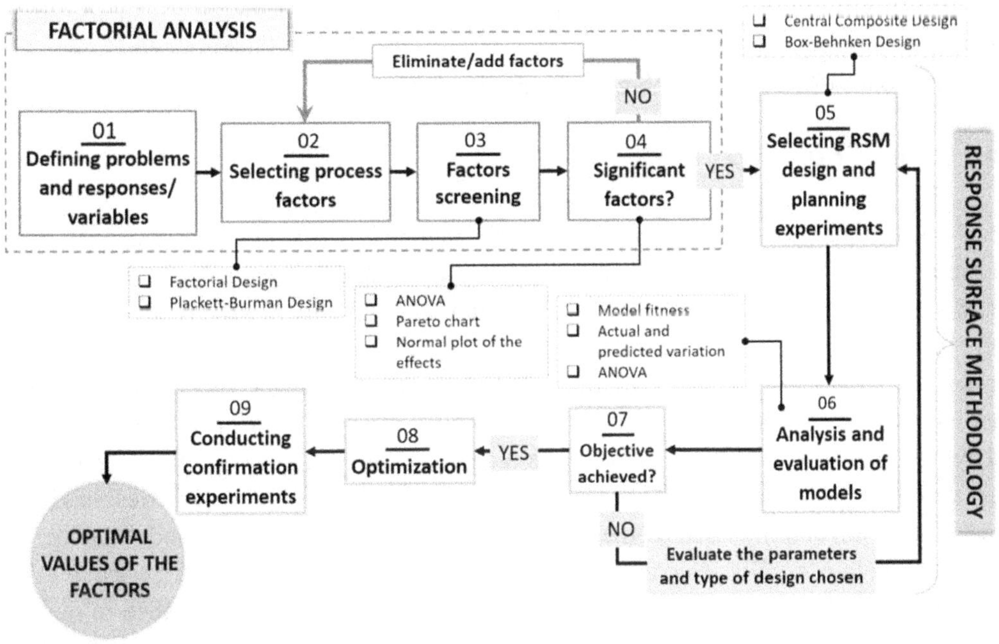

For the exploitation of metal ion biosorption by microorganisms is important to analyse the possibilities of improving the process effectiveness by changing only the values of the independent variables. This involves identifying the admissible region of space in which the variation of independent variables does not affect the maximum performance of the process (de Oliveira et al., 2019). According to Sadhukhan et al. (2016), the optimization of a process involves three main stages:

1. Performing the experiments that were statistically designed;
2. Estimation of the coefficients of the mathematical model that best fit experimental data;
3. Predicting the response and verifying the adequacy of the model.

One of the most effective ways of performing process analysis, by combining experimental design and analysis, modelling techniques and optimization methods is the Response Surface Methodology (RSM) (de Oliveira et al., 2019). Mathematical models of RSM can be applied to investigate the interactive effects of process variables and to build a mathematical model that accurately describes the process. This methodology is generally used in multivariate experimental design, statistical modelling and process optimization (Bayuo et al., 2019; Mourabet et al., 2017). The most common RSM designs are Central Composite Design (CCD) and Box-Behnken design. The CCD is frequently used as a statistical design to optimize biosorption process variables based on a multivariate non-linear model. This design allows the study of the interactions between independent variables that can have a significant influence on the process (Sadhukhan et al., 2016). The Box-Behnken design uses "*the twelve middle edge nodes and three center nodes to fit a 2nd order polynomial equation*" (Rao & Kumar, 2012) to generate higher order response surfaces using fewer required runs than a normal factorial technique (Veza et al., 2023). The RSM, as part of DOE, has been frequently applied to obtain a valid result with a minimum of resources, effort and time (Biswas et al., 2017). Furthermore, RSM is useful for mapping a response surface over a given region of interest, optimizing the response, or/and selecting the operating conditions to achieve the required response (Rajewski & Dobrzyńska-Inger, 2021).

Process Screening

In statistics, the adequacy of the mathematical model generated by RSM that describes the relationship between the response and the independent variables is evaluated by analysis of variance (ANOVA). ANOVA is a useful statistical technique that can be applied to any regression model and based on the parameters provided by it, the best model that describes the process can be established. The statistical parameters such as *p*-value, *F*-value, Variance Inflation Factor (VIF), coefficient of determination R^2, adjusted coefficient of determination (Adj. R^2) and predicted coefficient of determination (Pred. R^2) provided by ANOVA are those on which the magnitude of the independent variables effects are assessed and the method that best describes the process is established (Amini et al., 2008; Keskin Gündoğdu et al., 2016; Temel et al., 2022). Based on the R^2, adj. R^2 and pred. R^2 values the model that best fits with the experimental data is usually selected (Ghoniem et al., 2020; Maiti et al., 2022; Parmar et al., 2020; Tafakori et al., 2017; Temel et al., 2022). The R^2 value is the proportion of variance in the dependent variable that can be explained by the independent variable and the higher it is (closer to 100%) better the model fits with experimental data. At a value above 90% or 0.9, a high correlation is considered to exist between the model and the experimental data. The adj. R^2 coefficient explains the variance of the response as affected by the independent variables, while the pred. R^2 coefficient indicates how well the model predicts the response for the new experiments. Models with higher values of pred. R^2 predict the answer with a better accuracy (Ghoniem et al., 2020).

Another important parameter provided by ANOVA is the *p*-value, which indicates whether or not the effects of the independent variables or the interactions between them are significant for the process. The independent variables or interaction between them have a significant influence on the considered process when the related *p*-value is lower than a preselected *α* level (usually $\alpha = 0.05$) (Ghoniem et al., 2020). The *F*-value (*F* statistic) can be used to compare statistical models applied to the experimental data and, implicitly, to identify the model that best describes the process. The *F-value* of a model is calculated as the ratio of the mean squared dispersion to the square of the mean residual dispersion. This value is then compared to the *F* critical (*Fcrit*), which is obtained from the F-tables for a given degree of freedom.

If the F-value is higher than *Fcrit*, it indicates that the model is statistically significant (Ghoniem et al., 2020). The Variance Inflation Factor (VIF) provided by ANOVA indicates the intensity of collinearity between independent variables considered. Multicollinearity is a common issue that arises when independent variables in a regression model are highly correlated with each other. This can lead to problems in regression analysis by increasing the variance of the regression coefficients, making them sensitive to even minor changes in the data. The end result of multicollinearity is that the regression coefficient becomes unstable and difficult to interpret. Depending on the numerical value, the error dispersion inflation factor can be interpreted as follows (James et al., 2013):

- if VIF =1 is no correlation between factors;
- if VIF is between 1 and 10 is a moderate correlation; and
- if VIF > 10 is a high correlation between factors.

It is also considered that the higher the value of the VIF parameter, the less reliable the regression results will be (Daoud, 2017). For each model, ANOVA also provides the total degree of freedom (DF), the sum of the adjusted squares (Adj SS), the adjusted average of the squares (Adj MS), parameters that consolidate those confirmed by *p-value* and R^2 (Hlihor et al., 2023).

The individual and cumulative effects of the independent variables can be highlighted by the graphical representations of some regression results, which allow the observation of how the applied regression models fit with the experimental results. The most representative graphs are:

- Pareto chart of standardized effects,
- Normal probability plot of the standardized effects,
- Normal probability plot of residuals,
- Response surface and contour line plots of responses to different values of independent variables.

The Pareto diagram highlights the magnitude and importance of individual and cumulative effects of independent variables, and the normal probability plot of standardized effects helps in establish the direction of influence exerted by variables (negative or positive) (Bingol et al., 2010; Ponnusami et al., 2007). In statistics, the residual value is the difference between the experimental value and the corresponding value predicted by the model. Plotting residual values helps to determine if the model is appropriate and if regression assumptions are being met. Residual values also indicate how much a model explains variation in experimental data. Thus, the normal probability plot of residuals is a graphical technique used to check the adequacy of the chosen model with the experimental data and consists in the graphical representation of the residues compared to the expected values. If the plotted points are close to the reference diagonal and appear to be normally distributed, then the model fits with the experimental data well (Apostol et al., 2016; Hasan et al., 2009).

Response surface and contour line plots are two graphical representations of the regression equation obtained from applying the RSM to the experimental data. These plots are used to visualize the relationship between the response and experimental levels of each independent variable and are useful for establishing the desired response values and operating conditions (Biswas et al., 2019). In a contour line plot, the response surface is viewed as a two-dimensional plane in which all points that have the same response are connected to produce contour lines with constant responses. The response surface generally displays a three-dimensional view that can provide a clearer picture of the response (Antony, 2003).

Process Optimization

Mathematical optimization of a process implies the selection of the best item (concerning a criterion) from a set of available alternatives. Therefore, the typical optimization consists of adjusting the values of an input function (the independent variable) to minimize or maximize the output value (the dependent variable or the response). The function to be optimized is generally called the objective function or cost function, and a set of input values is called a solution (El Mourabit, 2016). In mathematical terms, the objective function expresses the objective that the decision maker is trying to optimize and is usually subject to equality or inequality constraints during the optimization process. The constraints define the set of alternatives that are truly options. A solution that satisfies the constraints is called a feasible solution, and the set of allowed input values is called the search space. The search space is the region that defines the set of all feasible solutions (El Mourabit, 2016). Given the proposed targets, process optimisation can be achieved by considering the following alternatives:

- response maximization,
- response minimization,
- reaching a certain imposed target - in the experimental domain, or
- setting sets of independent variables values to achieve a target response value (possibly minimizing the costs and environmental impacts) or a target desirability value.

Optimisation of biosorption involves finding the optimal solution by maximising the objective function (ion retention efficiency or uptake capacity). Determining the optimal (most favourable) solution requires setting a minimum and maximum level of each independent variable and associating it with a "weight" to adjust the shape of the objective function (desirability - D). The probability of the objective function reaching the desired value is zero outside the limits and 1 when the set objective is reached. Thus, when a response should be maximized, the probability of the predicted response (y) (desirability) is calculated using the equation (7) (Lazli & Hellal, 2020).

$$D = \begin{cases} 0 & if\ y < y_{min} \\ \dfrac{y - y_{min}}{y_{max} - y_{min}} & if\ y_{min} < y < y_{max} \\ 1 & if\ y > y_{max} \end{cases} \tag{7}$$

where y_{min} and y_{max} are the minimum and maximum values of the dependent variable.

Finding the optimal solution involves going step by step through the experimental range until the maximum value is predicted based on a best-fit model (Fertu et al., 2022). Since biosorption is mostly described by a second order polynomial equation, several variants of favourable solutions may exist. This can occur because this method does not have a constant curvature and varies according to the starting point and the slope covered step by step (Sadhukhan et al., 2016).

Analysing the studies available in the literature, it can be stated that RMS is the most used for modelling and optimizing the biosorption of metal ions by various types of biosorbents, the Box-Behnken design being the preferred one for optimization (Table 5). This model was used by Yuan et al. (2019)

to optimize the biosorption of Cd(II) ions by *Shewanella putrefaciens*, by Ghoniem et al. (2020) for the biosorption of Cu(II) by *Azotobacter nigricans* NEWG-1 or by Abigail et al. (2015) for the biosorption of Cr(VI) by *Penicillium griseofulvum* MSR1. CCD was used by Mathew & Krishnamurthy (2018) to assess the interactive effect of initial ion concentration and contact time on biosorption of Pb(II) by *Bacillus toyonensis* SCE1, *Bacillus toyonensis* SCE4, *Bacillus toyonensis* SCE5, *Bacillus anthracis* SCE2 and *Acinetobacter baumannii* SCE3. They found that an increase in contact time contributes to greater Pb(II) removal, with the highest effectiveness observed after 150 minutes of contact time. However, statistical modelling has shown that an extended contact time leads to desorption, and consequently, the efficiency of the process decreases.

Table 5. Screening and optimization of heavy metal biosorption by response surface methodology

Microorganisms	*Metal Ion*	*Purpose of Experiments*	*Independent Variables*	*Experimental Design*	*Results*	*References*
Aspergillus niger	Pb(II)	Screening of independent variables effects and process optimization	Initial solution pH, initial ion concentration and biosorbent dosage	Central composite design	Maximum removal = 96.21% at initial solution pH = 4.27, initial ion concentration = 30 mg/L, biosorbent dosage = 2.17 g/L, Second order polynomial equation R^2 = 0.9928, Adj R^2 = 0.9864, Pred. R^2 = 0.9695	(Amini et al., 2008)
Shewanella putrefaciens	Cd(II)	Process optimization	Contact time, initial solution pH and initial ion concentration	Box-Behnken design	Maximum removal = 86.54% at initial solution pH = 5, contact time = 4 days and initial ion concentration = 20 mg/L R^2 = 0.9281, Adj R^2 = 0.8356	(Yuan et al., 2019)
Aspergillus niger	Cr(VI)	Screening of independent variables effects and process optimization	Initial solution pH, biosorbent dosage, initial ion concentration, contact time	Box-Behnken design	Maximum removal = 87.00% at initial solution pH = 4.00, initial ion concentration = 33.33 mg/L, biosorbent dosage = 1.0 g/L and contact time = 48.45 min Second order polynomial equation R^2 = 0.9813, Adj R^2 = 0.9626, Pred. R^2 = 0.8075	(Mondal et al., 2017)
Rhodococcus sp. HX-2	Pb(II)	Screening of independent variables effects and process optimization	Biosorbent dose, pH, temperature and contact time	Box–Behnken design	Optimum biosorption capacity = 180.957 mg/g at biosorbent dose = 0.83 g/L, initial solution pH = 4, temperature = 24.42°C, contact time = 3 h Second order polynomial equation R^2 = 0.9844, Adj R^2 = 0.9688	(Hu et al., 2020)
Azotobacter nigricans NEWG-1	Cu(II)	Screening of independent variables effects and process optimization	Incubation time, initial culture pH and initial ion concentration	Box–Behnken design	Maximum removal = 80.56% at 200 mg/L Cu(II), incubation time = 4 days, initial solution pH = 8.5, R^2 = 98.74%, pred R^2 = 81.33%, adj. R^2 = 96.47%	(Ghoniem et al., 2020)
Aspergillus niger	Cd(II)	Process optimization	Solution pH, initial Cd(II) concentration, biomass dosage, temperature and contact time	Box–Behnken design	Maximum removal = 93% at initial solution pH =5.0, initial ion concentration = 100 mg/L, biomass dosage = 1.5 g/mL, temperature = 35 °C and contact time = 45 min Second order polynomial equation R^2 = 0.9313, Adj. R^2 = 0.8763	(Bhateria & Dhaka, 2019)
Escherichia coli	Cd(II)	Process optimization	pH, temperature and biomass dosage	Central Composite Design	Optimum biosorption capacity = 284.697 nmol/mg at initial solution pH = 8, temperature = 40°C and biosorbent dosage = 0.01 g/L after 1 hour contact time Second order polynomial equation R^2 = 0.98, Adj R^2 = 0.96, Pred. R^2 = 0.93	(Tafakori et al., 2017)

Continued on following page

Table 5. Continued

Microorganisms	*Metal Ion*	*Purpose of Experiments*	*Independent Variables*	*Experimental Design*	*Results*	*References*
Penicillium griseofulvum MSR1	Cr(VI)	Process optimization	Biosorbent dosage, initial ion concentration and contact time	Box–Behnken design	Maximum removal = 79.9% at 2 g/L biosorbent dosage, 67.8 mg/L initial Cr(VI) concentration and at contact time of 37.5 min Second-order polynomial quadratic equation $R^2 = 0.9894$, Adj $R^2 = 0.9758$, Pred. $R^2 = 0.8303$	(Abigail. et al., 2015)
Vibrio alginolyticus PBR1	Cd(II)	Process optimization	pH, biosorbent dosage and metal concentration	Box–Behnken design	Maximum removal = 59.78% at initial solution pH = 5.7, biosorbent dosage = 2.0 g/L and metal concentration = 50 mg/L Second-order polynomial quadratic equation $R^2 = 0.9947$, Adj-$R^2 = 0.9879$, Pred-$R^2 = 0.9387$	(Parmar et al., 2020)
	Pb(II)				Maximum removal = 82.20% at initial solution pH = 5.7, biosorbent dosage = 2.0 g/L and metal concentration = 50 mg/L Second-order polynomial quadratic equation $R^2 = 0.9115$, Adj-$R^2 = 0.7977$, Pred-$R^2 = 0.0518$	
Trichoderma viride	Pb(II), Cd(II) Cu(II)	Screening of independent variables effects	Initial metal ion concentration, biosorbent dosage, temperature and pH	Box–Behnken design	The pH had the most significant effect on the biosorption process, the optimum value being around 5. The relationship between variables and response is described by second-order polynomial equation. $R^2 = 0.9716$, Adj-$R^2 = 0.9432$, Pred-$R^2 = 0.8442$ (Pb(II)) $R^2 = 0.9699$, Adj-$R^2 = 0.9397$, Pred-$R^2 = 0.8494$ (Cd(II)) $R^2 = 0.9982$, Adj-$R^2 = 0.9963$, Pred-$R^2 = 0.9900$ (Cu(II))	(Singh et al., 2010)
Alcaligenes faecalis strain U.B.I.	Pb(II)	Screening of independent variables effects and process optimization	Biosorbent dosage, temperature, contact time and pH	Complete Composite Design	Maximum removal = 81.60% at initial solution pH = 9, biosorbent dosage = 0.3 g/L, temperature = 25°C after 24 min. Second-order polynomial quadratic equation $R^2 = 0.9397$, Adj-$R^2 = 0.9014$, Pred-$R^2 = 0.9031$	(Ibrahim et al., 2023)
Aspergillus niger	Cr(VI)	Screening of independent variables effects and process optimization	Initial metal ion concentration, contact time, temperature and pH	Box–Behnken design	Maximum removal = 82.60% at initial solution pH = 5.0 with initial metal ion concentration = 103.55 ppm, temperature = 35°C after 60 minutes. Second-order polynomial quadratic equation $R^2 = 0.9959$, Pred-$R^2 = 0.9898$	(Shrivastava & Gupta, 2021)

Continued on following page

Table 5. Continued

Microorganisms	*Metal Ion*	*Purpose of Experiments*	*Independent Variables*	*Experimental Design*	*Results*	*References*
Shewanella putrefaciens	Cr(VI)	Screening of independent variables effects and process optimization	Biosorbent dosage, temperature, contact time and pH	Box–Behnken design	Maximum removal = 85.68% at initial solution pH = 8, biomass biosorbent dosage = 0.42 g/L, temperature of 38.44 °C and contact time of 16.57 h. Second-order polynomial quadratic equation $R^2 = 0.8112$, Adj-$R^2 = 0.6224$, Pred-$R^2 = 0.0185$	(Cheng et al., 2021)
Arthrobacter sp. 25	Pb(II)	Screening of independent variables effects and process optimization	Initial metal ion concentration, pH, contact time, agitation speed, temperature and biosorbent dosage	Plackett-Burman design for factor screening and Box–Behnken design for optimization	Maximum adsorption capacity = 9.6 mg/g at the initial ion concentration = 108.79 mg/L, initial solution pH = 5.75 and biosorbent dosage = 9.9 g/L Second-order polynomial quadratic equation $R^2 = 0.9548$, Adj-$R^2 = 0.8967$	(Jin et al., 2016)

CONCLUSION

The removal and recovery of heavy metals from industrial wastewater stand as a critical priority, addressing both the reduction of heavy metal pollution in water and the associated risks to human health, as well as the potential recovery of critical metals. The application of bioremediation (bioaccumulation and biosorption) to manage and mitigate environmental heavy metal pollution represents an increasingly explored alternative to traditional physico-chemical processes. Extensive research into the removal of heavy metal ions from aqueous solutions has identified a diverse array of microorganisms that show promise as biosorbents for heavy metals.

The challenges of scaling-up bioremediation processes from laboratory to industrial scale, with an emphasis on biosorption are significant, attributed to the intricate interactions between pollutants and biosorbents, coupled with the diverse factors influencing the process. Therefore, it is imperative, in the exploitation of biosorption, to thoroughly investigate the impact of solution pH, biosorbent dose, initial metal concentration, process temperature, and contact time between the biosorbent and metal ions. However, the scale-up of biosorption technology from laboratory to industrial scale requires the mathematical modelling of experimental data obtained at the laboratory level. Utilizing appropriate models proves to be crucial in understanding process mechanisms, analyzing experimental data, predicting responses under established operational conditions, and optimizing the process.

A comprehensive analysis of published studies on heavy metals biosorption reveals that the most valuable models provide insights into process kinetics, maximum biosorption capacity, and thermodynamics. Equally essential are those methods that investigate the interactive effects of process variables, enabling the construction of accurate mathematical models. Prominent among the widely applied mathematical models are the Langmuir, Freundlich, Hill, Dubinin-Radushkevitch, Redlich-Peterson, and Sips isotherm models, along with Lagergren, Ho, Weber, Morris, and Elovich kinetic models. Furthermore, models derived from the design of experiments (DOE) tools, including Full Factorial Design, Plackett-Burman Design, Central Composite Design, and Box-Behnken Design, play a key role in this context. The DOE tools can serve to establish experimental matrices, refine and evaluate extensive experimental data, optimize processes, and formulate predictive equations for desired outcomes under specific conditions.

Therefore, this paper can significantly contribute in understanding and application of biosorption modelling for heavy metal removal. It addresses key parameters, utilizes advanced mathematical models, and leverages design of experiments (DOE) tools, providing valuable insights for advancing biosorption technology. The perspectives opened by this work extended to improved strategies for heavy metal removal, enhanced industrial-scale processes, and the sustainable management of environmental contaminants.

REFERENCES

Abdi, O., & Kazemi, M. (2015). A review study of biosorption of heavy metals and comparison between different biosorbents. *Journal of Materials & Environmental Sciences*, *6*(5), 1386–1399.

Abigail, M. E. A., Samuel, M. S., & Chidambaram, R. (2015). Hexavalent chromium biosorption studies using *Penicillium griseofulvum* MSR1 a novel isolate from tannery effluent site: Box–Behnken optimization, equilibrium, kinetics and thermodynamic studies. *Journal of the Taiwan Institute of Chemical Engineers*, *49*, 156–164. doi:10.1016/j.jtice.2014.11.026

Ali, I. H., & Alrafai, H. A. (2016). Kinetic, isotherm and thermodynamic studies on biosorption of chromium(VI) by using activated carbon from leaves of *Ficus nitida. Chemistry Central Journal*, *10*(1), 1–7. doi:10.1186/s13065-016-0180-1 PMID:27252777

Amini, M., Younesi, H., Bahramifar, N., Lorestani, A. A. Z., Ghorbani, F., Daneshi, A., & Sharifzadeh, M. (2008). Application of response surface methodology for optimization of lead biosorption in an aqueous solution by *Aspergillus niger. Journal of Hazardous Materials*, *154*(1–3), 694–702. doi:10.1016/j.jhazmat.2007.10.114 PMID:18068898

Antony, J. (2003). A systematic methodology for Design of Experiments. In J. Antony (Ed.), *Design of Experiments for Engineers and Scientists* (pp. 29–43). Butterworth-Heinemann. doi:10.1016/B978-075064709-0/50005-3

Apostol, L. C., Ghinea, C., Alves, M., & Gavrilescu, M. (2016). Removal of Erythrosine B dye from water effluents using crop waste pumpkin seed hulls as adsorbent. *Desalination and Water Treatment*, *57*(47), 22585–22608. doi:10.1080/19443994.2015.1132477

Ayawei, N., Ebelegi, A. N., & Wankasi, D. (2017). Modelling and interpretation of adsorption isotherms. *Journal of Chemistry*, *2017*, 1–11. doi:10.1155/2017/3039817

Aydar, A. Y. (2018). Utilization of Response Surface Methodology in Optimization of Extraction of Plant Materials. In V. Silva (Ed.), *Statistical Approaches with Emphasis on Design of Experiments Applied to Chemical Processes*. IntechOpen. doi:10.5772/intechopen.73690

Barros, M. S. D., Arroyo, P. A., & Silva, E. A. (2013). General Aspects of Aqueous Sorption Process in Fixed Beds. In H. Nakajima (Ed.), *Mass Transfer - Advances in Sustainable Energy and Environment Oriented Numerical Modeling*. IntechOpen. doi:10.5772/51954

Batool, F., Akbar, J., Iqbal, S., Noreen, S., & Bukhari, S. N. A. (2018). Study of isothermal, kinetic, and thermodynamic parameters for adsorption of cadmium: An overview of linear and nonlinear approach and error analysis. *Bioinorganic Chemistry and Applications*, *2018*, e3463724. doi:10.1155/2018/3463724 PMID:30065756

Begum, S. A. S., Tharakeswar, Y., Kalyan, Y., & Naidu, G. R. (2015). Biosorption of Cd (II), Cr (VI) & Pb (II) from aqueous solution using *Mirabilis jalapa* as adsorbent. *Journal of Encapsulation and Adsorption Sciences*, *5*(2), 2. doi:10.4236/jeas.2015.52007

Beni, A. A., & Esmaeili, A. (2020). Biosorption, an efficient method for removing heavy metals from industrial effluents: A Review. *Environmental Technology & Innovation*, *17*, 100503. doi:10.1016/j.eti.2019.100503

Bhateria, R., & Dhaka, R. (2019). Optimization and statistical modelling of cadmium biosorption process in aqueous medium by *Aspergillus niger* using response surface methodology and principal component analysis. *Ecological Engineering*, *135*, 127–138. doi:10.1016/j.ecoleng.2019.05.010

Bingol, D., Tekin, N., & Alkan, M. (2010). Brilliant Yellow dye adsorption onto sepiolite using a full factorial design. *Applied Clay Science*, *50*(3), 315–321. doi:10.1016/j.clay.2010.08.015

Biswas, G., Kumari, M., Adhikari, K., & Dutta, S. (2017). Application of response surface methodology for optimization of biosorption of fluoride from groundwater using *Shorea robusta* flower petal. *Applied Water Science*, *7*(8), 4673–4690. doi:10.1007/s13201-017-0630-5

Biswas, S., Bal, M., Behera, S., Sen, T., & Meikap, B. (2019). Process optimization study of Zn^{2+} adsorption on biochar-alginate composite adsorbent by response surface methodology (RSM). *Water (Basel)*, *11*(2), 325. doi:10.3390/w11020325

Boshagh, F., & Rostami, K. (2020). A review of application of experimental design techniques related to dark fermentative hydrogen production. *Journal of Renewable Energy and Environment*, *7*(2), 27–42.

Bulgariu, L., Bulgariu, D., & Rusu, C. (2015). Marine Algae Biomass for Removal of Heavy Metal Ions. In S.-K. Kim (Ed.), *Springer Handbook of Marine Biotechnology* (pp. 611–648). Springer. doi:10.1007/978-3-642-53971-8_25

Chakraborty, S., Chowdhury, S., & Das Saha, P. (2012). Insight into biosorption equilibrium, kinetics and thermodynamics of crystal violet onto *Ananas comosus* (pineapple) leaf powder. *Applied Water Science*, *2*(2), 135–141. doi:10.1007/s13201-012-0030-9

Cheng, J., Gao, J., Zhang, J., Yuan, W., Yan, S., Zhou, J., Zhao, J., & Feng, S. (2021). Optimization of hexavalent chromium biosorption by *Shewanella putrefaciens* using the Box-Behnken Design. *Water, Air, and Soil Pollution*, *232*(3), 92. doi:10.1007/s11270-020-04947-7

Ciobanu, A.-A., Vasile, G., & Bulgariu, L. (2023). Biosorption of Cu(II) ions on algae biomass and derived biochar. *Environmental Engineering and Management Journal*, *22*(5), 899–907. doi:10.30638/eemj.2023.072

Cozma, P., Hlihor, R.-M., Rosca, M., Simion, I.-M., Apostol, M., & Gavrilescu, M. (2021). Progress in microbial bioremediation of industrial effluents using different bioreactors design: An Overview. *2021 International Conference on E-Health and Bioengineering (EHB)*, 1–4. 10.1109/EHB52898.2021.9657749

Daoud, J. I. (2017). Multicollinearity and regression analysis. *Journal of Physics: Conference Series, 949*, 012009. doi:10.1088/1742-6596/949/1/012009

de Oliveira, L. G., de Paiva, A. P., Balestrassi, P. P., Ferreira, J. R., da Costa, S. C., & da Silva Campos, P. H. (2019). Response surface methodology for advanced manufacturing technology optimization: Theoretical fundamentals, practical guidelines, and survey literature review. *International Journal of Advanced Manufacturing Technology, 104*(5), 1785–1837. doi:10.1007/s00170-019-03809-9

Desta, M. B. (2013). Batch sorption experiments: Langmuir and Freundlich isotherm studies for the adsorption of textile metal ions onto teff straw (*Eragrostis tef*) agricultural waste. *Journal of Thermodynamics, 2013*, e375830. doi:10.1155/2013/375830

Ebelegi, A. N., Ayawei, N., & Wankasi, D. (2020). Interpretation of adsorption thermodynamics and kinetics. *Open Journal of Physical Chemistry, 10*(3), 166–182. doi:10.4236/ojpc.2020.103010

El Mourabit, S. (2016). *Optimization of Concrete Beam Bridges: Development of Software for Design Automation and Cost Optimization* [Independent thesis Advanced level, KTH, School of Architecture and the Built Environment (ABE), Civil and Architectural Engineering, Structural Engineering and Bridges]. https://urn.kb.se/resolve?urn=urn:nbn:se:kth:diva-189273

Espinoza-Sánchez, M. A., Arévalo-Niño, K., Quintero-Zapata, I., Castro-González, I., & Almaguer-Cantú, V. (2019). Cr(VI) adsorption from aqueous solution by fungal bioremediation based using *Rhizopus sp. Journal of Environmental Management, 251*, 109595. doi:10.1016/j.jenvman.2019.109595 PMID:31561145

Fertu, D. I., Bulgariu, L., & Gavrilescu, M. (2022). Modeling and optimization of heavy metals biosorption by low-cost sorbents using response surface methodology. *Processes (Basel, Switzerland), 10*(3), 523. doi:10.3390/pr10030523

Filote, C., Roşca, M., Hlihor, R. M., Cozma, P., Simion, I. M., Apostol, M., & Gavrilescu, M. (2021). Sustainable application of biosorption and bioaccumulation of persistent pollutants in wastewater treatment: Current practice. *Processes (Basel, Switzerland), 9*(10), 1696. doi:10.3390/pr9101696

Filote, C., Roşca, M., Simion, I. M., & Hlihor, R. M. (2022). Continuous systems bioremediation of wastewaters loaded with heavy metals using microorganisms. *Processes (Basel, Switzerland), 10*(9), 1758. doi:10.3390/pr10091758

Gavrilescu, M. (2004). Removal of heavy metals from the environment by biosorption. *Engineering in Life Sciences, 4*(3), 219–232. doi:10.1002/elsc.200420026

Gebreslassie, Y. T. (2020). Equilibrium, kinetics, and thermodynamic studies of malachite green adsorption onto fig (*Ficus cartia*) leaves. *Journal of Analytical Methods in Chemistry, 7384675*, 1–11. Advance online publication. doi:10.1155/2020/7384675 PMID:32190403

Ghoniem, A. A., El-Naggar, N. E.-A., Saber, W. I. A., El-Hersh, M. S., & El-khateeb, A. Y. (2020). Statistical modeling-approach for optimization of Cu^{2+} biosorption by *Azotobacter nigricans* NEWG-1; characterization and application of immobilized cells for metal removal. *Scientific Reports*, *10*(1), 9491. doi:10.1038/s41598-020-66101-x PMID:32528020

Hajahmadi, Z., Younesi, H., Bahramifar, N., Khakpour, H., & Pirzadeh, K. (2015). Multicomponent isotherm for biosorption of Zn(II), Co(II) and Cd(II) from ternary mixture onto pretreated dried *Aspergillus niger* biomass. *Water Resources and Industry*, *11*, 71–80. doi:10.1016/j.wri.2015.07.003

Hasan, S. H., Srivastava, P., & Talat, M. (2009). Biosorption of Pb(II) from water using biomass of *Aeromonas hydrophila*: Central composite design for optimization of process variables. *Journal of Hazardous Materials*, *168*(2), 1155–1162. doi:10.1016/j.jhazmat.2009.02.142 PMID:19345004

Hlihor, R.-M., Cozma, P., Ghinea, C., Pavel, V. L., Robu, B. M., & Gavrilescu, M. (2010). Thermodynamic study of Cd(II) sorption on soil. *Scientific Papers Series A. Agronomy*, *54*(2), 92–95. http://www.uaiasi.ro/revagrois/PDF/2010_2_94.pdf

Hlihor, R. M., Diaconu, M., Fertu, D., Chelaru, C., Sandu, I., & Tavares, T. (2013). Bioremediation of Cr(VI) Polluted wastewaters by sorption on heat inactivated *Saccharomyces cerevisiae* biomass. *International Journal of Environmental Research*, *7*(3), 581–594. doi:10.22059/ijer.2013.638

Hlihor, R. M., Diaconu, M., Leon, F., Curteanu, S., Tavares, T., & Gavrilescu, M. (2015). Experimental analysis and mathematical prediction of Cd(II) removal by biosorption using support vector machines and genetic algorithms. *New Biotechnology*, *32*(3), 358–368. doi:10.1016/j.nbt.2014.08.003 PMID:25224921

Hlihor, R. M., Roşca, M., Drăgoi, E. N., Simion, I. M., Favier, L., & Gavrilescu, M. (2023). New insights into the application of fungal biomass for chromium(VI) bioremoval from aqueous solutions using Design of Experiments and Differential Evolution based neural network approaches. *Chemical Engineering Research & Design*, *190*, 233–254. doi:10.1016/j.cherd.2022.12.024

Hlihor, R. M., Roşca, M., Tavares, T., & Gavrilescu, M. (2017). The role of *Arthrobacter viscosus* in the removal of Pb(II) from aqueous solutions. *Water Science and Technology*, *76*(7), 1726–1738. doi:10.2166/wst.2017.360 PMID:28991789

Hu, X., Cao, J., Yang, H., Li, D., Qiao, Y., Zhao, J., Zhang, Z., & Huang, L. (2020). Pb^{2+} biosorption from aqueous solutions by live and dead biosorbents of the hydrocarbon-degrading strain *Rhodococcus* sp. HX-2. *PLoS One*, *15*(1), e0226557. doi:10.1371/journal.pone.0226557 PMID:31995615

Hubbe, M., Azizian, S., & Douven, S. (2019). Implications of apparent pseudo-second-order adsorption kinetics onto cellulosic materials: A review. *BioResources*, *14*(3), 7582–7626. doi:10.15376/biores.14.3.Hubbe

Ibrahim, U. B., Yusuf, I., Saleh, A., Fardami, A. Y., Yahaya, H. I., Jodi, A. M., Sanusi, Z. M., Kawo, A. H., Nataala, M. K., & Yahaya, S. (2023). Biomass assessment and optimization of *Alcaligenes faecalis* isolated from some Nigerian mining sites for heavy metal uptake using response surface methodology model. *UMYU Scientifica*, *2*(3), 128–141. doi:10.56919/usci.2323.019

James, G., Witten, D., Hastie, T., & Tibshirani, R. (2013). *An Introduction to Statistical Learning* (Vol. 103). Springer. doi:10.1007/978-1-4614-7138-7

Jankovic, A., Chaudhary, G., & Goia, F. (2021). Designing the design of experiments (DOE) – An investigation on the influence of different factorial designs on the characterization of complex systems. *Energy and Building*, *250*, 111298. doi:10.1016/j.enbuild.2021.111298

Jiang, L., Zhou, W., Liu, D., Liu, T., & Wang, Z. (2017). Biosorption isotherm study of Cd^{2+}, Pb^{2+} and Zn^{2+} biosorption onto marine bacterium *Pseudoalteromonas* sp. SCSE709-6 in multiple systems. *Journal of Molecular Liquids*, *247*, 230–237. doi:10.1016/j.molliq.2017.09.117

Jin, Y., Wang, X., Zang, T., Hu, Y., Hu, X., Ren, G., Xu, X., & Qu, J. (2016). Biosorption of lead(II) by *Arthrobacter* sp. 25: Process optimization and mechanism. *Journal of Microbiology and Biotechnology*, *26*(8), 1428–1438. doi:10.4014/jmb.1603.03074 PMID:27197671

Kajjumba, G. W., Emik, S., Öngen, A., Kurtulus Özcan, H., & Aydın, S. (2019). Modelling of Adsorption Kinetic Processes - Errors, Theory and Application. In S. Edebali (Ed.), *Advanced Sorption Process Applications*. IntechOpen. doi:10.5772/intechopen.80495

Keskin Gündoğdu, T., Deniz, İ., Çalışkan, G., Şahin, E. S., & Azbar, N. (2016). Experimental design methods for bioengineering applications. *Critical Reviews in Biotechnology*, *36*(2), 368–388. doi:10.3109/07388551.2014.973014 PMID:25373790

Kim, S. Y., Jin, M. R., Chung, C. H., Yun, Y.-S., Jahng, K. Y., & Yu, K.-Y. (2015). Biosorption of cationic basic dye and cadmium by the novel biosorbent *Bacillus catenulatus* JB-022 strain. *Journal of Bioscience and Bioengineering*, *119*(4), 433–439. doi:10.1016/j.jbiosc.2014.09.022 PMID:25454694

Konig-Peter, A., Kilar, F., & Pernyeszi, T. (2019). Copper(II) biosorption characteristics of lyophilized and thermally treated *Pseudomonas* cells. *Environmental Engineering and Management Journal*, *18*(2), 455–464. doi:10.30638/eemj.2019.043

Kulkarni, R., Shetty, K. V., & Srinikethan, G. (2014). Cadmium (II) and nickel (II) biosorption by *Bacillus laterosporus* (MTCC 1628). *Journal of the Taiwan Institute of Chemical Engineers*, *45*(4), 1628–1635. doi:10.1016/j.jtice.2013.11.006

Lanati, A., Poli, C., Imberti, M., Menegon, A., & Grohovaz, F. (2015). A Design of Experiment Approach to Optimize an Image Analysis Protocol for Drug Screening. In V. Zazzu, M. B. Ferraro, & M. R. Guarracino (Eds.), *Mathematical Models in Biology: Bringing Mathematics to Life* (pp. 65–84). Springer International Publishing. doi:10.1007/978-3-319-23497-7_5

Largitte, L., & Pasquier, R. (2016). A review of the kinetics adsorption models and their application to the adsorption of lead by an activated carbon. *Chemical Engineering Research & Design*, *109*, 495–504. doi:10.1016/j.cherd.2016.02.006

Lazli, W., & Hellal, A. (2020). Biosorption of phenol by modified dead leaves of *Posidonia oceanica* immobilized in calcium alginate beads: Optimal experimental parameters using central composite design. *Water S.A.*, *46*(2 April). Advance online publication. doi:10.17159//wsa/2020.v46.i2.8248

Lee, K.-M., & Gilmore, D. F. (2006). Statistical experimental design for bioprocess modeling and optimization analysis. *Applied Biochemistry and Biotechnology*, *135*(2), 101–115. doi:10.1385/ABAB:135:2:101 PMID:17159235

Limcharoensuk, T., Sooksawat, N., Sumarnrote, A., Awutpet, T., Kruatrachue, M., Pokethitiyook, P., & Auesukaree, C. (2015). Bioaccumulation and biosorption of Cd^{2+} and Zn^{2+} by bacteria isolated from a zinc mine in Thailand. *Ecotoxicology and Environmental Safety*, *122*, 322–330. doi:10.1016/j.ecoenv.2015.08.013 PMID:26300116

Liu, L., Xia, M., Hao, J., Xu, H., & Song, W. (2021). Biosorption of Pb(II) by the resistant *Enterobacter* sp.: Investigated by kinetics, equilibrium and thermodynamics. *Archives of Environmental Protection*, *47*(3), 28–36. doi:10.24425/aep.2021.138461

Liu, Y., & Liu, Y. J. (2008). Biosorption isotherms, kinetics and thermodynamics. *Separation and Purification Technology*, *61*(3), 229–242. doi:10.1016/j.seppur.2007.10.002

Maiti, S., Prasad, B., & Minocha, A. K. (2022). Optimization of Pb(II) removal from aqueous solutions by fly ash using Box-Behnken design. *Environmental Engineering and Management Journal*, *21*(1), 125–135. doi:10.30638/eemj.2022.012

Mathew, B. B., & Krishnamurthy, N. B. (2018). Screening and identification of bacteria isolated from industrial area groundwater to study lead sorption: Kinetics and statistical optimization of biosorption parameters. *Groundwater for Sustainable Development*, *7*, 313–327. doi:10.1016/j.gsd.2018.07.007

Meier, L. (2014). *Statistical and Numerical Methods for Chemical Engineers*. https://people.math.ethz.ch/~muellepa/2016Stat/Script/cheming-e-part-I.pdf

Misra, D. N. (1973). Jovanovich adsorption isotherm for heterogeneous surfaces. *Journal of Colloid and Interface Science*, *43*(1), 85–88. doi:10.1016/0021-9797(73)90350-0

Mohapatra, R. K., Parhi, P. K., Pandey, S., Bindhani, B. K., Thatoi, H., & Panda, C. R. (2019). Active and passive biosorption of Pb(II)using live and dead biomass of marine bacterium *Bacillus xiamenensis* PbRPSD202: Kinetics and isotherm studies. *Journal of Environmental Management*, *247*, 121–134. doi:10.1016/j.jenvman.2019.06.073 PMID:31238200

Mondal, N. K., Samanta, A., Dutta, S., & Chattoraj, S. (2017). Optimization of Cr(VI) biosorption onto *Aspergillus niger* using 3-level Box-Behnken design: Equilibrium, kinetic, thermodynamic and regeneration studies. *Journal of Genetic Engineering and Biotechnology*, *15*(1), 151–160. doi:10.1016/j.jgeb.2017.01.006 PMID:30647651

Montgomery, D. C. (2013). *Design and Analysis of Experiments* (8th ed.). John Wiley & Sons, Inc.

Ncibi, M. C. (2008). Applicability of some statistical tools to predict optimum adsorption isotherm after linear and non-linear regression analysis. *Journal of Hazardous Materials*, *153*(1), 207–212. doi:10.1016/j.jhazmat.2007.08.038 PMID:17900804

Obradovic, B. (2020). Guidelines for general adsorption kinetics modeling. *Hemijska Industrija*, *74*(1), 65–70. doi:10.2298/HEMIND200201006O

Pagnanelli, F. (2011). Equilibrium, Kinetic and Dynamic Modelling of Biosorption Processes. In P. Kotrba, M. Mackova, & T. Macek (Eds.), *Microbial Biosorption of Metals* (pp. 59–120). Springer Netherlands. doi:10.1007/978-94-007-0443-5_4

Palmer, P. B., & O'Connell, D. G. (2009). Regression analysis for prediction: Understanding the process. *Cardiopulmonary Physical Therapy Journal, 20*(3), 23–26. doi:10.1097/01823246-200920030-00004 PMID:20467520

Parmar, P., Shukla, A., Goswami, D., Patel, B., & Saraf, M. (2020). Optimization of cadmium and lead biosorption onto marine *Vibrio alginolyticus* PBR1 employing a Box-Behnken design. *Chemical Engineering Journal Advances, 4*, 100043. doi:10.1016/j.ceja.2020.100043

Ponnusami, V., Krithika, V., Madhuram, R., & Srivastava, S. N. (2007). Biosorption of reactive dye using acid-treated rice husk: Factorial design analysis. *Journal of Hazardous Materials, 142*(1), 397–403. doi:10.1016/j.jhazmat.2006.08.040 PMID:17011118

Rajewski, J., & Dobrzyńska-Inger, A. (2021). Application of Response Surface Methodology (RSM) for the optimization of chromium(III) synergistic extraction by supported liquid membrane. *Membranes, 11*(11), 854. doi:10.3390/membranes11110854 PMID:34832083

Rangabhashiyam, S., Anu, N., Giri Nandagopal, M. S., & Selvaraju, N. (2014). Relevance of isotherm models in biosorption of pollutants by agricultural byproducts. *Journal of Environmental Chemical Engineering, 2*(1), 398–414. doi:10.1016/j.jece.2014.01.014

Rao, J. S., & Kumar, B. (2012). 3D Blade root shape optimization. In *10th International Conference on Vibrations in Rotating Machinery* (pp. 173–188). Woodhead Publishing. 10.1533/9780857094537.4.173

Roşca, M., Diaconu, M., Hlihor, R.-M., Cozma, P., & Gavrilescu, M. (2020). Prediction of equillibrium sorption isotherm for cadmium biosorption by microorganisms: comparison of linear and nonlinear methods. *2020 International Conference on E-Health and Bioengineering (EHB)*, 1–4. 10.1109/EHB50910.2020.9280180

Rosca, M., Hlihor, R.-M., Cozma, P., Comăniţă, E.-D., Simion, I. M., & Gavrilescu, M. (2015). Potential of biosorption and bioaccumulation processes for heavy metals removal in bioreactors. *2015 E-Health and Bioengineering Conference (EHB)*, 1–4. doi:10.1109/EHB.2015.7391487

Roşca, M., Hlihor, R.-M., Cozma, P., Drăgoi, E. N., Diaconu, M., Silva, B., Tavares, T., & Gavrilescu, M. (2018). Comparison of *Rhodotorula* sp. and *Bacillus megaterium* in the removal of cadmium ions from liquid effluents. *Green Processing and Synthesis, 7*(1), 74–88. doi:10.1515/gps-2016-0218

Rosca, M., Hlihor, R. M., Cozma, P., Simion, I. M., Filote, C., Grecu, C., Stoleru, V., & Gavrilescu, M. (2021). Scaling-up strategies of heavy metals microbial bioremediation. *2021 International Conference on E-Health and Bioengineering (EHB)*, 1–4. 10.1109/EHB52898.2021.9657641

Rosca, M., Hlihor, R.-M., & Gavrilescu, M. (2019). Bioremediation of Persistent Toxic Substances: From Conventional to New Approaches in Using Microorganisms and Plants. In P. K. Arora (Ed.), *Microbial Technology for the Welfare of Society* (Vol. 17, pp. 289–312). Springer Singapore. doi:10.1007/978-981-13-8844-6_14

Roşca, M., Silva, B., Tavares, T., & Gavrilescu, M. (2023). Biosorption of hexavalent chromium by *Bacillus megaterium* and *Rhodotorula* sp. inactivated biomass. *Processes (Basel, Switzerland), 11*(1), 1. Advance online publication. doi:10.3390/pr11010179

Saadi, R., Saadi, Z., Fazaeli, R., & Fard, N. E. (2015). Monolayer and multilayer adsorption isotherm models for sorption from aqueous media. *Korean Journal of Chemical Engineering*, *32*(5), 787–799. doi:10.1007/s11814-015-0053-7

Sadhukhan, B., Mondal, N. K., & Chattoraj, S. (2016). Optimisation using central composite design (CCD) and the desirability function for sorption of methylene blue from aqueous solution onto *Lemna major*. *Karbala International Journal of Modern Science*, *2*(3), 145–155. doi:10.1016/j.kijoms.2016.03.005

Sedlakova-Kadukova, J., Kopcakova, A., Gresakova, L., Godany, A., & Pristas, P. (2019). Bioaccumulation and biosorption of zinc by a novel *Streptomyces* K11 strain isolated from highly alkaline aluminium brown mud disposal site. *Ecotoxicology and Environmental Safety*, *167*, 204–211. doi:10.1016/j.ecoenv.2018.09.123 PMID:30340085

Shamim, S. (2018). Biosorption of Heavy Metals. In J. Derco & B. Vrana (Eds.), *Biosorption*. IntechOpen. doi:10.5772/intechopen.72099

Shrivastava, S., & Gupta, M. (2021). Application of Box-Behnken model to study the biosorption of chromium Cr(VI) ion by *Aspergillus niger* isolated from polluted site. *International Journal of Current Microbiology and Applied Sciences*, *10*(7), 414–424. doi:10.20546/ijcmas.2021.1007.045

Singh, R., Chadetrik, R., Kumar, R., Bishnoi, K., Bhatia, D., Kumar, A., Bishnoi, N. R., & Singh, N. (2010). Biosorption optimization of lead(II), cadmium(II) and copper(II) using response surface methodology and applicability in isotherms and thermodynamics modeling. *Journal of Hazardous Materials*, *174*(1), 623–634. doi:10.1016/j.jhazmat.2009.09.097 PMID:19836883

Tafakori, V., Zadmard, R., Tabandeh, F., Amoozegar, M. A., & Ahmadian, G. (2017). Equilibrium isotherm, kinetic modeling, optimization, and characterization studies of cadmium adsorption by surface-engineered *Escherichia coli*. *Iranian Biomedical Journal*, *21*(6), 380. doi:10.18869/acadpub.ibj.21.6.380 PMID:28555492

Temel, F. A., Avci, E., & Turan, N. G. (2022). Investigation of copper(II), zinc(II) and lead(II) removal onto expanded perlite by adsorption from the wastes of metal casting industry: Statistical modeling and optimization. *Environmental Engineering and Management Journal*, *21*(5), 757–767. doi:10.30638/eemj.2022.070

Tinega, J., Omosa, I., & Letema, S. (2023). Potential removal of chromium (VI) ions by macadamia nutshell from steel industry wastewater. *Environmental Engineering and Management Journal*, *22*(3), 411–427. doi:10.30638/eemj.2023.032

Torkia, Y. B., Bouaziz, N., Al-Muhtaseb, S. A., & Lamine, A. B. (2014). Adsorption energy and pore-size distributions of activated carbons calculated using Hill's model. *Adsorption Science and Technology*, *32*(7), 571–590. doi:10.1260/0263-6174.32.7.571

Turner, N. H. (1975). Kinetics of chemisorption: An examination of the Elovich equation. *Journal of Catalysis*, *36*(3), 262–265. doi:10.1016/0021-9517(75)90035-4

Uysal, U. G., Demir, A., & Goren, N. (2022). Investigation of cobalt(II) adsorption from aqueous solution using *Genista albida* as a low-cost adsorbent: Optimization based upon response surface methodology. *Environmental Engineering and Management Journal*, *21*(4), 637–650. doi:10.30638/eemj.2022.059

Veza, I., Spraggon, M., Fattah, I. M. R., & Idris, M. (2023). Response surface methodology (RSM) for optimizing engine performance and emissions fueled with biofuel: Review of RSM for sustainability energy transition. *Results in Engineering*, *18*, 101213. doi:10.1016/j.rineng.2023.101213

Vishan, I., Saha, B., Sivaprakasam, S., & Kalamdhad, A. (2019). Evaluation of Cd(II) biosorption in aqueous solution by using lyophilized biomass of novel bacterial strain *Bacillus badius* AK: Biosorption kinetics, thermodynamics and mechanism. *Environmental Technology & Innovation*, *14*, 100323. doi:10.1016/j.eti.2019.100323

Volesky, B. (2007). Biosorption and me. *Water Research*, *41*(18), 4017–4029. doi:10.1016/j.watres.2007.05.062 PMID:17632204

Wang, J., & Guo, X. (2020a). Adsorption kinetic models: Physical meanings, applications, and solving methods. *Journal of Hazardous Materials*, *390*, 122156. doi:10.1016/j.jhazmat.2020.122156 PMID:32006847

Wang, J., & Guo, X. (2020b). Adsorption isotherm models: Classification, physical meaning, application and solving method. *Chemosphere*, *258*, 127279. doi:10.1016/j.chemosphere.2020.127279 PMID:32947678

Wierzba, S. (2015). Biosorption of lead(II), zinc(II) and nickel(II) from industrial wastewater by *Stenotrophomonas maltophilia* and *Bacillus subtilis*. *Polish Journal of Chemical Technology*, *17*(1), 79–87. doi:10.1515/pjct-2015-0012

Yuan, W., Cheng, J., Huang, H., Xiong, S., Gao, J., Zhang, J., & Feng, S. (2019). Optimization of cadmium biosorption by *Shewanella putrefaciens* using a Box-Behnken design. *Ecotoxicology and Environmental Safety*, *175*, 138–147. doi:10.1016/j.ecoenv.2019.03.057 PMID:30897412

Zohrehvandi, S., & Bashiri, M. (2014). Evaluation of full experimental design method against fractional design method and Taguchi design method in machining operation. *Journal of Control Science and Engineering*, *2*, 43–51. doi:10.17265/2328-2231/2014.01.006

KEY TERMS AND DEFINITIONS

Adsorption Isotherm: Is defined as the graphical representation of the relationship between the quantities of sorbate (heavy metal) adsorbed per adsorbent unit and sorbate remaining in the solution at equilibrium.

Analysis of Variance (ANOVA): Is a useful statistical technique that can be applied to any regression model and based on the parameters provided by it, the best model that describes the process can be established.

Biosorption: Is the process that involves the ability of different types of biomasses, in particular microorganisms (bacteria, fungi, yeasts), to retain the metal ions on the surface of the cell wall, mainly due to the interaction between metal ions and functional groups present on the cell wall.

Biosorption Kinetics: Can be described by different models based on equilibrium studies, which are used to correlate the experimental data obtained considering a given initial metal ion concentration, biosorbent dose, pH, temperature, and different values of contact time between biosorbent and metal ions in an aqueous solution.

Design of Experiments (DOE): A systematic and efficient statistical technique, that based on planning, conducting, analysing and interpreting of controlled tests, the relationship between several input and output variables is studied.

Process Optimization: Involves identifying the admissible region of space in which the variation of independent variables does not affect the maximum performance of the process.

Regression Analysis: Is a statistical technique that helps in determining the relationship between a single dependent variable (response) and one or more independent variables (influencing factors).

Thermodynamic Studies: Provides information on the appropriate temperature range for sorption and the nature of the sorbent and sorbate at equilibrium.

Chapter 6
Various Biosorbents and Their Mechanisms for the Removal of Heavy Metals From Wastewater

Indranee Changmai
Gauhati University, India

ABSTRACT

Heavy metals are threatening to human health as they are arduous to remove from water bodies. Many technologies like photocatalysis, reverse osmosis, precipitation, and adsorption are used to get rid of those heavy metals from water. Among all these methods, adsorption is one of the most efficient methods to remove those heavy metals. In this chapter, the adsorption of heavy metals by dead biomass like algae, fungi, bacteria, and other biomass from waste will be discussed elaborately. The mechanisms of adsorption by these biomasses will be discussed, and the factors affecting their adsorption efficiencies are also included. The newest approach and emerging techniques are also considered. The future aspects and limitations are also considered.

1. INTRODUCTION

As the population is growing day by day, industrialization is also spiking up. Industrial waste has now become one of the most common concerns in the world. Heavy metals are toxic, incessant, non-degradable, and detrimental to ecosystem health (Arief et. Al, 2008). Industrial activities produce an enormous number of pollutants like dyes, pharmaceutical products, heavy metals metalloids, etc. and their wastewater directly goes into the terrestrial and aquatic ecosystems as they have always had problems with the demolition of these pollutants (Volesky and Holant, 1995). The dissipation of heavy metals into the environment is detrimental to human and aquatic health. Heavy metals like Cr, Mn, Fe, Ni, Cu, Pb, As, Cd, Zn, Hg, Co, etc. are very injurious to environmental and human health as they can cause many health issues like cancer, growth inhibition, organ and nervous system damage, etc. (Javanbakht et al., 2014). According to the World Health Organization (WHO), the concentration of these heavy metals should not be greater than 10 ppm (Wase and Forster, 1997). Although the toxicity of these contaminants is very

DOI: 10.4018/979-8-3693-1618-4.ch006

brutal, however as, the concentration of these contaminants is very low in water bodies, it is very hard to detect this type of contamination. Among those heavy metals: lead, mercury, chromium, cadmium, and arsenic are considered the most toxic ones among them (Wase and Forster, 1997). According to the International Agency for Research on Cancer (IARC), these heavy metals are classified as probable or known human carcinogens (Wase and Forster, 1997). These heavy metals are becoming one of the biggest threads for the life of aquatic systems and therefore, the removal of these heavy metals is pivotal for the upcoming days. Some heavy metals and their effects are shown in the table

Table 1. Some heavy metal ions and their permissible concentration (ppm) and their effects on human health

Metal Ions	Allowed Value in Drinking Water (WHO) (ppm)	Health Effects
Copper	1.5	Liver damage, insomnia
Lead	0.05	Kidney diseases and the nervous system affect
Arsenic	0.05	skin infection and cancer
Mercury	0.001	Nervous system diseases and arthritis
Zinc	5	Vomiting. depression
Cadmium	.005	Acute and chronic intoxications
Iron	0.3	Diabetes, nausea
Chromium	0.05	Lung tumor, diarrhea

A schematic diagram is given to represent the different origins and sources of the heavy metals and the different pathways to enter the human body (Elnabi *et al,* 2023).

Figure 1. Schematic diagram heavy metals contamination from different sources and their impact on environmental health (Elnabi et al., 2023)

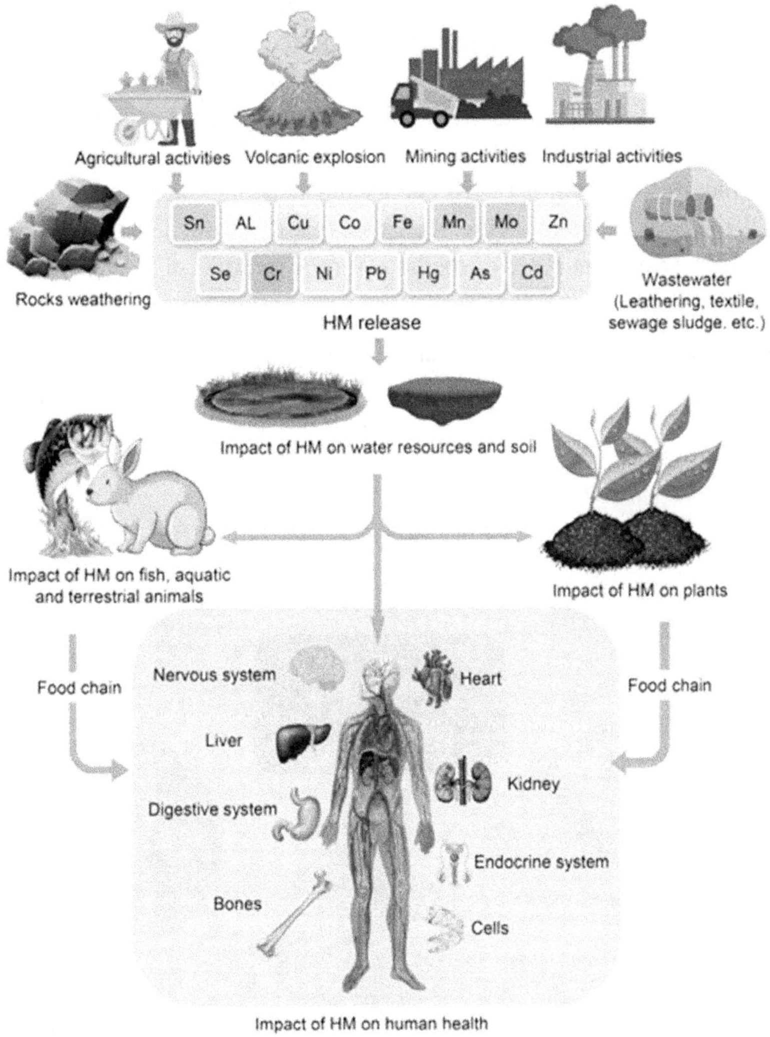

There are many types of methods for the removal of heavy metal from wastewater and those can be divided into physical, chemical, and biological methods. Physical methods include ion exchange, reverse osmosis, filtration, etc., chemical methods include precipitation, redox reaction, electrochemical, etc. The biological methods include the adsorption through micro-organisms (Hassan, 2010). Among those methods, adsorption is considered one of the most suitable methods as it is a very inexpensive and less energy-consuming method. Adsorption of heavy metals can be obtained by using different types of adsorbents, activated carbon is a well-known material for adsorption, however, its expensiveness is its limitation (Javanbakht et al., 2014). Therefore, among those methods, bio adsorption is selected as one of the most advantageous methods for the removal of heavy metals. Bioadsorption is a process, where non-living biological resources are used to immobilize and concentrate heavy metals and metalloids from aqueous solution (Wase and Forster, 1997). The biomasses are inexpensive, easily available, and environment friendly, and no hazardous by-products are produced by them (Javanbakht et al., 2014). Generally, when we speak about the adsorption method by biomasses, those bioadsorbents can be divided into two main categories- metabolism-independent and metabolism-dependent biomasses. Those which are living biomasses are called as metabolism dependent as their work efficiencies depend on the energy of the living particles. That adsorption is termed as bioaccumulation. On the other hand, biosorption that uses dead biomasses is termed energy or metabolism-independent bioabsorption. Dead biomasses are preferred over living (bioaccumulation) as no nutrition supply is needed for them and they are independent of surrounding conditions it is called energy independent and passive uptake of heavy metals method (Jinsong, 2014). Although these biosorbents are readily available but selectivity of these materials is very important. The different types of methods for the removal of heavy metals are shown below.

Figure 2. Different types of heavy metal removal methods

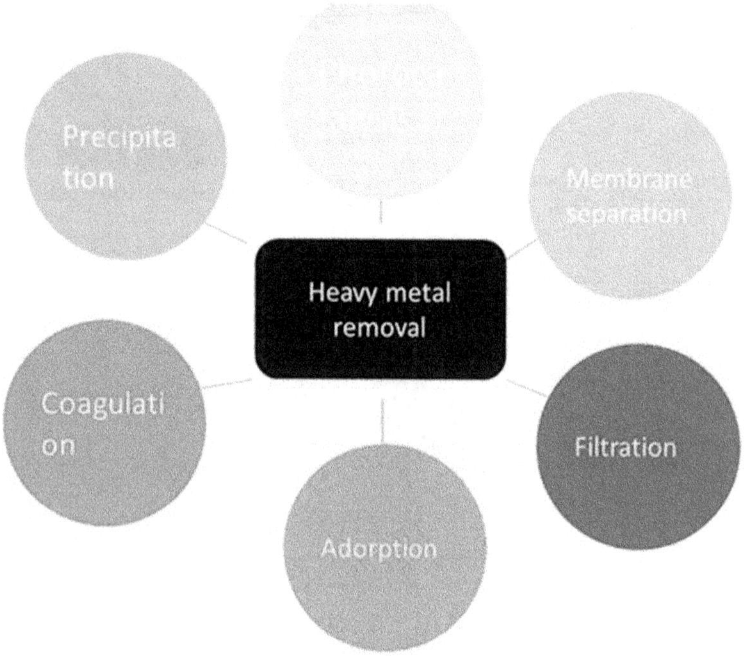

For the bio adsorption process, different types of microorganisms like Algae, fungi, bacteria, and yeast are used as bio adsorbents. Also, many low-cost biomaterials that are usually considered waste like peat, tree bark, peanut husk, etc. can be used for the adsorption of heavy metal ions (Wase and Forster, 1997). Those different microorganisms interacted differently with an aqueous metal solution. These bio adsorbents are very effective for the very low concentration of metal in the solution too. So, these bio adsorbents are very selective and effective. To know about their selectiveness and to make them more effective for the removal of heavy ions or metals we must know the mechanism of these adsorbents. On the other hand, the effectiveness of the adsorption depends on many factors, therefore to get the highest efficiencies from the adsorbents, the optimization of these factors for different biomaterials is very important. Bioadsorption through microorganisms is very cost-effective. To make the efficiencies more effective, the improvements of these biomasses also can be done by modification. The different types of bioadsorbents and their mechanisms are shown diagrammatically below.

Figure 3. Types of bioadsorbents

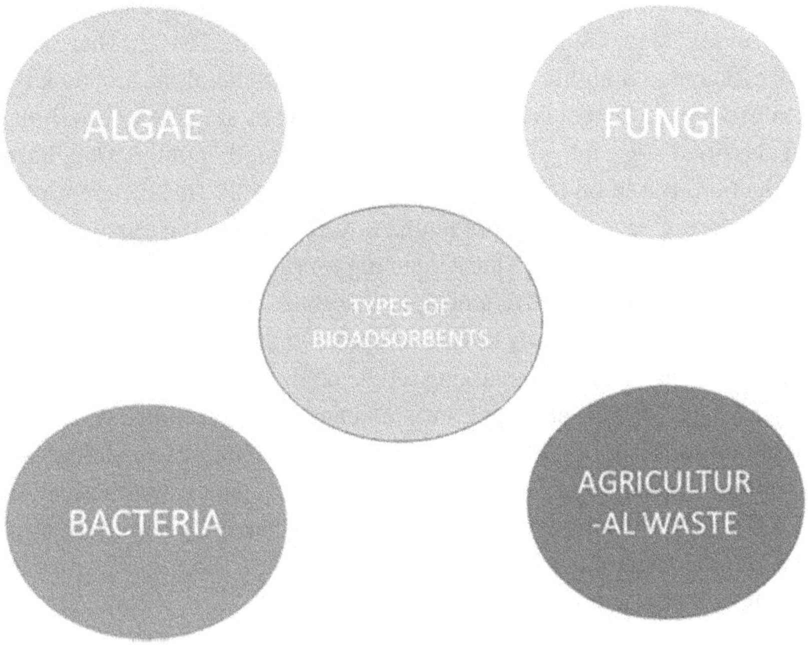

Figure 4. Mechanisms involved in the bioadsorption (Lau et al., 2003)

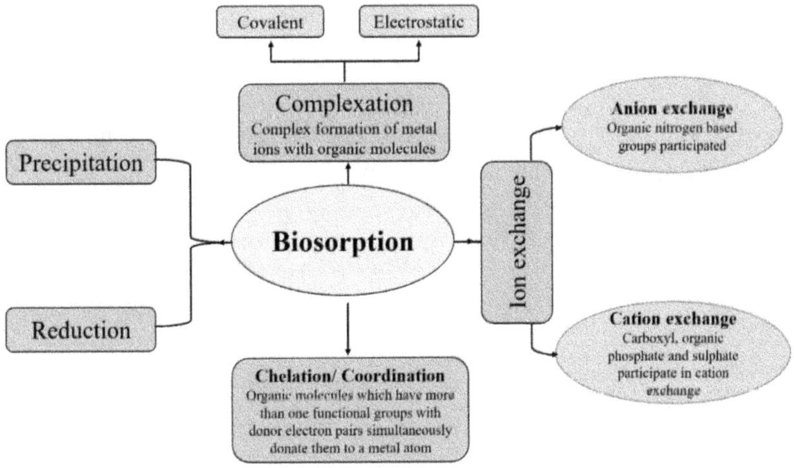

2. TYPES OF BIOADSORBENTS

The bioadsorbents are mainly divided into three broad categories and that are:

a. ***Algae***: Algae are one of the most significant organisms in the food chain for the environment. Algae are very cost-effective and abundant in nature. Almost over three million tons of algae are harvested annually (Jinsong et al., 2014). The adsorption rate of macroalgae and microalgae can reach up to 80% to 95% (Lau, et al., 2014, Ghimire,et al., 2008, Beetul et al., 2016,). That is why algal biomass is treated as a very efficient and highly efficient adsorbent. In general, in broad categories' algae can be categorized into two parts Macroalgae and Microalgae (Lin et al., 2020, Greeshma et al., 2022). In 1945, F. E. Fritsch classified algae into eleven classes based on pigmentation, flagella, thallus structure, and mode of reproduction. Based on the rate of efficiencies, mainly five different phyla are identified for adsorption and they are: Phaeophyta (Brown algae), Chlorophyta (Green algae), Rhodophyta (Red algae), Bacillariophyta and Cyanophyta (Blue-Green algae) (Lin et al., 2020, Greeshma et al., 2022,). Out of them, Bacillariophyta is the least efficient as an adsorbent. The low efficiencies are because of the cell wall structure of the photosynthetically produced lipid. On the other hand, Phaeophyta is marked as the highest efficient heavy metal adsorbent among all the phylum due to its multicellular body having a large surface area. This type of algae also contains small and acidic vacuole and mucus forms from the cell wall are favorable for adsorption purposes. Among the marine brown macroalgae, the *Sargassum* species is the best option (Jinsong et al., 2014). The adsorption capacities of these adsorbents generally are Pb > Cu = Ni > Cd > Zn (Jinsong et al., 2014). Also, other microalgae from the Chlorophyta group are also very efficient (Greeshma et al., 2022). The following chart Table 2 describes the different types of algae and their cell wall characteristics (Jinsong et al., 2014).

Table 2. Different types of algae and their cell wall structure

Phylum	Known Name	Cell Wall
Phaeophyta	Brown algae	Cellulose, non-sulphated phycocolloids, flagella present
Rhodophyta	Red algae	Cellulose, sulphated phycocolloids, flagella absent
Chlorophyta	Green algae	Cellulose, flagella present

The adsorption efficiencies of seaweed algae and microalgae for different heavy metals are listed below (Ordóñez et al., 2023, Mohammadi and Mahmoudnia, 2023)

Table 3. Different algae and their target metals for the removal (Ordóñez et al., 2023)

Type	Species	Metal
Brown algae	*Hydroclathrus clathratus*	Pb, Cr
	Cystoseira barbata	Pb, Cr
	Macrocystis pyrifera	Zn, Cd, Ni
	Fucus vesiculosus	Cd, Pb, Cu
	Sargassum filipendula	Cd, Zn
	Undaria pinnatifida	Hg
	Sargassum sp.	Pb, Cu
	Lessonia nigrescens	Cu
Red algae	*Gracilaria chilensis*	Cu
	Gracilaria fisheri	Cd, Cu
	Kappaphycus alvarezii	Cr, Ni, Cu
	Ceramium virgatum	Cd
	Callithamnion corymbosum	Cu, Co, Zn
Green algae	*Codium vermilara*	Cu
	Chlorella vulgaris	Cu

Table 4. Study of various microalgae for the removal of different metals with their optimum condition (Mohammadi and Mahmoudnia, 2023)

The Investigated Microalgae	Heavy Metal Investigated	pH	Contact Time (Minutes)	Absorbent Dose (gr/L)	Temperature
Chlamydomonas reinhardtii	Au(III)	2–3	70	0.1	26°C
Chlorella vulgaris	Cr(VI)	2	240	1	25°C
Cyanophyta & Chlorophyta	Ra226 U238	6–7 4–5	60	0.2	Environment temperature
Spirulina platensis	Cd2+	8	90	2	26°C
Scenedesmus obliquus	Cu(II)	5–7	60	0.03	Environment temperature
Scenedesmus quadricauda	Cd(II) & Pb(II)	5	60	0.2	Environment temperature

b. Fungi: Since the 1980s, fungi have been reported as biomass for the heavy metals' removal. Fungi are searched as adsorbent of heavy metal removal due to their cell material which is advantageous for metal bindings (Dhankhar and Hooda, 2011, Gadd, 1990). For different types of Fungi, the percentage of those protein lipids, and polysaccharides are different. Fungi are eukaryotic microorganisms and they are divided into three main classes: yeast, molds, and mushrooms (Paknikar et. Al., 1993, Ayele et al., 2021). The cell wall of fungi is made up of mainly polysaccharides and also proteins, lipids, etc. For the binding of metals, many functional groups like phosphate, amine hydroxyl, and sulphate, are found in those polymers. Those functional groups are excellent for the adsorption of heavy metals. Due to the high multiplicativity, high yield, and easily grown capacity, those adsorbents are highly advantageous. They are available as many industrial waste products also. Among those fungi genera, Rhizopus and Penicillium are established as bioadsorbents for heavy metal removal from wastewater (Paknikar et. Al., 1993, Ayele et al., 2021). After those other genera like Aspergillus, Streptoverticillum, and Saccharomyces are also explored (Carris et al., 2012, Mohmand et al., 2011, Siegel et al., 1990). A wide range of biosorption capacities from 0.012 to 1.979mmol/g for fungi has been reported. The different types of fungi species and their targeted metals are listed in Table 5.

Table 5. Biosorption by fungal biomass (mg g^{-1}) (Ahemad and K. Mulugeta, 2013)

Fungal Species	Metal Ions
A. Niger, Mucor rouxii	Au
Penicillium, Aspergillus, Rhizopus, Mucor	Ag, Cu, Cd, Pb
Phanerocheate chryosporium	Cd, Pb, Cu

The adsorption capacity of various fungal species for the removal of heavy metals is represented in the table with their experiment conditions (Ayele et al., 2021).

Table 6. Study of various fungi species and their adsorption capacity of heavy metals at different conditions (Ayele et al., 2021)

Fungal Species	Heavy Metal	pH	Initial Concentration (mg/L)	Optimum parameters Temperature (°C)	Bisorbent Dose (mg/L)	Agitation Speed (rpm)	Contact Time (min)	Adsorption Capacity (mg/g)
Streptomyces ciscaucasicus	Zn (II)	5	150	28	2000	90	480	54
Pleurotus ostreatus	Cr (VI)	5.6	— —	28 + 2	368.21	— —	22	— —
Pleurotus ostreatus	Cd (II)	6	10	26 ± 1	500	125	10	—
Mucor rouxii	Pb (II)	6	10	— —	50	125	300	53.75
Trametes versicolor	Cd (II)	5.5	600	25	25	400	60	120.6 ± 3.8
Aspergillus fumigatus	Pb (II) Cd (II)	5 5	10 10	28 ± 1 28 ± 1	0.04 0.04	150 150	280 280	3.651 0.83
Rhizopus arrhizus	Ni (II)	6	100	25	0.5	150	360	368.21
Lentinus edodes	Hg (II) Cd (II)	6 6	25–600 25–600	15–45 15–45	— —	—	— —	403.0 ± 2.9 274.3 ± 3.6
	Zn (II)	6	25–600	15–45	—	— —	—	57.1 ± 1.1
Aspergillus niger	Pb (II) Cd (II) Cu (II)	4 4 4	— — —	— — —	— —	— — —	120 120 144	7.24 3.43 2.66
	Ni (II)	5	—	—	—	—	192	0.96
Cladosporium resinae	Cd (II)	—	50	28	— —	120	672	—
Paecilomyces variotii	Cd (II)	—	50	28	—	120	792	—

c. ***Bacteria*****:** In the entire ecosystem, Bacteria occupy a large portion of it (Abbas et al., 2014, Priyadarshanee and Das, 2020). In general, Bacteria have a very simple cell structure. Mainly, Bacteria can be divided into two groups: Gram-positive and Gram-negative based on cell wall composition (Abbas et al., 2014, Priyadarshanee and Das, 2020). The gram-positive cell wall has a thick layer of peptidoglycan linked by amino acids and generally the thickness of these layers varies from 20 to 80 nm (Priyadarshanee and Das, 2020). On the other hand, Gram-negative bacteria have a thin cell wall of a single peptidoglycan layer, and the thickness of this layer is 10 to 15 nm. The gram-positive bacteria contain polymers joined by a phosphate group and those polymers are called teichuronic and teichoic acid. These acids create a negative charge on the surface by extending to the outer surface of peptidoglycan. The Gram-negative bacteria contain lipopolysaccharides (LPS), phospholipids, and proteins on the cell wall that make the cell surface negative. Mainly, the carboxyl and phosphoryl groups of peptide chains make the bacteria a negative charge and it will help to attract the metal cations present in the wastewater. The groups present in the cell wall materials, decide the selectivity and efficiencies of the bacteria. Among all phylum, three phyla are widely recognized for the removal of heavy metals and those are Actinobacteria, Proteobacteria, and Firmicutes (Abbas et al., 2014, Puranik and Paknikar, 1997). The adsorption capacity of bacterial biomasses ranges from 0.23 to 0.90 mmol/g (Eman, 2012). Bacteria are readily available anywhere and their size is very small and can grow under any harsh environment. Among different species Bacillus, Pseudomonas, Streptomyces, Escherichia, Micrococcus, etc., are marked as good adsorbent for heavy metal removal. A table is presented on the adsorption efficiencies of different bacteria on different heavy metals (Pham et al., 2022).

Table 7. Study on various bacteria and their heavy metal removal efficiencies (Pham et al., 2022)

Bacterial Biosorbents	Target Metals	Amount of Heavy Metals Uptake (mg/L)	Biosorption Efficiency (%)
Pseudomonas alcaliphila NEWG-2	Cr	200	96.6
Pseudomonas sp. *strain* DC-B3	Cr	55.35	41
Pseudomonas aeruginosa G12	Cr	10	93
Cellulosimicrobium funkei AR6	Cr	164.66	82.33
Stenotrophomonas maltophilia	Cr	19.84	99.2
Acinetobacter sp. WB-1	Cr	6.82	68.17
Cellulosimicrobium sp.	Cr	96.98	96.98
Stenotrophomonas sp.	Cr	270	90
Cellulosimicrobium sp.	Pb	200	84.62
Methylobacterium sp.	Pb	300	62.28
Aeribacillus pallidus MRP280	Pb	86.47	96.78
Bacillus sp. PZ-1	Pb	400	>90
Arthrobacter viscosus	Pb	100	97
Arthrobacter sp. 25	Pb	95.04	86.25
Pseudomonas sp. I3	Pb	49.48	98.96
Bacillus badius AK	Pb	60	60
Klebsiellap enumoniae	Cd	40.18	40.18
Rhodotorula sp.	Cd	40	80
Bacillus megaterium sp.	Cd	39.5	79
Bacillus sp. Q3	Cd	108.2	93.76
Pseudomonas aeruginosa FZ-2	Hg	10	99.7
Vibrio parahaemolyticus PG02	Hg	5	90
Pseudomonas aeruginosa	Cd, Pb	62.8 (Cd); 73.1 (Pb)	87 (Cd); 98.5 (Pb)
Saccharomyces cerevisiae	Pb, Cd	0.045 (Pb); 0.47 (Cd)	70.3 (Pb); 76.2 (Cd)
Desulfovibrio desulfuricans (immobilize on zeolite)	Zn	174	100
Micrococcus luteus DE2008	Pb, Cu	20.4 (Cu); 98.25 (Pb)	25.42 (Cu); 36.07 (Pb)
Bacillus sp.	Pb, Cu, Cd	990 (Cd); 970 (Cu); 200 (Pb)	>90 (Cd, Cu); 20 (Cd)
Oceanbacillus profundus	Pb, Zn	45 (Pb); 1.08 (Zn)	97 (Pb); 54 (Zn)
Staphylococcus epidermidis	Cr, Zn	118 (Zn); 112 (Cr)	59 (Zn), 56 (Cr)
Streptomyces sp.	Pb, Cd, Cu	1.43 (Cu); 0.91 (Pb) 3.66 (Cd)	Pb (83.4); Cu (74.5); Cd (68.4)
Klebsiella sp. USL2S	Hg, Pb, Cd, Ni	8500 (Hg); 10,000 (Pb); 1026 (Cd); 8479 (Ni)	85 (Hg); 97.13 (Pb); 73.33 (Cd) 86.06 (Ni)
Pseudomonas azotoformans JAW1	Cd, Cu, Pb	24.64 (Cd); 17.44 (Cu); 19.55 (Pb)	98.57 (Cd); 69.76 (Cu); 78.23 (Pb)

*d. **Plant-based byproducts***: Other than micro-organisms, plants-based byproducts mostly the waste from agricultural fields are used as adsorbents. For example, rice husk, tea waste, vegetable and fruit peels, sludge, starch, peat, wood, wool, sugarcane bagasse, etc. are very effective adsorbents (Calderón et al., 2020),. Different fruit peels like orange, lemon, and banana peels are used as adsorbents for lead, copper, and cadmium removal. Banana peels show the highest efficiencies for the removal of cadmium. Due to the lignocellulosic structure and surface negativity of the jackfruit peel it can adsorb Chromium and Nickel from wastewater (Calderón et al., 2020, Madeła and Skuza, 2021,). Due to the presence of amines, carboxyl, and H- bonded phenolic group, the aloe vera waste also can remove heavy metals above 90% efficiency in the pH range from 4.0 to 6.0 (Calderón et al., 2020). Also, those bioadsorbents are further modified by incorporating natural polymers like cellulose, chitosan, alginate, etc. The table for the other bioadsorbents from agricultural waste is listed below (Madeła and Skuza, 2021).

Table 8. Other bioadsorbents and their efficiencies for the removal of different types of dyes (Madeła and Skuza, 2021)

Metals	Biowaste	Sorption Capacity (mg/g)
Cd (II)	Agave bagasse Rice straw Citrus maxima peel Passion fruit shell Sugarcane bagasse Cabbage waste Agave bagasse (HCl) Agave bagasse (HNO3) Agave bagasse (NaOH)	13.27 13.90 135.2 86.75 23.23 22.12 12.50 13.50 18.32
Pb(II)	Agave bagasse Lentil husk Barley straws Citrus maxima peel Passion fruit shell Sugarcane bagasse Mango seed Maize silk Peach kernels Sunflower husks Cabbage waste Wheat bran Agave bagasse (HCl) Agave bagasse (HNO3) Agave bagasse (NaOH)	35.60 81.43 23.20 154.50 109.92 28.27 183.00 70.80 25.14 36.90 61.27 87.00 42.31 54.29 50.12
Cu(II)	Citrus maxima peel Passion fruit shell Sugarcane bagasse Mango peel Sunflower husks Beet pulp Gooseberry waste	83.7 30.09 16.09 46.09 34.89 31.4 24.0
Zn(II)	Agave bagasse Mango peel Cabbage waste Agave bagasse (HCl) Agave bagasse (HNO3) Agave bagasse (NaOH)	7.84 28.21 12.24 12.40 14.43 20.54
Ni(II)	Citrus maxima peel Passion fruit shell Sugarcane bagasse Mango peels Calamansi peels	70.79 20.64 12.04 39.75 11.00
Cr(VI)	Walnut shells Bean waste Bean shells Waste from tea Rice husk Passion fruit peel amino-riched	138.89 21.20 96.05 54.65 379.63 675.65
Cr(III)	Carrot residues Watermelon rinds	45.09 172.6
Fe(III)	Bean shells	66.63
As(V)	Orange peels	36.81

3. MECHANISM OF HEAVY METAL REMOVAL BY DIFFERENT ADSORBENTS

Understanding the adsorption process of the heavy metals from wastewater by biomasses is very difficult but at the same time, it is very crucial. the mechanisms of the adsorption cannot be explained by one procedure. The adsorption occurs due to the indulgence of many phenomena cumulatively. Also, the biomasses involved in the adsorption process are diverse. Mainly the bioabsorption process is a physiochemical process, however, biological interactions are also involved. Many processes like precipitation, ion exchange, electrostatic force, chelation, etc. are involved. The adsorption that takes place by dead biomasses is called a passive mechanism as it is independent of metabolism. In the adsorption mechanism, the process involves Reduction, Precipitation, and Complexation which is the formation of the complex by metal ions with organic molecules via electrostatic or covalent bonds, Ion Exchange i.e., anion and cation exchange between the metal ions and functional group present in the cell wall. Generally, in anion exchange, nitrogen groups are involved and for cation exchange, carboxyl, phosphate, and hydroxyl groups are involved (Eman, 2013, Wang, and Chen, 2009). Also, chelation happens when there is more than one functional group present in the cell wall. The total mechanisms involved in the heavy metal removal process are shown below (Madeła and Skuza, 2021).

Figure 5. Mechanisms involved in the bioadsorption process (Madeła and Skuza, 2021)

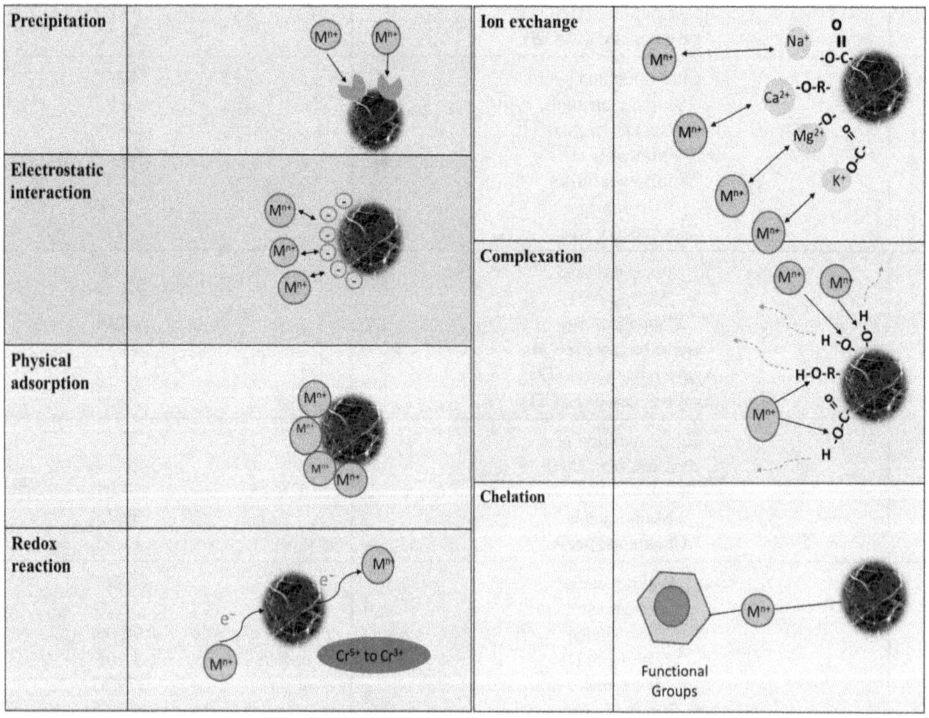

To generalize the mechanism process, it can be understood through the following steps (Eman, 2012, Wang, and Chen, 2009):

1. The heavy metal cations can bind to the cell surface of bioadsorbents.
2. The proteins of the cell structure can translocate the metal ions.
3. When heavy metals react with the functional groups of the cell wall, precipitation occurs.

Also, the adsorption of metal ions by those biomasses depends on the functional groups present in the cell wall as different meat ions prefer binding on selected ligands (Ahemad and Mulugeta, 2013, Wang and Chen, 2009). Class A metal groups like Lanthanides, Actinides, Li, Be, Na, Mg, K, Ca, Sc, Rb, Sr, La, and Ra are absorbed by mainly hydroxyl, carbonate, sulfate, nitrate, fluoride, carboxylate, dioxygen, phosphate ligands through oxygen. Class B metal groups like Pd, Rh, Ag, Pt, Au, Hg, and Pb are absorbed by cyanide ions, iodide ions, carbon monoxide, sulfide also by chloride, bromide, ammonia, etc. ligands. However, some metal ions named borderline metal ions can bind with all types of ligands at different preferences (Nieboe and D.H.S. Richardson, **1980**). The main constituents of the cell wall structure of different microorganisms are shown below (Ordóñez et al., 2023).

Figure 6. Different cell wall structures of bioadsorbents (Ordóñez et al., 2023)

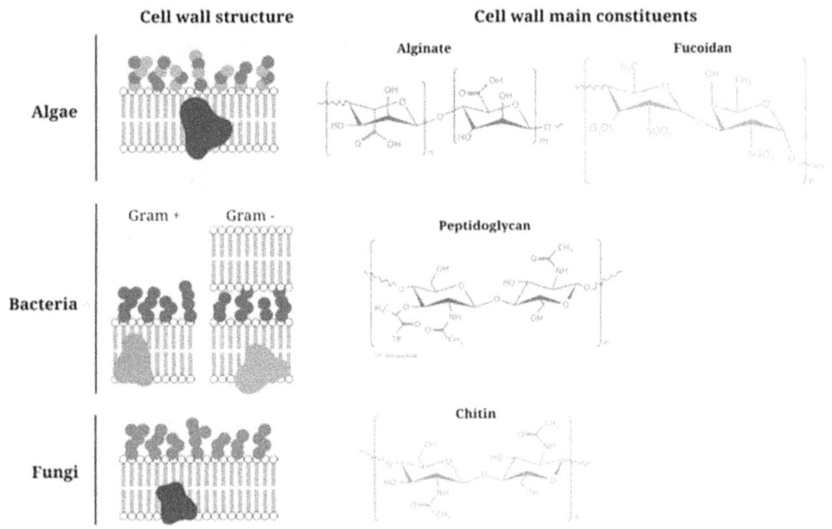

For the different microorganisms, the mechanisms are different, therefore the discussion of mechanism can be divided into three parts i.e., for bacteria, fungi, and algae.

a. **Mechanism of adsorption of Algal biomasses**:

For the adsorption by Algae mainly two principal mechanisms occur (Davis et al., 2000).

i) Ion exchange between the ions like sodium, potassium, and calcium present at the cell wall and the heavy metal ions.
ii) Complex formation between the functional group present in the cell wall and metal ions.

The Algal cell wall consists of polysaccharides like mannan, xylene, and chitin which provide different functional groups like carboxyl, phosphate, sulfate, amine, amino, imidiazole, etc. Those are the main factors for the ion exchange mechanism. Also, the adsorption can happen through van der Waal forces due to the presence of cellulose mesh in the cell wall. Although the maximum adsorption mechanism depends on the ionic exchange, however, some covalent bonding is also involved in the mechanism. This type of bonding can happen via carboxyl and amino groups (Ahmed et al, 2022). Algal biomass adsorption mainly consists of surface complexation, ion exchange, and redox reactions (Lin et al., 2020). The formation of surface complexation is preferred for the heavy ions lead copper and cadmium with two distal functional groups (Ahmed et al, 2022). Also, in some solutions metal is not present in cation form before the process. Therefore, redox reactions can transfer from one state to another state like Cr^{6+} to Cr^{3+}, As^{5+} to As^{3+}, and deduce the toxicity (Yang and Chen, 2008). Ion exchange usually occurs for the light ions. In algal biomass adsorption, lead (Pb) is the highest-adsorbed heavy metal (Lin et al., 2020). For Copper adsorption, the Cyanophyta group of Algae has a better adsorption capacity (Lin et al., 2020). In some studies, it is found that *Spirulina* platensis is tolerant to mercury and lead-containing environment. The mechanism is that the algae can conceal heavy metals by using extracellular polysaccharides and intracellular polyphosphates (Yang and Chen, 2008, Ahalya et al., 2003). Cadmium adsorption mainly happens via chelation and ion exchange is the main mechanism for Nickel ions (Ahalya et al., 2003, Javier et al., 2023). In some cases, multiple mechanisms contribute simultaneously like ion exchange, precipitation, and chelation happen in lead ions and Chromium adsorbed via electrostatic interaction (Javier et al., 2023, Sarojini et al., 2021, Raize et al., 2004, Chen et al., 2023). The bioadsorption mechanism is shown in the diagram given below (Chen et al., 2023).

Figure 7. Mechanisms for the bioadsorption by algae (Chen et al., 2023)

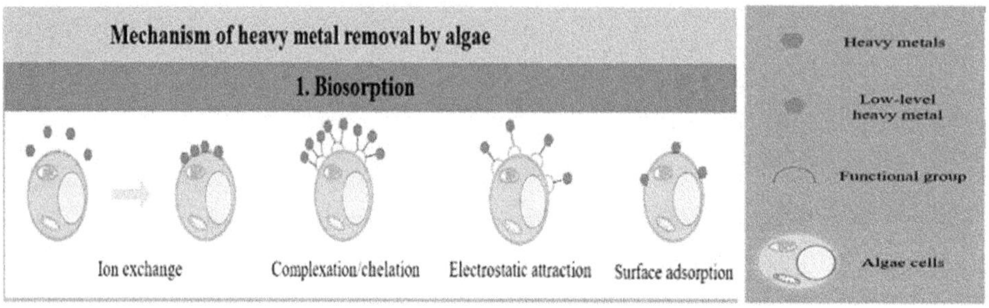

b. **Mechanism of adsorption of Fungal biomasses**

The biosorption of fungal biomasses also depends on the functional group present at the cell wall of the fungi. Ion exchange, complexation, adsorption of ionic species, and hydrolysis are observed. In short, the ion exchange mechanism is favored in fungal biomasses over complexation. The functional groups like amine, carboxyl, amide, and phosphate groups form complex metal ions (Kapoor, and Viraraghavan, 1995, Kapoor and Viraraghavan, 1997). Research suggests that the carboxylate and amine groups are important for the fungi A. niger (Kapoor, and Viraraghavan, 1995, Kapoor and Viraraghavan, 1997). Penicillium chrysogenum possesses zinc and lead complex properties via the phosphoryl group (Sarret et al, 1998). Amide groups contributed to the adsorption of Pb and Cu (Kiran et al, 2005, Wang, and Chen, 2006). The chitin and chitosan present in the cell walls of Fungi also participated in the adsorption process. The ions calcium, potassium, and magnesium present in the cell wall are exchanged with the heavy metal ions of the wastewater. In many research studies, the exchange of these ions with heavy metals can be observed (Kiran et al, 2005, Wang, and Chen, 2006). The glycoproteins of the cell wall contain a phosphate group and it is responsible for the negative charge on the surface of the cell wall above pH 3 (Tobin et al, 1984). The amino group of chitin and carboxylate contained by protein contributes to adsorption at pH 4 to 5. Chelation happens due to the presence of Aspartate, glutamate, and cysteine (Brady et al., 1994). The study of Tsezos and Volesky (B. Volesky and Tsezos, 1981) shows that thorium and uranium adsorption happen as the physical interaction via cell wall happens and then precipitation of uranyl hydroxide occurs. For chromium adsorption, electrostatic interaction happens (Kiran et al, 2005, Wang, and Chen, 2006). The chitin and chitosan present in the cell wall do not adsorb all metal ions and are selective in nature. They only adsorb the transition and post-transition metals, not adsorb the alkali and alkaline earth metals (Sağ, 2001). The mechanism is shown diagram in below (Gouda and Taha, 2023).

Figure 8. Mechanism for the bioadsorption by fungi (Gouda and Taha, 2023)

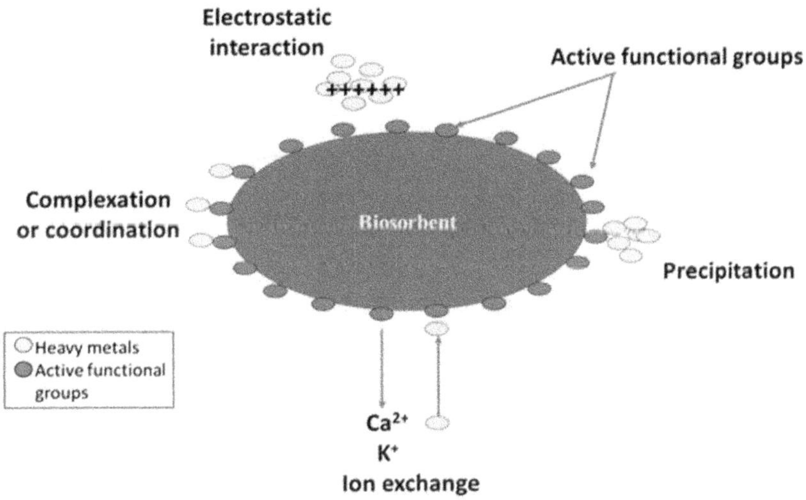

c. **Mechanism of adsorption of bacterial biomasses**

The adsorption of the heavy metals is initiated by the cell wall of the bacteria. Therefore, it plays a crucial role in bioadsorption. The metal ions are deposited on the surface and this process is called as extracellular method. The functional groups play a significant role in this extracellular process. The Carboxyl group is the dominant functional group in the bacterial cell wall, so the cations of heavy metals are attracted by this type of functional group which are negatively charged (Sağ, 2001). Other functional groups are also present there like hydroxyl, phosphate, and amine, etc. (van der Wal et al, 1997). On the other hand, the amine group can also absorb the anionic metals through hydrogen bonding or electrostatic interaction generally at low pH (Vijayaraghavan and Yun, 2008). All total it can be said that, with an increase of the pH, the cell wall becomes more negative, which supports the adsorption of the cation group of heavy metals and at the low pH environment, the protonation of the functional group helps the adsorption of anionic group of heavy metals. So, the adsorption of heavy metals by bacteria can be described in two steps. First is the interaction between the functional groups present in the cell wall with heavy metal ions and second is the deposition of heavy metals in higher concentrations (Vijayaraghavan and Yun, 2007, Vijayaraghavan and Yun, 2008). The mechanism of bio-absorption by bacteria is mainly due to the interaction between the cell wall components and heavy metals. The figure depicted the mechanism of bacteria for the heavy metal removal.

Figure 9. Different mechanisms for the adsorption process by bacterial biomass (Elnabi et al., 2023)

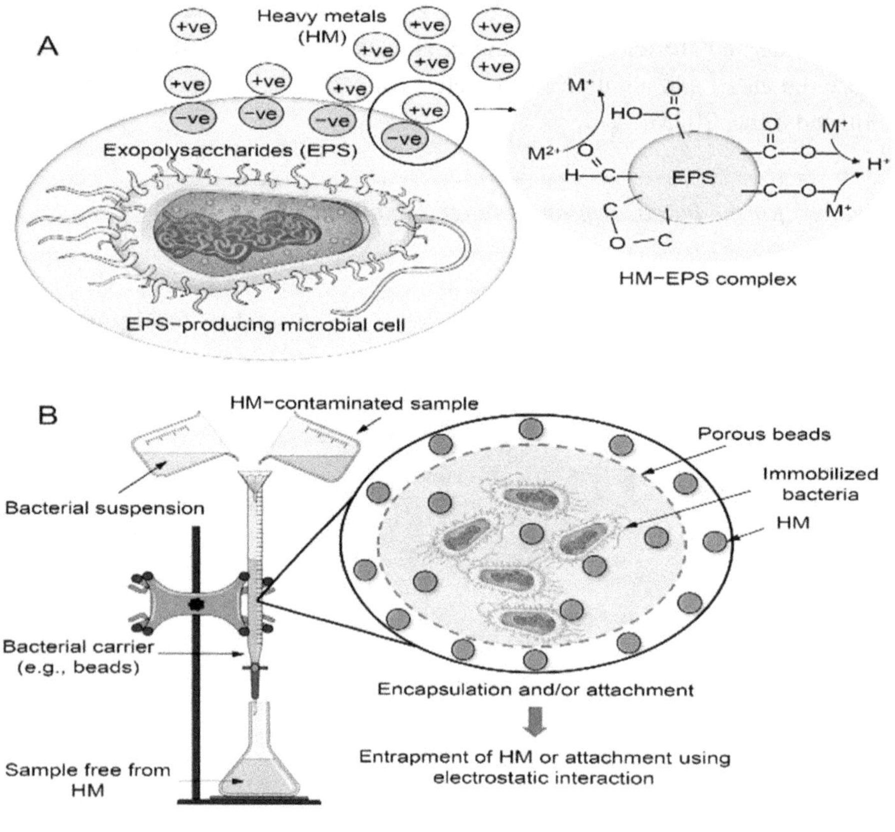

4. FACTORS AFFECTING HEAVY METALS ADSORPTION BY BIOADSORBENTS

The efficiencies of the adsorption are affected by many factors. Those factors cannot be ignored as those factors play a magnificent role in the mechanism of adsorption. the factors are explained below:

a. pH: Adsorption of the metals happens through ion exchange, so cation and anion adsorption occur. Many reports support that the higher value of the pH gives rise to metal precipitation, which can affect the potential of the adsorbent (Mosbah1 and Sahmoune, 2013, Selatnia et al, 2004, Iqbal and A. Saeed, 2007). At different pH values, the protonation and deprotonation of the function group present in the cell wall can occur. The pH of those adsorbs is totally opposite to each other. For cationic adsorption, the pH should be above 4.5 and anionic adsorption is favored when the pH is between 1.5 to 4 (Kratochvil and Volesky, 1998). The change in the pH can change the chemically active binding sites of the adsorbents.

If the active functional group is acidic or basic in nature, then the pH effect is huge. For the change of the pH, different adsorption sites got activated or deactivated. Also, in some cases when the pH values are at extreme levels, it can destroy the structure of the adsorbent, and desorption of the metal's ions can happen. For chitin possesses materials, where anion binds via electrostatic attraction, with the increase of pH the anions can be detached from the body of the adsorbent (Domard, 1987). At the solid-liquid interfaces, the adsorption is highly dependent on the pH value.

b. Initial Concentration: Both initial metal ions concentration and initial biosorbent concentration have an impact on the adsorption. In the low concentration of metal ions, the ratio of solute to surface area is low and becomes independent of the concentration. At higher concentration values, the sites for adsorption become comparatively low, so the adsorption becomes dependent on the initial concentration. Also, in some cases, the higher concentration of the meat ions can increase the amount of metal uptake (Mackova, 2011). The initial concentration of adsorbent is as important as the amount of solute uptake. Usually, as the amount of bioadsorbent increases the capacity of metal uptake increases for the high surface area and a large number of functional groups (Mackova, 2011). However, at a very high amount of bio adsorbent, the available solute is not sufficient to cover the exchangeable site of adsorbent and in return, the interference of those binding sites can lower the adsorption of heavy metals (Mackova, 2011).

c. Temperature: The temperature effect on the adsorption depends on the enthalpy change. The relation between the temperature and enthalpy can be explained by the equation below (Mackova, 2011, Ansari et al, 2011)

$$K_{int} = \exp(-\Delta H^0 + T\,\Delta S^0)/\,RT - = \exp(-\Delta H^0/RT) - \exp(-\Delta S^0/R)$$

Where K is the intrinsic adsorption equilibrium constant. ΔH is the enthalpy change, T is the temperature, ΔS is the entropy change, R is the gas constant. For different types of adsorption processes, the dependence of the temperature is different. For the chemical adsorption process, the change of enthalpy is positive i.e., the adsorption is endothermic in nature, so the adsorption process is better at the high temperature (Mackova, 2011). On the other hand, the change of enthalpy in the physical adsorption process is negative or exothermic in nature, therefore, the adsorption process is favored at low temperatures. As in bioadsorbent, the metal binding sites are numerous, therefore, the effect of the metal uptake is contributed by all sites. Although, the heat of the reaction is generally constant for the degree of site occupation in biomass adsorbent, however in some cases the temperature dependency plays a huge role. For example, in Cu adsorption by microbial biomasses, the heat of the reaction declined as the occupation site increased (Mackova, 2011). In total, it can be concluded that if the temperature range is not very large then it does not affect as much as other factors (Lin, 2020).

d. Ionic strength effect: The ionic strength of the metal in bioadsorption does not get too much attention. However, an increase in the ionic strength can decline the adsorption process and precipitation of metal can occur (Eman, 2012). This effect is significant in those cases where the adsorption mechanism is based on the electrostatic interaction. The study of the ionic strength effect is first done by (Lin, 2020)

e. Presence of other competing ions: In practical life, industrial wastewater consists of several heavy metals, and generally, for study of adsorption is done by only one type of heavy metal ion. In the presence of several ions, binding sites become less for the total amount of solute and therefore the uptake of metals becomes low as the cell walls are mostly nonspecific. The increase of concentration of other co-ions decreases the adsorption. for example, a study shows that in the presence of copper ions, the uptake of lead decreases, and further in addition of chromium ions in that solution, the adsorption of lead further declined (Sağ, 2001). The uptake of metals is highly dependent on the physicochemical properties of metals and the difference in the electronegativity of the atom's radius. It is reported that with the increase of the ionic radius, the metal uptake increases (Sağ, 2001). The Presence of lighter metals like sodium, and calcium has little impact on the adsorption of heavy metals (Vijayaraghavan and Yun, 2007, Vijayaraghavan and Yun, 2008). Also, the anion's presence can affect the efficiencies of adsorption (Sağ, 2001).

f. Contact time: The less contact time signifies the more efficiency of the adsorbent. In general, all the adsorption took place rapidly for the first few minutes and after that, it became slow due to a decrease of the adsorption site availability.

5. DIFFERENT ISOTHERM MODELS FOR THE EVALUATION OF BIOADSORPTION KINETICS

To know about the capacity of the adsorbent materials, the study of the Isotherm models is important. An adsorption isotherm is defined as the graphical representation of the relationship between the quantities of sorbate (heavy metal) adsorbed per unit of adsorbent and the sorbate that remains in the solution at equilibrium. This plot illustrates the distribution of adsorbable solute between the liquid and solid phases at different concentrations. Equilibrium is reached when the solute concentration remains constant in the solution (Desta, 2013). The Isotherm can give the kinetics of the adsorption and different models signifies the processes involved in the bioadsorption. For monolayer and multilayer adsorption, different models are explained. The given table describes the different models with their equations is given (Basu et al, 2022).

Table 9. Different models with their equations is given (Basu et al, 2022)

Isotherms	Equation	Applications	Parameters Involved
Langmuir	$R_L = \frac{1}{b \cdot C_0 + 1}$	Homogeneous substrate.	2
Freundlich	$q = k \cdot C_e^{1/n}$	Heterogeneous substrates.	2
Hill–Debboers	$q_e = \frac{q_m C_e^{n_H}}{K_D + C_e^{n_H}}$	Mobile adsorption. Interaction of adsorbed species.	2
Temkim	$q_e = \frac{RT}{B} \ln K_T C_e$	Indirect adsorbate/ adsorbate interaction	2
Sipps	$q_e = \frac{q_m a_s C_e^{Bs}}{1 + a_s C_e^{Bs}}$	Heterogeneous substrates with high adsorbate concentration	3
Redlich–Peterson	$q_e = \frac{A_R C_e}{1 + B_R C_e^{m_R}}$	Nonideal monolayer adsorption	3
Dubinin–Radushkevich	$q_e = Q_s \exp\left(-B_D \varepsilon^2\right)$ $\varepsilon = RT \ln\left(1 + \frac{1}{C_e}\right)$	Non-isothermal conditions.	3

6. CHALLENGES

Bioadsorption is receiving attention due to its minimal cost and no toxicity to the environment. However, the longevity and durability of the adsorbent is not appreciably good. The first challenge is the selection of different biomasses for different metal adsorption. The biomasses are large in number and exact mechanisms are not very clear to date. So, selecting the biomasses is quite a challenge. Another weakness is that they are lake of selectiveness property towards the metal ions and enhancing that property is quite a tedious job. Also, the reuse and recycle of these bioadsorbent are not much explored yet. The adsorption efficiencies of these biomasses haven't reached the level of the other adsorbent materials. The structure of the cell wall of those biomasses easily can get damaged in various environments. The recycling of these bioadsorbents materials is quite tough. As the heavy metals are attached to the cell wall of those biomasses, the structure of the cell wall and their applicability change after the adsorption, and therefore recycling got way harder. If the recycling of the biomasses cannot be done then after one use, it is not as effective as before. Another challenge is that although the bioadsorption has been known for many years the mechanisms of the process are not able to understand clearly to date. So, without a proper understanding of the mechanisms improvement in that area is not possible. The adsorbents are generally related to biological phenomena so the effects come from all sides, physical, chemical, and biological. Therefore, to recognize each and every phenomenon is not possible.

7. FUTURE PROSPECTS

The bioadsorbents need to be improved for practical application in different areas. The efficiencies and reusability are the major barriers to the application. The recent advancement of Algal, Fungal, Bacterial, and other biomasses made from wastes are discussed briefly.

a. **Algal biomasses**: Algal biomasses are highly cost-effective and easily grown products in any environment. the freshly immobilized Algae can remove heavy metals effectively (Raize et al, 2004) With the help of a new treatment of immobilized method, the performance of microalgae can be improved. Another method is that the adsorption process can be enhanced by genetically altered algae (Raize et al, 2004). The combination of more other bioadsorbents like fungi, and bacteria with Algal biomasses and co-immobilization can drastically improve the efficiencies. Genetic engineering of the algae for the addition of different proteins can enhance the selectivity of the metal uptakes and the desorption rate can be increased also.
b. **Fungal biomasses:** fungal biomasses can be used in different applications for the adsorption of different toxic metals. They help in the decomposition, nutrient cycling, and reduction of toxicity via different methods. The fermentation processes of food can provide the fungal biomasses and therefore, their product cost has become very cheap (Venkobachar, 1990). Without immobilization, fungi can be used in a column-like reactor. For the practical application, the mechanisms of the adsorption need to be clear, and different advanced technologies can be used. The genetic engineering of the cell wall can be done also. hybrid technologies can be combined with non-technological processes like electrochemical processes. Advanced computing technologies also be beneficial for the study of past and present trajectories and it will lead to the improvement in adsorption (Raize et al, 2004).
c. **Bacterial biomasses:** Just like the algal and fungal biomasses, for the improvement of the adsorbent immobilization hybrid technologies can be used. The parameters that affect the adsorption should be optimized and studied. To a better understanding of the metal and microbial interaction, the microbiological tool can be used to know the interaction at the molecular level (Kapoor and Viraraghavan, 1995, Ansari, 2011) Also, the reuse and recyclability of the adsorbent should be improved.

Also, research needs to focus on the process of bioadsorbent for commercial use and optimization of adsorbent dosage where the study does not focus much. The experiment should be at the industrial level also where multi-metal-containing wastewater resides. Those rigorous experiments and realization can be the main tool for practical application in the future.

8. CONCLUSION

The bioadorption process is quite advantageous to other techniques as it does not require any sophisticated instruments, the cost is very low due to the availability of bioadsorbent being very large, their growth does not require any controlled equation and many can be found as waste in agriculture and food industries. As industrial wastes are a huge barrier to the green environment, green techniques need to be implemented in the future. Therefore, the bioadsorbent which is non-toxic to the environment is realized to be a suitable candidate for the removal of heavy metals. The main three bioadsorbents are discussed in detail here. Their structure, mechanism, and the factors affecting their performance are discussed here. With the improvement of the research and technologies, the application of those adsorbents can further take to an upgraded level. The study of these bioadsorbents can give an idea of the performance of different bioadsorbents, their merit, and demits. The mechanisms study can help to analyze the drawback of those adsorbents and with the improvement of those sections can enhance the adsorption phenomenon. Other bioadsorbent like rice husk, wood, and pretreated biomasses are used for the adsorption. The adsorption of those dead masses is more effective than living biomasses. The adsorption rate is higher in dead biomasses and also as the dead biomasses are independent of metabolism, the adsorption can easily be manageable. Among the different algae, the brown macroalgae performed better than other algae. The green microalgae are also performed better. The improvement of algal biomasses can be done through immobilization. Generally, All the mechanisms of the adsorption, mostly ion exchange, chelation, precipitation, and complex formation is happened and electrostatic interaction, covalent bond, van der Waal force effect are there to some extent. The mechanism of these adsorption can be altered in the presence of the different pH. The pH value plays a huge role in the mechanism. From the functional group to the binding of light ions depends on the pH of the solution. According to the pH value, the anion and cation exchange can be altered. The cell wall outer surface charge negative or positive can be determined by the pH value. The adsorption is dependent on the no. of solute ions present in the solution and after that, the adsorption happens based on the ionic radius and electronegativity. Also, temperature does not have a high impact on the adsorption processes. Although the dosage of the biosorbent does not get much attention, however, this parameter has a high impact on the adsorption efficiencies. In the future, every parameter for the adsorption needs to be studied elaborately for the higher adsorption. Obviously, for the application of these biomasses in the practical platform, they are far behind, but for the enhancement of the green technology concept, this type of material is needed.

REFERENCES

Abbas, Ismail, Mostafa, & Abbas. (2014). Sulaymon Biosorption of Heavy Metals: A Review Salman. *Journal of Chemical Science and Technology*, *3*, 74–102.

Ahalya, N., Ramachandra, T. V., & Kanamadi, R. D. (2003). Biosorption of Heavy Metals. *Research Journal of Chemistry and Environment*, *7*, 71–79.

Ahemad, M., & Mulugeta, K. (2013). Recent Trends in Microbial Biosorption of Heavy Metals: A Review. *Biochemistry and Molecular Biology (New York, N.Y.)*, *1*(1), 19–26. doi:10.12966/bmb.06.02.2013

Ahmed, S. F. P., Kumar, P. S., Rozbu, M. R., Chowdhury, A. T., Nuzhat, S., Rafa, N., Mahlia, T. M. I., Ong, H. C., & Mofijur, M. (2022). Heavy metal toxicity, sources, and remediation techniques for contaminated water and soil. *Environmental Technology & Innovation*, *25*, 102114. doi:10.1016/j.eti.2021.102114

Ansari, M. I., Masood, F., & Malik, A. (2011). Bacterial Biosorption: A Technique for Remediation of Heavy Metals. *Microbes and Microbial Technology*, 283-319.

Arief, V. O., Trilestari, K., Sunarso, J., Indraswati, N., & Ismadji, S. (2008). Recent progress on Biosorption of Heavy Metals from Liquids Using Low Cost Biosorbents: Characterization, Biosorption Parameters and Mechanical Studies. *Clean (Weinheim), 36*(12), 937–962. doi:10.1002/clen.200800167

Ayele, A., Haile, S., Alemu, D., & Kamara, M. (2021). Comparative Utilization of Dead and Live Fungal Biomass for the Removal of Heavy Metal: A Concise Review. *TheScientificWorldJournal, 2021*, 10. doi:10.1155/2021/5588111 PMID:33927581

Basu, A., Ali, S. S., Hossain, S. K. S., & Asif, M. (2022). A Review of the Dynamic Mathematical Modeling of Heavy Metal Removal with the Biosorption Process. *Processes (Basel, Switzerland), 10*(6), 1154. doi:10.3390/pr10061154

Beetul, K., Gopeechund, A., Kaullysing, D., Moorgawa, S. M., Puchooa, D., & Bhagooli, R. (2016). Challenges and opportunities in the present era of marine algal applications. *Algae-Organisms for Imminent Biotechnology*, 40.

Brady, D., Stoll, A. D., Starke, L., & Duncan, J. R. (1994). Chemical and enzymatic extraction of Heavy metal binding polymers from isolated cell wall of Saccharomyces cerevisiae. *Biotechnology and Bioengineering, 44*(3), 297–302. doi:10.1002/bit.260440307 PMID:18618746

Calderón, O. A. R., Abdeldayem, O. M., Pugazhendhi, A., & Rene, E. R. (2020). Current Updates and Perspectives of Biosorption Technology: An Alternative for the Removal of Heavy Metals from Wastewater. *Current Pollution Reports*.

Carris, L. M., Little, C. R., & Stiles, C. M. (2012). Introduction to fungi. *The Plant Health Instructor*, 48.

Chang, J. S., & Hong, J. (1994). Biosorption of mercury by the inactivated cells of Pseudomonas aeruinosa. *Biotechnology and Bioengineering, 44*(8), 999–1006. doi:10.1002/bit.260440817 PMID:18618919

Chen, Z., Ahmed, I., Osman, D., Rooney, W., Oh, W., & Yap, P. (2023). Remediation of Heavy Metals in Polluted Water by Immobilized Algae: Current Applications and Future Perspectives. *Sustainability (Basel), 15*(6), 5128. doi:10.3390/su15065128

Davis, T. A., Volesky, B., & Vieira, R. (2000). Sargassum seaweed as biosorbent for heavy metals. *Water Research, 34*(17), 4270–4278. doi:10.1016/S0043-1354(00)00177-9

Desta, M. B. (2013). Batch sorption experiments: Langmuir and Freundlich isotherm studies for the adsorption of textile metal ions onto teff straw (*Eragrostis tef*) agricultural waste. *Journal of Thermodynamics, 2013*, 375830. doi:10.1155/2013/375830

Dhankhar, R., & Hooda, A. (2011). Fungal biosorption – an alternative to meet the challenges of heavy metal pollution in aqueous solutions. *Environmental Technology, 32*(5), 467–491. doi:10.1080/09593330.2011.572922 PMID:21877528

Domard, A. (1987). pH and c.d. measurements on a fully deacetylated chitosan: Application to CuII—polymer interactions. *International Journal of Biological Macromolecules, 9*(2), 98–104. doi:10.1016/0141-8130(87)90033-X

Doyle, R. J., Matthews, T. H., & Streips, U. N. (1980). Chemical basis for selectivity of metal ions by the Bacillus subtilis cell wall. *Journal of Bacteriology*, *143*(1), 471–480. doi:10.1128/jb.143.1.471-480.1980 PMID:6772632

Elnabi, M. K. A. (2023). Toxicity of Heavy Metals and Recent Advances in Their Removal: A Review. *Toxics*, *11*(7), 580. doi:10.3390/toxics11070580 PMID:37505546

Eman, Z. G. (2012). Production and characteristics of a heavy metals removing bioflocculant produced by Pseudomonas aeruginosa. *Polish Journal of Microbiology*, *61*(4), 281–289. doi:10.33073/pjm-2012-038 PMID:23484411

Gadd, G. M. (1990). Fungi and yeasts for metal binding. In H. Ehrlich & C. L. Brierley (Eds.), *Microbial Mineral Recovery* (pp. 249–275). McGraw-Hill.

Ghimire, K. N., Inoue, K., Ohto, K., & Hayashida, T. (2008). Adsorption study of metal ions onto crosslinked seaweed Laminaria japonica. *Bioresource Technology*, *99*(1), 32–37. doi:10.1016/j.biortech.2006.11.057 PMID:17267212

Gouda, S. A., & Taha, A. (2023). Biosorption of Heavy Metals as a New Alternative Method for Wastewater Treatment: A Review. *Egyptian Journal of Aquatic Biology & Fisheries*, *27*(2), 135–153. doi:10.21608/ejabf.2023.291671

Greeshma, K., Kim, H. S., & Ramanan, R. (2022). The emerging potential of natural and synthetic algae-based microbiomes for heavy metal removal and recovery from wastewaters. *Environmental Research*, *215*, 14238. doi:10.1016/j.envres.2022.114238 PMID:36108721

Gu, S., & Lan, C. Q. (2024). Mechanism of heavy metal ion biosorption by microalgal cells: A mathematic approach. *Journal of Hazardous Materials*, *463*, 132875. doi:10.1016/j.jhazmat.2023.132875 PMID:37918069

Hassan, S. H., Awad, Y. M., Kabir, M. H., Oh, S. E., & Joo, J. H. (2010). Bacterial Biosorption of heavy metals. *Biotechnology: Cracking New Pastures,* 79-110.

Iqbal, M., & Saeed, A. (2007). Production of an immobilized hybrid biosorbent for the sorption of Ni(II) from aqueous solution. *Process Biochemistry (Barking, London, England)*, *41*(2), 48–57. doi:10.1016/j.procbio.2006.07.022

Javanbakht, V., Alavi, S. A., & Zilou, H. (2014). Mechanisms of heavy metal removal using microorganisms as biosorbent. *Water Science and Technology*, *69*(9), 1775–1787. doi:10.2166/wst.2013.718 PMID:24804650

Javier, I., Ordóñez, I., Cortés, S., Maluenda, P., & Soto, I. (2023). Biosorption of Heavy Metals with Algae: Critical Review of Its Application in Real Effluents. *Sustainability (Basel)*, *15*(6), 5521. doi:10.3390/su15065521

Jinsong, H. J., Pau, J., & Chen, A. (2014). Comprehensive review on biosorption of heavy metals by algal biomass: Materials, performances, chemistry, and modelling simulation tools. *Bioresource Technology*, *160*, 67–78. doi:10.1016/j.biortech.2014.01.068 PMID:24630371

Kalita, N., & Baruah, P. P. (2023). Cyanobacteria as a potent platform for heavy metals biosorption: Uptake, responses and removal mechanisms. *Journal of Hazardous Materials Advances*, *11*, 100349. doi:10.1016/j.hazadv.2023.100349

Kapoor, A., & Viraraghavan, T. (1995). Fungal biosorption — an alternative treatment option for heavy metal bearing wastewaters: A review. *Bioresource Technology*, *53*, 195–206.

Kapoor, A., & Viraraghavan, T. (1997). Fungi as biosorbents. In J. Wase & C. Forster (Eds.), *Biosorbents for metal ions, 67–86.*

Kiran, I., Akar, T., & Tunali, S. (2005). Biosorption of Pb (II) and Cu (II) from aqueous solution by pretreated biomass of Neurospora crassa. *Process Biochemistry (Barking, London, England)*, *40*(11), 3550–3558. doi:10.1016/j.procbio.2005.03.051

Kratochvil, D., & Volesky, B. (1998). Advances in the biosorption of heavy metals. *Trends in Biotechnology*, *16*(7), 291–300. doi:10.1016/S0167-7799(98)01218-9

Lai, H., Ding, X., Cui, M., Zheng, J., Chen, Z., Pei, J., & Zhang, J. (2023). Mechanisms and influencing factors of biomineralization based heavy metal remediation: A review. *Biogeotechnics*, *1*(3), 100039. doi:10.1016/j.bgtech.2023.100039

Lau, T. C., Ang, P. O., & Wong, P. K. (2003). Development of seaweed biomass as a biosorbent for metal ions. *Water Science and Technology*, *47*(10), 49–54. doi:10.2166/wst.2003.0536 PMID:12862216

Lin, Z., Li, J., Luan, Y., & Dai, W. (2020). Application of algae for heavy metal adsorption: A 20-year meta-analysis. *Ecotoxicology and Environmental Safety*, *190*, 110089. doi:10.1016/j.ecoenv.2019.110089 PMID:31896472

Mackova, Macek, & Kotrba. (2011). *Microbial Biosorption of Metals*. Springer.

Madeła, M., & Skuza, M. (2021). Towards a Circular Economy: Analysis of the Use of Biowaste as Biosorbent for the Removal of Heavy Metals. *Energies*, *14*(17), 5427. doi:10.3390/en14175427

Mohmand, A. Q. K., Kousar, M. W., Zafar, H., Bukhari, K. T., & Khan, M. Z. (2011). Medical importance of fungi with special emphasis on mushrooms. *Isra Medical Journal*, *3*, 1–44.

Mosbah, R., & Sahmoune M. N. (2013). Biosorption of heavy metals by Streptomyces species - an overview. *Central European Journal of Chemistry*, *11*, 1412-1422.

Nieboe, E., & Richardson, D. H. S. (1980). The replacement of the no-descript term 'heavy metals' by a biologically and chemically significant classification of metal ions. *Environmental Pollution. Series B. Chemical and Physical*, *1*(1), 3–26. doi:10.1016/0143-148X(80)90017-8

Ordóñez, J. I., Cortés, S., Maluenda, P., & Soto, I. (2023). Biosorption of Heavy Metals with Algae: Critical Review of Its Application in Real Effluents. *Sustainability (Basel)*, *15*(6), 5521. doi:10.3390/su15065521

Paknikar, K. M., Palnitkar, U. S., & Puranik, P. R. (1993). In A. E. Torma, M. L. Apel, & C. L. Brierely (Eds.), *Biohydrometallurgical Technologies, V2* (pp. 229–236). The Minerals, Metals & Materials Society.

Pham, V. H. T., Kim, J., Chang, S., & Chung, W. (2022). Bacterial Biosorbents, an Efficient Heavy Metals Green Clean-Up Strategy: Prospects, Challenges, and Opportunities. *Microorganisms*, *10*(3), 610. doi:10.3390/microorganisms10030610 PMID:35336185

Priya, A. K., Gnanasekaran, L., Dutta, K., Rajendran, S., Balakrishnan, D., & Soto-Moscoso, M. (2022). Biosorption of heavy metals by microorganisms: Evaluation of different underlying mechanisms. *Chemosphere*, *307*, 135957. doi:10.1016/j.chemosphere.2022.135957 PMID:35985378

Priyadarshanee, M., & Das, S. (2020). Biosorption and removal of toxic heavy metals by metal tolerating bacteria for bioremediation of metal contamination: A comprehensive review. *Journal of Environmental Chemical Engineering*, *9*(1), 104686. doi:10.1016/j.jece.2020.104686

Puranik, P. R., & Paknikar, K. M. (1997). Biosorption of lead and zinc from solutions using Streptoverticillium cinnamoneum waste biomass. *Journal of Biotechnology*, *55*(2), 113–124. doi:10.1016/S0168-1656(97)00067-9 PMID:9232033

Raize, O., Argaman, Y., & Yannai, S. (2004). Mechanisms of biosorption of different heavy metals by brown marine macroalgae. *Biotechnology and Bioengineering*, *87*(4), 451–458. doi:10.1002/bit.20136 PMID:15286981

Raji, Z., Karim, A., Karam, A., & Khalloufi, S. (2023). Adsorption of Heavy Metals: Mechanisms, Kinetics, and Applications of Various Adsorbents in Wastewater Remediation—A Review. *Waste*, *1*(3), 775–805. doi:10.3390/waste1030046

Sağ, Y. (2001). biosorption of heavy metals by fungal biomass and modelling of fungal biosorption: A review. *Seperation and Purification Methods*, *30*, 1-48.

Sarojini, G., Babu, S. V., Rajamohan, N., Kumar, P. S., & Rajasimman, M. (2021). Surface modified polymer-magnetic-algae nanocomposite for the removal of chromium- equilibrium and mechanism studies. *Environmental Research*, *201*, 111626. doi:10.1016/j.envres.2021.111626 PMID:34217718

Sarret, G., Manceau, A., Spadini, L., Roux, J. C., Hazemann, J. L., Soldo, Y., Berard, L. E., & Menthonnex, J. J. (1998). Structural determination of Zn and Pb binding sites in Penicillium chrysogenum cell walls by EXAFS spectroscopy. *Environmental Science & Technology*, *32*(11), 1648–1655. doi:10.1021/es9709684

Selatnia, A., Boukazoula, A., Kechid, N., Bakhti, M. Z., & Chergui, A. (2004). Biosorption of Fe3+ from aqueous solution by a bacterial dead Streptomyces rimosus biomass. *Process Biochemistry (Barking, London, England)*, *39*(11), 1643–1651. doi:10.1016/S0032-9592(03)00305-4

Siegel, S. M. M., Galun, M., & Siegel, B. Z. (1990). Filamentous fungi as metal biosorbents: A review. *Water, Air, and Soil Pollution*, *53*(3-4), 335–344. doi:10.1007/BF00170747

Stumm, W., & Morgan, J. J. (1996). *Aquatic chemistry*. Wiley.

Tobin, J. M., Cooper, D. G., & Neufeld, R. J. (1984). Uptake of metal ions by Rhizopus Arrhizus Biomass. *Applied and Environmental Microbiology*, *74*(4), 821–824. doi:10.1128/aem.47.4.821-824.1984 PMID:16346521 van der Wal, A., Norde, W., Zehnder, A. J. B., & Lyklema, J. (1997). Determination of the total charge in the cell walls of gram-positive bacteria. *Colloids and Surfaces. B, Biointerfaces*, *9*(1-2), 81–100. doi:10.1016/S0927-7765(96)01340-9

Venkobachar, C. (1990). Metal Removal by Waste Biomass to Upgrade Wastewater Treatment Plants. *Water Science and Technology*, *22*(7-8), 319–320. doi:10.2166/wst.1990.0277

Vijayaraghavan, K., & Yun, Y. S. (2007). Utilization of fermentation waste (Corynebacterium glutamicum) for biosorption of Reactive Black 5 from aqueous solution. *Journal of Hazardous Materials*, *141*(1), 45–52. doi:10.1016/j.jhazmat.2006.06.081 PMID:16879915

Vijayaraghavan, K., & Yun, Y. S. (2008). Bacterial biosorbents and biosorption. *Biotechnology Advances*, *26*(3), 266–291. doi:10.1016/j.biotechadv.2008.02.002 PMID:18353595

Volesky, B., & Holant, Z. R. (1995). Biosorption of Heavy Metals. *Biotechnology Progress*, *11*(3), 235–250. doi:10.1021/bp00033a001 PMID:7619394

Volesky, B., & Tsezos, M. (1981). Biosorption of uranium and thorium. *Biotechnology and Bioengineering*, *23*(3), 583–604. doi:10.1002/bit.260230309

Wang, J., & Chen, C. (2006). Biosorption of heavy metals by Saccharomyces cerevisiae: A review. *Biotechnology Advances*, *24*(5), 427–445. doi:10.1016/j.biotechadv.2006.03.001 PMID:16737792

Wang, J., & Chen, C. (2009). Biosorbents for heavy metals removal and their future. *Biotechnology Advances*, *27*(2), 195–226. doi:10.1016/j.biotechadv.2008.11.002 PMID:19103274

Wase, J., & Forster, C. (1997). *Biosorbents for Metal Ions*. Taylor & Francis.

Yang, L., & Chen, J. P. (2008). Biosorption of hexavalent chromium onto raw and chemically modified Sargassum sp. *Bioresource Technology*, *99*(2), 297–307. doi:10.1016/j.biortech.2006.12.021 PMID:17336517

Zhang, P., Yang, M., Lan, J., Huang, Y., Zhang, J., Huang, S., Yang, Y., & Ru, J. (2023). Water Quality Degradation Due to Heavy Metal Contamination: Health Impacts and Eco-Friendly Approaches for Heavy Metal Remediation. *Toxics*, *11*(10), 828. doi:10.3390/toxics11100828 PMID:37888679

Chapter 7
Biogenic Sorption:
An Intelligent and Sustainable Approach to Reclaiming and Reutilizing Heavy Metals in the Modern Era

Periasamy Palanisamy
https://orcid.org/0000-0003-1069-5822
Nehru Institute of Engineering and Technology, Anna University, India

M. Kumaresan
Nehru Institute of Engineering and Technology, Anna University, India

M. Maheswaran
Nehru Institute of Engineering and Technology, Anna University, India

ABSTRACT

The treatment of heavy-metal-bearing wastewaters is a critical environmental challenge, as access to clean drinking water becomes increasingly difficult due to pollution and rising water treatment costs. Traditional adsorbents derived from fossil fuels are not sustainable and can contribute to secondary pollution. To address this issue, researchers have turned to biogenic sorbents made from modern biomass as a promising alternative. These biosorbents utilize biological waste materials that would otherwise contaminate water systems, promoting a circular economy and sustainable water treatment practices. This chapter explores the potential of biogenic sorbents for water decontamination, focusing on their use in the removal of heavy metals from contaminated waters. Further, this chapter delves into the preparation methods, adsorbent types, adsorption mechanisms, and regeneration techniques employed with these biosorbents.

DOI: 10.4018/979-8-3693-1618-4.ch007

1. INTRODUCTION TO BIOSORBENTS FOR HEAVY METAL REMOVAL

Heavy metal pollution is a substantial environmental distress, from the industries particularly mining, metallurgy, and energy harvesting units releases large quantities of toxic metals into the environment (Vardhan et al., 2019). Those heavy metals, including mercury (Hg), lead (Pb), chromium (Cr), and cadmium (Cd), cause a serious menace to living bodies' health and ecosystems(Vardhan et al., 2019). Conventional methods for eradicating heavy metal ions from aqueous solutions, such as chemical precipitation and electrochemical treatment, are habitually unproductive and produce large amounts of slurry. Therefore, different approaches are being required to address this persistent concern. One such methodology is biosorption, which utilizes biological materials to remove heavy metals from aqueous solution(Eccles, 1995; Sheng et al., 2004). Biosorption is the method in which living or dead biomass as well as cellular products like polysaccharides, sequesters metal ions through various physicochemical mechanisms(Veglio' & Beolchini, 1997; Volesky & Holan, 1995; Witek-Krowiak et al., 2011). It provides numerous advantages over traditional methods, comprising high efficiency, quick metal removal, and cost-effectiveness. Biosorption has multiplied its significant attention in recent years as a prospective solution for the treatment of high volume, low-concentration complex waste waters having heavy metals(Zouboulis et al., 2002).

The background mechanisms accountable for biosorption are not yet fully understood but are supposed to involve ion exchange, complexation, coordination, adsorption, electrostatic interaction, chelation, and micro precipitation. Biomass materials for example bacteria, fungi, yeast, and algae have been comprehensively experimented as potential biosorbents owing to their metal-binding capabilities. These biosorbents can meritoriously sequester heavy metal ions from dilute complex solutions, making them energetic candidates for the treatment of wastewater. Biosorption can take place through both metabolic and non-metabolic pathways(Reddy et al., 2012). Metabolic biosorption includes the active uptake of metal ions by living biomass. Non-metabolic biosorption, in contrast, take place through impulsive physicochemical processes and does not necessitate the expenditure of ATP. Both living and deceased biomass have been found to demonstration of biosorption properties which making them suitable for metal removal applications(Vijayaraghavan & Yun, 2008; Witek-Krowiak & Harikishore Kumar Reddy, 2013).

Figure 1. Biosorption method for heavy metal removal (adopted from Elgarahy et al., 2021 under creative common license)

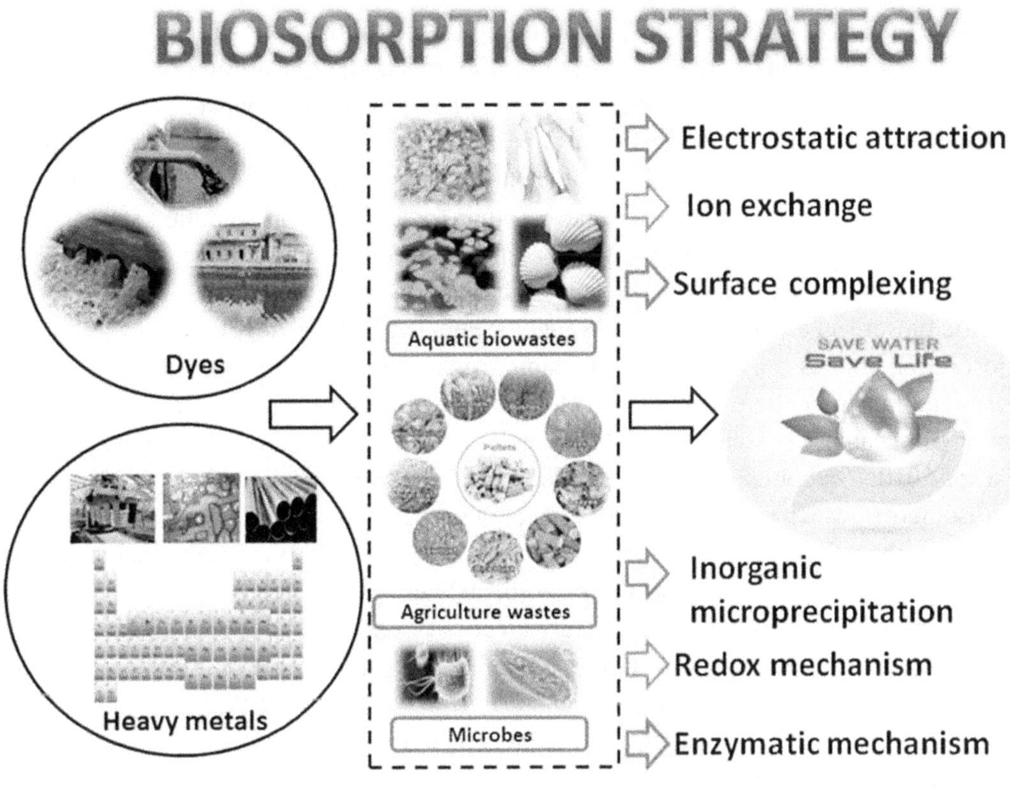

A several varieties of biomaterials has been studied for their biosorption capabilities. These biomaterials can be categorized into several categories, including bacteria, fungi, yeast, algae, industrial wastes, and agricultural wastes(Aksu, 2005; Tran et al., 2019; Vijayaraghavan & Yun, 2008). Bacteria, in particular, have been broadly experimented for their metal-binding capacities. Certain species of bacteria, such as Bacillus subtillis, have ascertained high metal sequestration abilities. Fungal biomass, including species like Rhizopus arrhizus, has also shown the capacity as a biosorbent for heavy metal removal(Behloul et al., 2017). Yeast, such as Saccharomyces cerevisiae, and algae, as well as brown algae like Sargasso seaweed, have likewise been studied for their metal-binding capabilities. Industrial and agricultural wastes, such as waste biomass from fermentation, food industries and corn cores, respectively, have been acknowledged as possible biosorbents due to their low cost and availability(Huang et al., 2020).

To boost the metal-binding capacity of biosorbents, several pre-treatment methods have been utilized. Physical treatments, such as heating, freezing, drying, and lyophilization, can adjust the cell wall of microbial biomass and improve metal uptake. Chemical treatments, comprising acid and alkali treatments, can also improve the metal-binding capabilities of biosorbents(Osman et al., 2023). Along with pre-treatment, the immobilization of biosorbents has been investigated for use in bioreactors. Immobilization imparts mechanical strength and resistance to degradation upon the biosorbents, allowing for their repeated use in metal removal processes. This approach is particularly advantageous for large-scale applications where the cost of biosorbent preparation must be taken into account(Osman et al., 2023).

The efficiency of biosorption processes can be influenced by various factors, including pH, temperature, contact time, initial metal concentration, biomass dosage, and agitation speed. The pH of the solution plays a crucial role in determining the metal-binding capacity of biosorbents. Different metals have different pH optima for biosorption, and adjusting the pH to the optimal range can significantly enhance the metal removal efficiency. Temperature also affects the biosorption process, with higher temperatures generally resulting in faster metal uptake. The contact time between the biosorbent and the metal solution is an important factor to consider, as longer contact times can lead to higher metal removal efficiencies(Hu et al., 2020). The initial metal concentration in the solution, as well as the dosage of biosorbent, should be optimized to achieve optimal metal removal. Agitation speed can impact the accessibility of metal ions to the biosorbent surface, and optimal agitation conditions should be determined to maximize metal removal efficiency. These optimization parameters can vary depending on the specific biosorbent and metal being targeted for removal(Fadel et al., 2017).

The biosorption process can be modeled and simulated using various mathematical models, including isotherm and kinetic models. Isotherm models describe the equilibrium relationship between the concentration of metal ions in solution and the amount of metal ions bound to the biosorbent(Michalak & Chojnacka, 2010; Zouboulis et al., 2002). Common isotherm models include the Langmuir and Freundlich isotherms, which can be used to estimate the maximum metal uptake capacity and the affinity of the biosorbent for the metal ions, respectively. Kinetic models, on the other hand, describe the rate at which metal ions are taken up by the biosorbent(Michalak & Chojnacka, 2010). Models such as the pseudo-first-order and pseudo-second-order kinetics can be used to determine the rate constants and predict the metal uptake behavior over time. Modeling and simulation of biosorption processes can provide valuable insights into the mechanisms and efficiency of metal removal, allowing for the optimization of process parameters. Advancements in biosorbent research are focused on the development of novel biosorbents with enhanced metal-binding capacities and selectivity. Researchers are exploring the modification of existing biosorbents through immobilization, encapsulation, or genetic engineering to improve their performance. The use of biopolymers, such as chitosan-based biosorbents, has shown promise in the removal of heavy metals. Furthermore, the application of biosorption extends beyond heavy metal removal. Biosorbents have also been utilized for the removal of other contaminants, including nutrients and organic pollutants, from wastewater. This versatility highlights the potential of biosorption as a comprehensive solution for wastewater treatment and environmental remediation(Aksu, 2001).

Biosorption offers a cost-effective and environmentally friendly approach to heavy metal removal from wastewater. The use of biological materials as biosorbents provides a sustainable and efficient method for metal sequestration. Ongoing research and advancements in biosorbent engineering and optimization will further enhance the effectiveness of biosorption processes. As the demand for effective wastewater treatment and metal recovery continues to grow, biosorption holds great potential as a viable solution. By harnessing the unique properties of living and deceased biomass, researchers and engineers can develop innovative and sustainable methods for the removal of heavy metals from industrial and municipal wastewater streams. With further research and development, biosorption can play a critical role in achieving regulatory standards and protecting the environment from the harmful effects of heavy metal pollution (Elgarahy et al., 2021).

Figure 2. Mechanisms involved in the biosorption process (Adopted from Elgarahy et al., 2021 under creative common license)

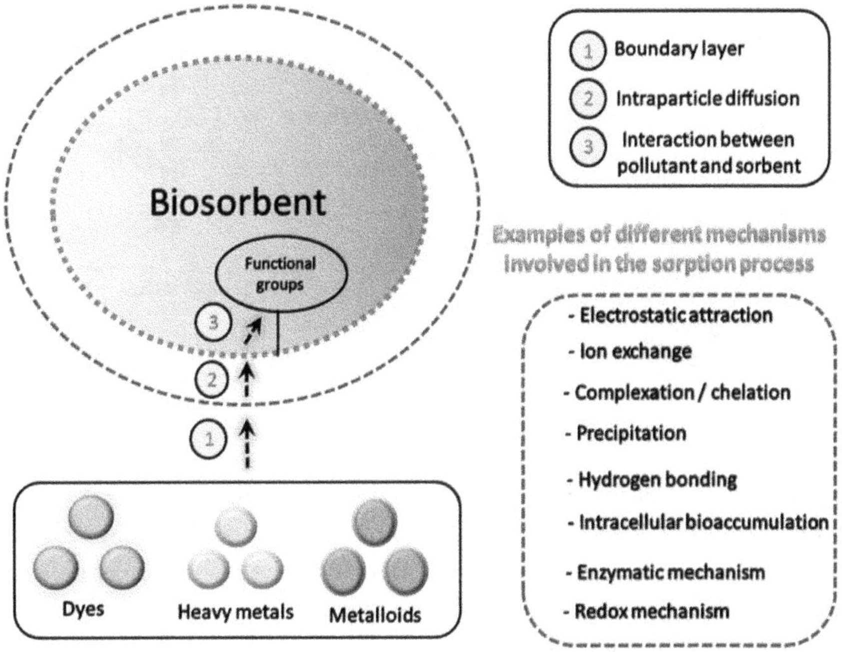

2. WOOD-BASED BIOSORBENTS

Wood-based biosorbents have attracted significant attention for their potential in removing heavy metals from water. These biosorbents are derived from various wood sources, such as sawdust and wood chips, and can be prepared through different methods, including co-precipitation and thermal decomposition(Osman et al., 2023). The porous structure of wood-based biosorbents provides a large surface area for adsorption, enhancing their metal removal capacity. Additionally, the presence of functional groups, such as hydroxyl and carboxyl groups, contributes to the sorptive interactions between the biosorbents and heavy metal ions(Adegoke et al., 2022). One notable example of a wood-based biosorbent is biochar, which is produced through the pyrolysis of biomass. Biochar exhibits excellent adsorption capabilities due to its high surface area and microporous structure. Studies have shown that biochar derived from wood waste can effectively remove heavy metals like copper (Cu) and nickel (Ni) from aqueous solutions. The adsorption mechanisms involve chemisorption and ion exchange, where the functional groups present on the biochar surface form strong bonds with the heavy metal ions, leading to their removal from the water(Yaashikaa et al., 2020).

Figure 3. Various processes for production of wood biochar (adopted from Z. Liu et al., 2022 under creative common licenses)

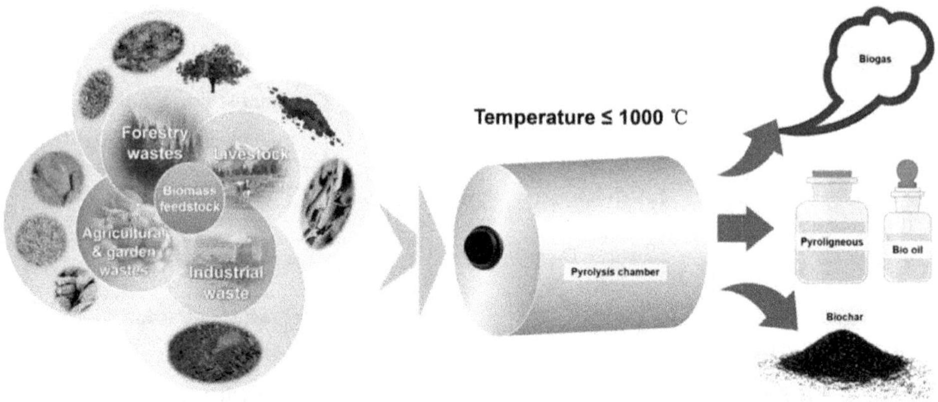

In recent years, there has been growing interest in non-conventional raw materials for the production of cellulose and other valuable materials. Wood-based biomass and agricultural waste, such as walnut shells and apricot kernels, have emerged as potential sources for the production of high-quality sorbents and ion-exchange materials. These materials offer advantages in terms of availability, cost-effectiveness, and sustainability. However, the delignification and chemical modification of non-wood cellulose from these sources are still relatively unexplored areas of research (Yelatontsev, 2023). The efficient processing of non-wood biomass into sorbents and ion-exchange materials requires the removal of lignin and the isolation of pure cellulose. Various chemical modification methods have been developed for this purpose, but they often involve complex disposal of waste solutions, limiting their industrial implementation. In contrast, the technology for producing sorbents from lignocellulosic waste is simpler and more cost-effective, as it eliminates the need for synthetic materials as a basis for the sorbent(Yelatontsev, 2023).

One promising approach is the utilization of walnut shells and fruit pits, which are abundant and inexpensive agricultural waste products. These materials possess sorption and ion-exchange properties due to the presence of surface groups such as carboxyl, hydroxyl, and phenolic groups. Chemical treatment of these raw materials can yield sorbents and ion-exchangers with characteristics comparable to industrial sorbents and activated carbon, at a much lower cost. This article aims to explore the development of biosorbents derived from non-wood biomass, with a focus on walnut shells and apricot kernels. It will discuss the pretreatment and delignification of the raw biomass, as well as the chemical modification processes involved in the synthesis of biosorbents. The physical and chemical characteristics of the resulting materials will be examined, along with their sorption capacities for various aqueous pollutants. The article will also highlight the potential applications and advantages of these biosorbents in environmental remediation and waste management(Albatrni et al., 2022).

2.1. Pretreatment of Raw Biomass

To improve the sorption kinetics and effectiveness of biosorbents derived from walnut shells and apricot kernels, the raw biomass undergoes pretreatment processes. This involves milling the biomass to a specific grain size and classifying it for further processing. The milled biomass is typically classified into different size fractions, with larger particles being recrushed and smaller particles sent for combustion or the production of activated carbon. The resulting grain size of 1.0-2.0 mm is optimal for biosorbent synthesis (Osman et al., 2023)(Shkliarenko et al., 2023).

2.2. Delignification of Biomass

Delignification is a crucial step in the production of biosorbents from walnut shells and apricot kernels. It involves the removal of lignin, which increases the specific surface area and sorption capacity of the biomass. Various delignification methods have been explored, including alkaline-acid-acid sulfate (AAS) delignification, which has shown promising results. AAS delignification can achieve a high product yield and a significant degree of delignification. The resulting cellulose intermediates have excellent physicochemical characteristics and can be further processed into biosorbents or ion-exchangers (Kildegaard et al., 2022)(Fadel et al., 2017).

2.3. Chemical Modification of Cellulose

Following delignification, the cellulose intermediates obtained from walnut shells and apricot kernels are subjected to chemical modification to enhance their sorption and ion-exchange properties. Phosphorylation is a common method used to introduce phosphate groups onto the cellulose structure, improving its selectivity for specific metal cations. The phosphorylation process involves the reaction of cellulose with phosphorus-containing reagents, resulting in the formation of cellulose phosphate. The synthesized cellulose phosphate biosorbents exhibit high adsorption capacities for various metal cations, including uranium, zirconium, and other heavy metals. The surface functional groups of the biosorbents, such as -COOH, -COH, and -OH, play a crucial role in their sorption and ion-exchange properties. The specific surface area and iodine number of the biosorbents are significantly increased through the phosphorylation process, resulting in improved sorption kinetics and capacities (Y. Liu & Shen, 2008; Sheng et al., 2004; Yelatontsev, 2023).

2.4. Physicochemical Characteristics of Biosorbents

The physicochemical characteristics of biosorbents derived from walnut shells and apricot kernels are important indicators of their performance and suitability for various applications. The specific surface area, pore volume, and bulk density of the biosorbents are key parameters that determine their sorption capacities and mechanical stability. Delignification significantly increases the specific surface area of the biomass, resulting in a higher adsorption capacity. The iodine number, which measures the adsorption capacity for iodine, is also enhanced through delignification. The biosorbents derived from walnut shells and apricot kernels exhibit excellent physicochemical properties, making them suitable for a wide range of environmental remediation applications (Ebelegi et al., 2023).

2.5. Sorption Capacities of Biosorbents

The biosorbents derived from walnut shells and apricot kernels demonstrate high sorption capacities for various aqueous pollutants. Their surface functional groups, combined with the increased specific surface area, allow for effective sorption of metal cations and other contaminants. The biosorbents show selectivity towards specific metal cations, such as uranium, zirconium, and heavy metals. The sorption capacities of the biosorbents can be further enhanced through optimization of the synthesis parameters, such as the concentration of phosphorus-containing reagents and the reaction conditions. The adsorption isotherms and kinetics of the biosorbents can be analyzed to determine their sorption mechanisms and optimize their performance for specific applications (De Gisi et al., 2016).

2.6. Potential Applications of Biosorbents

The biosorbents derived from walnut shells and apricot kernels have significant potential for various environmental remediation applications. Their high sorption capacities make them suitable for the removal of heavy metals, radioactive elements, and other pollutants from aqueous solutions. They can be used in wastewater treatment, industrial effluent management, and groundwater remediation. The low cost and availability of walnut shells and apricot kernels make these biosorbents attractive for large-scale applications. The waste products from the food industry, such as walnut shells and fruit pits, can be utilized effectively, reducing waste and providing a sustainable source of raw materials for the production of biosorbents (Karić et al., 2022).

2.7. Advantages of Biosorbents

The biosorbents derived from walnut shells and apricot kernels offer several advantages over conventional sorbents and ion-exchange materials. Their low cost, high sorption capacities, and selective sorption properties make them competitive alternatives to expensive industrial sorbents and activated carbon. The utilization of agricultural waste products as raw materials for biosorbent production also contributes to waste management and environmental sustainability (Kanamarlapudi et al., 2018). The waste-free technology for chemical processing of walnut shells and fruit pits provides a practical solution for the utilization of abundant and inexpensive raw materials. The development of biosorbents derived from non-wood biomass, specifically walnut shells and apricot kernels, offers promising opportunities for environmental remediation and waste management. The pretreatment and delignification of the raw biomass, followed by chemical modification, result in biosorbents with high sorption capacities for various aqueous pollutants. The physicochemical characteristics of the biosorbents, such as specific surface area, pore volume, and bulk density, play a crucial role in their performance (Aksu, 2005). The biosorbents demonstrate excellent sorption capacities and selectivity for metal cations, making them suitable for wastewater treatment, industrial effluent management, and groundwater remediation. The utilization of walnut shells and fruit pits as raw materials for biosorbent production provides a cost-effective and sustainable solution, reducing waste and promoting environmental sustainability. The waste-free technology for chemical processing of these raw materials offers a practical and efficient approach to the production of biosorbents with significant potential for various applications in environmental remediation (Kanamarlapudi et al., 2018).

3. BACTERIAL BIOSORBENTS

Bacterial biosorbents have shown great potential in the removal of heavy metals from contaminated waters. These biosorbents are typically derived from bacterial strains that possess metal-binding capabilities. The biosorption properties of bacterial biosorbents can be enhanced through various modification techniques, such as chemical reduction and micro-emulsion. The modification process can introduce additional functional groups or increase the surface area of the biosorbents, improving their metal removal efficiency (Vijayaraghavan & Yun, 2008). One example of a bacterial biosorbent is activated Pisum sativum peels embellished with triethylenetetramine (TETA). Activated TETA-Pisum sativum peels have demonstrated pH-dependent adsorption of Cr^{6+} ions, with maximum removal observed at pH 1.6. The biosorption capacity of the biosorbent was found to be 312.50 mg/g. The adsorption mechanism involves the interaction between the functional groups present on the biosorbent surface and the Cr6+ ions, leading to their sequestration from the water (Vijayaraghavan & Yun, 2008).

3.1. Algal Biosorbents

Algal biosorbents have shown promise in the removal of heavy metals from contaminated waters. These biosorbents are derived from various species of algae, which possess inherent metal-binding capabilities. Algal biosorbents can be prepared through different methods, including thermal decomposition and microwave irradiation. The composition of the algal biomass, including the presence of polysaccharides and proteins, contributes to its metal adsorption capacity. One notable example of an algal biosorbent is green algae, which has been shown to effectively remove metals like cadmium (Cd) from water. Green algae exhibit high selectivity for Cd, with removal efficiencies reaching nearly 100%. Other algal species, such as Chondracanthus chamissoi, have shown lower removal efficiencies for metals like copper (Cu). However, a combination of different algal species can enable the removal of a broad range of heavy metals, making algal biosorbents a versatile option for water decontamination (Wang & Chen, 2009; Mohammadi et al., 2023).

3.2. Preparation Methods for Biosorbents

3.2.1. Coprecipitation

Coprecipitation is a commonly used method for preparing biosorbents. In this process, metal ions are precipitated along with the biosorbent material, resulting in the formation of metal-loaded biosorbents. Coprecipitation can be achieved through various chemical reactions, such as the addition of a precipitating agent or the formation of complex compounds. This method allows for the incorporation of metal ions into the biosorbent structure, enhancing its metal removal capabilities.

3.2.2. Thermal Decomposition

Thermal decomposition involves subjecting the biosorbent material to high temperatures in an inert atmosphere. This process leads to the degradation of organic components and the formation of a porous structure. The resulting biosorbent exhibits increased surface area and pore volume, providing more sites for metal adsorption. Thermal decomposition can be conducted using different heating methods, such as microwave irradiation or conventional heating.

3.2.2. Chemical Reduction

Chemical reduction is a method used to modify biosorbents and enhance their metal adsorption properties. This process involves the addition of reducing agents, which react with metal ions to form elemental metals or metal compounds. The reduced forms of metals have higher affinity for the biosorbent material, resulting in improved metal removal efficiency. Chemical reduction can be performed through various chemical reactions, such as the use of sodium borohydride or sodium citrate as reducing agents.

3.2.3. Micro-Emulsion

Micro-emulsion is a preparation method that involves the formation of a stable dispersion of oil droplets in water. In the context of biosorbents, micro-emulsion can be used to incorporate metal ions into the biosorbent structure. The formation of micro-emulsion allows for the controlled deposition of metal ions onto the biosorbent material, enhancing its metal adsorption capacity. Micro-emulsion can be achieved through the use of surfactants and co-surfactants, which stabilize the oil droplets in the water phase.

3.2.4. Arc Discharge

Arc discharge is a technique used to generate plasma, which can be utilized to modify biosorbents for metal removal. In this process, an electric discharge is created between two electrodes submerged in a liquid medium containing the biosorbent material. The discharge generates a high-energy plasma, which leads to the formation of nanoparticles or surface modifications on the biosorbent material. These modifications can enhance the biosorbent's metal adsorption capabilities, making it more effective in removing heavy metals from water.

3.3. Adsorbent Types for Heavy Metal Removal

Activated carbon is a widely used adsorbent for the removal of heavy metals from water. It is derived from various carbonaceous materials, such as wood, coconut shells, or coal, through a process called activation. Activated carbon exhibits a highly porous structure with a large surface area, providing ample adsorption sites for heavy metal ions. The adsorption mechanism involves the physical attraction between the metal ions and the activated carbon surface. Biochar is a carbon-rich material produced through the pyrolysis of biomass. It possesses a highly porous structure and a large surface area, making it an effective adsorbent for heavy metals. Biochar can be prepared from different biomass sources, such as agricultural waste or wood chips. The adsorption mechanisms of biochar involve both physical and chemical interactions, including ion exchange, surface complexation, and precipitation (Yaashikaa et al., 2020).

Lignocellulosic waste, such as agricultural crop residues and agro-industrial waste, can be utilized as adsorbents for heavy metal removal. These materials are abundant and readily available, making them cost-effective options for water decontamination. Lignocellulosic waste can be prepared through various methods, including chemical treatments and physical modifications. The adsorption mechanisms of lignocellulosic waste involve the interaction between the functional groups present in the biomass and the heavy metal ions. Clays, such as bentonite and kaolinite, possess unique adsorption properties due to their layered structure and high cation exchange capacity. The interlayer spaces of clays can accommodate heavy metal ions through ion exchange processes. Additionally, the presence of functional groups on the clay surface contributes to the adsorption of metal ions. Clays can be modified through physical or chemical treatments to enhance their adsorption capabilities (Karić et al., 2022).

Zeolites are crystalline materials with a porous structure and high surface area. They have been extensively studied for their adsorption properties in various applications, including heavy metal removal. The framework structure of zeolites allows for the selective adsorption of certain metal ions based on their size and charge. Zeolites can be modified through ion exchange or surface modification techniques to enhance their metal adsorption capacity. Peat and humic soils are organic-rich materials that possess high sorption capacities for heavy metals. They contain humic substances, which are complex organic compounds with functional groups capable of binding metal ions. The adsorption mechanisms of peat and humic soils involve both surface complexation and ion exchange processes. These materials can be utilized as adsorbents for heavy metal removal, particularly in wetland treatment systems (Hong et al., 2019; Li et al., 2017).

3.4. Adsorption Isotherms and Kinetics

Adsorption isotherms provide insights into the equilibrium relationship between the concentration of metal ions in solution and the amount of metal ions adsorbed onto the biosorbent material. The most commonly used adsorption isotherms include the Langmuir and Freundlich isotherms. The Langmuir isotherm assumes monolayer adsorption on a homogeneous surface, while the Freundlich isotherm accounts for multilayer adsorption on a heterogeneous surface. Adsorption kinetics refers to the rate at which metal ions are adsorbed onto the biosorbent material. The kinetics of adsorption can be described using different models, such as the pseudo-first-order and pseudo-second-order models. The pseudo-first-order model assumes that the rate of adsorption is proportional to the difference between the initial concentration and the concentration at any given time. The pseudo-second-order model considers the adsorption rate to be proportional to the square of the concentration difference.

3.5. Regeneration of Spent Adsorbents

Thermal regeneration involves subjecting the spent biosorbent material to high temperatures to remove the adsorbed metal ions. The thermal regeneration process can be conducted through various heating methods, such as heating in an oven or using a furnace. The high temperatures cause desorption of the metal ions from the biosorbent surface, allowing for the reuse of the regenerated biosorbent. Chemical regeneration involves treating the spent biosorbent material with chemicals to remove the adsorbed metal ions. Different chemical regenerants can be used, such as acid solutions or complexing agents. The regenerant reacts with the metal ions, leading to their desorption from the biosorbent surface. Chemical regeneration can be an effective method for reusing the biosorbent material and recovering the metal ions (Zhang et al., 2023).

Supercritical fluid desorption is a novel technique for regenerating spent adsorbents. In this process, a supercritical fluid, such as carbon dioxide, is used to desorb the adsorbed metal ions from the biosorbent material. Supercritical fluids possess unique properties, such as high diffusivity and low viscosity, which allow for efficient desorption of the metal ions. This technique offers advantages in terms of energy efficiency and environmental sustainability. Effective management and disposal of exhausted adsorbents are essential to minimize environmental impacts. Depending on the characteristics of the spent biosorbent material and the adsorbed metal ions, different disposal methods can be employed. Options for managing exhausted adsorbents include landfill disposal, incineration, or recycling. The choice of disposal method should consider factors such as the potential release of contaminants and the overall environmental impact (Kravanja et al., 2022).

Biogenic sorbents offer a sustainable and cost-effective approach to the removal of heavy metals from contaminated waters. Wood-based biosorbents, bacterial biosorbents, and algal biosorbents have demonstrated promising metal removal capabilities. Various preparation methods can enhance the adsorption properties of biosorbents, including coprecipitation, thermal decomposition, and chemical reduction. Activated carbon, biochar, lignocellulosic waste, clays, zeolites, peat, and humic soils are among the types of adsorbents used for heavy metal removal. Adsorption isotherms and kinetics provide insights into the equilibrium and rate of metal adsorption. The regeneration of spent adsorbents can be achieved through thermal regeneration, chemical regeneration, or supercritical fluid desorption. Proper management and disposal of exhausted adsorbents are crucial for minimizing environmental impacts. By harnessing the potential of biogenic sorbents, we can contribute to a sustainable and clean water future (Osman et al., 2023).

Two commercially available heteropolysaccharides, xanthan and gellan, have been successfully used in remediation techniques (Balíková et al., 2022). These biodegradable and non-toxic fermentation products have shown promising sorptive performance in both native and modified forms. These green adsorbents are emerging as cost-effective and efficient alternatives for heavy metal-contaminated waters, making them a promising green-based material for water decontamination.

A biogenic FeS–kaolin composite (KL–FeS) was synthesized (Xu et al., 2022) using sulfate-reducing bacteria (SRB) mediation. The composite showed better removal efficiency for heavy metals (Cd(II), Pb(II), Cu(II), Zn(II), As(III) and Sb(III)) than other prepared composites. The adsorption process followed pseudo-second-order and Redlich-Peterson models, indicating a hybrid chemical reaction. The maximum adsorption capacities for Cd(II), Pb(II), Cu(II), Zn(II), As(III) and Sb(III) were 71.71, 133.54, 51.90, 54.41, 38.71, and 96.38 mg/g, respectively. The composite also displayed higher affinity for Pb(II), Sb(III) and Cu(II) in the multi-component system. A new hybrid adsorbent (HFO-BC) which was decorated with hydrous ferric oxide (HFO) within the pore of biochar was prepared (Li et al., 2020). This adsorbent shows excellent performance for heavy metal removal, with maximal sorption capacities of 29.9 mg/g for Cd(II) and 34.1 mg/g for Cu(II). It also shows satisfactory anti-interference ability, generating effluents equivalent to 774- and 1854-fold of its own weight. The exhausted HFO-BC can be effectively regenerated using $HCl–CaCl_2$ binary solution.

A hybrid (STpe-AgNP) (Akpomie & Conradie, 2020) was prepared using silver nanoparticles impregnated onto Solanum tuberosum peel (STpe) for ultrasonic assisted adsorption of bromophenol blue (BB) dye. The adsorbents were characterized using various techniques, and the Freundlich model was best fitted. The study found that competing metal ions interfered with BB adsorption. STpe and STpe-AgNP showed BB desorption of 91.3% and 88.5%, respectively. Ultra-sonication enhanced BB adsorption, but impregnation only slightly improved it. Iron plaque of Typha-latifolia and bacterial precipitated $Fe[OH]^{3-}$rich soil was used for the removal of Chromium [Cr^{+6}] and Copper [Cu^{2+}] from waste water (Brij Kishor, 2014). The adsorption capacity of these ions increased with decreasing media size, with maximum uptake values of 0.0229 mg Cr^{+6}/g and 0.048 mg Cu^{+2}/g at 298.15 K. The optimal media size for Chromium [Cr^{+6}] was 150 μ, and the maximum sorption efficiency was 86%.

The potential of ultrafine mesoporous magnetite (UFMNPs) for simultaneous removal of heavy metal ions from river water was investigated (Fato et al., 2019). The removal efficiencies of individual ions were 98, 87, 90, and 78%, respectively. The mixed ions showed efficiencies of 86, 80, 84, and 54%. This study highlights the potential of UFMNPs for wastewater treatment, particularly in removing multiple heavy metal ions from aqueous media. The adsorption and desorption of heavy metals by micro and nano-sized biogenic $CaCO_3$ induced by Bacillus subtilis and the pH effect on adsorption was invested (R. Liu et al., 2018). Results showed that biogenic $CaCO_3$ maintained high adsorption capability for heavy metals within wide pH ranges. The study also revealed that biogenic $CaCO_3$ and heavy metal ions were governed by a physical adsorption process, attributed to its large total specific surface area.

4. BACTERIAL BIOSORBENTS: A PROMISING SOLUTION FOR HEAVY METAL REMOVAL

Heavy metal contamination in water bodies is a significant threat to the environment and human health. Industries such as metallurgy, mining, electroplating, and agriculture contribute to the continuous discharge of untreated effluents containing toxic heavy metals. Conventional methods for heavy metal removal are expensive, inefficient, and generate large amounts of toxic waste. In recent years, microbial biosorption and bioaccumulation have emerged as promising alternatives for heavy metal removal. Bacteria, cyanobacteria, algae, and fungi have shown the ability to bind and accumulate heavy metals, offering a sustainable and environmentally friendly approach to wastewater treatment (Singh et al., 2022). Biosorption is a complex process involving various mechanisms such as complexation, chelation, coordination, ion exchange, precipitation, and reduction. Microorganisms possess functional groups on their cell surfaces, such as carboxyl, amino, and sulfhydryl groups, which can interact with heavy metal ions. These interactions result in the binding and immobilization of heavy metals, preventing their release into the environment. The specific mechanisms of biosorption vary depending on the microbial species and the characteristics of the heavy metal ions (Singh et al., 2022).

Complexation and chelation involve the formation of stable complexes between heavy metal ions and functional groups on the microbial cell surface. Functional groups such as carboxyl and amino groups can act as ligands, binding to heavy metal ions through coordination bonds. This process is driven by the affinity between the functional groups and the heavy metal ions, leading to the removal of heavy metals from the aqueous phase. Ion exchange occurs when heavy metal ions in the aqueous phase are exchanged with other ions present on the microbial cell surface. The exchange is driven by the difference in affinity between the heavy metal ions and the ions on the cell surface. This mechanism allows for the efficient removal of heavy metals from wastewater by microbial biosorbents (Eivazihollagh et al., 2019).

Precipitation and reduction involve the conversion of heavy metal ions into insoluble forms or lower oxidation states. Microorganisms can produce extracellular substances, such as polysaccharides, that facilitate the precipitation of heavy metal ions as metal sulfides or hydroxides. Additionally, some microorganisms possess enzymes capable of reducing heavy metal ions to less toxic forms, reducing their environmental impact. Several process parameters significantly influence the biosorption efficiency of microbial biosorbents. The pH of the solution, temperature, initial concentration of metal ions, biosorbent dose, and speed of agitation all play crucial roles in determining the success of heavy metal removal. Optimizing these parameters can enhance the biosorption capacity and selectivity of microbial biosorbents. The pH of the solution affects the charge of both the heavy metal ions and the functional groups on the microbial cell surface. Different heavy metal ions exhibit varying affinities for functional groups at different pH values. Adjusting the pH to an optimal range can enhance the biosorption efficiency of microbial biosorbents.

Temperature influences the metabolic activity and growth of microorganisms, which directly affects their biosorption capacity. Higher temperatures generally increase biosorption rates; however, extreme temperatures can denature proteins and reduce biosorption efficiency. Optimizing the temperature conditions is crucial for maximizing the performance of microbial biosorbents. The initial concentration of metal ions in the wastewater directly impacts the biosorption capacity of microbial biosorbents. Higher metal ion concentrations result in greater availability of target ions for binding to functional groups on the microbial cell surface. However, excessively high metal ion concentrations can saturate the biosorbent and reduce the overall removal efficiency (Timková et al., 2018).

The amount of biosorbent added to the wastewater, known as the biosorbent dose, significantly affects the biosorption capacity of microbial biosorbents. Increasing the biosorbent dose can increase the number of available binding sites for heavy metal ions, leading to higher removal efficiencies. However, an excessive biosorbent dose may result in aggregation and reduced accessibility of binding sites. The speed of agitation or mixing within the wastewater plays a crucial role in enhancing the contact between the biosorbent and the metal ions. Stirring or shaking the solution promotes mass transfer and facilitates the interaction between heavy metal ions and microbial biosorbents. Optimal agitation speeds should be determined to maximize the biosorption efficiency. Various physical and chemical treatments can be employed to modify the biomass of microbial biosorbents before their use in heavy metal removal. These modifications aim to enhance the biosorption capacity, selectivity, and stability of the biosorbents. Common modification strategies include heat treatment, chemical cross-linking, immobilization, and surface functionalization (Ahmadi et al., 2022).

Heat treatment involves subjecting the microbial biomass to elevated temperatures, which can induce structural changes and improve the biosorption properties. Heat treatment can enhance the stability and mechanical strength of the biosorbents, making them more suitable for practical applications. Chemical cross-linking involves the use of cross-linking agents to create covalent bonds between biomolecules, reinforcing the structure of the biosorbents. Cross-linking can improve the resistance of microbial biosorbents to harsh operating conditions and increase their capacity for heavy metal removal. Immobilization techniques aim to fix the microbial biomass in a solid support matrix, such as alginate beads or activated carbon. Immobilized biosorbents offer advantages such as improved mechanical stability, easy separation from the solution, and the potential for repeated use. Surface functionalization involves modifying the surface of microbial biosorbents with specific functional groups or nanoparticles to enhance their heavy metal binding capacity. Functionalized biosorbents can exhibit improved selectivity, higher binding affinity, and enhanced stability, making them more effective for heavy metal removal (Adegoke et al., 2022).

The economic feasibility of using microbial biosorbents for heavy metal removal is an essential consideration. The regeneration and reuse of biosorbents after heavy metal removal can significantly reduce the overall cost of the treatment process. Furthermore, the potential for metal recovery from spent biosorbents offers opportunities for resource recycling and economic benefits. Microbial biosorbents have found applications in various industries and environmental remediation projects. They can be employed for the treatment of industrial effluents, mine drainage, contaminated soils, and groundwater. The versatility and sustainability of microbial biosorbents make them a promising solution for heavy metal removal and environmental protection (Wang & Chen, 2009).

5. CONCLUSION

Bacterial biosorbents offer a sustainable and environmentally friendly approach to heavy metal removal from wastewater. The unique mechanisms of biosorption, including complexation, chelation, ion exchange, precipitation, and reduction, enable microorganisms to bind and immobilize heavy metal ions effectively. Optimizing process parameters and modifying the biomass of microbial biosorbents can further enhance their performance. The economic feasibility and diverse applications of microbial biosorbents make them a promising solution for addressing heavy metal contamination and mitigating the environmental impacts of industrial activities. The development and utilization of bacterial biosorbents provide a promising avenue for sustainable heavy metal removal and wastewater treatment. With ongoing research and advancements in materials science and biotechnology, microbial biosorbents have the potential to revolutionize the field of environmental pollution. healthier planet.

REFERENCES

Adegoke, K. A., Adesina, O. O., Okon-Akan, O. A., Adegoke, O. R., Olabintan, A. B., Ajala, O. A., Olagoke, H., Maxakato, N. W., & Bello, O. S. (2022). Sawdust-biomass based materials for sequestration of organic and inorganic pollutants and potential for engineering applications. *Current Research in Green and Sustainable Chemistry*, *5*, 100274. doi:10.1016/j.crgsc.2022.100274

Ahmadi, H., Hafiz, S. S., Sharifi, H., Rene, N. N., Habibi, S. S., & Hussain, S. (2022). Low cost biosorbent (Melon Peel) for effective removal of Cu (II), Cd (II), and Pb (II) ions from aqueous solution. *Case Studies in Chemical and Environmental Engineering*, *6*, 100242. doi:10.1016/j.cscee.2022.100242

Akpomie, K. G., & Conradie, J. (2020). Biogenic and chemically synthesized Solanum tuberosum peel–silver nanoparticle hybrid for the ultrasonic aided adsorption of bromophenol blue dye. *Scientific Reports, 10*(1), 1–18. doi:10.1038/s41598-020-74254-y

Aksu, Z. (2001). Equilibrium and kinetic modelling of cadmium(II) biosorption by C. vulgaris in a batch system: Effect of temperature. *Separation and Purification Technology*, *21*(3), 285–294. doi:10.1016/S1383-5866(00)00212-4

Aksu, Z. (2005). Application of biosorption for the removal of organic pollutants: A review. *Process Biochemistry (Barking, London, England)*, *40*(3–4), 997–1026. doi:10.1016/j.procbio.2004.04.008

Albatrni, H., Qiblawey, H., & Al-Marri, M. J. (2022). Walnut shell based adsorbents: A review study on preparation, mechanism, and application. *Journal of Water Process Engineering*, *45*, 102527. doi:10.1016/j.jwpe.2021.102527

Balíková, K., Farkas, B., Matúš, P., & Urík, M. (2022). Prospects of Biogenic Xanthan and Gellan in Removal of Heavy Metals from Contaminated Waters. *Polymers*, *14*(23), 5326. Advance online publication. doi:10.3390/polym14235326

Behloul, M., Lounici, H., Abdi, N., Drouiche, N., & Mameri, N. (2017). Adsorption study of metribuzin pesticide on fungus Pleurotus mutilus. *International Biodeterioration & Biodegradation*, *119*, 687–695. doi:10.1016/j.ibiod.2016.07.005

De Gisi, S., Lofrano, G., Grassi, M., & Notarnicola, M. (2016). Characteristics and adsorption capacities of low-cost sorbents for wastewater treatment: A review. *Sustainable Materials and Technologies*, *9*, 10–40. doi:10.1016/j.susmat.2016.06.002

Ebelegi, A. N., Elijah, N. T., Godwin, J., Ebelegi, A. N., Elijah, N. T., & Godwin, J. (2023). Evaluation of Physicochemical Parameters of Biosorbents Produced from Groundnut Hull Using Microwave Assisted Irradiation Method. *Open Journal of Physical Chemistry*, *13*(1), 1–12. doi:10.4236/ojpc.2023.131001

Eccles, H. (1995). Removal of heavy metals from effluent streams - Why select a biological process? *International Biodeterioration & Biodegradation*, *35*(1–3), 5–16. doi:10.1016/0964-8305(95)00044-6

Eivazihollagh, A., Svanedal, I., Edlund, H., & Norgren, M. (2019). On chelating surfactants: Molecular perspectives and application prospects. *Journal of Molecular Liquids*, *278*, 688–705. doi:10.1016/j.molliq.2019.01.076

Elgarahy, A. M., Elwakeel, K. Z., Mohammad, S. H., & Elshoubaky, G. A. (2021). A critical review of biosorption of dyes, heavy metals and metalloids from wastewater as an efficient and green process. *Cleaner Engineering and Technology*, *4*, 100209. doi:10.1016/j.clet.2021.100209

Fadel, M., Hassanein, N. M., Elshafei, M. M., Mostafa, A. H., Ahmed, M. A., & Khater, H. M. (2017). Biosorption of manganese from groundwater by biomass of Saccharomyces cerevisiae. *HBRC Journal*, *13*(1), 106–113. doi:10.1016/j.hbrcj.2014.12.006

Fato, F. P., Li, D. W., Zhao, L. J., Qiu, K., & Long, Y. T. (2019). Simultaneous Removal of Multiple Heavy Metal Ions from River Water Using Ultrafine Mesoporous Magnetite Nanoparticles. *ACS Omega*, *4*(4), 7543–7549. doi:10.1021/acsomega.9b00731

Hong, M., Yu, L., Wang, Y., Zhang, J., Chen, Z., Dong, L., Zan, Q., & Li, R. (2019). Heavy metal adsorption with zeolites: The role of hierarchical pore architecture. *Chemical Engineering Journal*, *359*, 363–372. doi:10.1016/j.cej.2018.11.087

Hu, X., Cao, J., Yang, H., Li, D., Qiao, Y., Zhao, J., Zhang, Z., & Huang, L. (2020). Pb2+ biosorption from aqueous solutions by live and dead biosorbents of the hydrocarbon-degrading strain Rhodococcus sp. HX-2. *PLoS One*, *15*(1), e0226557. Advance online publication. doi:10.1371/journal.pone.0226557

Huang, D., Li, B., Ou, J., Xue, W., Li, J., Li, Z., Li, T., Chen, S., Deng, R., & Guo, X. (2020). Megamerger of biosorbents and catalytic technologies for the removal of heavy metals from wastewater: Preparation, final disposal, mechanism and influencing factors. *Journal of Environmental Management*, *261*, 109879. doi:10.1016/j.jenvman.2019.109879

Kanamarlapudi, S. L. R. K., Kumar Chintalpudi, V., Muddada, S., Kanamarlapudi, S. L. R. K., Kumar Chintalpudi, V., & Muddada, S. (2018). *Application of Biosorption for Removal of Heavy Metals from Wastewater*. Biosorption. doi:10.5772/intechopen.77315

Karić, N., Maia, A. S., Teodorović, A., Atanasova, N., Langergraber, G., Crini, G., Ribeiro, A. R. L., & Đolić, M. (2022). Bio-waste valorisation: Agricultural wastes as biosorbents for removal of (in)organic pollutants in wastewater treatment. *Chemical Engineering Journal Advances*, *9*, 100239. doi:10.1016/j.ceja.2021.100239

Kildegaard, G., Balbi, M. D. P., Salierno, G., Cassanello, M., De Blasio, C., & Galvagno, M. (2022). A Cleaner Delignification of Urban Leaf Waste Biomass for Bioethanol Production, Optimised by Experimental Design. *Processes, 10*(5), 943. doi:10.3390/pr10050943

Kishor, B. (2014). Removal of Heavy Metals from Wastewater using Iron Plaque of Typha-latifolia & Bacterial Precipitated Fe (OH) 3 Rich Soil. *International Journal of Engineering Research & Technology (Ahmedabad), 3*(5), 26–31.

Kravanja, K. A., Finšgar, M., Knez, Ž., & Knez Marevci, M. (2022). Supercritical Fluid Technologies for the Incorporation of Synthetic and Natural Active Compounds into Materials for Drug Formulation and Delivery. *Pharmaceutics, 14*(8), 1670. Advance online publication. doi:10.3390/pharmaceutics14081670

Li, Y., Gao, L., Lu, Z., Wang, Y., Wang, Y., & Wan, S. (2020). Enhanced Removal of Heavy Metals from Water by Hydrous Ferric Oxide-Modified Biochar. *ACS Omega, 5*(44), 28702–28711. doi:10.1021/acsomega.0c03893

Li, Y., Li, L., & Yu, J. (2017). Applications of Zeolites in Sustainable Chemistry. *Chem, 3*(6), 928–949. doi:10.1016/j.chempr.2017.10.009

Liu, R., Guan, Y., Chen, L., & Lian, B. (2018). Adsorption and desorption characteristics of Cd2+ and Pb2+ by micro and nano-sized biogenic CaCO3. *Frontiers in Microbiology, 9*(JAN), 294049. doi:10.3389/fmicb.2018.00041

Liu, Y., & Shen, L. (2008). A general rate law equation for biosorption. *Biochemical Engineering Journal, 38*(3), 390–394. doi:10.1016/j.bej.2007.08.003

Liu, Z., Xu, Z., Xu, L., Buyong, F., Chay, T. C., Li, Z., Cai, Y., Hu, B., Zhu, Y., & Wang, X. (2022). Modified biochar: Synthesis and mechanism for removal of environmental heavy metals. *Carbon Research, 1*(1), 1–21. doi:10.1007/s44246-022-00007-3

Michalak, I., & Chojnacka, K. (2010). Interactions of metal cations with anionic groups on the cell wall of the macroalga Vaucheria sp. *Engineering in Life Sciences, 10*(3), 209–217. doi:10.1002/elsc.200900039

Mohammadi, A., Mahmoudnia, F., Mohammadi, A., & Mahmoudnia, F. (2023). *Biological Treatment of Heavy Metals with Algae*. Heavy Metals - Recent Advances., doi:10.5772/intechopen.110301

Osman, A. I., El-Monaem, E. M. A., Elgarahy, A. M., Aniagor, C. O., Hosny, M., Farghali, M., Rashad, E., Ejimofor, M. I., López-Maldonado, E. A., Ihara, I., Yap, P. S., Rooney, D. W., & Eltaweil, A. S. (2023). Methods to prepare biosorbents and magnetic sorbents for water treatment: a review. *Environmental Chemistry Letters, 21*(4), 2337–2398. doi:10.1007/s10311-023-01603-4

Reddy, D. H. K., Lee, S. M., & Seshaiah, K. (2012). Biosorption of toxic heavy metal ions from water environment using honeycomb biomass-an industrial waste material. *Water, Air, and Soil Pollution, 223*(9), 5967–5982. doi:10.1007/s11270-012-1332-0

Sheng, P. X., Ting, Y. P., Chen, J. P., & Hong, L. (2004). Sorption of lead, copper, cadmium, zinc, and nickel by marine algal biomass: Characterization of biosorptive capacity and investigation of mechanisms. *Journal of Colloid and Interface Science, 275*(1), 131–141. doi:10.1016/j.jcis.2004.01.036

Shkliarenko, Y., Halysh, V., & Nesterenko, A. (2023). Adsorptive Performance of Walnut Shells Modified with Urea and Surfactant for Cationic Dye Removal. *Water (Basel), 15*(8), 1536. doi:10.3390/w15081536

Singh, A., Sharma, A., Verma, R. K., Chopade, R. L., Pandit, P. P., Nagar, V., Aseri, V., Choudhary, S. K., Awasthi, G., Awasthi, K. K., Sankhla, M. S., Singh, A., Sharma, A., Verma, R. K., Chopade, R. L., Pandit, P. P., Nagar, V., Aseri, V., Choudhary, S. K., … Sankhla, M. S. (2022). Heavy Metal Contamination of Water and Their Toxic Effect on Living Organisms. *The Toxicity of Environmental Pollutants*. doi:10.5772/intechopen.105075

Timková, I., Sedláková-Kaduková, J., & Pristaš, P. (2018). Biosorption and Bioaccumulation Abilities of Actinomycetes/Streptomycetes Isolated from Metal Contaminated Sites. *Separations, 5*(4), 54. doi:10.3390/separations5040054

Tran, H. N., Nguyen, H. C., Woo, S. H., Nguyen, T. V., Vigneswaran, S., Hosseini-Bandegharaei, A., Rinklebe, J., Kumar Sarmah, A., Ivanets, A., Dotto, G. L., Bui, T. T., Juang, R. S., & Chao, H. P. (2019). Removal of various contaminants from water by renewable lignocellulose-derived biosorbents: A comprehensive and critical review. *Critical Reviews in Environmental Science and Technology*, *49*(23), 2155–2219. doi:10.1080/10643389.2019.1607442

Vardhan, K. H., Kumar, P. S., & Panda, R. C. (2019). A review on heavy metal pollution, toxicity and remedial measures: Current trends and future perspectives. *Journal of Molecular Liquids*, *290*, 111197. doi:10.1016/j.molliq.2019.111197

Veglio', F., & Beolchini, F. (1997). Removal of metals by biosorption: A review. *Hydrometallurgy*, *44*(3), 301–316. doi:10.1016/S0304-386X(96)00059-X

Vijayaraghavan, K., & Yun, Y. S. (2008). Bacterial biosorbents and biosorption. *Biotechnology Advances*, *26*(3), 266–291. doi:10.1016/j.biotechadv.2008.02.002

Volesky, B., & Holan, Z. R. (1995). Biosorption of Heavy Metals. *Biotechnology Progress*, *11*(3), 235–250. doi:10.1021/bp00033a001

Wang, J., & Chen, C. (2009). Biosorbents for heavy metals removal and their future. *Biotechnology Advances*, *27*(2), 195–226. doi:10.1016/j.biotechadv.2008.11.002

Witek-Krowiak, A., & Harikishore Kumar Reddy, D. (2013). Removal of microelemental Cr(III) and Cu(II) by using soybean meal waste - Unusual isotherms and insights of binding mechanism. *Bioresource Technology*, *127*, 350–357. doi:10.1016/j.biortech.2012.09.072

Witek-Krowiak, A., Szafran, R. G., & Modelski, S. (2011). Biosorption of heavy metals from aqueous solutions onto peanut shell as a low-cost biosorbent. *Desalination*, *265*(1–3), 126–134. doi:10.1016/j.desal.2010.07.042

Xu, R., Li, Q., Yang, Y., Jin, S., Liao, L., Wu, Z., Yin, Z., Xu, B., Nan, X., He, Y., Zhu, B., & Jiang, T. (2022). Removal of heavy metal(loid)s from aqueous solution by biogenic FeS–kaolin composite: Behaviors and mechanisms. *Chemosphere*, *299*, 134382. doi:10.1016/j.chemosphere.2022.134382

Yaashikaa, P. R., Kumar, P. S., Varjani, S., & Saravanan, A. (2020). A critical review on the biochar production techniques, characterization, stability and applications for circular bioeconomy. *Biotechnology Reports (Amsterdam, Netherlands)*, *28*, e00570. doi:10.1016/j.btre.2020.e00570

Yelatontsev, D. (2023). Production of versatile biosorbent via eco-friendly utilization of non-wood biomass. *Chemical Engineering Journal*, *451*, 138811. doi:10.1016/j.cej.2022.138811

Zhang, T., Yang, Y., Li, X., Zhou, Z., & Wei, B. (2023). Ultrasonic–Thermal Regeneration of Spent Powdered Activated Carbon. *Sustainability (Basel)*, *15*(11), 9060. Advance online publication. doi:10.3390/su15119060

Zouboulis, A. I., Lazaridis, N. K., & Matis, K. A. (2002). Removal of toxic metal ions from aqueous systems by biosorptive flotation. *Journal of Chemical Technology and Biotechnology*, *77*(8), 958–964. doi:10.1002/jctb.663

Chapter 8
Biosorption Processes for Heavy Metal Removal in Aqueous Media

Joan Nyika
https://orcid.org/0000-0001-8300-6990
University of Johannesburg, South Africa

Megersa Olumana Dinka
https://orcid.org/0000-0003-3032-7672
University of Johannesburg, South Africa

ABSTRACT

Biosorption technique is a growing alternative to existent conventional treatment techniques used in heavy metal remediation and associated with the production of toxic sludge. This chapter demonstrated the use of the technique and its processes, the biosorbents used, and the sorbates that they target. Evidently, many plant by-products, bacteria, fungi, and algae species are used in the process in their living or dead forms to take up a variety of heavy metals. The process is regulated by factors such as biosorbent dosage, temperature, pH, and initial metal concentration. There is need to advance research on the technique to optimize its metal removal efficacy.

INTRODUCTION

Heavy metals describe elements of the periodic table elements (metalloids and metals) whose atomic density is five times greater than that of water or more than 4 g/cm^3 (Nyika & Dinka, 2023a). Examples of heavy metals include arsenic (As), cadmium (Cd), chromium (Cr), copper (Cu), iron (Fe), mercury (Hg), manganese (Mn), nickel (Ni), lead (Pb), thallium (Tl) and antimony (Sb) among others. The metals occur naturally in the earth's crust but can be released during various anthropic activities such as mining, industrial activities and from solid waste and wastewater constituents (Priyadarshanee & Das, 2021). Continuous accretion of heavy metals has also resulted from the manufacture of paints, explora-

DOI: 10.4018/979-8-3693-1618-4.ch008

tion of gas and oil, coal combustion and unscientific application of agricultural inputs (fertilizers and agrochemicals). The situation is more pronounced in contemporary society characterized with a rise in industrialization and the quest for economic development globally.

Heavy metals have a persistent, non-destroyable and non-biodegradable nature, which makes them to accumulate in environmental compartments, enter and get transferred in food chains to become health and environmental hazards (Nyika & Dinka, 2023b). Apart from occurring in rocks, the elements occur in compartments of the biosphere including the soils and water. The metals can accumulate on land and enter in soil systems to disrupt their ecosystem functions. They are then transferred to water resources to render them non-palatable and unsuitable for consumptive uses. From soils and water, they can be taken up the roots to accumulate in plants and food before their transfer from primary to secondary consumers. In trace amounts, some metals such as Cu, Zn and Fe can be useful in various physiological functions of the body though in higher concentrations they are toxic to the body. Other heavy metals apart from these essential ones are toxic to organisms even in minute levels. Their toxicity potential to organisms is dependent on the health state, genetics, gender and age of the organisms as well as the specific chemical species involved, the exposure route and the dose of the heavy metal ingested or inhaled (Pham . 2022). In higher animals for instance, ingestion of high concentrations of heavy metals is associated with immunological toxicity, hepatoxicity, carcinogenicity, nephrotoxicity, neurotoxicity and toxicity of reproductive systems (Nyika & Dinka, 2023a). Mercury and Cr ingestion by humans causes carcinogenic effects (Mitra et al., 2022). The poisoning of Pb in children results to intellectual dysfunctions among children while Cd and Hg toxicity have been associated with the Itai-itai and Minamata diseases, respectively. The main route of toxicity is through cytotoxicity (damaging and eventual death of cells), production of reactive oxygen species (ROS) to induce oxidative stress and the destruction of genetic material (DNA) in the organisms.

Heavy metals cannot be metabolized but they can be transformed to lesser toxic forms. As such, it is key to manage and control their release to the biosphere. Such advances require the application of chemical and physical techniques to treat the metals in polluted environs. Examples of physical techniques employed include reverse osmosis, membrane technology, ion exchange and filtration while some chemical techniques include redox reactions, electrochemical treatment and chemical precipitation. Although the techniques are effective in heavy metal remediation to some degree, they are sometimes labour and energy intensive, non-selective and non-specific to heavy metals, inefficient and expensive (Shamim, 2018; Redha, 2020). Some of the techniques also have low efficacy and produce secondary toxic waste and sludge as byproducts leading to extensive pollution (Priyadarshanee & Das, 2021). Owing to these challenges new advances on the use of biological techniques that employ microorganisms are being advocated in the scientific field (Shamim, 2018; Nyika & Dinka, 2023 a, b, c). Bioremediation techniques are preferred since they have a high recovery efficiency, are natural, safe, greener, less polluting, and cheaper compared to other conventional techniques (Pham, Kim, Chang, & Chung, 2022). This chapter explored the application of biosorption, a type of bioremediation technique for heavy metal removal from various environmental compartments and prospects, which could facilitate the wider application of the technique to alleviate pollution at field scale.

The Principles of Biosorption and Types of Biosorbents

Biosorption refers to a physicochemical and biological process in which microbes and plant biomass adsorb and absorb contaminants such as dyes and heavy metals in their ionic forms (Redha, 2020). The process is orchestrated by two phases: the sorbate and the biosorbent. The latter is the solid form and occurs as fungi, bacteria, algae and plant biomass while the former, is the liquid/aqueous phase, which contains the contaminant ions (Ali, Sihabudeen, & Hussain, 2016). In heavy metal biosorption, plant biomass, fungi, bacteria and algae are the commonest biosorbents whereby, absorption or adsorption of the contaminants occurs in their cell wall (Tsezos, Remoundaki, & Hatzikioseyian, 2014; Nyika & Dinka, 2023b). Through functional groups such as the hydroxyl, carboxyl, amino, sulfhydryl, carbonyl and phosphate moieties found in the carbohydrates, alcohols proteins, nucleic acids, polysaccharides and genetic material of specific biosorbents, metal ions append on the cell wall to become bio-adsorbed (Javanbakht, Alavi, & Zilouei, 2014). The interaction between biosorbents and sorbates through functional groups can be studied using X-ray absorption fine structure spectroscopy, energy-dispersive X-ray spectroscopy, Raman spectroscopy, infrared spectroscopy, titration, and X-ray photoelectron spectroscopy among other advanced techniques (Redha, 2020; Nyika & Dinka, 2023c). Processes such as coordination, chelation, ion exchange and surface complexation of metal cations with microbial or biomass functional groups drive the biosorption process using microbe cells that are either alive or dead. The microbes could be either fungi, algae, or bacteria since they have functional groups that complex the heavy metals and convert them to less noxious substances.

Plants as Biosorbents

Agricultural by-products such as peels, rind and cortexes have a growing potential to be used as biosorbents (Nyika & Dinka, 2023c). Such wastes contain fermented sugars, lignin, cellulose and hemicellulose in their cell wall that are hosts to functionals groups used in taking up heavy metal ions. The preference in the use of agricultural wastes as biosorbents is associated with their ease in availability and the ability to channel such by-products into a circular economy through waste valorisation (Nyika & Dinka, 2023c). Using the hydroxyl, carbonyl and carboxyl moieties in aloe vera wastes cadmium and uranium cations were adsorbed effectively from wastewater (Noli, Kapashi, & Kapnisti, 2019). Peels from banana, pomegranate, pineapples, tomato, lemon, orange as well as watermelon rind and cortexes from tangerine and kiwi have been used to adsorb heavy metals such as Cd, Cr, Mn, Pb, Sb and Zn from aqueous solutions (Bharat & Manchanda, 2017; Abbas & Nussrat, 2020; Nyika & Dinka, 2023c). Plant biomass from pokeweed, sawdust from the Durian tree, peanut shells and pruning from olive tree have been used in the biosorption of lead from aqueous media effectively (Redha, 2020).

The success of using plant biosorbents varies based on the targeted metal and the source of the biomass. The use of tangerine and kiwi cortes in remediation of Cd, Cr and Zn from wastewater reported 35, 36 and 25% efficacy rate, respectively when the adsorbent dose was 2 g, the initial concentration of each of the metals was 100 mg/l and the contact time was 3 hours (Al-Qahtani, 2016). In another study, orange peels recorded 85% uptake level for antimony (Sb) from wastewater at a pH of 6 and in a 150-minute contact time (Hasan, Al-Tameemi, & Abbas, 2021). Using the watermelon rind biomass, 80% of Zn was removed from synthetic water at an average temperature of 30°C, 450 mg/l initial concentration of the metal, 3-hour contact time and an adsorbent dosage of 1.5 mg/l (Altowayti et al., 2021).

Fungi as Biosorbents

Fungal species have a rich cell wall composition, which suggests the presence of multiple functional groups and high potential for use as biosorbents (Benila & Sumithra, 2017). According to Dhankhar and

Hood (2011), the cell wall of fungal species contains chitosans, α and β- glucans, cellulose, chitin, lipids, glucoproteins, polyuranides, pigments and inorganic salts with multiple functional groups for metals ions to be bio-adsorbed. These components make the cell wall negatively charged to enable electrostatic attraction of contaminant metal ions as well as complexes of heavy metals and N and O components of chitosan and chitin (Velkova et al., 2018). A study using Fourier -transform infrared spectroscopy (FTIR) established that *Mucor rouxii* fungi had amine and phosphate functional groups that were useful in biosorption of Cd, Ni, Pb and Zn cations (Yan & Viraraghavan, 2008). Other fungal species used in metal biosorption include the *Trichoderma, Rhizopus* and *Aspergillus* species since their biomass can be produced using cost effective culture media and they grow rapidly (Velkova et al., 2018). Table 1 shows some of the fungal species used in biosorption and the metals they target.

Table 1. Fungal species used in biosorption of heavy metals

Species	Targeted Heavy Metals	Efficacy (%)	Reference
Trichoderma viride	Cr (VI) Ni (II) Zn (II)	40.1 75 53	Kumar et al., 2011
Trichoderma asperellum	Cu (II)	100	Tan &Ting, 2012
Trichoderma harzianium	U (VI)	98.1-99.3	Akhtar et al., 2009
Penicillium citrinum	Cu (II)	-	Verma et al., 2013
Penicillium fellutinum	Ni (II) Zn (II)	-	Rashid et al., 2016
Aspergillus niger	U (VI)	93.09	Ding et al., 2012
A consortium of *Penicillium, Fusarium* and *Aspergillus* species	Cr (III) Cr (VI)	- -	Mahmoud et al., 2015
Aspergillus ustus	Cd (II)	97	Mahmoud et al., 2011
Rhizopus cohnii	Cr (VI)	100	Li et al., 2008
Trametes versicolor	Cu (II)	-	Akar et al., 2009

Algae as Biosorbents

Macro- and micro-algae species have also been applied in biosorption of heavy metals from aqueous solutions. Macroalgae, which are generally seaweeds classified based on their pigmentation occur as green, red and brown are usually multi-cellular in nature (Anostopoulos & Kyzas, 2015). Microalgae on the other hand are unicellular and grow in both fresh and saline water and occur based on their morphology, photosynthetic membrane arrangement and pigmentation (just as macroalgae). The cell wall of algae has lipids, proteins, polysaccharides and chitin that contain the functional moieties useful in heavy metal biosorption (Ivanova et al., 2012; He & Chen, 2014). Of the different types of algae, the brown algae are the most preferred in biosorption due to their high efficacy in effluents that are dilute, high surface area to volume ratio and the ability to recover and regenerate the adsorbed metal ions (Akbari et al. 2015). About 121.95 mg/g and 111.1 mg/g of Pb (II) and Cd (II) ions were bio-adsorbed by the blue green algae (*Anabaena sphaerica*) (Abdel-Aty et al., 2013). The carbonyl, hydroxyl, carboxyl and amino functional groups enabled the biosorption of the heavy metals. *Cystoseira indica* (Moghaddam, Fatemi,

& Keshtkar, 2013) and *Sargassum sp.* (Abdi & Kazemi, 2015) algal biomass were used in biosorption of Cd, Pb and U cations through esterification of the carboxylic functional groups on the microbes' cell wall.

Bacteria as Biosorbents

Bacterial species have an intricate cell wall with high adsorptive capacity and multiple binding sites, which enable them to act as selective sorbents. The cell wall thickness and composition of gram positive and negative is different. The cell wall of gram-positive bacteria has a thicker peptidoglycan layer, consists of teichuronic and teichoic acids and hence, has a higher sorption potential compared to the gram-negative one (Zyoud et al., 2019). In a number of studies (Table 2), bacteria have been involved in remediation of heavy metal ions from aqueous solutions using their phosphoric, sulfhydryl, amine and hydroxyl functional groups (Javanbakht et al., 2014).

Table 2. Bacteria species used in biosorption of heavy metals from aqueous solutions

Bacterial Species	Targeted Heavy Metal	Efficacy (%)	Reference
Aeromonas hydrophila	Pb (II)	-	Javanbakht et al. 2014
Bacillus amyloliquefaciens	U (VI)	-	Liu et al., 2019
Bacillus subtilis	Cd (II)	50.7	Ahmad et al., 2014
Anthrobacter sp.	Cu (II)	89.56	Hasan & Srivastava, 2009
Bacillus drentensis	Cd, Cu, Fe, Zn	93	Kim, Lee & Wang, 2014
Bacillus strain CR-7	Cd, Co, Cu, Fe, Mn, Ni, Pb, Zn	-	Xu et al., 2011
Escherichia coli	Pt (IV)	90.2	Mao et al., 2015
	Pd (II)	-	Cho et al., 2016
Staphylococcus aureus	Cu Ni Pb	83.0-89.5 92.6-97.5 99.4-100	Mahmoud et al., 2016
Bacillus subtilis	Cd, Cu, Pb, Zn	99	Michalak, Chojnacka, & Witek-krowiak, 2013
B. cereus	Cu (II) and Pb (II)	-	Priyadarshanee & Das, 2021
Pseudomonas stuzeri	Cd, Cu and Pb	-	
Streptomyces pilosus	Cu	-	
P. plecoglossicida	Cd	-	
Cyanobacteria mycrocystis	Sb	-	
Rhodococcus opacus	Cr, Cu and Pb	-	
P. azotoformans	Cd Cu Pb	24.64 17.44 19.55	Pham et al. 2022
Klebsiella sp USL2S	Hg Cd Pb Ni	85 100 26.1 84.7	
Pseudomonas alcaliphila NEWG-2	Cr	96.6	
Cellulosimicrobium sp.	Pb	84.62	
Klebsiella enumoniae	Cd	80	
P. aeruginosa FZ-2	Hg	99.7	

A number of advantages are associated with the use of bacteria biosorbents in biosorption (Filote et al., 2021). The biosorbents are cheap and simple to use in the form of biomass; they take up many metals without additional requirement, can treat large volumes of aqueous solution and can be regenerated. The biosorbents can withstand severe conditions such as high salinity, pH and temperatures in polluted environs and hence have high efficiency in decreasing the environmental effects of both solid waste and wastewater (Pham et al. 2022). The microbes however can be high energy consuming, lead to incomplete remediation of heavy metals due to saturation of binding sites and their biomass production and maintenance can be exorbitant (Igiri et al., 2018).

Evidently, different biosorbents are able to take up heavy metal cations from polluted environs but the biosorption capacity differs based on various modifications and pre-treatment measures applied on the biosorbents, various environmental factors and the specific biosorption mechanism applied.

Mechanisms of Biosorption

Biosorption can occur via metabolism dependent and/or metabolism independent mechanisms (Figure 1). The metabolism dependent mechanism occurs in living organisms (bacteria, algae and fungi) and involves chelation and physical adsorption (Shamim, 2018). Chelation is the process of binding of molecules and ions to metals through the presence or formation of a single central atom and polydentate ligand coordinate bond. Physical adsorption involves the actual attaching of molecules, ions and atoms from a dissolved solid, liquid or gas to a surface. In the metabolism dependents mechanism, an adsorbate film is created on the adsorbent's surface. The result of adsorption is covalent bonding of the biosorbent and sorbate, precipitation of metal ions, their electrostatic attraction and /or complexation. Most heavy metal compounds serve as coordination complexes and undergo reduction during biosorption (Shamim, 2018). All metabolism dependent biosorption processes use energy in the form of adenosine triphosphate (ATP) through ligands of the hydroxyl, sulfhydryl, carbonyl, carboxyl and phosphoryl moieties.

Figure 1. Mechanisms of biosorption (Pham et al., 2022)

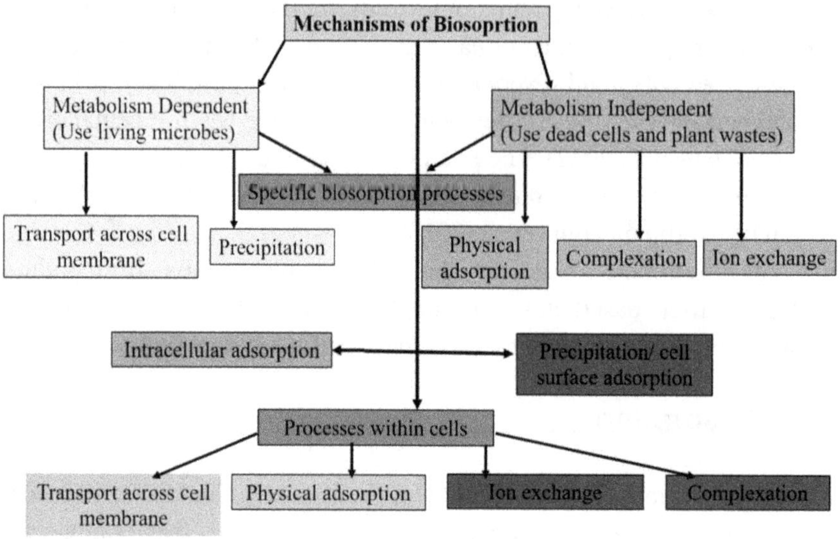

The metabolism independent biosorption mechanism involves the use of biomass of dead cells. In this case, heavy metal ions attach on the dead cells via physicochemical biosorption or ionic interactions. Using dead cells is preferred due to its cost effectiveness, the low demand for pure microbial cultures and nutrient medium, high efficacy over a wide range of pH, ease in biomass regeneration and high rate of heavy metal sorption followed by easy desorption (Vijayaraghavan & Yun, 2008). Unlike the living cells, dead cells do not self-replenish and have no continuous ability to take up metals (Shamim, 2018). Adsorption occurs rapidly followed by a period of accumulation that is energy-dependent and slow. In living or dead cells, the fate of a heavy metal ions could be bioaccumulation, detoxification and/or efflux depending on the microbial species used in biosorption.

Biosorption mechanisms are also classified as intracellular accumulative, cell surface sorption orchestrated and extracellular depending on the location cell when heavy metal remediation occurs (Pham et al., 2022). Intracellular accumulative processes are bio-accumulative in nature, involve either transport in adjacent cells or precipitation. Extracellular precipitation occurs via transport of metal ions across membranes while cell surface sorption occurs via physical adsorption, complexation, diffusion and via ion exchange (Redha, 2020).

Factors that Influence the Process

A number of factors including the pH, temperature, contact time, biomass dosage, initial metal concentration and biosorbent's surface area influence biosorption process. pH influences the adsorption and absorption potential of a sorbent by regulating the chemistry of cell wall functional groups (Wei et al., 2016). Increase in biosorption is directly proportional to pH increase up until the optimum pH whereby biosorption is at the peak and beyond the point, metals precipitate and form insoluble hydroxide complexes with metals and other anions (Abdi & Kazemi, 2015; Bilal et al., 2018). The biosorption of Cd, Co, Cu, Ni and Zn at acidic conditions is reduced compared to higher pH because of competition between the hydronium (H_3O+) and metal cations for the biosorbent's binding sites (Gadd, 2009). Increase in pH enable the interaction of phosphoryl, hydroxyl and carboxyl groups with metal ions following deprotonation and hence increased biosorption (Bilal et al., 2018). In addition to influencing the chemical composition of functional groups, pH influences dominant metal ions in an aqueous solution. In a study assessing the biosorption capacity of Cr (III) by olive stones, it was established that the metal exists in five forms $Cr(OH)_3$, $Cr(OH)_4^-$, $Cr(OH)^{2+}$, $Cr(OH)_2^+$ and Cr^{3+}, which are dominant based on the pH levels (Blazquez et al., 2009). The forms have different affinities to functional groups and hence influence the biosorption process, mobility and solubility of metal cations.

The activity on the cell wall of biosorbents is dependent on temperature. This depends on the nature of the involved process (if it is exothermic or endothermic). In exothermic biosorption, increasing the temperature reduces metal ion remediation in aqueous solutions and vice versa for endothermic reactions (Tasar, Kaya, & Ozer, 2014). Increasing the temperature in a Pb (II) biosorption process using peanut shells resulted to decreased capacity alluding that to the process was exothermic. At temperatures ranging 20-35 °C, biosorption rises steadily (Redha, 2020).

Although the contact time between the sorbate and biosorbent does not directly influence biosorption, it can be a limiting factor as established in many experimental reactions. Increasing the contact time is directly proportional to biosorption capacity so long as the binding sites are available. Once saturation has occurred and no binding sites are available, increasing contact time has no influence on the process (Hajahmadi et al. 2015). Effects of the initial metal concentration on biosorption process has the same effect as the contact time. In this case, the true effect of initial metal concentration can only be felt upon increasing it up to the saturation level when all binding sites are unavailable. In many experiments 5-200 mg/l is the preferred initial metal concentration (Redha, 2020).

The specific uptake of contaminant metal ions at low biosorbent dosages is shown to increase (Sao, 2014). However, using such doses in complex aqueous solutions could enhance competition for binding sites at the cell wall and hence reduce biosorption capacity. Increasing the biosorbent dosage translates to increase in binding sites and more adsorption and absorption of the metal. Usually, the commonest published biosorbent dosage is 0.5- 0,6 g/l (Redha, 2020).

The biosorbent's surface area also affects biosorption process in a directly proportional relationship. For instance, the use of powdered lettuce leaves in biosorption of Cu, Pb and Zn ions showed higher sorption capacity compared to dried leaves due to a high surface area for the former compared to the latter (Shartooh, 2013). The presence of multiple metals as biosorption targets could influence the process. This is because the metals compete for binding sites and have varied affinity for such sites (Sao, 2014). Chemical treatment of biosorbents also promote their biosorption capacity (Calero et al., 2013)

Applications of Biosorption in Specific Heavy Metal Removal in Aqueous Media

The preference to the biosorption technique in heavy metal remediation of aqueous solutions is growing due to its associated advantages (Pham et al. 2022). The technique is simple in demand for equipment and their operation, desorption of contaminants is easy and their adsorption rate is high. Additionally, the technique does not increase the chemical oxygen demand (COD) and a wide range of biosorbents are available to facilitate the process (Nyika & Dinka, 2023d). In this section, biosorption of some of the noxious heavy metals including Pb, Cd and Cr from aqueous environments is discussed.

Lead is a common heavy metal discharged from battery industries, mining and oil sectors to the tune of 5-66, 0.02-2.5 and 125-150 mg/l, respectively (Tasar et al., 2014). The metal is lethal to living organisms and causes organ damage, mental dysfunction, cognitive and behavioural impairments in humans, for instance (Nadeem et al., 2016; Morosanu et al., 2017; Nyika & Dinka, 2023a). It is therefore key to remediate the metal in wastewater before releasing it to the environment to reduce the noxious effects of the metal. A number of biosorption studies aimed at removal of Pb from wastewater, the biosorbent used and the efficacy rates are as shown in Table 3.

Table 3. Biosorbents used in removal of Pb, Cr and Cd from wastewater and their efficacy rates

Biosorbent	Adsorption Capacity (mg/g)	Reference
	Lead	
Anabaena sphaerica algal biomass	121.95	Abdel-Aty et al., 2013
Pruning of olive tree treated with sodium hydroxide	121.06	Calero et al., 2013
Pruning of olive tree treated with nitric acid	85.09	
Pruning of olive tree treated with sulfuric acid	65.62	
Cyclosorus interruptus	46.25	Zhou et al., 2015
Peanut shell	39	Tasar et al., 2014
Coconut coir	37.04	Yusoff et al., 2014
Durian tree sawdust	20.37	
Empty oil palm fruit bunch	37.59	
Pokeweed treated with nitric acid	14.51	Wang et al., 2018
Chromium		
Pleutrotus sajor-caju	122.36	Hanif, 2011
Agaricus bitorquis	152.74	
Ganoderma lucidum	149.58	
Chlorella vulgaris	161.41	Sibi, 2016
Portulaca oleracea	54.95	Mishra, Dubey, & Shinghal, 2015
Watermelon rind	172.6	Reddy, Lakshmipathy, & Sarada, 2014
Opuntia cladodes	18.5	Fernandez -Lopez, Angosto, & Aviles, 2014
Opuntia ectodermis	16.4	
Aspergillus flavus	257	Azhar & Iram, 2014
Cadmium		
Klebsiella sp.	170.4	Zhou et al., 2015
Anabaena sphaerica	111.1	Abdel-Aty et al. 2013
Aloe vera waste	104.2	Noli et al., 2019
Saccharomyces cerevisiae	68.78	Farhan & Khadom, 2015
Streptomyces rimosus	64.9	Sahmoune, 2018
Ulva lactuca	29.1	Ghoneim et al., 2014

Chromium is a teratogenic, carcinogenic, toxic and mutagenic heavy metal, which exists in its hexavalent (Cr (VI) and trivalent (Cr (III) forms (Rezaei, 2016). It is a common pollutant in both surface- and ground-water resources with the hexavalent form being more toxic (500 times than trivalent Cr) due to its high solubility and mobility in aqueous environs (Mishra et al. 2015). Just like Pb, Cr causes hepatoxicity, immunological dysfunction, kidney damage and respiratory complications (Nyika & Dinka, 2023a). The metal is discharged from industrial activities such as tannery activities, in steel and iron foundries, during inorganic chemical, textile and electroplating industrial processes (Redha, 2020). Owing to this toxicity potential, it is key to remediate wastewater off the metal prior to discharge. Several agricultural by-products, algal and fungal biosorbents have been applied to remove different Cr forms from aqueous media as shown in Table 3.

Cadmium is another toxic heavy metal, which accumulates in fish and plants to be transferred to human beings (Al-Homaidan et al., 2015). Usually, the metal binds on amino acid ligands of aspartate, histidine, glutamate and cysteine to cause iron deficiency and ultimately, anaemia in humans, nephrotoxicity and hepatoxicity (Jaishankar et al., 2014). Sources that discharge the metal to natural resources include alloy, mining, fertilizer, battery, paint, pigment, smelting and electroplating industries. Some of the studies involving the bioremediation of Cd via biosorption and their removal rates are as shown in Table 3.

Challenges and Future Prospects of Biosorption

Biosorption in remediation of heavy metals from aqueous solution is a viable technique owing to its cost-effective production and use in addition to good adsorption efficacy (Wei et al. 2016). A number of biosorbents living or dead and sourced from microbes or agricultural waste only require fermentation to be effective biosorbents. This makes the technique applicable and affordable especially to developing countries (Abdi & Kazemi, 2015). Additionally, most biosorbents can be regenerated following desorption and the adsorbed metals ions can be recovered for reuse making the technique an enhancement to circularity and ultimately, sustainable environmental management (Redha, 2020). For this reason, biosorption has a future towards a cleaner environment.

Evaluation of biosorption studies showed its capability for use in removal of single or several heavy metals at experimental level. Although the adsorption capacities are high, they do not reflect the intricate real-life aqueous environs whereby an assortment of metal ions is evident mixed with organic pollutants and at extreme environmental conditions such as temperature and pH occur. Such conditions hinder complete bioremediation of metal contaminants. To advance the process and its field application, it is imperative to use actual contaminated water in these experiments in environs whose factors such as pH mimic actual situations (Nyika & Dinka, 2023c). Such research advances also investigate the selectivity and specificity of biosorbents to mixtures of heavy metals and provides clues on how to improve biosorption efficacy (Gupta, Nayak, & Agarwal, 2015).

The reported studies in this chapter showed biosorption as having the potential to detoxify and bioaccumulate heavy metals through microbes though at laboratory scale. To the best of our knowledge, none of the evaluated studies were at field scale. Furthermore, the widescale application of biosorption has been hampered by incomplete understanding of the metal uptake processes from aqueous solution. Most of the evaluated biosorption studies focused on ion exchange, redox and chemical precipitation experiments (Gadd, 2009). To this end, there is need to extend research on the technique, its mechanism and how its use can be commercialized and transferred from laboratory scale to field scale application.

Biosorption effectiveness is also affected by biosorbent saturation in the binding sites leading to less heavy metal remediation removal rate and efficacy. As such, evaluating the biosorption capacity of organic pollutants is essential because it influences the uptake of heavy metals by competing for binding sites and regulating other environmental factors such as the pH (Nyika & Dinka, 2023c). Optimising the process to reduce the frequency of biosorbent saturation and eventual need to replace it is also essential and can be advanced through further experimentation and research on the technique.

Treatment of biosorbents and application of genetic engineering technologies in biosorption is a growing advancement towards optimisation of the process. Such modifications improve the tolerance of microbes to heavy metals and hence ability to better adsorb and accumulate them (Wu et al., 2021; Zhu, Zhang & Yu, 2020). Some of the genetic techniques used to improve the biosorption of heavy metals from aqueous solutions include the use of next generation sequencing to study transcriptomes and whole genomes useful in the process (Mosa et al., 2012). Proteogenomics, which is based on combining both proteomics and genomics principles is also being researched on to enhance the tolerance of microbes to heavy metals, the contaminants' hyperaccumulation in microbes and biomass and eventually, higher biosorption efficacy (Mosa et al., 2012). Giese et al. (2020) and Velkova et al. (2018) also suggested that the immobilization of microbial biomass on an appropriate carrier improves the biosorbents' porosity, chemical and physical stability, which enhances biosorption. The mechanism in which immobilization improves the biosorption process is barely understood and hence the need for more research. The use of extremophiles in biosorption of heavy metals from aqueous solutions should also be exploited due to their high removal efficacy as Pham et al. (2022) advised.

To remain green and environmentally cautious during the process of biosorption, it is key to plan on the approaches to reuse and/or dispose the resultant by-products. The metals can be desorbed and rechannelled back for various industrial processes (Benila & Sumithra, 2017; Redha, 2020). The biosorbent biomass waste can be reused for gasification especially the plant-sourced one due to its calorific values (Pathak, Mandavgane & Kulkarni, 2015). Such considerations are key in promoting environmental cleanness during biosorption as well as ensuring sustainability of the process.

CONCLUSION

In this study, the potential for biosorption technique in removal of heavy metals from aqueous environments is demonstrated. The method is preferred due to its ease to implement, the ubiquitous nature of biosorbents and the environmentally friendly nature of the technique compared to conventional methods. Biosorption is however regulated by factors such as the initial metal concentration, the biosorbent dose, the contact time between the sorbate and biosorbent, temperature and pH of polluted environs. For this reason, different heavy metal uptake levels are realized based on the targeted metal and environment in which they are situated. The chapter noted of the need to advance biosorption experiments from laboratory to field-scale to enable real-life application of the technique. Such advances require optimization of the process through chemical treatment of biosorbents, their immobilization using suitable carriers and genetic modifications to enhance their ability to take up heavy metal contaminants.

REFERENCES

Abbas, M., & Nussrat, T. (n.d.). Statistical analysis of experimental data for adsorption process of cadmium by watermelon rinds in continuous packed bed column. *Innovation Journal of Creative Change, 13*(3), 124-1121.

Abdel-Aty, M., Ammar, S., Abdel Ghafar, H., & Ali, K. (2013). Biosorption of cadmium and lead from aqueous solution by fresh water alga *Anabaena sphaerica* biomass. *Journal of Advanced Research*, *4*(4), 367–374. doi:10.1016/j.jare.2012.07.004

Abdi, O., & Kazemi, M. (2015). A review study of biosorption of heavy metals and comparison between different biosorbents. *Journal of Materials & Environmental Sciences*, *6*(5), 1386–1399.

Ahmad, F., Haydar, S., Bhatti, A., & Bari, J. (2014). Application of artificial neural network for the prediction of biosorption capacity of immobilized *Bacillus subtilis* for the removal of cadmium ions from aqueous solution. *Biochemical Engineering Journal*, *84*, 83–90. doi:10.1016/j.bej.2014.01.004

Akar, T., Akar, T., Kaynak, Z., Anilan, B., Cabuk, A., Tabak, O., Demir, T., & Gedikbey, T. (2009). Removal of copper(II) ions from synthetic solution and real wastewater by the combined action of dried *Trametes versicolor* cells and montmorillonite. *Hydrometallurgy*, *97*(1-2), 98–104. doi:10.1016/j.hydromet.2009.01.009

Akbari, M., Hallajisani, A., Keshtkar, R., Shahbeig, H., & Ali, S. (2015). Equilibrium and kinetic study and modelling of Cu(II) and Co(II) synergistic biosorption from Cu(II)-Co(II) single and binary mixtures on brown algae *C. indica*. *Journal of Environmental Chemical Engineering*, *3*(1), 140–149. doi:10.1016/j.jece.2014.11.004

Akhtar, K., Khalid, M., Akhtar, W., & Ghauri, M. (2009). Removal and recovery of uranium from aqueous solutions by Ca-alginate immobilized *Trichoderma harzianum*. *Bioresource Technology*, *100*(20), 4551–4558. doi:10.1016/j.biortech.2009.03.073

Al-Homaidan, A., Alabdullatif, A., Al-Hazzani, A., Al-Ghanayem, A., & Alabbad, F. (2015). Adsorptive removal of cadmium ions by *Spirulina platensis* dry biomass. *Saudi Journal of Biological Sciences*, *22*(6), 795–800. doi:10.1016/j.sjbs.2015.06.010

Al-Qahtani, K. (2016). Water purification using different waste fruit cortexes for removal of heavy metals. *Journal of Taibah University for Science : JTUSCI*, *10*(5), 700–708. doi:10.1016/j.jtusci.2015.09.001

Ali, A., Sihabudeen, M., & Hussain, Z. (2016). Biosorption of heavy metals by *Pseudomonas* bacteria. *International Research Journal of Engineering and Technology*, *3*(8), 1446–1450.

Altowayti, W., Othman, N., Al-Gheethi, A., Dzahir, N., Asharuddin, S., Alshalif, A., Nasser, I., Tajarudin, H., & AL-Towayti, F. A. H. (2021). Adsorption of Zn^{2+} from synthetic wastewater using dried watermelon rind (D-WMR): An overview of nonlinear and linear regression and error analysis. *Molecules (Basel, Switzerland)*, *26*(20), 6176. doi:10.3390/molecules26206176

Anastopoulos, I., & Kyzas, G. Z. (2015). Progress in batch bio-sorption of heavy metals onto algae. *Journal of Molecular Liquids*, *209*, 77–86. doi:10.1016/j.molliq.2015.05.023

Azhar, W., & Iram, S. (2014). kinetics, equilibrium and thermodynamics studies on biosorption of heavy metals by fungal biomass. *International Journal of Engineering Sciences & Emerging Technologies*, *7*(3), 693–700. doi:10.1007/s11356-022-22860-w

Benila, M., & Sumithra, P. (2017). Optimization of Chromium Biosorption by Fungal Adsorbent, *Trichoderma sp*. BSCR02 and its Desorption Studies. *Journal of Biosciences, 24*(2), 65–71. https://doi.org/.hjb.2017.08.005 doi:10.1016/j

Bharat, S., & Manchanda, D. (2017). Efficient bio-adsorbents for removal of heavy metals from water: A review. *International Journal of Chemical Studies*, *5*(4), 1691–1694.

Bilal, M., Rasheed, T., Sosa-Hernandez, J., Raza, A., Nabeel, F., & Iqbal, H. (2018). Biosorption: An interplay between marine algae and potentially toxic elements—A review. *Marine Drugs*, *16*(2), 1–16. doi:10.3390/md16020065

Blazquez, G., Hernainz, F., Calero, M., Martın-Lara, M.A., & Tenorio, G. (2009). The effect of pH on the biosorption of Cr (III) and Cr (VI) with olive stone. *Chemical Engineering Journal, 148*(2-3), 473–479. https://doi.org/.2008.09.026 doi:10.1016/j.cej

Calero, M., Perez, A., Blazquez, G., Ronda, A., & Martin-Lara, M. (2013). Characterization of chemically modified biosorbents from olive tree pruning for the biosorption of lead. *Ecological Engineering*, *58*, 344–354. doi:10.1016/j.ecoleng.2013.07.012

Cho, W., Kang, S. B., Kim, S., Yun, Y., & Won, S. (2016). Reusable polyethylenimine-coated polysulfone/bacterial biomass composite fibre biosorbent for recovery of Pd(II) from acidic solutions. *Chemical Engineering Journal*, *302*, 545–551. doi:10.1016/j.cej.2016.05.091

Dhankhar, R., & Hooda, A. (2011). Fungal biosorption-an alternative to meet the challenges of heavy metal pollution in aqueous solutions. *Environmental Technology*, *32*(5), 467–491. doi:10.1080/09593330.2011.572922

Ding, X., Tan, X., Hu, N., Li, Y., Wang, Y.-D., & Tan, Y. (2012). Removal and recovery of uranium(VI) from aqueous solutions by immobilized *Aspergillus niger* powder beads. *Bioprocess and Biosystems Engineering*, *35*(9), 1567–1576. doi:10.1007/s00449-012-0747-8

Farhan, N., & Khadom, A. (2015). Biosorption of heavy metals from aqueous solutions by *Saccharomyces Cerevisiae*. *International Journal of Industrial Chemistry*, *6*(2), 119–130. doi:10.1007/s40090-015-0038-8

Fernandez-Lopez, A., Angosto, M., & Aviles, D. (2014). Biosorption of hexavalent chromium from aqueous medium with opuntia biomass. *TheScientificWorldJournal*, *4*, 1–8. doi:10.1155/2014/670249

Filote, C., Rosca, M., Hlihor, R., Cozma, P., Simion, I., Apostol, M., & Gavrilescu, M. (2021). Sustainable application of biosorption and bioaccumulation of persistent pollutants in wastewater treatment: Current Practice. *Processes (Basel, Switzerland)*, *9*(10), 1696. doi:10.3390/pr9101696

Gadd, G. (2009). Biosorption: Critical review of scientific rationale, environmental importance and significance for pollution treatment. *Journal of Chemical Technology and Biotechnology*, *84*(1), 13–28. doi:10.1002/jctb.1999

Ghoneim, M., El-Desoky, S., El-Moselhy, M., Amer, A., Abou El-Naga, H., Mohamedein, I., & Al-Prol, E. (2014). Removal of cadmium from aqueous solution using marine green algae, *Ulva lactuca*. *Egyptian Journal of Aquatic Research*, *40*(3), 235–242. doi:10.1016/j.ejar.2014.08.005

Giese, E., Silva, D., Costa, A., Almeida, S., & Dussan, K. (2020). Immobilized microbial nanoparticles for biosorption. *Critical Reviews in Biotechnology*, *40*(5), 653–666. doi:10.1080/07388551.2020.1751583

Gupta, V., Nayak, A., & Agarwal, S. (2015). Bio-adsorbents for remediation of heavy metals: Current status and their future prospects. *Environmental Engineering Research*, *20*(1), 1–8. doi:10.4491/eer.2015.018

Hajahmadi, Z., Younesi, H., Bahramifar, N., Khakpour, H., & Pirzadeh, K. (2015). Multicomponent isotherm for biosorption of Zn(II), CO(II) and Cd(II) from ternary mixture onto pretreated dried *Aspergillus niger* biomass. *Water Resources and Industry*, *11*, 71–80. doi:10.1016/j.wri.2015.07.003

Hanif, M. (2011). Biosorption of Cr (III) and Cr (VI) by Newly Isolated White Rot Fungi: Batch and Column Studies. *Asian Journal of Chemistry*, *23*(8), 3375–3383.

Hasan, H., & Srivastava, P. (2009). Batch and continuous biosorption of Cu(II) by immobilized biomass of *Arthrobacter sp. Journal of Environmental Management*, *90*(11), 3313–3332. doi:10.1016/j.jenvman.2009.05.005

Hasan, M., Al-Tameemi, I., & Abbas, M. (2021). Orange peels as a sustainable material for treating water polluted with antimony. *Journal of Ecological Engineering*, *22*(2), 25–35. doi:10.12911/22998993/130632

He, J., & Chen, J. (2014). A comprehensive review on biosorption of heavy metals by algal biomass: Materials, performances, chemistry, and modelling simulation tools. *Bioresource Technology*, *160*, 67–78. doi:10.1016/j.biortech.2014.01.068

Igiri, B., Okoduwa, S., Idoko, G., Akabuogu, E., Adeyi, A., & Ejiogu, I. (2018). Toxicity and bioremediation of heavy metals contaminated ecosystem from tannery wastewater: A Review. *Journal of Toxicology*, *27*, 2568038. doi:10.1155/2018/2568038

Ivanova, D., Kadukova, J., Kavulicova, J., & Horvathova, H. (2012). Determination of the Functional Groups in Algae *Parachlorella Kessleri* by Potentiometric Titrations. *Nova Biotechnologica et Chimica*, *11*(2), 93–99. doi:10.2478/v10296-012-0010-3

Jaishankar, M., Tseten, T., Anbalagan, N., Mathew, B. B., & Beeregowda, K. N. (2014). Toxicity, mechanism and health effects of some heavy metals. *Interdisciplinary Toxicology*, *7*(2), 60–72. doi:10.2478/intox-2014-0009

Javanbakht, V., Alavi, S. A., & Zilouei, H. (2014). Mechanisms of heavy metal removal using microorganisms as biosorbent. *Water Science and Technology*, *69*(9), 1775–1787. doi:10.2166/wst.2013.718

Kim, I., Lee, M., & Wang, S. (2014). Heavy metal removal in groundwater originating from acid mine drainage using dead *Bacillus drentensis* sp. immobilized in polysulfone polymer. *Journal of Environmental Management*, *146*, 568–574. doi:10.1016/j.jenvman.2014.05.042

Kumar, R., Bhatia, D., Singh, R., Rani, S., & Bishnoi, N. (2011). Sorption of heavy metals from electroplating effluent using immobilized biomass *Trichoderma viride* in a continuous packed-bed column. *International Biodeterioration & Biodegradation*, *65*(8), 1133–1139. doi:10.1016/j.ibiod.2011.09.003

Li, H., Li, Z., Liu, T., Xiao, X., Peng, Z., & Deng, L. (2008). A novel technology for biosorption and recovery hexavalent chromium in waste water by bio-functional magnetic beads. *Bioresource Technology*, *99*(14), 6271–6279. doi:10.1016/j.biortech.2007.12.002

Liu, L., Liu, J., Liu, X., Dai, C., Zhang, Z., Song, W., & Chu, Y. (2019). Kinetic and equilibrium of U(VI) biosorption onto the resistant bacterium *Bacillus amyloliquefaciens. Journal of Environmental Radioactivity*, *203*, 117–124. doi:10.1016/j.jenvrad.2019.03.008

Mahmoud, E., Abdou, H., Mohamed, S., & Osman, M. (2016). Engineered *Staphylococcus aureus* via immobilization on magnetic Fe3O4-phthalate nanoparticles for biosorption of divalent ions from aqueous solutions. *Journal of Environmental Chemical Engineering*, *4*(4), 3810–3824. doi:10.1016/j.jece.2016.08.022

Mahmoud, E., Yakout, A., Abdel-Aal, H., & Osman, M. (2011). Enhanced biosorptive removal of cadmium from aqueous solutions by silicon dioxide nano-powder, heat inactivated and immobilized *Aspergillus ustus. Desalination*, *279*(1-3), 291–297. doi:10.1016/j.desal.2011.06.023

Mahmoud, E., Yakout, A., Hany, A., & Osman, M. (2015). Speciation and selective biosorption of Cr(III) and Cr(VI) using nanosilica immobilized-fungi biosorbents. *Journal of Environmental Engineering*, *141*(4), 1–9. doi:10.1061/(ASCE)EE.1943-7870.0000899

Mao, J., Kim, S., Wu, X. H., Kwak, I., Zhou, T., & Yun, Y. (2015). A sustainable cationic chitosan/*E. coli* fibre biosorbent for Pt(IV) removal and recovery in batch and column systems. *Separation and Purification Technology*, *143*, 32–39. doi:10.1016/j.seppur.2015.01.023

Michalak, I., Chojnacka, K., & Witek-krowiak, A. (2013). State of the art for the biosorption process- a review. *Applied Biochemistry and Biotechnology*, *170*(6), 1389–1416. doi:10.1007/s12010-013-0269-0

Mishra, A., Dubey, A., & Shinghal, S. (2015). Biosorption of chromium(VI) from aqueous solutions using waste plant biomass. *International Journal of Environmental Science and Technology*, *12*(4), 1415–1426. doi:10.1007/s13762-014-0516-0

Mitra, S., Chakraborty, A., Tareq, A., Emran, T., Nainu, F., Khusro, A., Idris, A., Khandaker, M., Osman, H., Alhumaydhi, F., & Gandara, S. (2022). Impact of heavy metals on the environment and human health: Novel therapeutic insights to counter the toxicity. *Journal of King Saud University. Science*, *34*(3), 101865. doi:10.1016/j.jksus.2022.101865

Moghaddam, R., Fatemi, S., & Keshtkar, A. (2013). Adsorption of lead (Pb^{2+}) and uranium (UO_2^{2+}) cations by brown algae; experimental and thermodynamic modelling. *Chemical Engineering Journal*, *231*, 294–303. doi:10.1016/j.cej.2013.07.037

Morosanu, I., Teodosiu, C., Paduraru, C., Ibanescu, D., & Tofan, L. (2017). Biosorption of lead ions from aqueous effluents by rapeseed biomass. *New Biotechnology*, *39*, 110–124. doi:10.1016/j.nbt.2016.08.002

Mosa, K., Saadoun, I., Kumar, K., Helmy, M., & Dhankher, O. (2016). Potential biotechnological strategies for the cleanup of heavy metals and metalloids. *Frontiers in Plant Science*, *7*, 303. doi:10.3389/fpls.2016.00303

Nadeem, R., Manzoor, Q., Iqbal, M., & Nisar, J. (2016). Biosorption of Pb(II) onto immobilized and native *Mangifera indica* waste biomass. *Journal of Industrial and Engineering Chemistry*, *35*, 185–194. https://doi.org/.jiec.2015.12.030 doi:10.1016/j

Noli, F., Kapashi, E., & Kapnisti, M. (2019). Biosorption of uranium and cadmium using sorbents based on Aloe vera wastes. *Journal of Environmental Chemical Engineering*, *7*(2), 102985. doi:10.1016/j.jece.2019.102985

Nyika, J., & Dinka, M. (2023a). *Global industrial impacts of heavy metal pollution in sub-Saharan Africa*. IGI Global. doi:10.4018/978-1-6684-7116-6

Nyika, J., & Dinka, M. (2023b). Factors and mechanisms regulating heavy metal phycoremediation in polluted water. *Discover Water*, *3*(1), 14. doi:10.1007/s43832-023-00041-1

Nyika, J., & Dinka, M. (2023c). (in press). The potential of reusing fruit bio-adsorbents for water purification: A minireview. *Materials Today: Proceedings*. Advance online publication. doi:10.1016/j.matpr.2023.07.285

Nyika, J., & Dinka, M. (2023d). The application of microorganism-derived biosorbents in the removal of heavy metals and dyes. In K. Ukhurebor, U. Aigble, & R. Onyancha (Eds.), *Adsorption applications for environmental sustainability*. IOP Publishing Limited. doi:10.1088/978-0-7503-5598-8ch8

Pathak, P., Mandavgane, S., & Kulkarni, B. (2015). Fruit peel waste as a novel cost bioadsorbent. *Reviews in Chemical Engineering*, *31*(4), 361–381. doi:10.1515/revce-2014-0041

Pham, V., Kim, J., Chang, S., & Chung, W. (2022). Bacterial Biosorbents, an efficient heavy metals green clean-up strategy: Prospects, challenges, and opportunities. *Microorganisms*, *10*(3), 610. doi:10.3390/microorganisms10030610

Priyadarshanee, M., & Das, S. (2021). Biosorption and removal of toxic heavy metals by metal tolerating bacteria for bioremediation of metal contamination: A comprehensive review. *Journal of Environmental Chemical Engineering*, *9*(1), 104686. doi:10.1016/j.jece.2020.104686

Rashid, A., Bhatti, H. N., Iqbal, M., & Noreen, S. (2016). Fungal biomass composite with bentonite efficiency for nickel and zinc adsorption: A mechanistic study. *Ecological Engineering*, *91*, 459–471. doi:10.1016/j.ecoleng.2016.03.014

Reddy, A., Lakshmipathy, R., & Sarada, N. C. (2014). Application of *Citrullus lanatus* rind as biosorbent for removal of trivalent chromium from aqueous solution. *Alexandria Engineering Journal*, *53*(4), 969–975. doi:10.1016/j.aej.2014.07.006

Redha, A. (2020). Removal of heavy metals from aqueous media by biosorption. *Arab Journal of Basic and Applied Sciences*, *27*(1), 183–193. doi:10.1080/25765299.2020.1756177

Rezaei, H. (2016). Biosorption of chromium by using *Spirulina sp. Arabian Journal of Chemistry*, *9*(6), 846–853. doi:10.1016/j.arabjc.2013.11.008

Sahmoune, M. (2018). Performance of *Streptomyces rimosus* biomass in biosorption of heavy metals from aqueous solutions. *Microchemical Journal*, *141*(5), 87–95. doi:10.1016/j.microc.2018.05.009

Sao, K. (2014). A review on heavy metals uptake by plants through biosorption. *International Proceedings of Economics Development and Research*, *75*(17), 78–83. doi:10.7763/IPEDR

Shamim, S. (2018). *Biosorption of heavy metals*. Intech Open Publishers. doi:10.5772/intechopen.72099

Shartooh, S. M. (2013). Lettuce Leaves as Biosorbent Material to Remove Heavy Metal Ions from Industrial Wastewater. *Journal of Baghdad for Science*, *11*(3), 1164–1170.

Sibi, G. (2016). Biosorption of chromium from electroplating and galvanizing industrial effluents under extreme conditions using *Chlorella vulgaris*. *Green Energy & Environment*, *1*(2), 172–177. doi:10.1016/j.gee.2016.08.002

Tan, S., & Ting, Y. (2012). Efficacy and reusability of alginate-immobilized live and heat-inactivated *Trichoderma asperellum* cells for Cu (II) removal from aqueous solution. *Bioresource Technology*, *123*, 290–295. doi:10.1016/j.biortech.2012.07.082

Tasar, S., Kaya, F., & Ozer, A. (2014). Biosorption of lead (II) ions from aqueous solution by peanut shells: Equilibrium, thermodynamic and kinetic studies. *Journal of Environmental Chemical Engineering*, *2*(2), 1018–1026. doi:10.1016/j.jece.2014.03.015

Tsezos, M., Remoundaki, E., & Hatzikioseyian, A. (2014). Biosorption - Principles and Applications for Metal Immobilization from waste-water streams. *Clean Production and Nano Technologies*, 23–33.

Velkova, Z., Kirova, G., Stoytcheva, M., Kostadinova, S., Todorova, K., & Gochev, V. (2018). Immobilized microbial biosorbents for heavy metal removal. *Engineering in Life Sciences*, *18*(12), 871–881. doi:10.1002/elsc.201800017

Verma, A., Shalu, Singh, A., Bishnoi, N. R., & Gupta, A. (2013). Biosorption of Cu(II) using free and immobilized biomass of *Penicillium citrinum*. *Ecological Engineering*, *61*, 486–49. doi:10.1016/j.ecoleng.2013.10.008

Vijayaraghavan, K., & Yun, S. (2008). Bacterial biosorbents and biosorption. *Biotechnology Advances*, *26*(3), 266–291. doi:10.1016/j.biotechadv.2008.02.002

Wang, G., Zhang, S., Yao, P., Chen, Y., Xu, X., Li, T., & Gong, G. (2018). Removal of Pb(II) from aqueous solutions by *Phytolacca americana L.* biomass as a low cost biosorbent. *Arabian Journal of Chemistry*, *11*(1), 99–110. doi:10.1016/j.arabjc.2015.06.011

Wei, W., Wang, Q., Li, A., Yang, J., Ma, F., Pi, S., & Wu, D. (2016). Biosorption of Pb (II) from aqueous solution by extracellular polymeric substances extracted from *Klebsiella sp.* J1: Adsorption behaviour and mechanism assessment. *Scientific Reports*, *6*(1), 31575. doi:10.1038/srep31575

Wu, C., Li, F., Yi, S., & Ge, F. (2021). Genetically engineered microbial remediation of soils co-contaminated by heavy metals and polycyclic aromatic hydrocarbons: Advances and ecological risk assessment. *Journal of Environmental Management*, *296*, 113185. doi:10.1016/j.jenvman.2021.113185

Xu, J., Song, C., Zhang, Q., Pan, H., Liang, Y., Fan, X., & Li, Y. (2011). Characterization of metal removal of immobilized *Bacillus strain CR-7* biomass from aqueous solutions. *Journal of Hazardous Materials*, *187*(1-3), 450–458. doi:10.1016/j.jhazmat.2011.01.047

Yan, G., & Viraraghavan, T. (2008). Mechanism of biosorption of heavy metals by *Mucor rouxii*. *Engineering in Life Sciences*, *8*(4), 363–371. doi:10.1002/elsc.200820237

Yusoff, S. N. M., Kamari, A., Putra, W. P., Ishak, C. F., Mohamed, A., Hashim, N., & Isa, I. M. (2014). Removal of Cu(II), Pb(II) and Zn(II) Ions from Aqueous Solutions Using Selected Agricultural Wastes: Adsorption and Characterisation Studies. *Journal of Environmental Protection*, *05*(04), 289–300. doi:10.4236/jep.2014.54032

Zhou, K., Yang, Z., Liu, Y., & Kong, X. (2015). Kinetics and equilibrium studies on biosorption of Pb(II) from aqueous solution by a novel biosorbent: *Cyclosorus interruptus*. *Journal of Environmental Chemical Engineering*, *3*(3), 2219–2228. doi:10.1016/j.jece.2015.08.002

Zhu, N., Zhang, B., & Yu, Q. (2020). Genetic engineering-facilitated co-assembly of synthetic bacterial cells and magnetic nanoparticles for efficient heavy metal removal. *ACS Applied Materials & Interfaces*, *12*(20), 22948–22957. doi:10.1021/acsami.0c04512

Zyoud, A., Alkowni, R., Yousef, O., Salman, M., Hamdan, S., Helal, M., Jaber, S. F., & Hilal, H. S. (2019). Solar light-driven complete mineralization of aqueous gram-positive and gram-negative bacteria with ZnO photocatalyst. *Solar Energy*, *180*, 351–359. doi:10.1016/j.solener.2019.01.034

KEY TERMS AND DEFINITIONS

Chelation: Application of chelation agents to bind on metal ions and form a ring structure (chelate).

Complexation: This is the association between a metal ion and ligands (mononuclear complex) and that of many metal ions centrally located and carrying a negative or positive charge with respect to specific ligands involved.

Coordination: The interaction of heavy metal cations and non-metal atoms on the cell wall through covalent bonding.

Ion Exchange: Exchange of binary metal ions (either cations or anions) with ions founds on the cell surface.

Precipitation: Formation of insoluble components between cell wall functional groups and contaminant ions which remain on the surface or are taken in by the biosorbent.

Reduction: The process of electron transfer following the interaction of a metal with the functional moieties on the cell wall so that the resultant product has a lower oxidation state.

Chapter 9
Potential of Agricultural Wastes as Alternative Biosorbents

Poorva Derashri
Amity Institute of Biotechnology, Amity University, Maharashtra, India

Sagar Barage
Amity Institute of Biotechnology, Amity University, Maharashtra, India

Shashi Bala Singh
Amity Institute of Biotechnology, Amity University, Maharashtra, India

ABSTRACT

Agricultural wastes have excellent potential as substitute biosorbents to reduce pollution in the environment. These include organic contaminants, pesticides, dyes, heavy metals, and other impurities from soil and water. In the chapter, the authors focused on the adsorption capacity, remediation mechanism, and possible uses of agricultural wastes as biosorbents as well as their viability and efficacy. The biosorption capabilities of a variety of agricultural wastes, including crop residues, fruit peels, rice husks, and sugarcane bagasse, has been extensively studied. The processes of biosorption generally encompass ion exchange, complexation, physical adsorption, and chemical interactions between the target pollutants and the functional groups on the biosorbents. This study emphasizes the use of agricultural waste as an efficient alternative for artificial absorbents and promotes a circular economy by addressing environmental issues. Thus, the development of effective biosorption techniques using agricultural wastes is a pressing need for achieving environmental sustainability.

1. INTRODUCTION

Agricultural practices are integral to human society, providing sustenance and raw materials for various industries (Hemathilake & Gunathilake, 2022). However, they generate a substantial amount of waste that, when mismanaged, poses significant environmental challenges. In a world grappling with escalating

DOI: 10.4018/979-8-3693-1618-4.ch009

pollution and dwindling natural resources, the potential of agricultural waste as alternative biosorbents for environmental remediation emerges as a compelling and underexplored subject (SUD et al., 2008).

This chapter delves into the realm of agricultural waste and its untapped capabilities as biosorbents, offering a sustainable solution for removing organic contaminants and heavy metals from soil and water. At the intersection of environmental science and agricultural innovation, this chapter explores the adsorption capacity, remediation mechanisms, and potential applications of agricultural waste as biosorbents, critically examining their viability and efficacy (F. A. Ahmad, 2023).

Agricultural waste comprises a wide array of materials, from the remnants of harvested crops to fruit and vegetable peelings left behind in food processing. The disposal of such waste can lead to air and water pollution, soil degradation, and the release of harmful chemicals. Furthermore, improper disposal practices can contribute to the emission of greenhouse gases, exacerbating the global challenge of climate change. In light of these challenges, the need for sustainable solutions to manage and repurpose agricultural waste becomes evident (Sadh et al., 2018).

Biosorbents, in the context of agricultural waste, offer a promising solution to this multifaceted problem. Biosorbents are biological materials with the capacity to adsorb pollutants from the environment, offering an environmentally friendly, readily available, and renewable means of pollution cleanup(Karić et al., 2022). They can target a wide range of pollutants, including pesticides, dyes, heavy metals, and various other impurities in aqueous solutions (Elgarahy et al., 2021). The mechanisms behind biosorbents' efficacy in pollutant removal are diverse, encompassing ion exchange, complexation, physical adsorption, and chemical interactions between the target pollutants and the biosorbent's functional groups (Dai et al., 2018).

Biosorbents present a dual solution, as they not only address agricultural waste problems but also combat environmental pollution (J. Wang & Chen, 2009). Repurposing agricultural waste as biosorbents promotes a circular economy, reducing environmental concerns while increasing the overall value derived from agricultural outputs (F. Ahmad & Zaidi, 2021). This approach aligns seamlessly with sustainable and green chemistry principles, offering an economical and environmentally beneficial remedy for soil and water pollution. Moreover, it reduces the need for artificial adsorbents, making the development of effective biosorption techniques using agricultural waste a contemporary imperative for environmental sustainability (F. A. Ahmad, 2023).

The potential of agricultural waste as alternative biosorbents, it begins by scrutinizing various agricultural wastes, such as crop residues, fruit peels, rice husks, and sugarcane bagasse, which have been extensively researched for their potential in sustainable pollution cleanup (Sen, 2023). Each of these waste materials is examined for its unique characteristics, adsorption capabilities, and potential applications in environmental remediation.

The core of this exploration lies in understanding how these biosorbents function in pollutant removal. Their efficacy in eliminating a wide range of pollutants, from organic contaminants to heavy metals, demands a thorough examination of the underlying mechanisms. These mechanisms, including ion exchange, complexation, physical adsorption, and chemical interactions, are elucidated to provide a comprehensive view of the biosorption processes at play (Raji et al., 2023).

Furthermore, this chapter underscores the significance of agricultural waste as an efficient alternative to artificial absorbents and traditional remediation processes. The synthesis of research findings, case studies, and practical applications demonstrates the real-world potential of harnessing agricultural waste for environmental remediation while addressing the broader context of sustainability.

The exploration of agricultural waste as alternative biosorbents aligns with contemporary imperatives. As there is an urgent need to address environmental pollution, and advancing sustainable practices, the role of agriculture in this pursuit becomes pivotal. By reimagining agricultural waste as a valuable resource in the fight against pollution, we not only address environmental concerns but also enhance the overall sustainability of agricultural systems (Harshala & Wagh, 2022; Noor & Khan, 2023).

In conclusion, the quest to harness the potential of agricultural waste as alternative biosorbents is a pressing concern in the context of environmental sustainability. By repurposing these abundant and renewable resources, we offer a sustainable, green, and economically viable solution to the challenges of soil and water pollution. The chapters that follow will delve into the specific aspects of this multifaceted exploration, offering a comprehensive understanding of the potential, mechanisms, and real-world applications of agricultural waste as biosorbents, contributing to the paradigm shift required for a more sustainable coexistence of agriculture and the environment.

1.1 Types of Agricultural Waste

Agricultural practices give rise to a diverse array of waste materials, each presenting unique challenges and opportunities for sustainable management (Duque-Acevedo et al., 2020). Among the primary categories of agricultural waste are crop residues, food processing waste, and livestock waste (Raut et al., 2023) Figure 1. Understanding the characteristics and potential of these waste streams is pivotal for harnessing their biosorbent capabilities effectively.

Figure 1. Types of Agricultural Waste

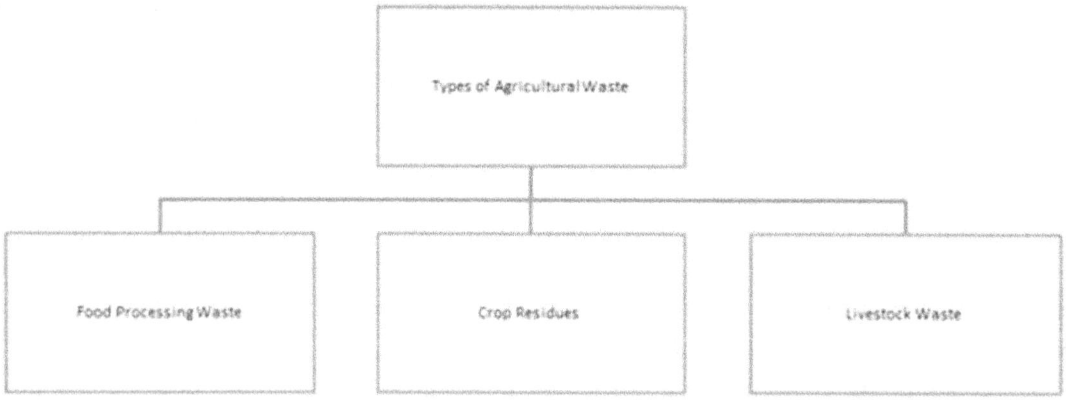

1.2 Crop Residues

Crop residues constitute a substantial portion of agricultural waste, encompassing the non-edible parts of plants left behind in the field after harvesting. These residues, including stalks, leaves, and stems, are conventionally perceived as waste. However, they hold immense potential as biosorbents due to their rich organic content and structural composition. The characteristics of crop residues vary widely based on the type of crop and agricultural practices (Sarkar et al., 2020). For instance, rice straw differs from corn stalks in composition. The organic matter in crop residues contains functional groups such as hydroxyl, carboxyl, and phenolic groups, which are crucial for biosorption processes. Studies have demonstrated that crop residues exhibit significant biosorption capabilities, particularly in removing heavy metals from aqueous solutions. The functional groups on the surface of crop residues play a pivotal role in adsorbing pollutants through mechanisms like ion exchange and physical adsorption (Yang et al., 2019). The applications of crop residues as biosorbents extend to diverse areas, from water treatment in agricultural runoff to the remediation of industrial effluents.

1.3 Food Processing Waste

Food processing industries produce a significant amount of waste, which includes anything from peelings and cores to the leftovers of processed grains, fruits, and vegetables (H. Kumar et al., 2020). Although disposing of this waste causes environmental issues, it also offers a chance to turn these leftovers into useful biosorbents. Waste from food processing is varied and contains both bioactive chemicals and organic components that are high in nutrients (Georganas et al., 2020). The waste is a diverse resource for biosorption applications since it frequently contains proteins, lipids, and polysaccharides. The biosorption capability of food processing waste is enhanced by its organic character. By using processes like complexation and physical adsorption, these materials are capable of efficiently absorbing pollutants, such as colors and organic toxins (De Gisi et al., 2016). A circular economy is supported by the use of food processing waste as biosorbents. By repurposing waste from food processing industries, not only are pollution concerns addressed, but a valuable resource is redirected towards environmental remediation (Bhatia ct al., 2023).

1.4 Livestock Waste

Livestock farming is a significant source of waste, primarily in the form of manure and other byproducts. While managing livestock waste is crucial for preventing environmental pollution, exploring its biosorbent potential offers a sustainable approach to waste utilization. Livestock waste is rich in organic matter, nutrients, and microbial content (Xie et al., 2023). The composition varies based on the type of livestock, diet, and management practices. Manure, for instance, contains organic compounds, ammonia, and phosphorus. Livestock waste, particularly manure, has demonstrated biosorption capabilities, especially in removing nutrients like phosphorus and heavy metals from water. The organic content and microbial activity in manure contribute to effective biosorption mechanisms (Ojuederie & Babalola, 2017). The use of livestock waste as biosorbents contributes to nutrient recovery and water purification. By harnessing the biosorption potential of manure, it is possible to mitigate the environmental impact of livestock farming while simultaneously addressing water quality concerns (R. R. Kumar et al., 2013).

The exploration of different types of agricultural waste opens avenues for sustainable waste management and environmental remediation. Crop residues, food processing waste, and livestock waste, when viewed through the lens of biosorption, transform from mere byproducts to valuable resources (Duque-Acevedo et al., 2020). As the chapters progress, a deeper understanding of the specific applications, challenges, and innovations related to each type of agricultural waste will unfold, contributing to the overall narrative of sustainable agricultural practices and environmental stewardship.

2. TECHNIQUES FOR BIO-SORBENTS PREPARATION FROM AGRICULTURAL WASTE

2.1 Preparation of Crop Residues Biosorbents

In the realm of environmental sustainability, the preparation of biosorbents from crop residues stands as a testament to the innovative utilization of agricultural waste (Peng et al., 2023). This section explores diverse techniques (Figure 2) employed in transforming crop residues into effective biosorbents, emphasizing their potential applications in pollutant removal.

Figure 2.

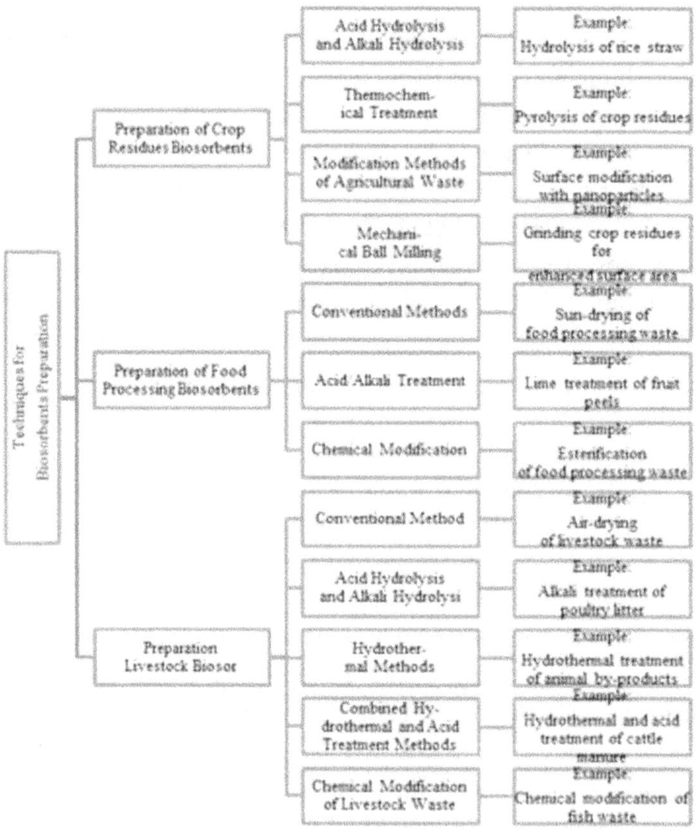

2.1.1. Acid Hydrolysis and Alkali Hydrolysis

Acid hydrolysis and alkali hydrolysis are pivotal techniques in the preparation of biosorbents from crop residues. Acid hydrolysis involves the use of strong acids like sulfuric acid or hydrochloric acid to break down complex structures within crop residues (Osman et al., 2023). This process cleaves chemical bonds, depolymerizing polysaccharides and lignocellulosic materials. For instance, applying acid hydrolysis to corn stover releases bioactive components such as cellulose and hemicellulose. The resulting material exhibits an increased surface area and enhanced sorption capabilities, making it a potent biosorbent for applications ranging from water treatment to industrial effluent remediation (Amalina et al., 2022).

On the other hand, alkali hydrolysis utilizes alkaline solutions, often sodium hydroxide or potassium hydroxide, to break down crop residues. This method is particularly effective in extracting hemicellulose, as alkalis react with acetyl and uronic acid groups in hemicellulose, leading to its solubilization (Groen et al., 2006). Wheat straw subjected to alkali hydrolysis undergoes structural modifications, resulting in improved adsorption properties. The altered chemical characteristics and increased porosity make the treated crop residues adept at adsorbing contaminants in a targeted and efficient manner (De Gisi et al., 2016).

2.1.2. Thermochemical Treatment

Thermochemical treatment represents a paradigm shift in biosorbent preparation, i.e. involving subjecting crop residues to high temperatures in the presence of chemicals (Adewuyi, 2020). Processes such as pyrolysis or gasification are employed, exemplifying the versatility of these techniques. Rice husks, when subjected to pyrolysis at elevated temperatures, undergo a transformation into biochar or activated carbon. This process enhances the porosity of the material, providing a multitude of active sites for sorption. The resulting biosorbent, characterized by stability and a high absorption capacity, exemplifies the effectiveness of thermochemical treatment in repurposing crop residues for environmental remediation (Amalina et al., 2022). This method not only tackles the issue of agricultural waste but also aligns with the broader goal of sustainable resource utilization.

2.1.3. Modification Methods of Agricultural Waste

To further enhance the sorption properties of agricultural waste, various modification methods come into play. Techniques such as esterification and etherification introduce deliberate alterations to the surface chemistry of the biosorbent (Bădescu et al., 2018). Esterification involves the reaction of hydroxyl groups with carboxylic acid, forming ester bonds. Sugarcane bagasse, for instance, can undergo esterification with acetic acid, resulting in the creation of ester linkages that alter the surface chemistry and enhance sorption properties (Hoang et al., 2021).

Etherification, another modification method, entails the substitution of hydrogen atoms with ether groups. By subjecting corn residues to etherification, the resultant biosorbent showcases improved capabilities in sequestering pollutants owing to the introduced functional groups (SUD et al., 2008). These modification methods offer a tailored approach to biosorbent development, allowing researchers to fine-tune the properties of agricultural waste for optimal pollutant removal (Kwikima et al., 2021). The versatility of these methods emphasizes the potential for customization, ensuring that biosorbents can be optimized for specific environmental challenges.

2.1.4. Mechanical Ball Milling

Mechanical ball milling introduces a physical method that utilizes mechanical force to reduce the particle size of crop residues. Employing ball mills, this process involves the impact and abrasion caused by milling balls, breaking down the material into smaller particles (Neikov, 2009). Rice straw, when subjected to mechanical ball milling, undergoes a reduction in particle size, leading to increased surface area and porosity (Gao et al., 2017). This finely ground biosorbent exhibits enhanced sorption capacity, making it a valuable tool in pollutant removal applications where finer particles offer increased adsorption sites (Kanamarlapudi et al., 2018).

The dynamic nature of the preparation of biosorbents from crop residues reflects the ongoing quest for sustainable solutions to environmental challenges (Osman et al., 2023). These innovations not only address the issue of agricultural waste management but also contribute to a circular economy where these residues are transformed into valuable resources, promoting sustainability and resilience in environmental stewardship (Bakan et al., 2022).

2.2 Preparation of Food Processing Biosorbents

In the pursuit of sustainable environmental solutions, the utilization of food processing waste as biosorbents offers a promising avenue. The preparation of these biosorbents involves employing various techniques, each contributing to the enhancement of sorption capabilities. From conventional methods that capitalize on the inherent properties of food processing waste to sophisticated chemical modifications, this section explores the diverse approaches (Figure 2), emphasizing their potential impact on pollutant removal (Roy et al 2023).

2.2.1. Conventional Methods

Conventional methods provide a straightforward and resource-efficient approach to preparing biosorbents from food processing waste (Razzak et al., 2022). The inherent bioactive compounds and fibers present in residues from fruits, vegetables, and other food processing by-products lay the foundation for their sorption capabilities. Citrus peels, for example, a common by-product of juice extraction, are rich in pectin and cellulose. These natural components endow citrus peels with the ability to adsorb heavy metals like lead and cadmium directly from aqueous solutions, showcasing the efficacy of this uncomplicated yet powerful method (Akinhanmi et al., 2020).

The simplicity of conventional methods lies in their ability to capitalize on the natural composition of food processing waste without requiring extensive pre-treatment. This not only makes the process cost-effective but also aligns with the principles of sustainability by minimizing resource inputs while maximizing output in the form of efficient biosorbents.

2.2.2. Acid/Alkali Treatment

Moving beyond conventional approaches, acid and alkali treatments emerge as transformative techniques for optimizing the sorption characteristics of food processing waste. Analogous to methods applied to crop residues, acid or alkali treatments aim to break down complex structures within the waste material (Bala et al., 2023). By doing so, the accessibility of bioactive components is enhanced, consequently improving the biosorption capacity of the material.

Consider apple pomace, a residual by-product of cider production. Through alkali treatment, the surface chemistry of apple pomace can be modified, exposing more hydroxyl groups (Perussello et al., 2017). This alteration results in an increased capacity to adsorb organic pollutants, such as dyes from wastewater. The acid/alkali treatment not only facilitates the modification of surface characteristics but also amplifies the versatility of food processing waste as biosorbents by expanding their applicability to a wider range of pollutants (F. Ahmad & Zaidi, 2021).

2.2.3. Chemical Modification

Chemical modification techniques introduce a layer of precision and tailoring to the preparation of food processing biosorbents (Adewuyi, 2020). Through methods like impregnation and functionalization, researchers gain the ability to customize the material for targeted sorption applications (R. Girish & ., 2018). This involves the introduction of specific functional groups or alterations to the surface chemistry of the waste material.

Take, for instance, rice bran, a by-product of rice milling. Chemical modification can entail functionalizing its surface with amino groups. This deliberate alteration enhances its affinity for heavy metal ions, showcasing improved selectivity and efficiency in removing targeted pollutants (Sapwarobol et al., 2021). Chemical modification not only elevates the biosorption properties of the material but also allows for a nuanced approach to environmental remediation by addressing specific pollutant types with tailored solutions.

2.3 Preparation of Livestock Biosorbents

Livestock waste, a substantial by-product of animal husbandry, has emerged as a promising resource for the development of biosorbents—a critical facet of sustainable environmental solutions (Koul et al., 2022). This section delves into diverse methodologies for converting livestock waste into effective biosorbents (Figure 2), showcasing their versatility and potential in addressing environmental pollution challenges.

2.3.1. Conventional Method

The conventional method involves utilizing livestock waste directly as biosorbents, with requiring minimal pre-treatment. Residues from animal husbandry activities, including manure and by-products, are harnessed for their inherent sorption potential (Jayathilakan et al., 2012). An illustrative example is the utilization of poultry litter, a common by-product rich in organic materials like cellulose and proteins. By leveraging these natural components, poultry litter serves as a direct biosorbent, minimizing the need for extensive pre-treatment processes (Gaspareto et al., 2020). This straightforward approach aligns with sustainability principles by repurposing waste into a valuable resource for pollutant removal.

2.3.2. Acid Hydrolysis and Alkali Hydrolysis

Applying acid and alkali hydrolysis methods to livestock waste represents a strategic approach to unlocking its sorption potential. Acid hydrolysis involves using strong acids like sulfuric acid or hydrochloric acid to break down complex structures within the waste (Becker et al., 2021). For instance, treating cattle manure with acid hydrolysis releases peptides and amino acids, contributing to enhanced sorption properties (Saady et al., 2021). Similarly, alkali hydrolysis utilizes alkaline solutions, such as sodium hydroxide or potassium hydroxide, to break down livestock waste (Arias et al., 2018). The choice between acid and alkali hydrolysis depends on the specific composition of the waste, emphasizing the importance of tailoring methods to maximize sorption efficiency.

2.2.3. Hydrothermal Methods

Hydrothermal treatment stands out as an innovative method for utilizing livestock waste. This technique involves subjecting waste to elevated temperatures and pressures in the presence of water. The process leads to the breakdown of complex structures within the waste, releasing valuable components for sorption (C. Wang et al., 2022). In practice, hydrothermal treatment of swine manure extracts bioactive compounds like phosphorus, enhancing the sorption capabilities of the resulting biosorbent. The efficacy of hydrothermal methods lies not only in their ability to extract valuable materials but also in their contribution to the overall sustainability of the waste utilization process (Zhang et al., 2022).

2.2.4. Combined Hydrothermal and Acid Treatment Methods

Combining hydrothermal and acid treatments represents a synergistic approach to maximize the sorption properties of livestock waste bio-sorbents (Mahato et al., 2020). Hydrothermal treatment opens up the structure of the waste, making it more accessible, while subsequent acid treatment further breaks down complex molecules, releasing additional bioactive components. This dual approach maximizes the extraction of valuable materials, contributing to the efficiency of the bio-sorption process. For example, employing a combination of hydrothermal and acid treatments in dairy manure processing enhances the release of organic compounds, amplifying the sorption capabilities of the resulting biosorbent (Liu et al., 2022). This integrated methodology showcases the potential for sophisticated techniques to optimize waste utilization.

2.2.5. Chemical Modification of Livestock Waste

Chemical modification techniques add a layer of precision to the development of livestock waste biosorbents (Adewuyi, 2020). Processes such as esterification and amidation modify the surface chemistry of the biosorbent, introducing functional groups that enhance sorption properties. By subjecting poultry litter to esterification, the surface can be tailored to increase the biosorbent's affinity for heavy metal ions (Raji et al., 2023). This exemplifies how chemical modification can be employed to fine-tune livestock waste for targeted pollutant removal applications. The versatility of these modification methods positions them as valuable tools in tailoring biosorbents to specific environmental challenges.

The diverse techniques for preparing bio-sorbents from agricultural waste offer a range of options for sustainable water and wastewater treatment. The selection of a specific method depends on the characteristics of the raw material, the desired properties of the bio-sorbent, and the targeted application. Continued research and development in this field are crucial for advancing the effectiveness and applicability of bio-sorbents in environmental remediation.

3. CHARACTERIZATION OF BIOSORBENTS BY VARIOUS TECHNIQUES

Biosorbents, derived from agricultural wastes and organic materials, have emerged as promising tools for environmental remediation. Their efficiency in adsorbing contaminants from aqueous solutions makes them valuable candidates for sustainable pollution cleanup. However, understanding the characteristics and properties of these biosorbents is crucial for optimizing their performance. This chapter explores the comprehensive characterization of biosorbents, delving into various advanced techniques employed to unveil their intricate structural, chemical, and morphological features.

3.1. Fourier Transform Infrared (FTIR) Spectroscopy

In the intricate landscape of biosorbent characterization, Fourier Transform Infrared (FTIR) spectroscopy emerges as a beacon, providing a profound exploration into the chemical composition of biosorbents derived from agricultural waste (Saxena et al., 2017). The technique is adept at identifying crucial functional groups such as hydroxyl (-OH), carboxyl (-COOH), and amino (-NH2) groups present in cellulose, hemicellulose, and lignin—the primary building blocks of agricultural waste-derived biosorbents (F. Ahmad & Zaidi, 2021). By relying on the absorption of infrared light by these functional groups, FTIR generates a spectrum that serves as a molecular fingerprint, revealing the unique chemical constituents and potential adsorption sites on the biosorbent surface (Tiquia-Arashiro et al., 2023).

FTIR's molecular fingerprinting capability becomes instrumental in tailoring biosorbents for enhanced pollutant adsorption, aligning with the overarching theme of utilizing agricultural waste as alternative biosorbents (Diale et al., 2023). The technique thus not only identifies the chemical makeup but also provides a roadmap for optimizing biosorbent functionality.

3.2. Scanning Electron Microscopy (SEM) and Transmission Electron Microscopy (TEM)

The structural intricacies of agricultural waste-based biosorbents come to life through the lenses of Scanning Electron Microscopy (SEM) and Transmission Electron Microscopy (TEM) (Mandal et al., 2017). These complementary techniques offer visualizations at both macro and nanoscales, crucial for understanding surface morphology—a parameter intricately linked to biosorbent adsorption capacity. SEM unveils macroscopic features such as pores, irregularities, and surface roughness, which directly influence the biosorbent's ability to adsorb pollutants (Adewuyi, 2020). On the other hand, TEM, with its higher resolution, provides a deeper dive into nanoscale details, elucidating internal structures and the distribution of nanoparticles on the biosorbent surface (Blaga et al., 2022).

3.3. X-ray Diffraction (XRD) Analysis

In the quest to harness the potential of agricultural waste as alternative biosorbents, X-ray Diffraction (XRD) analysis emerges as a crucial tool for unraveling the crystalline structures embedded within biosorbent matrices. For cellulose-rich materials like crop residues, XRD becomes instrumental in identifying the arrangement of cellulose microfibrils and crystalline regions within the biosorbent (He et al., 2022). This knowledge is paramount as it unveils how structural changes induced by chemical treatments or pollutant loading manifest in alterations to the XRD pattern. Understanding the crystalline nature of agricultural waste-based biosorbents not only contributes to characterizing their structural response but also guides strategies for optimizing their remediation efficiency.

3.4. Brunauer-Emmett-Teller (BET) Analysis

The optimization journey of agricultural waste-based biosorbents delves into the quantification of surface area and porosity through Brunauer-Emmett-Teller (BET) analysis. As the biosorbent landscape expands, understanding surface area becomes pivotal, directly correlating with the availability of active sites for adsorption (Sen, 2023). BET analysis, with its focus on nitrogen adsorption-desorption isotherms, becomes the compass guiding researchers to calculate surface area, pore volume, and pore size distribution (Tan et al., 2012).

3.5. Energy-Dispersive X-Ray Spectroscopy (EDS)

Complementing the visual insights provided by SEM, Energy-Dispersive X-ray Spectroscopy (EDS) takes center stage in unmasking the elemental composition of agricultural waste-derived biosorbents (Wyroba et al., 2015). This technique unveils the presence and distribution of elements on the biosorbent surface, offering valuable information about impurities or contaminants adsorbed during pollutant removal experiments. EDS not only contributes to comprehensive characterization but also aids in correlating elemental composition with adsorption performance, ensuring the effectiveness of biosorbent applications (Amalina et al., 2022).

3.6. Thermogravimetric Analysis (TGA)

The transformative journey of agricultural waste into biosorbents necessitates an evaluation of thermal stability, a task entrusted to Thermogravimetric Analysis (TGA) (Gómez et al., 2007) . For biosorbents derived from agricultural waste, understanding the temperature at which components decompose or volatilize becomes paramount. TGA serves as the thermal compass, guiding researchers to identify temperature ranges suitable for activation or modification processes (Díaz et al., 2021). This information ensures that the thermal integrity of agricultural waste-based biosorbents is maintained, aligning with the sustainable utilization theme and optimizing their functionality for environmental remediation.

3.7. X-Ray Photoelectron Spectroscopy (XPS)

Surface chemistry takes the spotlight as X-ray Photoelectron Spectroscopy (XPS) probes the chemical intricacies of agricultural waste-derived biosorbents (Hartmann & Lamb, 1997). XPS, being a surface-sensitive technique, identifies the oxidation states of elements on the biosorbent surface and offers insights into the presence of functional groups. As biosorbents undergo chemical treatments or interact with pollutants, XPS becomes the lens through which researchers comprehend alterations in binding energies, contributing to a nuanced understanding of surface properties (Jana et al., 2016). This detailed insight into surface chemistry aligns with the central theme of transforming agricultural waste into a resource for sustainable environmental remediation.

3.8. Nuclear Magnetic Resonance (NMR) Spectroscopy

In the quest to utilize agricultural waste for environmental remediation, Nuclear Magnetic Resonance (NMR) spectroscopy unravels the molecular structure of biosorbents (N. F. E. I. Nestle & Kimmich,

2000). This technique provides invaluable insights into the arrangement of atoms and the connectivity of molecular groups within agricultural waste-derived biosorbents. As biosorbents undergo chemical modifications or interact with adsorbates, NMR spectroscopy becomes the lens through which researchers comprehend changes in molecular structure (N. Nestle, 2002). This detailed perspective contributes to a comprehensive understanding of the structural characteristics of biosorbents, aligning with the overarching theme of optimizing their efficiency for environmental remediation.

3.9. Surface Charge and Zeta Potential Measurements

The electrochemical identity of agricultural waste-derived biosorbents comes to the forefront as surface charge and zeta potential measurements unveil crucial insights. In the context of biosorbents from agricultural waste, these measurements offer a window into the surface charge and the potential to attract or repel charged species in aqueous solutions. Understanding the zeta potential is essential for assessing the stability of biosorbent suspensions and predicting their interactions with pollutants (Zinicovscaia et al., 2020). These electrochemical insights contribute to a comprehensive characterization, guiding the optimization of agricultural waste-derived biosorbents for efficient pollutant removal—a pivotal step towards the sustainable and effective utilization of agricultural waste in environmental remediation (Karim et al., 2023).

4. FACTORS INFLUENCING ADSORPTION PROPERTIES

Adsorption, a crucial process in environmental science and remediation, is governed by a multitude of factors that intricately shape the interaction between bio-sorbents, derived from agricultural waste, and adsorbates. This chapter provides a comprehensive exploration of four key factors: pH and temperature, particle size and surface area, concentration of adsorbent, and exposure time. A detailed analysis of these factors aims to offer profound insights for optimizing adsorption processes utilizing bio-sorbents from agricultural waste.

4.1. pH and Temperature

4.1.1. pH

The pH of a solution profoundly influences adsorption efficiency by affecting the surface charge of both the bio-sorbent and adsorbate (Albatrni et al., 2022). In the context of bio-sorbents derived from agricultural waste, such as those from crop residues, the pH of the solution impacts the ionization state of surface functional groups. For instance, activated carbon, a common bio-sorbent, possesses carboxyl and hydroxyl groups whose ionization is pH-dependent. Recognizing the point of zero charge (PZC) or the pH of maximum adsorption is crucial for optimizing electrostatic interactions and overall adsorption capacity (Adam, 2016).

4.1.2. Temperature

The role of temperature in adsorption processes is complex and system specific. While adsorption is generally exothermic, higher temperatures can enhance adsorption for some bio-sorbents derived from agricultural waste (Sen, 2023). Understanding the temperature dependence is crucial for optimizing adsorption processes, especially in scenarios where thermodynamics play a key role. In agricultural waste bio-sorbents, variations in temperature can impact the kinetic energy, influencing molecular collisions and adsorbate-adsorbent interactions.

4.2. Particle Size and Surface Area

4.2.1. Particle Size

The particle size of bio-sorbents from agricultural waste significantly impacts the available surface area for adsorption. In the context of crop residues or food processing by-products, finer particles generally offer increased surface area, providing more active sites for adsorption (Holliday et al., 2022). However, the relationship between particle size and adsorption efficiency is intricate. Smaller particles may agglomerate, reducing effective surface area, while larger particles may limit accessibility to active sites. Striking a balance is crucial for optimizing the adsorption process ("Particle Size Reduction and Enlargement," 2019).

4.2.2. Surface Area

The specific surface area of the bio-sorbent is a fundamental parameter influencing adsorption capacity. For agricultural waste bio-sorbents, like activated carbon derived from crop residues, techniques like BET analysis provide quantitative insights into surface area. These materials often exhibit high surface areas due to their porous nature, enhancing accessibility and contributing to efficient adsorption of pollutants. Strategies to increase surface area, such as modifying particle size or employing hierarchical structures, become critical for enhancing the efficiency of agricultural waste bio-sorbents (Liu et al., 2022; Muhammad et al., 2022).

4.3. Concentration of Adsorbent

4.3.1. Adsorbent Dosage

The concentration of the adsorbent, known as adsorbent dosage, is a critical factor governing adsorption efficiency. For bio-sorbents from agricultural waste, understanding the dosage-response relationship is vital. While increasing the amount of adsorbent generally enhances pollutant removal, there exists an optimal dosage beyond which additional adsorbent may not significantly improve performance. Considerations such as agglomeration of particles or overcrowding of active sites must be taken into account for cost-effective and efficient adsorption processes (Chang & Li, 2019).

4.4. Exposure Time

4.4.1. Kinetics of Adsorption

Exposure time, or the duration of contact between the bio-sorbent and adsorbate, directly influences the kinetics of adsorption. Agricultural waste-derived bio-sorbents often exhibit unique kinetic behaviors based on factors such as the nature of the adsorbent and adsorbate, concentration, and temperature. Understanding the kinetics is essential for optimizing exposure times and predicting the efficiency of the process under different scenarios. Kinetic modeling can provide valuable insights into the adsorption mechanism (Raji et al., 2023).

4.4.2. Equilibrium and Pseudo-Equilibrium

Achieving equilibrium in adsorption refers to the point where the rate of adsorption equals the rate of desorption. In the context of agricultural waste bio-sorbents, understanding the dynamics of achieving equilibrium is crucial. Pseudo-equilibrium may be observed when the process appears to reach equilibrium, but subtle changes in concentration still occur. Studying these dynamics aids in determining the appropriate exposure time for practical applications, ensuring that the adsorption process is optimized for real-world scenarios (Siva Kumar et al., 2019).

5. ADSORBENT CLASSIFICATION AND THEIR APPLICATION

Bio-sorbents derived from agricultural waste represent a paradigm shift in environmental remediation, offering a sustainable and eco-friendly approach to address pollution challenges (Ngeno et al., 2022). These bio-sorbents are derived from diverse agricultural residues, such as crop remnants and livestock waste, showcasing the potential of transforming waste into valuable resources. This chapter comprehensively explores the classification of bio-sorbents based on origin, physical structure, chemical nature, and adsorption mechanisms, providing an in-depth understanding of their applications in real-life scenarios.

5.1 Natural and Synthetic Bio-Sorbents

5.1.1 Natural Bio-Sorbents From Agricultural Waste

The utilization of agricultural waste as natural bio-sorbents is a testament to the eco-friendly practices emerging in environmental science. Crop residues, including rice husks, sugarcane bagasse, and corn cobs, serve as natural adsorbents due to their inherent properties. For instance, rice husk biochar, a by-product of rice milling, has gained attention for its porous structure and adsorption capabilities (Juela et al., 2021). These materials, sourced directly from agricultural processes, leverage their natural composition for effective pollutant removal.

5.1.2 Engineered Bio-Sorbents From Agricultural Waste

Advances in materials science have led to the engineering of bio-sorbents from agricultural waste. Through processes like chemical modification or activation, these materials are tailored to enhance their adsorption capabilities. Engineered bio-sorbents may include activated carbon derived from agricultural residues or modified crop residues with increased surface area for improved adsorption (Adegoke et al., 2023). This synthesis of engineered materials reflects the evolving landscape of bio-sorbent development, where agricultural waste serves as a precursor for customized solutions.

5.2 Porous and Non-Porous Bio-Sorbents

5.2.2 Porous Bio-Sorbents From Agricultural Waste

The porous nature of certain agricultural waste materials is harnessed to create bio-sorbents with extensive internal surfaces. Crop residues subjected to thermochemical treatments, such as pyrolysis, result in biochar with remarkable porosity. This property enhances the availability of active sites for adsorption, making them particularly effective in capturing pollutants from aqueous solutions. The porous bio-sorbents derived from agricultural waste contribute to the sustainable utilization of natural resources for environmental benefit (F. A. Ahmad, 2023).

5.2.3 Non-Porous Bio-Sorbents From Agricultural Waste

While some agricultural waste materials inherently lack intricate internal structures, they may still exhibit non-porous characteristics suitable for adsorption. Certain fibrous crop residues, when utilized directly as bio-sorbents, demonstrate specific selectivity despite lacking extensive internal surfaces (Karim et al., 2023). This diversity in material design showcases the adaptability of agricultural waste for adsorption purposes, even in its natural form.

5.3 Inorganic and Organic Bio-Sorbents

5.3.1 Inorganic Bio-Sorbents From Agricultural Waste

Inorganic components present in agricultural waste, such as ash or mineral content, contribute to the inorganic nature of certain bio-sorbents. These materials, after suitable processing, can exhibit specific cation exchange capacities, enabling selective adsorption of ions based on their chemical composition (Sadh et al., 2018). Zeolites derived from agricultural waste are a notable example, where the aluminosilicate structure contributes to ion exchange capabilities, making them effective in the removal of heavy metals from water (Y. Li et al., 2017).

5.3.2 Organic Bio-Sorbents From Agricultural Waste

The majority of bio-sorbents from agricultural waste fall under the category of organic materials. Crop residues, rich in organic compounds like cellulose and lignin, serve as valuable precursors (S & P, 2019). Through various processing techniques, these materials can be harnessed as organic bio-sorbents, providing a platform for surface modification and precise control over adsorption characteristics. Activated carbon derived from agricultural by-products, with its porous structure and high surface area, stands out as a versatile organic bio-sorbent with applications in water and air purification.

5.4 Chemo-Sorbents and Physio-Sorbents

5.4.1 Chemo-Sorbents From Agricultural Waste

Certain bio-sorbents from agricultural waste engage in chemical interactions with adsorbate molecules, forming covalent or ionic bonds. This chemisorption mechanism leads to effective and often irreversible binding of pollutants. Modified crop residues with introduced functional groups, for example, can act as chemo-sorbents. This category aligns with the principles of green chemistry, utilizing agricultural waste in a sustainable manner for applications where irreversible adsorption is advantageous.

5.4.2 Physio-Sorbents From Agricultural Waste

The majority of bio-sorbents from agricultural waste rely on physical interactions like Van der Waals forces and hydrogen bonding. Crop residues processed through hydrothermal methods or activated carbon derived from agricultural by-products fall into the category of physio-sorbents. Their adsorption is typically reversible without significant chemical changes, making them suitable for scenarios where regeneratibility is a crucial aspect of the adsorption process (Agboola & Benson, 2021).

5.5 Applications With Real-Life Examples

5.5.1 Water Treatment

Rice Husk Biochar for Heavy Metal Removal: In regions with heavy metal contamination in water sources, rice husk biochar has been employed as a bio-sorbent. The porous structure of biochar facilitates the effective adsorption of heavy metals such as lead and cadmium, contributing to water purification efforts in a sustainable manner (Z. Li et al., 2023).

Activated Carbon from Crop Residues for Organic Contaminant Removal: Activated carbon derived from agricultural residues, such as coconut shells or sugarcane bagasse, has found application in the removal of organic contaminants from water. The high surface area and porosity of activated carbon make it highly effective in adsorbing pollutants like pesticides and dyes, providing a green alternative for water treatment (Demiral & Gündüzoğlu, 2010).

5.5.2 Air Quality Improvement

Modified Livestock Waste for Ammonia Adsorption: Livestock waste, when subjected to modification processes, has been explored as a bio-sorbent for ammonia adsorption in air purification systems. The tailored surface chemistry enhances its affinity for ammonia molecules, offering a sustainable solution for mitigating air pollution in agricultural settings (Dey et al., 2021).

5.5.3 Waste Minimization in Agriculture

Utilization of Crop Residues for Nutrient Recovery: Crop residues, often considered agricultural waste, can be repurposed as bio-sorbents to recover excess nutrients from agricultural runoff. Adsorbing nutrients like phosphorus and nitrogen, these bio-sorbents contribute to sustainable agriculture practices, minimizing environmental impact and promoting circular economy principles (Karić et al., 2022).

5.6 Drawbacks of Adsorbent Classes and Future Directions

The exploration of bio-sorbents derived from agricultural waste has illuminated their potential as environmentally friendly solutions for water and wastewater treatment. While these bio-sorbents have demonstrated significant promise, it is crucial to comprehensively analyze their drawbacks to refine and optimize their applications. This section delves into the nuanced limitations associated with distinct classes of adsorbents derived from agricultural waste and outlines in detail potential future directions to overcome these challenges.

5.6.1 Drawbacks of Adsorbent Classes

5.6.1.1 Crop Residues Bio-Sorbents

Crop residues-based bio-sorbents present challenges that include limited selectivity, variable composition, and slow adsorption kinetics. The heterogeneous nature of crop residues introduces variations in adsorption sites, impacting the material's efficacy in removing specific pollutants. Moreover, the diverse composition stemming from factors such as crop type and cultivation practices makes it difficult to produce consistent bio-sorbents. The relatively sluggish adsorption kinetics may hinder their practical application, especially in scenarios requiring rapid and efficient pollutant removal.

5.6.1.2 Food Processing Waste Bio-Sorbents

Bio-sorbents derived from food processing waste encounter several challenges, including high moisture content, limited regeneration potential, and a complex matrix. The presence of moisture within these bio-sorbents can compromise sorption capacity by engaging in competition with adsorbate molecules. Some materials may face challenges in being regenerated effectively, impacting their overall lifespan and sustainability. Furthermore, the intricate matrix of food processing waste, consisting of components like fats and proteins, can lead to fouling issues in water treatment systems, necessitating additional maintenance.

5.6.1.3 Livestock Waste Bio-Sorbents

Livestock waste bio-sorbents, particularly those derived from manure, exhibit drawbacks such as high organic content, the release of ammonia, and limited mechanical strength. The substantial organic content in these materials raises concerns about leaching organic matter into treated water, potentially causing secondary pollution. Additionally, the release of ammonia during adsorption processes poses environmental risks. The materials may also exhibit limited mechanical strength, presenting practical challenges, particularly in fixed-bed systems.

6. FUTURE DIRECTIONS

6.1 Crop Residues Bio-Sorbents

To overcome the limitations associated with crop residues-based bio-sorbents, future research should prioritize tailored modification techniques, nanostructuring for improved kinetics, and the development of standardized production protocols. Tailoring the surface chemistry through advanced functionalization offers an avenue to enhance selectivity, while nanostructuring can address the challenge of slow adsorption kinetics by increasing the surface area of the bio-sorbent. Standardized production protocols will contribute to ensuring the reproducibility and reliability of these bio-sorbents.

6.1.1 Food Processing Waste Bio-Sorbents

Future endeavors for bio-sorbents derived from food processing waste should focus on developing efficient drying and pre-treatment strategies, regeneration studies, and integration with other treatment processes. Effective drying methods and pre-treatments can mitigate challenges associated with high moisture content. Regeneration studies are essential to extend the lifespan of these materials, while integration with other treatment processes, such as membrane filtration or biological treatment, can provide synergistic effects and overcome the complex matrix issue.

6.1.2 Livestock Waste Bio-Sorbents

In the context of livestock waste bio-sorbents, future research should emphasize nutrient recovery technologies, ammonia capture and valorization, and enhanced mechanical stability. Nutrient recovery technologies can address concerns related to the high organic content by capturing valuable nutrients from the waste. Strategies for capturing and valorizing released ammonia can transform what could be an environmental pollutant into a valuable resource. Additionally, improving the mechanical stability of these bio-sorbents is crucial for their practical application, ensuring durability and efficacy in water treatment systems.

While bio-sorbents derived from agricultural waste represent a sustainable avenue for water treatment, a thorough understanding of their drawbacks is imperative for their effective implementation. Future research endeavors should strategically focus on the detailed recommendations provided, ensuring that these bio-sorbents evolve into reliable and efficient materials capable of addressing environmental pollution and fostering sustainable water management practices.

7. CONCLUSION

This chapter embarked on a journey through the untapped potential of agricultural waste as alternative biosorbents. Beginning with an insightful introduction, we navigated the landscape of crop residues, food processing waste, and livestock by-products. The meticulous exploration of techniques for biosorbent preparation revealed a diverse toolkit, from acid hydrolysis to chemical modifications, showcasing adaptability and efficiency.

Characterization techniques provided a crucial understanding of structural and chemical compositions, laying the groundwork for tailored applications. The examination of factors influencing adsorption properties highlighted the nuanced interplay between pH, temperature, particle size, surface area, concentration, and exposure time. Biosorbent classification based on source and their applications underscored versatility, offering solutions from wastewater treatment to soil remediation. Despite the promising strides, identified drawbacks emphasize the ongoing nature of this research, calling for continued innovation.

In essence, this chapter illuminates the transformative potential of agricultural waste-derived biosorbents, merging agriculture, waste management, and environmental sustainability. As we forge ahead, the chapter stands as a guide towards a greener and more sustainable future, marked by the resilient integration of nature's own resources into our environmental solutions.

REFERENCES

Adam, O. (2016). Removal of Resorcinol from Aqueous Solution by Activated Carbon: Isotherms, Thermodynamics and Kinetics. *American Chemical Science Journal*, *16*(1), 1–13. doi:10.9734/ACSJ/2016/27637

Adegoke, K. A., Akinnawo, S. O., Adebusuyi, T. A., Ajala, O. A., Adegoke, R. O., Maxakato, N. W., & Bello, O. S. (2023). Modified biomass adsorbents for removal of organic pollutants: A review of batch and optimization studies. *International Journal of Environmental Science and Technology*, *20*(10), 11615–11644. doi:10.1007/s13762-023-04872-2

Adewuyi, A. (2020). Chemically Modified Biosorbents and Their Role in the Removal of Emerging Pharmaceutical Waste in the Water System. *Water (Basel)*, *12*(6), 1551. doi:10.3390/w12061551

Agboola, O. D., & Benson, N. U. (2021). Physisorption and Chemisorption Mechanisms Influencing Micro (Nano) Plastics-Organic Chemical Contaminants Interactions: A Review. *Frontiers in Environmental Science*, *9*, 678574. Advance online publication. doi:10.3389/fenvs.2021.678574

Ahmad, F., & Zaidi, S. (2021). Potential Use of Agro/Food Wastes as Biosorbents in the Removal of Heavy Metals. In *Emerging Contaminants*. IntechOpen. doi:10.5772/intechopen.94175

Ahmad, F. A. (2023). The use of agro-waste-based adsorbents as sustainable, renewable, and low-cost alternatives for the removal of ibuprofen and carbamazepine from water. *Heliyon*, *9*(6), e16449. doi:10.1016/j.heliyon.2023.e16449 PMID:37292321

Akinhanmi, T. F., Ofudje, E. A., Adeogun, A. I., Aina, P., & Joseph, I. M. (2020). Orange peel as low-cost adsorbent in the elimination of Cd(II) ion: Kinetics, isotherm, thermodynamic and optimization evaluations. *Bioresources and Bioprocessing*, *7*(1), 34. doi:10.1186/s40643-020-00320-y

Albatrni, H., Qiblawey, H., & Al-Marri, M. J. (2022). Walnut shell based adsorbents: A review study on preparation, mechanism, and application. *Journal of Water Process Engineering*, *45*, 102527. doi:10.1016/j.jwpe.2021.102527

Amalina, F., Razak, A. S. A., Krishnan, S., Zularisam, A. W., & Nasrullah, M. (2022). A comprehensive assessment of the method for producing biochar, its characterization, stability, and potential applications in regenerative economic sustainability – A review. *Cleaner Materials*, *3*, 100045. doi:10.1016/j.clema.2022.100045

Arias, J. Z., Reuter, T., Sabir, A., & Gilroyed, B. H. (2018). Ambient alkaline hydrolysis and anaerobic digestion as a mortality management strategy for whole poultry carcasses. *Waste Management (New York, N.Y.)*, *81*, 71–77. doi:10.1016/j.wasman.2018.09.049 PMID:30527045

Bădescu, I. S., Bulgariu, D., Ahmad, I., & Bulgariu, L. (2018). Valorisation possibilities of exhausted biosorbents loaded with metal ions – A review. *Journal of Environmental Management*, *224*, 288–297. doi:10.1016/j.jenvman.2018.07.066 PMID:30055461

Bakan, B., Bernet, N., Bouchez, T., Boutrou, R., Choubert, J.-M., Dabert, P., Duquennoi, C., Ferraro, V., García-Bernet, D., Gillot, S., Mery, J., Rémond, C., Steyer, J.-P., Trably, E., & Tremier, A. (2022). Circular Economy Applied to Organic Residues and Wastewater: Research Challenges. *Waste and Biomass Valorization*, *13*(2), 1267–1276. doi:10.1007/s12649-021-01549-0

Bala, S., Garg, D., Sridhar, K., Inbaraj, B. S., Singh, R., Kamma, S., Tripathi, M., & Sharma, M. (2023). Transformation of Agro-Waste into Value-Added Bioproducts and Bioactive Compounds: Micro/Nano Formulations and Application in the Agri-Food-Pharma Sector. *Bioengineering (Basel, Switzerland)*, *10*(2), 152. doi:10.3390/bioengineering10020152 PMID:36829646

Becker, M., Ahn, K., Bacher, M., Xu, C., Sundberg, A., Willför, S., Rosenau, T., & Potthast, A. (2021). Comparative hydrolysis analysis of cellulose samples and aspects of its application in conservation science. *Cellulose (London, England)*, *28*(13), 8719–8734. doi:10.1007/s10570-021-04048-6 PMID:34316103

Bhatia, L., Jha, H., Sarkar, T., & Sarangi, P. K. (2023). Food Waste Utilization for Reducing Carbon Footprints towards Sustainable and Cleaner Environment: A Review. *International Journal of Environmental Research and Public Health*, *20*(3), 2318. doi:10.3390/ijerph20032318 PMID:36767685

Blaga, A. C., Tanasă, A. M., Cimpoesu, R., Tataru-Farmus, R.-E., & Suteu, D. (2022). Biosorbents Based on Biopolymers from Natural Sources and Food Waste to Retain the Methylene Blue Dye from the Aqueous Medium. *Polymers*, *14*(13), 2728. doi:10.3390/polym14132728 PMID:35808773

Chang, C., & Li, R. (2019). Agricultural waste. *Water Environment Research*, *91*(10), 1150–1167. doi:10.1002/wer.1211 PMID:31433884

Dai, Y., Sun, Q., Wang, W., Lu, L., Liu, M., Li, J., Yang, S., Sun, Y., Zhang, K., Xu, J., Zheng, W., Hu, Z., Yang, Y., Gao, Y., Chen, Y., Zhang, X., Gao, F., & Zhang, Y. (2018). Utilizations of agricultural waste as adsorbent for the removal of contaminants: A review. *Chemosphere*, *211*, 235–253. doi:10.1016/j.chemosphere.2018.06.179 PMID:30077103

De Gisi, S., Lofrano, G., Grassi, M., & Notarnicola, M. (2016). Characteristics and adsorption capacities of low-cost sorbents for wastewater treatment: A review. *Sustainable Materials and Technologies*, *9*, 10–40. doi:10.1016/j.susmat.2016.06.002

Demiral, H., & Gündüzoğlu, G. (2010). Removal of nitrate from aqueous solutions by activated carbon prepared from sugar beet bagasse. *Bioresource Technology*, *101*(6), 1675–1680. doi:10.1016/j.biortech.2009.09.087 PMID:19854640

Dey, S., Haripavan, N., Basha, S. R., & Babu, G. V. (2021). Removal of ammonia and nitrates from contaminated water by using solid waste bio-adsorbents. *Current Research in Chemical Biology*, *1*, 100005. doi:10.1016/j.crchbi.2021.100005

Diale, P. P., Hildebrandt, D., Glasser, D., Matambo, T. S., & Makgato, S. S. (2023). An analysis of the processes, kinetics and equilibrium of iron's biosorption on immobilized green microalgae. *South African Journal of Chemical Engineering*, *45*, 210–220. doi:10.1016/j.sajce.2023.05.007

Díaz, M. J., Ruiz-Montoya, M., Palma, A., & de-Paz, M.-V. (2021). Thermogravimetry Applicability in Compost and Composting Research: A Review. *Applied Sciences (Basel, Switzerland)*, *11*(4), 1692. doi:10.3390/app11041692

Duque-Acevedo, M., Belmonte-Ureña, L. J., Cortés-García, F. J., & Camacho-Ferre, F. (2020). Agricultural waste: Review of the evolution, approaches and perspectives on alternative uses. *Global Ecology and Conservation*, *22*, e00902. doi:10.1016/j.gecco.2020.e00902

Elgarahy, A. M., Elwakeel, K. Z., Mohammad, S. H., & Elshoubaky, G. A. (2021). A critical review of biosorption of dyes, heavy metals and metalloids from wastewater as an efficient and green process. *Cleaner Engineering and Technology*, *4*, 100209. doi:10.1016/j.clet.2021.100209

Gao, C., Xiao, W., Ji, G., Zhang, Y., Cao, Y., & Han, L. (2017). Regularity and mechanism of wheat straw properties change in ball milling process at cellular scale. *Bioresource Technology*, *241*, 214–219. doi:10.1016/j.biortech.2017.04.115 PMID:28570886

Gaspareto, T. C., Scapini, T., Venturin, B., Tápparo, D. C., Camargo, A. F., Di Luccio, M., Valerio, A., Favretto, R., Antes, F. G., Steinmetz, R. L. R., Treichel, H., & Kunz, A. (2020). Pre-treatment Strategies for Value Addition in Poultry Litter. *Frontiers in Bioengineering and Biotechnology*, *8*, 477. Advance online publication. doi:10.3389/fbioe.2020.00477 PMID:32523940

Georganas, A., Giamouri, E., Pappas, A. C., Papadomichelakis, G., Galliou, F., Manios, T., Tsiplakou, E., Fegeros, K., & Zervas, G. (2020). Bioactive Compounds in Food Waste: A Review on the Transformation of Food Waste to Animal Feed. *Foods*, *9*(3), 291. doi:10.3390/foods9030291 PMID:32150822

Girish, C. (2018). Various Impregnation Methods Used for the Surface Modification of the Adsorbent: A Review. *International Journal of Engineering & Technology*, *7*(4.7), 330. doi:10.14419/ijet.v7i4.7.20571

Gómez, X., Cuetos, M. J., García, A. I., & Morán, A. (2007). An evaluation of stability by thermogravimetric analysis of digestate obtained from different biowastes. *Journal of Hazardous Materials*, *149*(1), 97–105. doi:10.1016/j.jhazmat.2007.03.049 PMID:17448599

Groen, J. C., Maldonado, L., Moulijn, J. A., & Pérez-Ramírez, J. (2006). *On the role of iron in preparation of mesoporous Fe-MFI zeolites via desilication*. doi:10.1016/S0167-2991(06)80916-X

Harshala, K., & Wagh, N. D. (2022). Use of Agricultural Waste-Based Biosorbents for the Removal of Heavy Metals from Aqueous Solution: A Review. *Nature Environment and Pollution Technology*, *21*(3), 1003–1014. doi:10.46488/NEPT.2022.v21i03.007

Hartmann, A. J., & Lamb, R. N. (1997). X-ray photoemission spectroscopy of thin films. *Current Opinion in Solid State and Materials Science*, *2*(5), 511–516. doi:10.1016/S1359-0286(97)80038-1

He, Q., Bai, Y., Lu, Y., Cui, B., Huang, Z., Yang, Q., Jiang, D., & Shao, D. (2022). Isolation and characterization of cellulose nanocrystals from Chinese medicine residues. *Biomass Conversion and Biorefinery*. Advance online publication. doi:10.1007/s13399-022-03380-6 PMID:36259074

Hemathilake, D. M. K. S., & Gunathilake, D. M. C. C. (2022). Agricultural productivity and food supply to meet increased demands. In *Future Foods* (pp. 539–553). Elsevier. doi:10.1016/B978-0-323-91001-9.00016-5

Hoang, M. T., Pham, T. D., Pham, T. T., Nguyen, M. K., Nu, D. T. T., Nguyen, T. H., Bartling, S., & Van der Bruggen, B. (2021). Esterification of sugarcane bagasse by citric acid for Pb2+ adsorption: Effect of different chemical pretreatment methods. *Environmental Science and Pollution Research International, 28*(10), 11869–11881. doi:10.1007/s11356-020-07623-9 PMID:31953762

Holliday, M. C., Parsons, D. R., & Zein, S. H. (2022). Agricultural Pea Waste as a Low-Cost Pollutant Biosorbent for Methylene Blue Removal: Adsorption Kinetics, Isotherm And Thermodynamic Studies. *Biomass Conversion and Biorefinery*. Advance online publication. doi:10.1007/s13399-022-02865-8

Jana, A., Bhattacharya, P., Sarkar, S., Majumdar, S., & Ghosh, S. (2016). An ecofriendly approach towards remediation of high lead containing toxic industrial effluent by a combined biosorption and microfiltration process: A total reuse prospect. *Desalination and Water Treatment, 57*(12), 5498–5513. doi:10.1080/19443994.2015.1004596

Jayathilakan, K., Sultana, K., Radhakrishna, K., & Bawa, A. S. (2012). Utilization of byproducts and waste materials from meat, poultry and fish processing industries: A review. *Journal of Food Science and Technology, 49*(3), 278–293. doi:10.1007/s13197-011-0290-7 PMID:23729848

Juela, D., Vera, M., Cruzat, C., Alvarez, X., & Vanegas, E. (2021). Adsorption properties of sugarcane bagasse and corn cob for the sulfamethoxazole removal in a fixed-bed column. *Sustainable Environment Research, 31*(1), 27. doi:10.1186/s42834-021-00102-x

Kanamarlapudi, S. L. R. K., Chintalpudi, V. K., & Muddada, S. (2018). Application of Biosorption for Removal of Heavy Metals from Wastewater. In *Biosorption*. InTech. doi:10.5772/intechopen.77315

Karić, N., Maia, A. S., Teodorović, A., Atanasova, N., Langergraber, G., Crini, G., Ribeiro, A. R. L., & Đolić, M. (2022). Bio-waste valorisation: Agricultural wastes as biosorbents for removal of (in)organic pollutants in wastewater treatment. *Chemical Engineering Journal Advances, 9*, 100239. doi:10.1016/j.ceja.2021.100239

Karim, A., Raji, Z., Karam, A., & Khalloufi, S. (2023). Valorization of Fibrous Plant-Based Food Waste as Biosorbents for Remediation of Heavy Metals from Wastewater—A Review. *Molecules (Basel, Switzerland), 28*(10), 4205. doi:10.3390/molecules28104205 PMID:37241944

Koul, B., Yakoob, M., & Shah, M. P. (2022). Agricultural waste management strategies for environmental sustainability. *Environmental Research, 206*, 112285. doi:10.1016/j.envres.2021.112285 PMID:34710442

Kumar, H., Bhardwaj, K., Sharma, R., Nepovimova, E., Kuča, K., Dhanjal, D. S., Verma, R., Bhardwaj, P., Sharma, S., & Kumar, D. (2020). Fruit and Vegetable Peels: Utilization of High Value Horticultural Waste in Novel Industrial Applications. *Molecules (Basel, Switzerland), 25*(12), 2812. doi:10.3390/molecules25122812 PMID:32570836

Kumar, R. R., Park, B. J., & Cho, J. Y. (2013). Application and environmental risks of livestock manure. *Journal of the Korean Society for Applied Biological Chemistry, 56*(5), 497–503. doi:10.1007/s13765-013-3184-8

Kumar, S., & Asif, P. (2019). Equilibrium and Kinetic Studies of Biosorptive Removal of 2,4,6-Trichlorophenol from Aqueous Solutions Using Untreated Agro-Waste Pine Cone Biomass. *Processes (Basel, Switzerland)*, *7*(10), 757. doi:10.3390/pr7100757

Kwikima, M. M., Mateso, S., & Chebude, Y. (2021). Potentials of agricultural wastes as the ultimate alternative adsorbent for cadmium removal from wastewater. A review. *Scientific African*, *13*, e00934. doi:10.1016/j.sciaf.2021.e00934

Li, Y., Li, L., & Yu, J. (2017). Applications of Zeolites in Sustainable Chemistry. *Chem*, *3*(6), 928–949. doi:10.1016/j.chempr.2017.10.009

Li, Z., Zheng, Z., Li, H., Xu, D., Li, X., Xiang, L., & Tu, S. (2023). Review on Rice Husk Biochar as an Adsorbent for Soil and Water Remediation. *Plants*, *12*(7), 1524. doi:10.3390/plants12071524 PMID:37050150

Liu, G., Dai, Z., Liu, X., Dahlgren, R. A., & Xu, J. (2022). Modification of agricultural wastes to improve sorption capacities for pollutant removal from water – a review. *Carbon Research*, *1*(1), 24. doi:10.1007/s44246-022-00025-1

Mahato, N., Sharma, K., Sinha, M., Baral, E. R., Koteswararao, R., Dhyani, A., Hwan Cho, M., & Cho, S. (2020). Bio-sorbents, industrially important chemicals and novel materials from citrus processing waste as a sustainable and renewable bioresource: A review. *Journal of Advanced Research*, *23*, 61–82. doi:10.1016/j.jare.2020.01.007 PMID:32082624

Mandal, A., Singh, N., & Nain, L. (2017). Agro-waste biosorbents: Effect of physico-chemical properties on atrazine and imidacloprid sorption. *Journal of Environmental Science and Health. Part B, Pesticides, Food Contaminants, and Agricultural Wastes*, *52*(9), 671–682. doi:10.1080/03601234.2017.1331677 PMID:28679066

Muhammad, S., Abdul Khalil, H. P. S., Abd Hamid, S., Albadn, Y. M., Suriani, A. B., Kamaruzzaman, S., Mohamed, A., Allaq, A. A., & Yahya, E. B. (2022). Insights into Agricultural-Waste-Based Nano-Activated Carbon Fabrication and Modifications for Wastewater Treatment Application. *Agriculture*, *12*(10), 1737. doi:10.3390/agriculture12101737

Neikov, O. D. (2009). Mechanical Crushing and Grinding. In *Handbook of Non-Ferrous Metal Powders* (pp. 47–62). Elsevier. doi:10.1016/B978-1-85617-422-0.00002-1

Nestle, N. (2002). NMR studies on heavy metal immobilization in biosorbents and mineral matrices. *Reviews in Environmental Science and Biotechnology*, *1*(3), 215–225. doi:10.1023/A:1021255727484

Nestle, N. F. E. I., & Kimmich, R. (2000). NMR imaging of heavy metal absorption in alginate, immobilized cells, and kombu algal biosorbents. *Biotechnology and Bioengineering*, *51*(5), 538–543. doi:10.1002/(SICI)1097-0290(19960905)51:5<538::AID-BIT5>3.0.CO;2-D PMID:18629817

Ngeno, E. C., Mbuci, K. E., Necibi, M. C., Shikuku, V. O., Olisah, C., Ongulu, R., Matovu, H., Ssebugere, P., Abushaban, A., & Sillanpää, M. (2022). Sustainable re-utilization of waste materials as adsorbents for water and wastewater treatment in Africa: Recent studies, research gaps, and way forward for emerging economies. *Environmental Advances*, *9*, 100282. doi:10.1016/j.envadv.2022.100282

Noor, A., & Khan, S. A. (2023). Agricultural Wastes as Renewable Biomass to Remediate Water Pollution. *Sustainability (Basel), 15*(5), 4246. doi:10.3390/su15054246

Ojuederie, O., & Babalola, O. (2017). Microbial and Plant-Assisted Bioremediation of Heavy Metal Polluted Environments: A Review. *International Journal of Environmental Research and Public Health, 14*(12), 1504. doi:10.3390/ijerph14121504 PMID:29207531

Osman, A. I., El-Monaem, E. M. A., Elgarahy, A. M., Aniagor, C. O., Hosny, M., Farghali, M., Rashad, E., Ejimofor, M. I., López-Maldonado, E. A., Ihara, I., Yap, P.-S., Rooney, D. W., & Eltaweil, A. S. (2023). Methods to prepare biosorbents and magnetic sorbents for water treatment: A review. *Environmental Chemistry Letters, 21*(4), 2337–2398. doi:10.1007/s10311-023-01603-4

Particle Size Reduction and Enlargement. (2019). *Coulson and Richardson's Chemical Engineering*. Elsevier. doi:10.1016/B978-0-08-101098-3.00006-8

Peng, X., Jiang, Y., Chen, Z., Osman, A. I., Farghali, M., Rooney, D. W., & Yap, P.-S. (2023). Recycling municipal, agricultural and industrial waste into energy, fertilizers, food and construction materials, and economic feasibility: A review. *Environmental Chemistry Letters, 21*(2), 765–801. doi:10.1007/s10311-022-01551-5

Perussello, C. A., Zhang, Z., Marzocchella, A., & Tiwari, B. K. (2017). Valorization of Apple Pomace by Extraction of Valuable Compounds. *Comprehensive Reviews in Food Science and Food Safety, 16*(5), 776–796. doi:10.1111/1541-4337.12290 PMID:33371603

Raji, Z., Karim, A., Karam, A., & Khalloufi, S. (2023). Adsorption of Heavy Metals: Mechanisms, Kinetics, and Applications of Various Adsorbents in Wastewater Remediation—A Review. *Waste, 1*(3), 775–805. doi:10.3390/waste1030046

Rangabhashiyam, S., & Balasubramanian, P. (2019). The potential of lignocellulosic biomass precursors for biochar production: Performance, mechanism and wastewater application—A review. *Industrial Crops and Products, 128*, 405–423. doi:10.1016/j.indcrop.2018.11.041

Raut, N. A., Kokare, D. M., Randive, K. R., Bhanvase, B. A., & Dhoble, S. J. (2023). Introduction: fundamentals of waste removal technologies. In 360-Degree Waste Management, Volume 1 (pp. 1–16). Elsevier. doi:10.1016/B978-0-323-90760-6.00009-6

Razzak, S. A., Faruque, M. O., Alsheikh, Z., Alsheikhmohamad, L., Alkuroud, D., Alfayez, A., Hossain, S. M. Z., & Hossain, M. M. (2022). A comprehensive review on conventional and biological-driven heavy metals removal from industrial wastewater. *Environmental Advances, 7*, 100168. doi:10.1016/j.envadv.2022.100168

Roy, P., Mohanty, A. K., Dick, P., & Misra, M. (2023). A Review on the Challenges and Choices for Food Waste Valorization: Environmental and Economic Impacts. *ACS Environmental Au, 2*(2), 58–75. doi:10.1021/acsenvironau.2c00050 PMID:36941850

Saady, N. M. C., Rezaeitavabe, F., & Ruiz Espinoza, J. E. (2021). Chemical Methods for Hydrolyzing Dairy Manure Fiber: A Concise Review. *Energies, 14*(19), 6159. doi:10.3390/en14196159

Sadh, P. K., Duhan, S., & Duhan, J. S. (2018). Agro-industrial wastes and their utilization using solid state fermentation: A review. *Bioresources and Bioprocessing*, *5*(1), 1. doi:10.1186/s40643-017-0187-z

Sapwarobol, S., Saphyakhajorn, W., & Astina, J. (2021). Biological Functions and Activities of Rice Bran as a Functional Ingredient: A Review. *Nutrition and Metabolic Insights*, *14*, 117863882110585. doi:10.1177/11786388211058559 PMID:34898989

Sarkar, S., Skalicky, M., Hossain, A., Brestic, M., Saha, S., Garai, S., Ray, K., & Brahmachari, K. (2020). Management of Crop Residues for Improving Input Use Efficiency and Agricultural Sustainability. *Sustainability (Basel)*, *12*(23), 9808. doi:10.3390/su12239808

Saxena, A., Bhardwaj, M., Allen, T., Kumar, S., & Sahney, R. (2017). Adsorption of heavy metals from wastewater using agricultural–industrial wastes as biosorbents. *Water Science*, *31*(2), 189–197. doi:10.1016/j.wsj.2017.09.002

Sen, T. K. (2023). Agricultural Solid Wastes Based Adsorbent Materials in the Remediation of Heavy Metal Ions from Water and Wastewater by Adsorption: A Review. *Molecules (Basel, Switzerland)*, *28*(14), 5575. doi:10.3390/molecules28145575 PMID:37513447

Sud, D., Mahajan, G., & Kaur, M.SUD. (2008). Agricultural waste material as potential adsorbent for sequestering heavy metal ions from aqueous solutions – A review. *Bioresource Technology*, *99*(14), 6017–6027. doi:10.1016/j.biortech.2007.11.064 PMID:18280151

Tan, Y. H., Davis, J. A., Fujikawa, K., Ganesh, N. V., Demchenko, A. V., & Stine, K. J. (2012). Surface area and pore size characteristics of nanoporous gold subjected to thermal, mechanical, or surface modification studied using gas adsorption isotherms, cyclic voltammetry, thermogravimetric analysis, and scanning electron microscopy. *Journal of Materials Chemistry*, *22*(14), 6733. doi:10.1039/c2jm16633j PMID:22822294

Tiquia-Arashiro, S., Li, X., Pokhrel, K., Kassem, A., Abbas, L., Coutinho, O., Kasperek, D., Najaf, H., & Opara, S. (2023). Applications of Fourier Transform-Infrared spectroscopy in microbial cell biology and environmental microbiology: Advances, challenges, and future perspectives. *Frontiers in Microbiology*, *14*, 1304081. Advance online publication. doi:10.3389/fmicb.2023.1304081

Wang, C., Wang, Z., Wang, X., Li, N., Tao, J., Zheng, W., Yan, B., Cui, X., Cheng, Z., & Chen, G. (2022). A Review on the Hydrothermal Treatment of Food Waste: Processing and Applications. *Processes (Basel, Switzerland)*, *10*(11), 2439. doi:10.3390/pr10112439

Wang, J., & Chen, C. (2009). Biosorbents for heavy metals removal and their future. *Biotechnology Advances*, *27*(2), 195–226. doi:10.1016/j.biotechadv.2008.11.002 PMID:19103274

Wyroba, E., Suski, S., Miller, K., & Bartosiewicz, R. (2015). Biomedical and agricultural applications of energy dispersive X-ray spectroscopy in electron microscopy. *Cellular & Molecular Biology Letters*, *20*(3). Advance online publication. doi:10.1515/cmble-2015-0028 PMID:26208393

Xie, S., Tran, H.-T., Pu, M., & Zhang, T. (2023). Transformation characteristics of organic matter and phosphorus in composting processes of agricultural organic waste: Research trends. *Materials Science for Energy Technologies*, *6*, 331–342. doi:10.1016/j.mset.2023.02.006

Yang, X., Wan, Y., Zheng, Y., He, F., Yu, Z., Huang, J., Wang, H., Ok, Y. S., Jiang, Y., & Gao, B. (2019). Surface functional groups of carbon-based adsorbents and their roles in the removal of heavy metals from aqueous solutions: A critical review. *Chemical Engineering Journal*, *366*, 608–621. doi:10.1016/j.cej.2019.02.119 PMID:34522159

Zhang, X., Qin, Q., Sun, X., & Wang, W. (2022). Hydrothermal treatment: An efficient food waste disposal technology. *Frontiers in Nutrition*, *9*, 986705. Advance online publication. doi:10.3389/fnut.2022.986705 PMID:36172524

Zinicovscaia, I., Yushin, N., Grozdov, D., Vergel, K., Popova, N., Artemiev, G., & Safonov, A. (2020). Metal Removal from Nickel-Containing Effluents Using Mineral–Organic Hybrid Adsorbent. *Materials (Basel)*, *13*(19), 4462. doi:10.3390/ma13194462 PMID:33050087

Chapter 10
Microbial Architects of Environmental Restoration:
Exploring the Frontiers of Genetically Engineered Bacteria in Remediation Strategies

Anjali Rajpoot
Banasthali Vidyapith, India

Veena Sharma
Banasthali Vidyapith, India

Tanya Aggarwal
Banasthali Vidyapith, India

ABSTRACT

Environmental pollution poses a significant threat to ecosystems and human health worldwide. Traditional methods of environmental remediation often fall short in addressing the complexities of contemporary pollutants. The emergence of genetic engineering technologies has opened new avenues for innovative solutions to environmental challenges. This chapter aims to explore the potential of genetically engineered bacteria as a cutting-edge tool for environmental remediation. By delving into the principles, applications, and concerns surrounding this technology, the chapter seeks to contribute to the understanding of its feasibility and sustainability. However, as we navigate the frontiers of this innovative approach, it is imperative to acknowledge and address the environmental concerns and conduct thorough risk assessments associated with the deployment of genetically engineered bacteria. Ecological impact, potential risks to human health, and ethical considerations necessitate careful consideration and mitigation strategies to ensure responsible and sustainable implementation.

DOI: 10.4018/979-8-3693-1618-4.ch010

INTRODUCTION

In recent years, the field of environmental science has witnessed a paradigm shift with the advent of genetically engineered bacteria as innovative tools for addressing pollution and contamination challenges. This book chapter explores the transformative potential of genetically modified bacteria in environmental remediation, drawing insights from influential review papers that have significantly shaped our understanding of this evolving field. The initial section provides a comprehensive overview of the current state of environmental degradation, setting the stage for an in-depth exploration of the intricate details of genetic engineering applied to bacteria. This includes an elucidation of key mechanisms and molecular tools employed to enhance the environmental detoxification capabilities of these genetically modified organisms. Through a meticulous examination of leading review papers, the chapter navigates the diverse applications of genetically engineered bacteria across various environmental matrices. Furthermore, the chapter critically analyzes the ecological implications, ethical considerations, and regulatory frameworks surrounding the deployment of genetically modified bacteria, offering a balanced perspective on the socio-scientific landscape. It delves into existing regulatory frameworks and advocates for international collaboration to enhance guidelines governing the use of genetically engineered organisms in environmental applications. The chapter underscores the need for robust risk management strategies and emphasizes the importance of public engagement, highlighting the necessity of a comprehensive and adaptive regulatory framework.

Although microbial communities in the environment possess the ability to degrade various pollutants, the degradation of many pollutants occurs slowly, leading to their accumulation and posing severe hazards. Microorganisms often lack the necessary pathways to degrade novel pollutants due to their complex chemical structures, which inhibits complete transformation. This limitation may stem from the novelty of these pollutants, hindering microbial evolution to develop appropriate catabolic pathways. Additionally, populations of microorganisms responsible for degradation might not be sufficiently large or active, or complex mixtures of pollutants may resist degradation.

To address these challenges, one strategy involves bioaugmentation, where exogenous microorganisms are introduced to enhance indigenous populations' degradation capabilities. This can include naturally occurring microorganisms with the required genes or genetically engineered microorganisms (GEMs). However, concerns arise regarding the stability of inserted genetic material in GEMs and the potential negative perception of genetic material transfer to indigenous organisms.

Advancements in molecular biology have facilitated the development of novel microbial strains tailored for bioremediation. These advancements encompass various aspects such as constructing and regulating pathways, expanding substrate ranges, and improving metabolic efficiency. For instance, genetic engineering techniques can introduce genes encoding bacterial hemoglobin (VHb) to enable microbial growth and catabolic activities under hypoxic conditions.

This chapter focuses on the transformative landscape of environmental bioremediation facilitated by genetically engineered bacteria, with a specific emphasis on the remediation of heavy metal pollutants. It accentuates the ecological and health advantages over conventional physico-chemical strategies, advocating for a synergy of microbiological, ecological, and biochemical insights coupled with innovative field engineering designs. The chapter delves into the intricate realm of microbial metal bioremediation, highlighting its significance as a cost-effective and eco-friendly strategy for addressing heavy metal pollutants. This approach represents a paradigm shift from conventional physico-chemical methods, offering unique advantages in terms of sustainability and environmental compatibility.

The exploration of current challenges and future directions in the field concludes the chapter, shedding light on the evolving nature of genetically engineered bacteria as a tool for environmental remediation. While genetically engineered bacteria offer eco-friendly alternatives, concerns regarding genetic pollution and risk mitigation demand rigorous attention, necessitating a careful balance between environmental benefits and potential risks. The chapter highlights the molecular genetics contribution to risk mitigation, particularly in addressing environmental and public health concerns.

In essence, this chapter serves as a beacon, illuminating the transformative potential of genetically engineered bacteria in shaping the future of environmental remediation. Its significance lies not only in deciphering the scientific nuances but also in catalyzing collective awareness of the ethical, ecological, and societal implications associated with the deployment of genetically engineered bacteria. As we navigate the complex landscape of environmental challenges, this chapter stands as an indispensable guide, steering the discourse towards a sustainable and harmonious coexistence between technology and the environment. Serving as an informative and forward-looking contribution, this chapter is a valuable resource for researchers, practitioners, and policymakers navigating the dynamic intersection of genetic engineering and environmental science. It advocates for the conscientious and sustainable assimilation of genetic engineering technologies into environmental policies.

Scope of Genetically Engineered Bacteria

Microbial bioremediation is a promising approach to address environmental pollution. Genetically engineered bacteria (GEB) have emerged as a powerful tool for bioremediation. GEB can be designed to degrade specific pollutants, accumulate heavy metals, or detoxify hazardous chemicals. This technology involves the modification of bacterial genomes to enhance their bioremediation capabilities. The scope of genetically engineered bacteria for environmental remediation includes a comprehensive overview of the genetic modifications involved in engineering bacteria for environmental remediation and the diversity of bacterial species used and their specific applications in addressing different types of pollutants.

According to Singh (2011), GEB is an emerging tool for environmental remediation and future research. Singh *et al.,* (2011) discusses the challenges and future perspectives of using GEB for bioremediation. The deliberate release of GEB into the environment is a subject of considerable public concern, and the potential risks associated with the release of GEB into the environment have led to the construction of biological containment systems by which bacteria are killed in a controlled suicide process. Active biological containment systems usually consist of two different components, a killing element designed to induce cell death and a control element that modulates the expression of the killing function Garbisu (1999).

GEB can be used to treat mercury-polluted environments, as discussed in Kiyono (2006). The authors genetically engineered Escherichia coli to express a mercury transport system and organo mercurial lyase enzyme simultaneously, and over express polyphosphate, a strong chelator of essential divalent metals. The mercury accumulation system was designed so that over expressed polyphosphate would serve as a mercury accumulator; the mercury transport system would make the bacterial cell specifically accumulate mercury; and the intracellular accumulation process would allow the bioaccumulation system to be less sensitive to ambient conditions.

GEB is a promising tool for environmental remediation. The genetic modifications involved in engineering bacteria for environmental remediation and the diversity of bacterial species used and their specific applications in addressing different types of pollutants are important aspects of this technology.

Mechanism of Action

The mechanism of action for genetically engineered bacteria in environmental remediation involves the molecular mechanisms underlying their ability to degrade environmental pollutants and the interactions between engineered bacteria and contaminants. The importance of specificity and selectivity in these interactions is crucial for effective remediation strategies. This section discusses various strategies utilized by these bacteria to degrade environmental pollutants.

Molecular Mechanisms: Genetically engineered bacteria can degrade various environmental pollutants through different mechanisms, such as:

- *Genetic Modifications:* Researchers often engineer bacteria to enhance their pollutant degradation capabilities through genetic modifications. DNA fragments containing one or more nucleotides can be precisely incorporated, removed, or substituted within an organism's genome using advanced scientific techniques, enabling targeted genetic engineering at both the global (genome) and local (gene) levels (Soutullo et al., 2016). Among the commonly used tools for gene editing are transcription activator-like effector nucleases (TALEN) and clustered regularly interspaced short palindromic repeat (CRISPR)-associated nucleases (Sharma and Shukla, 2022). CRISPR-Cas, an efficient and user-friendly gene editing technology, has been developed, offering potential benefits for bioremediation. The DNA-binding module of TALEN exhibits sequence specificity to the host genome (Priya et al., 2022), leading to the creation of double-stranded breaks (DSBs) upon binding to DNA, resulting in stable sticky ends. ZFNs have a 30-amino acid DNA-binding domain, and the Fok 1 cleavage domain introduces DSBs at specific locations within the host genome.

A novel approach has emerged, incorporating hybrid nucleases such as TALENs and ZFNs, which has broadened the capabilities of the CRISPR-Cas system, allowing simultaneous editing of multiple genes with high precision (Gao et al., 2022). The CRISPR-Cas system, originating from Streptococcus pyogenes, provides viral immunity, where guide RNA (gRNA) combines crisper-derived RNA (crRNA) and trans-acting antisense RNA (trcRNA) to form gRNA, directing the Cas9 enzyme to induce DSBs at target DNA sequences recognized by the gRNA (Sarma et al., 2021). Bioremediation research stands to benefit from the introduction of these gene editing technologies, facilitating knock-in and knock-out strategies (Liu et al., 2021).

Recent studies have primarily utilized the CRISPR-Cas system in model organisms such as Pseudomonas or Escherichia coli (Okoli et al., 2021), and advancements in CRISPR tools and gRNA design have expanded applications to non-model organisms like Achromobacter sp. HZ01 and Comamonas testosteroni for bioremediation (Zhang et al., 2021). Bacteria inhabiting contaminated environments present ideal candidates for metabolic engineering and genome editing due to their adaptation to diverse toxic pollutants (Patel et al., 2022b).

Understanding metabolic pathways is crucial for microbial bioremediation, enabling the breakdown of toxic contaminants through enzymes such as haloalkane dehalogenases and fenpropathrin biodegradation mechanisms in Bacillus sp. DG-02 (Kurade et al., 2021). Metabolic engineering enhances bioremediation processes by modifying existing pathways, utilizing enzymes like oxidases, esterases, monooxygenases, oxidoreductases, and phenoloxidases (Mehta et al., 2021). Enzyme-based bioremediation offers numerous benefits as a green approach, with recombinant enzymes produced through genetic techniques and extracellular enzymes facilitating the breakdown of pollutants such as PAHs, TNT, and PCBs (Gallo et al., 2021). Even partial degradation of contaminants can be beneficial, as they may serve as intermediates in biological pathways or energy sources for microbes (Marycz et al., 2022).

Genetically modified microorganisms have shown effectiveness in bioremediating organophosphates and organochlorines, with metabolic engineering enabling the breakdown of persistent substances by bacteria like Pseudomonas sp. WBC-3 and Sphingobium japonicum (Xu et al., 2022). CRISPR technology enables precise genome editing, with CRISPR-Cas systems types I, II, and III allowing targeted DNA cleavage by Cas9 and guided RNA (gRNA), facilitating gene manipulation for enhanced bioremediation (Zhang et al., 2022). Deletion of the yvmC gene in Bacillus licheniformis using CRISPR-Cas9 improved biotransformation efficiency to 100%, showcasing the potential for developing microbial strains with enhanced remediation capabilities (Qian et al., 2022a).

- *Recombination by Natural Methods*: Several bacteria acquire novel genetic traits through processes like conjugation, transformation, and transduction. In laboratory settings, these methods of genetic material exchange are frequently employed to introduce desired genes into bacterial cells, thereby imparting specific advantages to the strain. Conjugation necessitates close physical contact between donor and recipient strains to transfer genetic material. This method can be utilized to confer specialized degradation characteristics by transferring degradation genes to indigenous bacteria in polluted environments. For instance, through conjugation, a strain of Pseudomonas putida capable of degrading camphor, octane, salicylate, and naphthalene was engineered. This strain harbored a hybrid plasmid containing XYL and NAH genes resulting from the recombination of CAM and OCT genes (Glick and Patten, 2022). Consequently, the enhanced strain exhibited rapid growth on crude oil due to its superior ability to metabolize hydrocarbons compared to wild-type strains with single plasmid systems.

Transformation occurs when bacterial cells actively uptake free DNA and integrate it into their genomes, representing a key aspect of horizontal transfer of genetic material observed in populations of competent bacteria that develop spontaneously (Deckers et al., 2022). It has recently gained recognition as a potent approach for introducing desired traits in strain enhancement research. Evidence of natural transformation in both bacteria and archaea indicates its presence before the last common ancestor of these domains. Transformation has been instrumental in studying the molecular processes involved in the degradation of pollutants such as trichloroethylene (TCE) and polychlorinated biphenyls (PCBs).

Transduction is a conventional technique for strain enhancement, often employed in conjunction with other natural mechanisms of gene transfer like conjugation and transformation to generate significantly improved strains with diverse genetic backgrounds. It involves the transfer of pieces of the host genome bundled into capsids. These capsid delivery agents facilitate the transfer of desired genes to specified recipients (Sendhil et al., 2022).

- *Gene cloning for the degradation of synthetic pollutants:* Gene cloning offers a viable approach to enhance the degradation of synthetic pollutants. Cloning specific genes enables the production of multiple identical copies, thereby increasing the desired product within a strain. In bioremediation applications, genes encoding degradation-specific enzymes can be cloned and expressed. In a study utilizing Cosmid pCP13, the 4-chlorobiphenyl-degrading gene of Pseudomonas was cloned, and the degradation potential of the cloned 808 gene was evaluated. Enzymes such as dioxygenase, dehydrogenase, and hydrolase contribute to the full mineralization of 4-chlorobiphenyl. Similarly, the cloning and expression of TOD genes in the chromosome of Dinococcus radiodurans facilitate the degradation of toluene, chlorobenzene, indole, and 3,4-dichloro-1-butene.

Genes encoding carbofuran hydrolase, an enzyme involved in carbamate pesticide breakdown, demonstrate broad activity. Achromobacter sp. WM11 successfully cloned and isolated the hydrolase gene responsible for methyl carbamate breakdown. Furthermore, the cloning and expression of serine alkaline protease in Shewanella strain AC10 resulted in increased output. For the bioremediation of nitrile chemicals, the gene encoding nitrile hydrolase (NHase) was extracted and utilized, facilitating the transformation of nitrile into valuable amides. Genes atzA and atzB, responsible for atrazine degradation in boreal groundwater, have been cloned and expressed.

Hydroquinone dioxygenase, an enzyme with diverse functions, is produced by one of the genes involved in p-nitrophenol bioremediation. The mpd gene and its clones have been cloned, studied, and utilized for chlorpyrifos decomposition in polluted areas. Overall, cloning desired bacterial and fungal genes has proven effective in exploring the capabilities of a wide array of bioremediation-specific genes. These genes can be transferred into model bacterial cells and utilized in treating widespread pollution in soil and water systems (Duque et al., 2022).

- *Utilizing Catabolic Pathways through Engineering*: Aromatic compounds present a wide array of contaminants in soil and water, making them prime targets for bioremediation using genetically engineered bacteria. Reineke (1998) provided an overview of recombinant strain engineering through patchwork assembly, aiming at complete chloroaromatic degradation. This method involves amalgamating pathways from different bacteria via conjugation into a single recombinant host, resulting in an integrated pathway capable of mineralizing specific compounds. For instance, Hrywna et al. (1999) successfully cloned and expressed the ohb operon from Pseudomonas aeruginosa and the fcb operon from Arthrobacter globiformis in Comamonas testosteroni strain VP44. These operons encode enzymes for chlorobenzoic acid (CBA) metabolism. By introducing plasmids containing the ohb and fcb operons into the host strain, a bacterium capable of complete monochlorobiphenyl mineralization was engineered.

Monti et al. (2005) engineered the genes responsible for 2,4-dinitrotoluene (2,4-DNT) degradation from Burkholderia sp. strain DNT into Pseudomonas fluorescens ATCC 17400. This recombinant strain efficiently degraded 2,4-DNT when provided as the sole nitrogen source, with resultant carbon utilized co-metabolically. Notably, this engineered strain exhibited superior DNT degradation capability at lower temperatures compared to Burkholderia and displayed reduced toxicity to certain plants under specific conditions. Similarly, Wu et al. (2006) utilized Escherichia coli as a host for cloning and expressing genes from Comamonas sp. strain CNB1, encoding a novel partial reductive pathway for 4-chloronitrobenzene and nitrobenzene.

Another intriguing example involves cloning and expressing Burkholderia cepacia toluene o-monooxygenase genes in Pseudomonas fluorescens 2-79 (Yee et al., 1998). Coating wheat seeds with the resulting recombinant strain significantly enhanced the removal of trichloroethylene (TCE) in the rhizosphere compared to controls, despite TCE not being an aromatic compound. Whether transferring one or multiple genes from various organisms to the target microorganism, the objective is to confer functional capabilities that enhance contaminant bioremediation. The aforementioned examples illustrate the diverse approaches in engineering microorganisms to optimize their catabolic pathways for specific contaminants.

- *Mutation:* Mutation serves as a well-recognized mechanism for altering the genetic characteristics of an organism, contributing to strain formation. Genetic changes at the molecular level influence the phenotype of an organism. Mutations can arise spontaneously or be induced deliberately, resulting in either detrimental or advantageous effects on the organism, with natural selection eliminating less favorable mutations (Patel et al., 2022). Degradation-specific genes are found in various genetic elements such as plasmids, integrons, and transposons. Bacteria exhibit enhanced survival capabilities in polluted environments due to mutations, offering potential for site-specific mutations to target resistant SPs in polluted soil and water. Bioremediation strategies harness microbial metabolism, which has demonstrated benefits from mutations targeted at specific loci. Mutant enzymes facilitate the breakdown of SPs, enhancing catalytic processes (Mutanda et al., 2022).

Genetically modified organisms (GMOs) are organisms subjected to genetic alterations, leading to the creation of new strains with improved genetic features, while less-fit populations are selectively eliminated. Microorganisms adapt their genomes in response to exposure to polluted environments, enabling indigenous microbial populations to adapt to changing conditions (Chakdar et al., 2022). Mutation can be induced by exposure to mutagens, which are either chemical or physical agents. Mutagens, including physical agents like x-rays and chemical agents like base analogs and intercalating agents, have the capacity to induce mutations in cells (Xue et al., 2022). Mutagenesis is commonly employed in industrial settings to modulate the production of desired metabolites, thereby utilizing molecular techniques to enhance bioremediation. Random mutation has emerged as a molecular approach for generating microbial strains capable of efficiently degrading SPs. UV irradiation, ethyl methyl sulphate, and ethidium bromide (EtBr) are employed as mutagens to induce mutations. For instance, Gopinath et al. utilized random mutation to accelerate the biodegradation of Congo Red, while Joshi and Thagard (2013) developed a UV-irradiated mutant of Pseudomonas sp. 1 F for the biodegradation of the sulfonated azo dye Green HE4B (GHE4B). The mutant strain exhibited a 25% reduction in the time required to fully degrade the resistant dye GHE4B compared to the wild-type strain. UV irradiation resulted in increased enzymatic activity of lignin peroxidase, laccase, veratryl alcohol, NADH-(dichlorophenol-indophenol) reductase, and oxidase (Téllez et al., 2022).

Interactions with Contaminants: The interactions between genetically engineered bacteria and contaminants play a crucial role in the effectiveness of remediation strategies. Key aspects of these interactions include:

- *Specificity:* Engineered bacteria should be specifically designed to target and degrade specific contaminants, ensuring that they do not negatively impact the environment or human health.
- *Selectivity:* The ability of genetically engineered bacteria to differentiate between target contaminants and non-target molecules is essential for effective remediation. This selectivity ensures that the bacteria do not degrade essential nutrients or introduce new contaminants into the environment.
- *Horizontal Gene Transfer:* The potential for horizontal gene transfer between genetically engineered bacteria and native microflora should be carefully considered, as this transfer can lead to the spread of antibiotic resistance genes or other undesirable traits (Singh *et al.,* 2011).

The mechanism of action for genetically engineered bacteria in environmental remediation involves their molecular mechanisms for degrading pollutants and the interactions between engineered bacteria and contaminants, with a focus on specificity and selectivity. These factors are crucial for the successful implementation of genetically engineered bacteria in remediation strategies and ensuring their safety and efficacy in the environment.

Case Studies and Success Stories

The case studies and success stories of this topic examine the successful applications of genetically engineered bacteria in real-world environmental remediation projects and analyze the efficiency and durability of these engineered bacteria in comparison to traditional remediation methods.

According to a study conducted by Singh (2011), genetically engineered bacteria are an emerging tool for environmental remediation and future research. The study highlights the wide metabolic and physiological versatility of microorganisms that can be used to degrade many pollutants. Nowadays, microbial systems are employed in bioremediation programs, generally in the treatment of soils and waters contaminated with organic pollutants. There are instances where natural populations are not suitable for use in the remediation of polluted sites, and therefore the utilization of genetically engineered microorganisms (GEMs) is being considered for in-situ bioremediation of contaminated ecosystems.

Another study conducted by Singh et al. (2011), highlights that the use of genetically engineered bacteria in field conditions may offer more efficient bioremediation of contaminated sites and make the bioremediation more viable and eco-friendly technology. However, many challenges are to be addressed concerning the release of genetically engineered bacteria in field conditions. There are possible risks associated with the use of GE bacteria in field conditions, with particular emphasis on ways in which molecular genetics could contribute to the risk mitigation. Both environmental as well as public health concerns need to be addressed by the molecular biologists.

Genetically engineered bacteria have shown promise in environmental remediation, particularly in the biodegradation of heavy metals and pollutants (Singh *et al.,* 2011). However, there are still critical gaps in our knowledge of engineering bacteria for oil-spill bioremediation, such as the performance of enzymes compared to native bacteria, the degradation of petroleum hydrocarbons, and the environmental effects of engineered bacteria on native soil populations (French *et al.,* 2020).

The use of genetically engineered bacteria in environmental remediation projects is a promising approach. However, the potential risks associated with the release of genetically engineered bacteria in field conditions need to be addressed, and policies and safeguards need to be developed to prevent accidents and eliminate potential harmful effects of GMOs in the environment.

ENVIRONMENT CONCERNS AND RISK ASSESSMENT

Ecological Impact

- *Assessing the potential ecological consequences of releasing genetically engineered bacteria into the environment:* This involves evaluating the potential impact of GE bacteria on ecosystems, including the effects on native microorganisms, plant and animal populations, and the potential for horizontal gene transfer to wild-type bacteria (Bittihn *et al.,* 2020).

- *Exploring methods for mitigating unintended ecological effects and promoting the safe application of this technology:* This includes developing containment strategies and policies to prevent accidents and eliminate potential adverse effects of GE bacteria on the environment, such as sharing antibiotic genes with native microorganisms.

The use of genetically engineered bacteria for environmental remediation offers promising prospects for bioremediation of contaminated sites and making the bioremediation process more efficient and eco-friendly. However, there are concerns and challenges associated with the release of GE bacteria in field conditions, including potential risks to both environmental and public health (Singh *et al.,* 2011). As a result, it is crucial to conduct thorough risk assessments and develop strategies for safe and responsible application of this technology.

Human Health Risk

Investigating the potential risks to human health associated with the use of genetically engineered bacteria in environmental remediation: It is crucial to understand the potential risks to human health when using genetically engineered bacteria for environmental remediation. This includes assessing the bio-safety of these bacteria and the factors related to genetic pollution (Singh *et al.,* 2011). Strategies for minimizing human exposure and ensuring the safety of communities near remediation sites: To address both environmental and public health concerns, it is essential to develop strategies that minimize human exposure to genetically engineered bacteria and ensure the safety of communities living near remediation sites (Singh *et al.,* 2011). Advances in molecular biology, such as CRISPR genome editing, have enabled the development of synthetic biology solutions for environmental remediation. For example, genetically engineered bacteria have been used to decontaminate oil sands tailings ponds, which are a significant source of environmental concern. However, it is essential to consider the potential risks to human health and the environment when implementing these technologies (Rafeeq *et al.,* 2023).

Ethical Considerations

Manipulating Microbial Communities for Environmental Benefit

- The manipulation of microbial communities for environmental benefit raises several important ethical concerns, including the potential impact on ecosystems, biodiversity, and human health (Ma *et al.,* 2018) (Peixoto *et al.,* 2022).
- It is crucial to consider the consequences of altering natural microbial communities and the potential risks associated with the introduction of genetically engineered bacteria in ecosystems (Building Synthetic Microbial Communities for Biology, Mitigating Climate Change, Sustainability, and Biotechnology (Synthetic Communities) (Nsf22607) | NSF - National Science Foundation).

Public Perception, Consent, and Transparency

- *Public perception:* The deployment of genetically engineered bacteria in environmental remediation may have implications for public perception and acceptance of such technologies (Ma *et al.*, 2018). It is essential to engage with the public and address concerns related to the use of genetically engineered bacteria in the environment.
- *Consent:* The involvement of local communities and stakeholders in the decision-making process is crucial for ensuring the responsible development and implementation of genetically engineered bacteria in environmental remediation (Ma *et al.*, 2018). Their consent and support are necessary for the successful adoption and integration of these technologies.
- *Transparency:* The deployment of genetically engineered bacteria should be transparent and well-regulated to ensure the safety of both human and the environment (Ma *et al.*, 2018)(Peixoto *et al.*, 2022). This includes the disclosure of information about the use of these bacteria, their potential risks, and the monitoring of their impact on ecosystems and human health.

To address these concerns, it is essential to develop a framework for the responsible development and implementation of genetically engineered bacteria in environmental remediation (Peixoto *et al.*, 2022). This framework should consider the potential risks and benefits, engage with local communities, and ensure transparency in the decision-making process.

REGULATORY FRAMEWORK AND GUIDELINES

International Perspectives

The regulatory framework and guidelines for the use of genetically engineered organisms in environmental applications are of paramount importance to ensure the safety of the environment and human health. Here is an overview of international regulations governing the use of genetically engineered organisms in environmental applications and the identification of gaps and challenges in the current regulatory frameworks, along with potential improvements.

- The regulations governing the use of genetically modified organisms (GMOs) and products containing, composed of, or derived from GMOs are a key aspect of the international regulatory framework.
- These regulations encompass various aspects, including their use in closed systems, intentional introduction into the environment, handling, transportation, and labeling.
- Different countries and regions have diverse regulatory processes for products of genetic engineering, reflecting broader social, legal, and cultural differences among nations (National Academies Press (US), 2016). For instance, the European Union (EU) has established a regulatory process for the assessment and approval of genetically engineered foods, triggering assessment and approval based on the presence of specific traits introduced into organisms.

Public Engagement in Regulatory Decisions

Public engagement is crucial in shaping regulatory decisions and ensuring the responsible deployment of genetically engineered bacteria in environmental remediation. Here are some ways to involve the public in regulatory decisions:

- *Transparency:* The regulatory process should be transparent, and information about the use of genetically engineered bacteria should be disclosed to the public. This includes information about the potential risks and benefits of these bacteria and their impact on the environment and human health.
- *Engagement:* Local communities and stakeholders should be involved in the decision-making process to ensure that their concerns are addressed and their consent is obtained.
- *Education:* Public education and awareness campaigns should be developed to inform the public about the use of genetically engineered bacteria in environmental remediation and the potential risks and benefits of these technologies (Peixoto *et al.,* 2022).

The responsible use of genetically engineered bacteria in environmental remediation requires a comprehensive risk assessment and risk management strategies. Additionally, public engagement is crucial in shaping regulatory decisions and ensuring the safe and responsible deployment of these technologies.

Potential Improvements

- To address the identified gaps and challenges, potential improvements include enhancing transparency in the regulatory process, promoting international cooperation and harmonization of regulations, and ensuring that risk assessment and decision-making are based on the best available scientific evidence (National Academies Press (US), 2016; Movement of Certain Genetically Engineered Organisms, 2020).
- Furthermore, there is a need to develop flexible, risk-based permitting procedures that can accommodate the import, interstate movement, and environmental release of regulated genetically engineered organisms (Movement of Certain Genetically Engineered Organisms, 2020).

Gaps and Challenges

- The international regulatory framework for genetically engineered organisms in environmental applications is complex and multifaceted. Addressing the identified gaps and challenges and implementing potential improvements are essential for ensuring the safe and responsible use of these organisms while facilitating international trade and environmental protection.
- One of the challenges in the current regulatory frameworks is the need to maintain transparency, apply the scientific method, and adhere to the precautionary principle in the assessment of the potential adverse effects of genetically engineered organisms on the environment and human health (National Academies Press (US), 2016).

- Another challenge is the coordination of regulatory efforts across different agencies and jurisdictions, particularly in the context of international trade standards and the movement of genetically engineered organisms across borders (Rafeeq *et al.,* 2023).

Risk Management Strategies

The responsible use of genetically engineered bacteria in environmental remediation requires a thorough risk assessment and risk management strategies. Here are some suggestions for these strategies:

- *Risk assessment:* The risk assessment process should include a comprehensive evaluation of the potential risks associated with the use of genetically engineered bacteria in environmental remediation. This includes assessing the potential risks to human health, the environment, and ecosystems (National Academies Press (US), 2016; Rafeeq *et al.,* 2023).
- *Risk management:* Risk management strategies should be developed to minimize the potential risks associated with the use of genetically engineered bacteria in environmental remediation. This includes developing protocols for the safe handling, transportation, and disposal of genetically engineered bacteria, as well as monitoring their impact on the environment and human health (Rafeeq *et al.,* 2023).

APPLICATIONS IN BIOREMEDIATION

While the deployment of Genetically Engineered Microorganisms (GEMs) in practical scenarios has been restricted due to concerns regarding the potential risks associated with horizontal gene transfer, the findings obtained remain significant in evaluating both the advantages and challenges related to their use in bioremediation. This understanding becomes increasingly vital given the potential future release of GEMs into controlled environments for the purpose of degrading contaminants.

An exemplary case is illustrated by Pseudomonas fluorescens HK44, the first genetically engineered bacterium sanctioned by the U.S. Environmental Protection Agency for soil bioremediation (Ripp et al., 2000). This strain harbors the naphthalene catabolic plasmid (pUTK21) containing the lux gene, which triggers bioluminescence in response to PAH exposure. Consequently, the lux protein (luciferase) production facilitated real-time monitoring of host cell levels in PAH-contaminated soil inoculated with HK44. Furthermore, encapsulated HK44 cells in a biosensor provided real-time PAH level measurements in the same soil sample. Utilizing this monitoring system, HK44 demonstrated survival in the field for a period of 2 years while effectively degrading PAHs.

In a similar vein, a genetically engineered E. coli strain was engineered as a bioreporter for benzene, toluene, and related compounds by incorporating the luc gene under the control of the Pu promoter and transcription activator XylR from the TOL plasmid of Pseudomonas putida mt-2 (Willardson et al., 1998). However, unlike the HK44 system, these recombinant cells were not introduced into contaminated soil or water, providing measurements of toluene-like compounds in field-collected samples but not real-time data during bioremediation.

In another noteworthy application, Escherichia coli engineered to express the atrazine chlorohydrolase gene, but chemically cross-linked to render it non-viable, was employed for in situ soil remediation of atrazine-contaminated soil. Concurrent supplementation with phosphate led to a significant 97% reduction in atrazine concentration (Strong et al., 2000).

The engineering of bacterial hemoglobin, vgb/VHb has demonstrated efficacy across a diverse range of prokaryotic and eukaryotic organisms, resulting in enhanced growth and production of valuable products. Examples include the synthesis of recombinant and non-recombinant proteins in bacteria (Khosravi et al., 1990; Kallio and Bailey, 1996), antibiotics in fungi (DeModena et al., 1993; Magnolo et al., 1991), and tissue plasminogen activator in mammalian cells (Pendse and Bailey, 1994). While the beneficial effects of VHb are commonly observed, they are not universal (Liu et al., 1995; Wei et al., 1998b).

The expression of VHb in recombinant hosts has been found to impact respiration rates, leading to either increases (up to 2–3 fold) or decreases (up to 2–10 times lower) compared to cells lacking vgb (Liu et al., 1995; Tsai et al., 1996; Patel et al., 2000; Urgun-Demirtas et al., 2004). These variations may stem from VHb-enhanced respiration or increased oxygen utilization efficiency. VHb presence in heterologous bacterial hosts has also shown variable effects on metabolism (Birnbaum and Bailey, 1991; Kallio et al., 1994; Liu et al., 1996; Wei et al., 1998b), with microarray studies indicating significant gene regulation in VHb-producing recombinant E. coli (Roos et al., 2004).

In the realm of bioremediation studies, vgb was introduced into several soil bacteria to investigate its potential for enhancing growth and aromatic compound degradation. Initial experiments with Xanthomonas maltophillia using benzoic acid as a model compound yielded only modest improvements (Liu et al., 1996). However, more practical applications were explored by transforming Burkholderia strain DNT with pSC160 to enhance 2,4-dinitrotoluene (2,4-DNT) degradation (Patel et al., 2000). The presence of VHb increased 2,4-DNT degradation by approximately 50–60%, possibly by augmenting respiration and growth, as well as facilitating oxygen-dependent steps in the degradation pathway.

Further investigations with plasmid pJS39 in E. coli JM109 highlighted VHb's potential to stimulate DNT dioxygenase directly. Under limited aeration, VHb-enhanced DNT dioxygenase levels and/or activity were observed, resulting in increased conversion rates compared to strains lacking VHb (Fish et al., 2001). Purified VHb was also shown to stimulate the disappearance of 2,4-DNT, suggesting a direct interaction with dioxygenase (Lin et al., 2003). These findings underscore VHb's promising role in enhancing bioremediation processes and merit further exploration in environmental cleanup efforts.

FUTURE DIRECTIONS AND RECOMMENDATIONS

Research and Development

To enhance the efficiency, specificity, and safety of genetically engineered bacteria in remediation strategies, the following research directions and recommendations are proposed:

1. *Efficiency*: Develop novel techniques to improve the growth rate and survival of genetically engineered bacteria in contaminated environments. Explore the potential of synthetic biology and synthetic genomics to optimize bacterial performance in bioremediation processes (Ayub *et al.,* 2022; Fan *et al.,* 2022).
2. *Specificity:* Investigate the mechanisms behind bacterial interactions with target contaminants, such as heavy metals or organic pollutants, to develop more selective and effective remediation strategies (Ayub *et al.,* 2022). Evaluate the potential of bioaugmentation, which involves the addition

of specific microorganisms to contaminated sites, for enhancing the specificity of bioremediation processes (Ayub *et al.,* 2022).

3. *Safety:* Assess the potential risks associated with the use of genetically engineered bacteria in environmental remediation, such as the possibility of gene escape or unintended impacts on ecosystems. Develop robust regulatory frameworks to ensure the safe and responsible use of genetically engineered bacteria in bioremediation applications (Ayub *et al.,* 2022).

Interdisciplinary Collaboration

- It is crucial to encourage collaboration between biotechnologists, ecologists, and regulatory experts to advance the field of microbial-assisted environmental restoration. This collaboration can lead to the development of more effective and sustainable remediation strategies that consider the complex interactions between genetically engineered bacteria, their target contaminants, and the surrounding ecosystem (Ayub *et al.,* 2022; Lemke & DeSalle, 2023).
- Foster communication and knowledge exchange among these disciplines to better understand the complex interactions between genetically engineered bacteria, their target contaminants, and the surrounding ecosystem (J. S. Singh *et al.,* 2011; Singh, 2011).
- By promoting interdisciplinary collaboration, researchers can work together to address the challenges of environmental remediation and develop innovative solutions that consider the unique challenges and opportunities presented by genetically engineered bacteria.
- This approach will ultimately lead to more efficient, specific, and safe remediation strategies for contaminated sites, benefiting both the environment and human health (Rafeeq *et al.,* 2023).

Integration With Traditional Remediation Approaches

- It is essential to emphasize the integration of genetically engineered bacteria with traditional remediation approaches to maximize efficacy and the importance of a holistic and adaptive environmental cleanup approach. This integration holds great potential for advancing remediation strategies and achieving more comprehensive and sustainable results.
- The use of genetically engineered bacteria in environmental remediation is an emerging and promising tool, as highlighted in various research perspectives. For instance, a review on genetically engineered bacteria as an emerging tool for environmental remediation and future research perspectives emphasizes the potential of these organisms in sustainable environmental practices (J. Singh, 2011; Singh *et al.,* 2011).
- Additionally, genetically engineered microorganisms have been studied for their potential in environmental remediation, and their future prospects and recommendations have been summarized, indicating the high level of interest and research in this area (Ayub *et al.,* 2022).
- Furthermore, it is important to consider the potential of genetically engineered bacteria in driving ecological restoration. Microbial ecology plays a fundamental role in the natural world, influencing processes such as nutrient cycling, energy flow, and disease. Therefore, integrating genetically

engineered bacteria with traditional remediation methods can have a significant impact on ecological restoration.

- By integrating genetically engineered bacteria with traditional remediation approaches, a more comprehensive and effective strategy can be developed, contributing to the advancement of environmental restoration efforts. This integration aligns with the ongoing research and the future prospects of using genetically engineered bacteria for environmental remediation, emphasizing the importance of staying updated with the latest developments and integrating them into practical remediation strategies.

Potential Improvements

- To address the identified gaps and challenges, potential improvements include enhancing transparency in the regulatory process, promoting international cooperation and harmonization of regulations, and ensuring that risk assessment and decision-making are based on the best available scientific evidence (National Academies Press (US), 2016; Movement of Certain Genetically Engineered Organisms, 2020).
- Furthermore, there is a need to develop flexible, risk-based permitting procedures that can accommodate the import, interstate movement, and environmental release of regulated genetically engineered organisms (Movement of Certain Genetically Engineered Organisms, 2020).

CONCLUSION

In conclusion, the chapter "Microbial Architects of Environmental Restoration: Exploring the Frontiers of Genetically Engineered Bacteria in Remediation Strategies" has illuminated the transformative potential of genetically engineered bacteria as a pioneering tool in the realm of environmental remediation. Through a comprehensive exploration of the principles, applications, and concerns surrounding this emerging technology, it becomes evident that these microbial architects hold great promise in addressing the complex challenges posed by environmental pollutants. The discussed case studies and success stories underscore the tangible impact of genetically engineered bacteria in real-world remediation projects, showcasing their efficiency and adaptability across diverse environmental contexts. Looking ahead, the recommendations presented in this chapter underscore the importance of continued research and development to refine the efficiency, specificity, and safety of genetically engineered bacteria. The integration of these innovative solutions with traditional remediation approaches emerges as a promising avenue, emphasizing a holistic and synergistic approach to environmental cleanup.

In essence, "Microbial Architects of Environmental Restoration" serves as a foundational exploration into the potential and challenges of genetically engineered bacteria in environmental remediation. As we stand at the intersection of biology and environmental science, this chapter invites further interdisciplinary collaboration, critical inquiry, and responsible innovation to unlock the full potential of microbial architects in the ongoing quest for a sustainable and restored environment.

REFERENCES

Armstrong, R. (2023). Towards the microbial home: An overview of developments in next-generation sustainable architecture. *Microbial Biotechnology*, *16*(6), 1112–1130. doi:10.1111/1751-7915.14256 PMID:37070748

Ayub, ur Rehman, Umar, Zuhra, & Shabaan. (2022). Role of legumes in phytoremediation of heavy metals. In Advances in Legumes for Sustainable Intensification (pp. 345-360). Academic Press.

Azad, M. A. K., Amin, L., & Sidik, N. M. (2014). Genetically engineered organisms for bioremediation of pollutants in contaminated sites. *Chinese Science Bulletin*, *59*(8), 703–714. doi:10.1007/s11434-013-0058-8

Bittihn, P., Didovyk, A., Tsimring, L. S., & Hasty, J. (2020). Genetically engineered control of phenotypic structure in microbial colonies. *Nature Microbiology*, *5*(5), 697–705. doi:10.1038/s41564-020-0686-0 PMID:32284568

Carpi, A. (Ed.). (2011). *Progress in molecular and environmental bioengineering: from analysis and modeling to technology applications*. BoD–Books on Demand.

Cases & de Lorenzo. (2005). *Genetically modified organisms for the environment: stories of success and failure and what we have learned from them.* Academic Press.

Davison, J. (2005). Risk mitigation of genetically modified bacteria and plants designed for bioremediation. *Journal of Industrial Microbiology & Biotechnology*, *32*(11-12), 639–650. doi:10.1007/s10295-005-0242-1 PMID:15973534

Deckers, M., De Loose, M., Papazova, N., Deforce, D., Fraiture, M.-A., & Nancy, H. C. (2022). Roosens. "First monitoring for unauthorized genetically modified bacteria in food enzymes from the food market.". *Food Control*, *135*, 108665. doi:10.1016/j.foodcont.2021.108665

Dua, M., Singh, A., Sethunathan, N., & Johri, A. (2002). Biotechnology and bioremediation: Successes and limitations. *Applied Microbiology and Biotechnology*, *59*(2-3), 143–152. doi:10.1007/s00253-002-1024-6 PMID:12111139

Duque, E., Udaondo, Z., Molina, L., de la Torre, J., Godoy, P., & Ramos, J. L. (2022). Providing octane degradation capability to Pseudomonas putida KT2440 through the horizontal acquisition of oct genes located on an integrative and conjugative element. *Environmental Microbiology Reports*, *14*(6), 934–946. doi:10.1111/1758-2229.13097 PMID:35651318

Fan, J.-X., Niu, M.-T., Qin, Y.-T., Sun, Y.-X., & Zhang, X.-Z. (2022). Progress of engineered bacteria for tumor therapy. *Advanced Drug Delivery Reviews*, *185*, 114296. doi:10.1016/j.addr.2022.114296 PMID:35439571

French, K. E., Zhou, Z., & Terry, N. (2020). French, Katherine E., Zhongrui Zhou, and Norman Terry. "Horizontal 'gene drives' harness indigenous bacteria for bioremediation.". *Scientific Reports*, *10*(1), 15091. doi:10.1038/s41598-020-72138-9 PMID:32934307

Gao, J.-J., Wang, B., Li, Z.-J., Xu, J., Fu, X.-Y., Han, H.-J., Wang, L.-J., Zhang, W.-H., Deng, Y.-D., Wang, Y., Gong, Z.-H., Tian, Y.-S., Peng, R.-H., & Yao, Q.-H. (2022). Metabolic engineering of Oryza sativa for complete biodegradation of thiocyanate. *The Science of the Total Environment*, *820*, 153283. doi:10.1016/j.scitotenv.2022.153283 PMID:35066037

Garbisu, C., & Alkorta, I. (1999). Utilization of genetically engineered microorganisms (GEMs) for bioremediation. *Journal of Chemical Technology and Biotechnology*, *74*(7), 599–606. doi:10.1002/(SICI)1097-4660(199907)74:7<599::AID-JCTB82>3.0.CO;2-G

Kiyono, M., & Pan-Hou, H. (2006). Genetic engineering of bacteria for environmental remediation of mercury. *Journal of Health Science*, *52*(3), 199–204. doi:10.1248/jhs.52.199

Kumar, R. S., Singh, D., Bose, S. K., & Trivedi, P. K. (2020). Biodegradation of environmental pollutant through pathways engineering and genetically modified organisms approaches. In *Microorganisms for Sustainable Environment and Health* (pp. 137–165). Elsevier. doi:10.1016/B978-0-12-819001-2.00007-3

Lemke, M., & DeSalle, R. (2023). The next generation of microbial ecology and its importance in environmental sustainability. *Microbial Ecology*, *85*(3), 781–795. doi:10.1007/s00248-023-02185-y PMID:36826587

Liu, J., Zhou, X., Wang, T., Fan, L., Liu, S., Wu, N., Xu, A., Qian, X., Li, Z., Jiang, M., Zhou, J., & Dong, W. (2022). Construction and comparison of synthetic microbial consortium system (SMCs) by non-living or living materials immobilization and application in acetochlor degradation. *Journal of Hazardous Materials*, *438*, 129460. doi:10.1016/j.jhazmat.2022.129460 PMID:35803189

Liu, L., Bilal, M., Duan, X., & Hafiz, M. N. (2019). Iqbal. "Mitigation of environmental pollution by genetically engineered bacteria—Current challenges and future perspectives.". *The Science of the Total Environment*, *667*, 444–454. doi:10.1016/j.scitotenv.2019.02.390 PMID:30833243

Lloyd, J. R., Lovley, D. R., & Macaskie, L. E. (2003). Biotechnological application of metal-reducing microorganisms. *Advances in Applied Microbiology*, *53*, 85–128. doi:10.1016/S0065-2164(03)53003-9 PMID:14696317

Ma, Y., Chen, H., Lan, C., & Ren, J. (2018). Help, hope and hype: Ethical considerations of human microbiome research and applications. *Protein & Cell*, *9*(5), 404–415. doi:10.1007/s13238-018-0537-4 PMID:29675808

Mutanda, I., Sun, J., Jiang, J., & Zhu, D. (2022). Bacterial membrane transporter systems for aromatic compounds: Regulation, engineering, and biotechnological applications. *Biotechnology Advances*, *59*, 107952. doi:10.1016/j.biotechadv.2022.107952 PMID:35398204

Offiong, N.-A. O., Edet, J. B., Shaibu, S. E., Akan, N. E., Atakpa, E. O., Sanganyado, E., Okop, I. J., Benson, N. U., & Okoh, A. (2023). Metagenomics: An emerging tool for the chemistry of environmental remediation. *Frontiers in Environmental Chemistry*, *4*, 7. doi:10.3389/fenvc.2023.1052697

Patel, H., Shakhreliya, S., Maurya, R., Pandey, V. C., Gohil, N., Bhattacharjee, G., Alzahrani, K. J., & Singh, V. (2022). CRISPR-assisted strategies for futuristic phytoremediation. In *Assisted phytoremediation* (pp. 203–220). Elsevier. doi:10.1016/B978-0-12-822893-7.00006-9

Peixoto, R. S., Voolstra, C. R., Sweet, M., Duarte, C. M., Carvalho, S., Villela, H., Lunshof, J. E., Gram, L., Woodhams, D. C., Walter, J., Roik, A., Hentschel, U., Thurber, R. V., Daisley, B., Ushijima, B., Daffonchio, D., Costa, R., Keller-Costa, T., Bowman, J. S., ... Berg, G. (2022). Harnessing the microbiome to prevent global biodiversity loss. *Nature Microbiology*, *7*(11), 1726–1735. doi:10.1038/s41564-022-01173-1 PMID:35864220

Pieper, D. H., & Reineke, W. (2000). Engineering bacteria for bioremediation. *Current Opinion in Biotechnology*, *11*(3), 262–270. doi:10.1016/S0958-1669(00)00094-X PMID:10851148

Priya, A., Dutta, K., & Daverey, A. (2022). A comprehensive biotechnological and molecular insight into plastic degradation by microbial community. *Journal of Chemical Technology and Biotechnology*, *97*(2), 381–390. doi:10.1002/jctb.6675

Rafeeq, H., Afsheen, N., Rafique, S., Arshad, A., Intisar, M., Hussain, A., Bilal, M., & Hafiz, M. N. (2023). Iqbal. “Genetically engineered microorganisms for environmental remediation.”. *Chemosphere*, *310*, 136751. doi:10.1016/j.chemosphere.2022.136751 PMID:36209847

. Rafeeq, Hamza, Nadia Afsheen, Sadia Rafique, Arooj Arshad, Maham Intisar, Asim Hussain, Muhammad Rajendran, P., J. Muthukrishnan, and P. Gunasekaran. “Microbes in heavy metal remediation.” (2003).

Resources, N.National Academies of Sciences, Engineering, and Medicine. (2016). Regulation of current and future genetically engineered crops. In *Genetically engineered crops: Experiences and prospects*. National Academies Press.

Sayler, G. S., & Ripp, S. (2000). Field applications of genetically modified bacteria for bioremediation processes. *Current Opinion in Biotechnology*, *11*, 286–289. doi:10.1016/S0958-1669(00)00097-5 PMID:10851144

Sharma, B., & Shukla, P. (2022). Futuristic avenues of metabolic engineering techniques in bioremediation. *Biotechnology and Applied Biochemistry*, *69*(1), 51–60. doi:10.1002/bab.2080 PMID:33242354

Shukla, K. P., Singh, N. K., & Sharma, S. (2010). Bioremediation: Developments, current practices and perspectives. *Genetic Engineering and Biotechnology Journal*, *3*, 1–20.

Singh, J. S., Abhilash, P. C., Singh, H. B., Singh, R. P., & Singh, D. P. (2011). Genetically engineered bacteria: An emerging tool for environmental remediation and future research perspectives. *Gene*, *480*(1-2), 1–9. doi:10.1016/j.gene.2011.03.001 PMID:21402131

Téllez, A. A. (2022). Prediction of the impact induced by Cd in binary interactions with other divalent metals on wild-type and Cd-resistant strains of Dictyosphaerium chlorelloides. *Environmental Science and Pollution Research International*, 1–11.

Urgun-Demirtas, M., Stark, B., & Pagilla, K. (2006). Use of genetically engineered microorganisms (GEMs) for the bioremediation of contaminants. *Critical Reviews in Biotechnology*, *26*(3), 145–164. doi:10.1080/07388550600842794 PMID:16923532

Wood, T. K. (2008). Molecular approaches in bioremediation. *Current Opinion in Biotechnology*, *19*(6), 572–578. doi:10.1016/j.copbio.2008.10.003 PMID:19000765

Chapter 11
The Potential of Agricultural Waste Chars as Low-Cost Adsorbents for Heavy Metal Removal From Water

Bothwell Nyoni
https://orcid.org/0000-0001-8371-6116
Nelson Mandela University, South Africa & National University of Science and Technology, Zimbabwe

Bienvenu Gael Fouda-Mbanga
Nelson Mandela University, South Africa

Bongibethu Msekeli Hlabano-Moyo
Iowa State University, USA

Yvonne Boitumelo Nthwane
https://orcid.org/0000-0003-0730-9521
Tshwane University of Technology, South Africa & University of Johannesburg, South Africa

Bongani Yalala
National University of Science and Technology, Zimbabwe

Zikhona Tywabi-Ngeva
https://orcid.org/0000-0002-5290-6942
Nelson Mandela University, South Africa

Percy Hlangothi
https://orcid.org/0000-0002-6590-4356
Nelson Mandela University, South Africa

ABSTRACT

Ensuring that every human being has access to clean and safe drinking water is a major global concern. The challenge of heavy metal contamination in water poses a serious risk to human health. This chapter aims to explore an eco-friendly and cost-effective solution to address water contamination issues. The focus is on using pyrolysis char derived from agricultural waste as a low-cost adsorbent to efficiently remove heavy metals from water. A thorough review of relevant literature, emphasising the advantages of pyrolysis char and highlighting recent advancements in the field, is performed. Additionally, an outline of potential challenges and future research direction in this promising area is given.

DOI: 10.4018/979-8-3693-1618-4.ch011

INTRODUCTION

With the increasing anthropogenic activities attributed to technological advancements within chemical manufacturing, mining and agriculture, the deterioration of global water quality is imminent (Singh et al., 2023). This increase in demand for scarce water resources can best be abated through effective management and use of much more benign chemical processes, re-use efforts and pollution control. Water quality on the global platform varies significantly with developing countries generally bearing the brunt due to much more lenient regulations and inadequate pollution control attributed to limited infrastructure among other constraints. Despite stringent regulations, developed countries face high water treatment costs due to intense industrialisation. This highlights the urgent need for more cost-effective water treatment technologies. Industrial wastewater, containing pollutants like perfluoroalkyl substances and heavy metals, jeopardizes drinking water. Contamination makes water unsuitable for consumption, with treatment costs potentially soaring. Unchecked pollution leads to severe health issues in humans and wildlife, disrupting ecosystems and posing a significant threat to food security by altering predator-prey ratios. Carbon based adsorption materials have been at the forefront of water treatment and purification research in the past decade, these include activated carbon, carbon nanotubes, carbon aerogels and carbon black (Cheng et al., 2019). With agricultural activity on a global scope generating over 140 billion tonnes of agricultural waste annually (United Nations, 2009) which if pyrolysed could amount to above 70 billion tonnes of pyrolysis char based on reported char yields, biomass char becomes a readily available carbon-based adsorbent. Whilst activated carbon and other carbon-based materials have been utilised distinctly as adsorbents, recent advances have coupled the adsorptive capabilities of these materials with electro-motive techniques such as capacitive deionisation which allow for benign regeneration of the adsorbent and thus limiting the environmental burden (Cheng et al., 2019, Botha et al., 2021). Such a development also limits material handling and associated costs, thus paving a way for cheaper alternative water treatment methods. Pyrolysis biochar costs far much less than common commercial adsorbents. Commercial adsorbents are characterised by BET specific surface areas in the excess of $1500m^2/g$ (Yang et al., 2018) whilst pyrolysis char typically exhibits less BET specific surface area ($< 500\ m^2/g$) thus limited adsorption sites (Tomczyk et al., 2020). However, despite specific surface area concerns, similar adsorptive mechanisms borne by these materials and the low cost of pyrolysis char are seen as attributes that would set the stage for a shift towards the widespread use of biomass char for adsorption tasks.

This chapter aims to position pyrolysis char as a viable alternative to current carbonaceous materials in water treatment. It provides a comprehensive overview of using pyrolysis char as an adsorbent, emphasising the optimal agricultural waste types for pyrolysis, process details, and the sustainability of this approach. The chapter also delves into sources of heavy metal contamination, associated risks, and existing mitigation technologies for treating contaminated waters. Material characteristics of pyrolysis char relevant to contaminant adsorption are explored, along with discussions on adsorption mechanisms and influencing factors. A comparative analysis of the effectiveness of various adsorbents against pyrolysis char is conducted to identify commonalities and strengths, aiming to pave the way for composite adsorption materials that balance cost and efficiency. Case studies highlighting the successful use of pyrolysis char in heavy metal removal, along with performance metrics, offer guidance for addressing challenges and advancing the technology. The chapter concludes by summarising key aspects discussed throughout.

AGRICULTURAL WASTE PYROLYSIS: A SUSTAINABLE APPROACH

Pyrolysis is a high-temperature decomposition process that occurs in the absense of oxygen, transforms organic compounds into gases, liquids, and solid char in three phases: drying, pyrolytic decomposition, and char formation. It offers on-farm utility, converting agricultural biomass into chemicals, char, and energy. The liquids serve as renewable energy, biofuel feedstock, or chemicals (Nyoni & Hlangothi, 2023). Solid char acts as soil amendment (biochar) or low-cost adsorbent (Sikder & Joardar, 2019). Figure 1 depicts a typical process of converting agricultural waste into biochar and other valuable products via pyrolysis. Recovered liquids are known as pyroligneous liquid. Syngas, rich in hydrogen and carbon compounds, serves in heat, power generation, or as a feedstock for chemicals and fuels. Solid residue, char or biochar, becomes a carbon-rich adsorbent for water treatment (Bhurachevskaya et al., 2023).

Figure 1. Pyrolysis route for the synthesis of biochar, liquid chemicals and fuels, and syngas

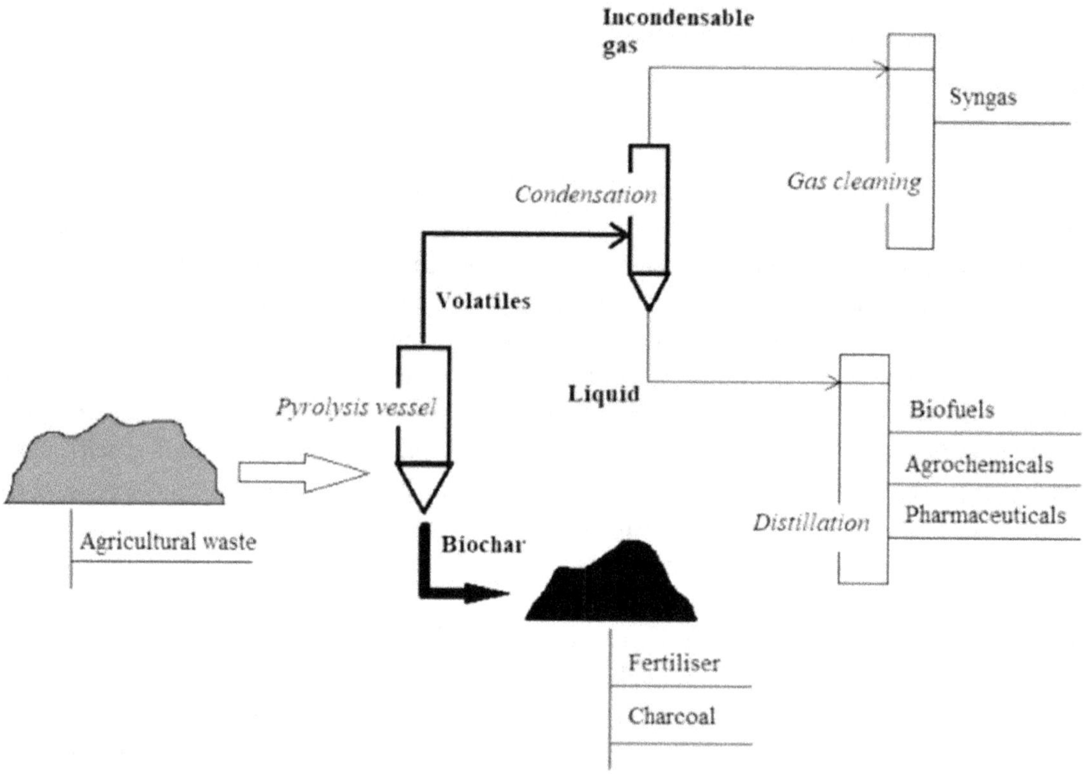

Types of Agricultural Waste

Agricultural waste refers to discarded material associated with various farming practices and food production processes (Obi et al., 2016; Pattanaik et al., 2019; Rahimi et al., 2022). Sometimes referred to as agro-waste, these materials have gained recognition for their potential in sustainable resource management. For example, developing countries such as India have the potential of generating above 600 million tonnes of agricultural waste per year, of which 25-30% is used for animal feed and energy generation (Rahimi et al., 2022). In this chapter, some common types of agricultural waste will be explored following the classification given in Fig. 2.

Figure 2. Classification of agricultural waste

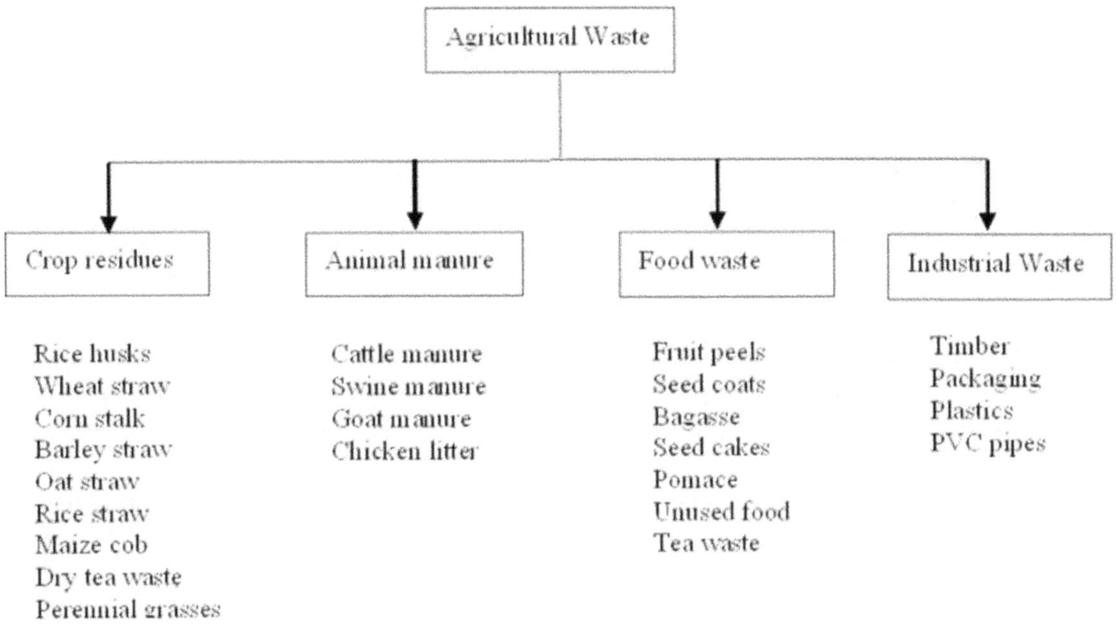

Crop Residues: Crop residues encompass the plant parts left behind after harvesting crops. These residues, such as corn stalks, wheat and barley straw, and rice husks, are rich in organic matter. The global production of crop residues of various types is in the magnitudes of billions of tonnes per year (Li et al., 2023). They can be effectively utilised as livestock feed, bedding material, or as inputs for biochar and bioenergy production.

Animal Manure: Animal manure, produced from livestock farming, presents a significant agricultural waste stream. Researchers have managed to show that adsorbents based on animal manure, especially cow dung, are efficient in the removal of dyes and heavy metals from aqueous solutions (Iwuozor et al., 2022).

Food Waste: Food processing generates substantial waste, including fruit and vegetable peels, seed coats, de-oiled seed cakes, pomace, sugarcane bagasse and discarded parts of fish or poultry (Ajala et al., 2021). The annual world production of sugarcane bagasse has been reported to be approximately 279 million metric tonnes (Chandel et al., 2012). These waste materials can be transformed into animal feed or converted into biochar or bioenergy.

Industrial Waste: Agricultural activities, notably wood manufacturing, generate substantial waste, with Germany producing over 10 million tonnes annually (Pandey, 2022). Stringent regulations in timber industries mitigate deforestation concerns (Ramage et al., 2017). Modern agriculture's agrochemical use results in packaging waste, emphasising the need for proper collection and disposal to prevent environmental contamination.

Types of Pyrolysis

Pyrolysis technologies can be classified based on the rate of heating and removal of pyrolysis gases from the reaction atmosphere. As such, pyrolysis has been classified as slow, intermediate, fast, and flash (Dhyani & Bhaskar, 2017; Dermibas, 2009; Bridgwater et al., 2007). It has been shown that in any pyrolysis process, slow heating rates promotes the production of char (Bridgwater et al., 2007).

Slow pyrolysis: Slow pyrolysis is typically carried out as a batch operation whereby the material is subjected to low to moderate temperatures, low particle heating rates and longer residence times (Dhyani & Bhaskar, 2017). A typical slow pyrolysis process occurs at a temperature of 500 °C and a heating rate range of 0.1 to 2 °C/s (Chaiwong et al., 2013; Bridgwater et al., 2007; Marshall, 2013; Onay & Kockar, 2004). Slow pyrolysis's distinctive feature is its extended residence time, facilitating prolonged interaction between biomass and the pyrolysis environment, yielding larger molecules in char, bio-oil, and gases. Typically performed in a fixed bed reactor, biomass undergoes pyrolysis in a stationary container, producing continuous extraction of gaseous and liquid products while solid char remains in the bed.

Intermediate pyrolysis: Intermediate pyrolysis occurs at temperatures ranging from 400 to 600 °C. It is a compromise between slow pyrolysis and fast pyrolysis in terms of reaction time and heating rate. The process is relatively faster than slow pyrolysis but slower than fast pyrolysis, allowing for better control over the reaction conditions and product quality (Nyoni & Hlangothi, 2022). Overall, intermediate pyrolysis is a promising process that offers a balance between product yield, reaction time, and energy efficiency, making it suitable for the conversion of agricultural waste into biochar.

Fast pyrolysis: In fast pyrolysis, the material is subjected to moderate to high temperatures depending on the desired product. Low to moderate temperatures result in maximum liquid yield and high temperatures favours maximum gas production (Li et al., 2004; Ellens & Brown, 2012). In both cases, high heat transfer rates and very short residence times are required. The rapid heating of the solid material and cooling of the gases minimises further cracking of large hydrocarbon molecules and thereby maximise the liquid yields (Dhyani & Bhaskar, 2017; Gable & Brown, 2016). Typical conditions for the fast pyrolysis of biomass are a temperature of 600°C, heating rate of 200 – 1000 °C/s and residence times of less than 2s (Demirbas, 2009; Bridgwater et al., 2007).

Flash pyrolysis: Flash pyrolysis sometimes referred to as very fast pyrolysis, is a more extreme form of pyrolysis, characterised by high temperatures, very high heating rates and extremely short residence times. To achieve flash pyrolysis, the material must be fed into a pyrolysis reactor in the form of powder (Singh et al., 2016). In flash pyrolysis, the temperature typically ranges from 700 to 1000 °C, and the heating rates are greater than 1000 °C/s. Flash pyrolysis, with high temperatures and rapid heating, swiftly decomposes materials into diverse products. Its brief residence times, typically under 0.5 seconds, limit material-reactor interaction (Goyal et al., 2008; Onay & Kockar, 2003). Despite challenges in reactor design and product handling, ongoing research aims to optimize this method for efficient and scalable biomass conversion.

Sustainable Aspects of Using Pyrolysis for Char Production

Pyrolytic conversion of agricultural waste into biochar is a sustainable solution, managing waste and curbing pollution. Biochar improves soil fertility, water retention, and structure. The process also yields bioenergy (syngas, bio-oil), supporting renewable energy. This closed-loop system aligns with circular economy principles, ensuring resource efficiency.

HEAVY METAL CONTAMINATION IN DRINKING WATER

Heavy metals are a group of metallic elements with a high atomic weight and density, i.e., > 20 and 5 g/cm^3, respectively. These are essentially transition elements and some metalloids such as arsenic (Raychaudhuri et al., 2021; Fouda-Mbanga et al., 2023). Common heavy metals in drinking water, such as lead, arsenic, mercury, cadmium, and chromium, can pose health risks. Excessive copper and zinc, essential in small amounts, also present concerns. Understanding sources, health effects, regulations, and mitigation is vital for ensuring safe drinking water, requiring collaborative efforts to monitor and address heavy metal presence and protect public health.

Sources of Heavy Metals in Drinking Water

The presence of heavy metals in drinking water can be attributed to natural occurrences, industrial and sewage effluent discharge, agricultural industry, illegal garbage disposal, and leakage of leachate from waste disposal sites (Dilebo et al., 2023).

Natural sources: Heavy metals can naturally leach into groundwater from geological formations, such as rock and soil, particularly in regions with high mineral content through the process of weathering (Srivastava et al., 2017). The existence of subterranean rocks and volcanoes beneath the water bodies could also be a source of some heavy metal containing salts (Fouda-Mbanga et al., 2023).

Industrial activities: Industrial processes like mining, metal plating, leather, textile, pulp and papermaking release heavy metals into the environment. These contaminants can infiltrate groundwater and surface water sources. The common heavy metals discharged from industrial activities are arsenic, cadmium, chromium, copper, lead, mercury, silver, nickel and zinc (Das & Poater, 2021).

Agricultural activities: The use of heavy metal-based fertilizers, pesticides, and herbicides in agriculture can contaminate water bodies through runoff and percolation (Srivastava et al., 2017). It is common practice to improve soil fertility by the application of sewage sludge, however, sewage sludge introduces heavy metals such as zinc, copper, lead and cadmium (Srivastava et al., 2017). Furthermore, the use of several pesticides introduces heavy metals into soils and plants (Singh et al., 2017).

Plumbing systems: Aging infrastructure and metal-based plumbing materials can introduce heavy metals into drinking water as it travels from the treatment plant to the end-user (Gonzalez et al. 2013; Ling et al., 2011). The most common heavy metals used in plumbing systems are iron, copper and lead. The length of water distribution networks, especially in large metropolitan areas can reach approximately 4000 km, for example in the city of Barcelona (Gonzalez et al., 2013). Therefore, water quality may be altered during transportation when interactions occur between the distributed water and the pipe material; these interactions may lead to dissolution of metals into drinking water.

Waste disposal: Improper disposal of electronic waste, batteries, and other hazardous materials can lead to the release of heavy metals into the environment (Das & Poater, 2021). The global use of electronic gadgets such as cell phones, computers, television, cooking utensils is increasing. Due to the short life of most electronic gadgets, the disposal of electronic waste is fast becoming a major concern. Electronic waste disposal introduces heavy metals such as lead, copper, silver and zinc (Nekouei et al., 2019).

Health Effects of Heavy Metals in Drinking Water

Exposure to elevated levels of heavy metals in drinking water can have serious health consequences. The specific health effects vary depending on the type of heavy metal and the duration and level of exposure. To ensure the safety of drinking water, regulatory agencies like the World Health Organisation (WHO), Department of Water and Sanitation (DWS) in South Africa, Environmental Protection Agency (EPA) in the United States have established maximum allowable levels for heavy metals in drinking water. These regulations set strict limits on the concentration of heavy metals in drinking water to protect public health. Table 1 presents heavy metals of major concern found in drinking water, their health effects, and maximum allowable limits.

Table 1. Heavy metals of major concern, their sources, health effects, and maximum allowable limits

Heavy Metal	Sources	Health Effects	Maximum Allowable Limits	References
Lead	Corrosion of lead-based water distribution pipes and fittings, Industrial effluent	Exposure can impair cognitive development in children, cause high blood pressure, kidney damage, and neurological disorders in adults	0.01 mg/L, WHO 0.015 mg/L, EPA	(Ab Razak et al., 2015; Chiririwa et al., 2017).
Mercury	Erosion of natural deposits, Illegal gold mining activities	Exposure can result in neurological damage, developmental issues in children, and cardiovascular problems	0.01 mg/L, WHO 0.02 mg/L, EPA	(Chen et al., 2016; Mapombere et al., 2022).
Arsenic	Erosion of natural deposits Runoff from orchards	Arsenic is a known carcinogen and can lead to various cancers, skin lesions, and cardiovascular problems	0.01 mg/L, WHO 0.01 mg/L, EPA	(Dixit & Hering, 2003; Ling et al., 2011).
Cadmium	Industrial effluent from battery manufacturers Water stagnating in household pipes Application of phosphate fertilisers	Exposure is associated with kidney damage, bone disease, and an increased risk of cancer	0.03 mg/L, WHO	(Ling et al., 2011; Izah et al., 2016; Kubier et al., 2019).
Chromium	Effluent from electroplating plants Leaching of rocks Slow oxidation of stainless-steel pipes	Hexavalent chromium, a toxic form of chromium, is known to cause cancer and respiratory problems	0.05 mg/L, WHO 0.1 mg/L, EPA	(Dube et al., 2016; Zhitkovich, 2019).
Nickel	Corrosion of nickel-based water distribution alloy pipes and fittings Electronic waste such as batteries Weathering of nickel-containing rocks	May cause free radicals in human tissue that result in DNA complications	0.02 mg/L, WHO 0.1 mg/L, EPA	(Forgacs et al., 2012; Dube et al., 2016).
Iron	Corrosion of steel-based water distribution systems	Intake has positive effects; however, an overload can cause diabetes, hemochromatosis, stomach problems, and nausea	0.3 mg/L, WHO 0.3 mg/L, EPA	(Vasudevan et al., 2009; Chaturvedi & Dave, 2012)
Copper	Corrosion of copper-based water distribution systems Mining operations Electronic waste	Gastrointestinal distress, liver or kidney damage	2.0 mg/L, WHO 1.3 mg/L, EPA	(Ling et al., 2011; Manne et al., 2022).

Existing Water Treatment Methods for Heavy Metal Removal

Ensuring safe drinking water involves diverse methods to minimize or eliminate heavy metals. Detection techniques identify metal presence, prompting crucial removal due to associated health hazards. Water treatment plants often combine methods based on specific metals, concentrations, and water characteristics, with ongoing monitoring assessing effectiveness. Additional options include electrochemical treatment, biological treatment, advanced oxidation processes, and membrane filtration.

Chemical Precipitation Method

This is one of the most popular methods for removing heavy metals from wastewater. In this technique, chemical agents like lime or sodium hydroxide are added to water to form insoluble metal hydroxides by interacting with the heavy metal ions in water. To remove harmful metals from water, insoluble precipitates are removed by sedimentation or filtration (Hualpa-Cutipa et al., 2022).

Adsorption Method

Adsorption, a widely used method for removing metal ions from water, employs an adsorbent to attract and bind them to its surface. Its popularity stems from low cost, simple design, and reliable operation. Common adsorbents include activated carbon, biomaterials, layered double hydroxides, and carbon nanotube-based materials (Gupta et al., 2021; Hualpa-Cutipa et al., 2022). Adsorption is also effective in removing organic contaminants.

Ion Exchange Method

Due to the rapidness and good removal efficiencies of ion exchange materials, ion exchange is a valuable method for the removal of harmful heavy metal ions from wastewater. In this method, ion exchange resins are used to replace heavy metal ions with less harmful ions like sodium or potassium. Both synthetic and natural ion exchange resins are used for this purpose, with synthetic resins being preferred due to their better ability to extract heavy metals from solutions (Gupta et al., 2021). This method is effective for removing specific heavy metals, but the resin needs regeneration or replacement over time.

Coagulation and Flocculation

Coagulation and flocculation involve the addition of chemical compounds (coagulants) to bind particles together with metal ions until they are heavy enough to ultimately settle down (Srivatsav et al., 2020). Chemical coagulants, such as alum or ferric chloride, are added to water to form flocs. Flocs attract and bind with heavy metal particles, allowing them to settle and be removed during sedimentation.

Reverse Osmosis

In this process, water is forced through a semi-permeable membrane, allowing only water molecules to pass while blocking heavy metal ions. Reverse osmosis is effective for removing a wide range of contaminants, including heavy metals (El-Dessouky & Ettouney, 2002).

PYROLYSIS CHAR AS AN ADSORBENT

Biochar is a carbon-rich residue produced by subjecting biomass to a thermochemical process (200-900 °C), and it typically has a high carbon content and porous structure (Yin et al., 2017). Agricultural and forest residues, animal waste, food processing waste, paper mill waste, municipal solid waste, and sludge are some of the most often used biomass raw sources. There is considerable research focus on the use of biochar for water remediation technologies because of several reasons. Firstly, biochar has high specific surface area which improves its removal efficiency (Levesque et al., 2021); secondly, it has been shown that the cost of production and regeneration is considerably lower than other commercial adsorbents (Ahmed et al., 2016) and thirdly, the use of biochar is in line with global efforts of improving sustainable development.

Properties of Biochar

The ability of biochar to adsorb heavy metals is dependent on its properties. In turn, pysico-chemical properties of biochar are dependent on the method of production and conditions used. The properties of biochar synthesised via pyrolysis of agricultural waste are dependent on the pyrolysis process parameters such as temperature and heating rate (Bavariani et al., 2019). The various properties of biochar that makes it a potential candidate as an adsorbent for heavy metal ions are discussed in the following subsections.

Elemental Composition

The elemental compositions of biochars synthesised from different agricultural wastes are given in Table 2. It is evident that chars synthesised from agricultural wastes have considerably high elemental carbon and oxygen. Elemental composition of biochar is dependent on pyrolysis temperature and heating rate. Ahmed et al., (2016) showed that the carbon content increases with increased temperature of pyrolysis. Conversely, hydrogen and oxygen contents were found to decrease with an increase in pyrolysis temperature. This observation is evident from Table 2. Increasing pyrolysis temperature consistently reduces hydrogen-to-carbon (H/C) and oxygen-to-carbon (O/C) ratios in biochar, indicating fewer alkyl and oxygenated groups, respectively (Nyoni et al., 2020). Higher H/C ratios in low-temperature poultry manure char suggest more alkyl groups, while elevated O/C ratios indicate increased oxygenated groups, contributing to a desirable negative surface charge that enhances biochar's heavy metal adsorption capabilities.

Specific Surface Area

Various studies indicate that the specific surface area (SSA) of biochar ranges between 1.5 to 600 m^2/g for biochars synthesised at pyrolysis temperatures ranging up to 1000 °C (Ahmed et al., 2016; Li et al., 2018). Temperature increase within a specific range correlates with a rise in biochar surface area, as lower temperatures fill pores with byproducts, decreasing area, while higher temperatures release gases, increasing pore quantity (Ahmed et al., 2016; Li et al., 2018). Beyond a critical temperature, surface area declines due to microporous destruction (Ahmed et al., 2016; Li et al., 2018). The specific surface areas of biochars from sugarcane and coconut shell have been reported to be 253.2 m^2/g and 25.8 m^2/g, respectively (Kajina & Rousset, 2018).

Pore Volume

Pore volume refers to the total space occupied by pores within the biochar structure. These pores can be categorised into different types based on their size: micropores (< 0.002 μm in diameter), mesopores (0.002 – 0.05 μm in diameter), and macropores (> 50 μm in diameter, normally > 1 cm^3/g in pore volume) (Khamkeaw et al., 2020). Pore volume is a crucial characteristic when evaluating biochars for adsorption applications. Biochars, which are carbonaceous materials derived from the pyrolysis of biomass, possess a complex porous structure consisting of various-sized pores. Pore volume is directly linked to surface area and the greater the pore volume, the more adsorption sites are available, enhancing the material's capacity to capture and retain various substances.

pH

Biochars are generally basic in nature, with pH ranging from 4 – 12 since they are mostly composed of inorganic metal carbonates and phosphates, as well as ash generated during pyrolysis and carbonisation (Yuan et al., 2011; Levesque et al., 2021). This is also affected by factors such as raw material and carbonisation temperature. When the pyrolysis temperature is adjusted, the pH normally rises owing to the degradation of functional groups with acidic properties such as carboxylic and phenolic groups and the volatilisation of organic acids (Zhao et al., 2015).

Water-Retaining Ability

Biochar, derived from varied sources through pyrolysis, shows differing water retention. Elevated pyrolysis temperatures enhance aromatisation and hydrophobicity, reducing functional groups and surface polarity, resulting in a gradual decline in water-holding capacity (Yu et al., 2013; Levesque et al., 2021). Most biochars are hydrophobic in nature and those with a large number of polar groups display an improved water retention efficiency (Batista et al., 2018).

Table 2. Properties of biochar synthesised from different materials and pyrolysis conditions

Biochar Starting Material	Pyrolysis Temperature (°C)	Elemental Analysis (%)					H/C Ratio	O/C Ratio	SSA (m^2/g)	PV (cm^3/g)	References
		C	H	N	S	O					
Rice straw	250	45.1	4.62	0.86	0.28	-	1.23	0.32	1.14	3.57	Yang et al. (2021)
Rice straw	450	50.4	2.75	0.83	0.27	-	0.65	0.28	7.65	5.31	Yang et al. (2021)
Rice straw	650	39.8	1.73	0.71	0.36	-	0.52	0.37	14.3	4.78	Yang et al. (2021)
Peanut shells	400	47.5	5.28	-	-	37.1	1.33	0.59	-	-	Hadey et al. (2022)
Peanut shells	500	48.2	5.06	-	-	35.7	1.26	0.56	-	-	Hadey et al. (2022)
Sugarcane bagasse	400	45.3	5.09	-	-	36.1	1.35	0.60	-	-	Hadey et al. (2022)
Sugarcane bagasse	500	47.1	5.03	-	-	34.9	1.28	0.56	-	-	Hadey et al. (2022)
Coconut husk	400	47.9	3.50	1.34	0.17	47.9	0.88	0.75	39.6	-	Suman & Gautam (2017)
Coconut husk	600	58.4	3.01	1.02	0.16	37.4	0.62	0.48	120.7	-	Suman & Gautam (2017)
Coconut husk	800	67.3	2.67	0.87	0.17	29.0	0.48	0.32	256.6	-	Suman & Gautam (2017)
Coconut husk	1000	74.1	2.50	0.50	0.17	22.7	0.40	0.23	590.4	-	Suman & Gautam (2017)
Wheat straw	450	64.8	3.10	0.80	0.10	23.0	0.57	0.36	-	-	Nanda et al. (2013)
Pinewood	450	75.5	3.70	0.20	0.10	17.0	0.59	0.20	-	-	Nanda et al. (2013)
Timothy grass	450	63.7	3.60	1.90	0.04	23.1	0.68	0.36	-	-	Nanda et al. (2013)
Canola stalk	250	46.2	4.98	1.07	0.68	-	1.29	0.74	0.66	3.50	Yang et al. (2021)
Canola stalk	450	61.9	3.02	1.01	1.06	-	0.59	0.58	2.50	5.63	Yang et al. (2021)
Canola stalk	650	41.7	1.86	1.02	1.96	-	0.54	0.89	7.94	6.23	Yang et al. (2021)
Cattle manure	250	52.2	-	1.90	0.30	-	-	-	1.4	0.003	Pituello et al. (2015)
Cattle manure	350	60.7	-	2.60	0.30	-	-	-	2.2	0.008	Pituello et al. (2015)
Cattle manure	450	63.2	-	2.20	0.30	-	-	-	2.9	0.014	Pituello et al. (2015)
Cattle manure	550	65.9	-	2.20	0.50	-	-	-	58.6	0.065	Pituello et al. (2015)
Peanut hull	400	74.8	4.5	2.7	0.09	9.7	0.01	0.72	0.52		Novak et al. (2009)
Peanut hull	500	81.8	2.9	2.7	0.10	3.3	0.03	0.42	1.22		Novak et al. (2009)

Stability

Biochar, like any other organic substance, eventually decays, but the main consideration is its rate of degradation. One of the key causes for biochar's slow rate of decay is the existence of aromatic structures with varied characteristics (Leng et al., 2019). The stability of biochar is a significant consideration for dye and heavy metal removal in water treatment facilities (Chen et al., 2019). Biochars with less amounts of oxygen containing functional groups exhibit increased stability (Srivatsav et al., 2020).

Grindability

Material grindability in industrial size reduction processes is vital for its usability. Carbonisation induces mechanical stability changes, making char more brittle and increasing grindability compared to the raw material. Abdullah & Wu (2009) highlight a substantial correlation between biochar grindability and pyrolysis temperature. This underscores the importance of considering pyrolysis conditions in industrial processes, particularly in biochar production and processing sectors, for informed decision-making.

MECHANISMS OF HEAVY METAL ADSORPTION BY PYROLYSIS CHAR

Adsorption Mechanisms

Adsorption mechanisms comprise three key stages: the transfer of metal ions from the aqueous solution to the adsorbate, their adsorption onto the solid surface, and the intra-particle movement.

Surface Complexation

Heavy metal adsorption onto the adsorbate is primarily governed by surface complexation, influenced by solution pH, metal speciation, and the nature of functional groups on biomass pyrolysis char surfaces (Goswami et al., 2022). The surface functional groups on biomass pyrolysis char, such as carboxyl (–COOH), hydroxyl (–OH), and phenolic (–Ph), play a pivotal role in facilitating these reactions (Liu et al., 2022; Adewuyi, 2020). Understanding these surface complexation mechanisms is critical for optimising the heavy metal removal process using biomass pyrolysis char and designing effective water treatment systems.

Ion Exchange

Ion exchange replaces positively charged metal ions with similarly charged ions on the adsorbate surface. Its efficacy depends on factors like solution pH, temperature, ionic strength, and the adsorbent's ion abundance (Wang et al., 2023). Biomass pyrolysis char, with high surface area and abundant exchangeable ions (mainly cations like Na^+ and K^+), excels in ion exchange processes. These ions result from incomplete carbonisation, offering ample active sites for heavy metal adsorption through ion exchange mechanisms, enhancing biochar's effectiveness in water treatment applications (Ahmaruzzaman, 2010).

Precipitation

Precipitation arises from pH changes and chemical interactions between metal ions and char surfaces, forming insoluble metal hydroxides or carbonates. Particularly effective under alkaline conditions, it aids heavy metal removal, especially for metals with low solubility products. Biomass pyrolysis char acts as an adsorbent surface, aiding precipitation by providing nucleation sites for the growth of metal hydroxides or sulphides (Wang et al., 2013). Its porous structure and functional groups serve as nucleation sites. Surface chemistry modifications enhance biochar's heavy metal removal through precipitation. Additionally, biochar can function as a pH buffer, optimising the precipitation process in certain cases (Faheem et al., 2018).

Chelation

Chelation in adsorption forms metal ion-chelating agent complexes, enhancing adsorption with multiple binding sites. Commonly used, chelation, involving organic compounds like ethylenediaminetetraacetic acid (EDTA) or functional groups on the adsorbent, improves selectivity. It creates stable complexes, preventing released metal ions from re-entering the solution and facilitating their removal (Raji et al., 2023).

Electrostatic Interaction

There is possibility of electrostatic attractions between positively charged metal ions and negatively charged functional groups of adsorbents to promote the adsorption. Additionally, attractive forces such as hydrophobic interactions, van der Waal forces, and hydrogen bonding could be involved in the process of metal adsorption on the surface of biosorbents (Huang et al., 2020; Raji et al., 2023).

Enzymatic Transfer

If the binding of heavy metal ions by living organisms is connected to the enzymes found within biological cells, the typical explanation for this process often involves the enzymatic mechanism, as highlighted by Huang et al. (2020). In support of this, Yin et al. (2008) conducted a study showcasing enzymatic adsorption. They utilised yeast strains, specifically Candida tropicalis and Candida lipolytica, to demonstrate the effective removal of chromium and nickel from solution.

Figure 3 outlines mechanisms contingent on the adsorbate material's structure. Surface complexing is likely for char adsorbate material, while ion exchange is expected for algal biomass (Huang et al., 2020).

Figure 3. Schematic description of adsorption mechanisms (Huang et al., 2020; Raji et al., 2023)

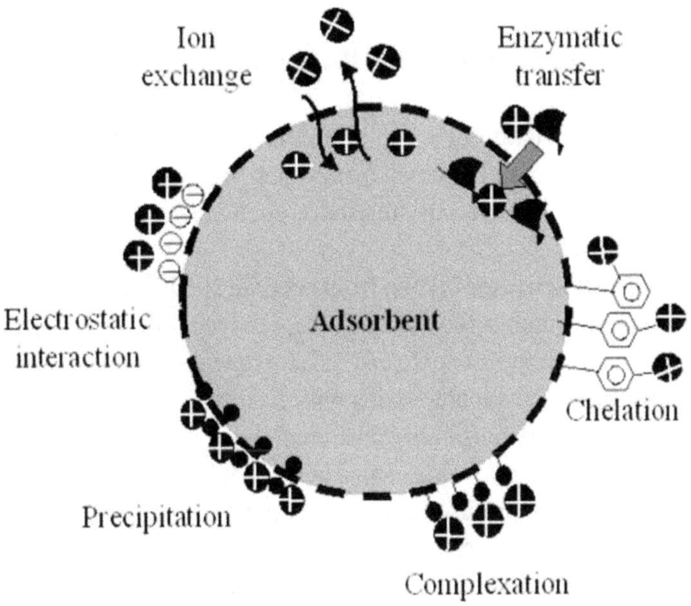

Factors Influencing Adsorption Efficiency

Several factors influence the efficiency of adsorption processes, which play a crucial role in removing contaminants such as heavy metals from water. The pH level of the solution is a significant factor, as it can affect the surface charge of both the adsorbent and the adsorbate. In a study by Cozmuta et al. (2012), it was reported that as the pH of the solution decreased, the maximum monolayer adsorption capacity also decreased. Thus, if, for instance, the pH decreases from 4 to 1, the maximum adsorption capacity decreases from 0.3569 to 0.1604 mol/kg.

The nature of the adsorbent material, its particle size, surface area and pore volume greatly influence adsorption efficiency; materials with larger surface areas typically exhibit higher adsorption capacities. This effect of surface area was studied by Muller (2010) in the adsorption of albumin-bonded bilirubin on different sized granular activated carbons with different BET surface areas of 913, 1173 and 1413 m^2/g. The best bilirubin sorption results were obtained by using a high activated carbon (BET surface area of 1413 m^2/g; benzene mesopore volume of about 0.18 cm^3/g) with an average particle diameter of about 82 μm. 58% of initially available bilirubin was adsorbed within 1 h adsorption time in batch experiments.

Temperature plays a role, with adsorption often being more effective at lower temperatures. The concentration of the adsorbate in the solution, as well as the contact time between the adsorbent and the solution, are critical factors. Horsfall & Spiff (2004) studied the influence of temperature on the adsorption of Pb^{2+} and Cd^{2+} from aqueous solution using Wild Cocoyam biomass. It was reported that the most suitable sorption temperature was 40 °C with maximum sorption capacities of 49.02 mg/g and 52.63 mg/g for Pb^{2+} and Cd^{2+} respectively.

Other physical and chemical properties of the adsorbate, such as solubility, and chemical speciation, also impact adsorption efficiency. Additionally, the presence of competing ions in the water can hinder the adsorption process. Understanding and optimising these factors are essential for designing effective adsorption systems in water treatment processes.

Biochar as a Low-Cost Adsorbent and Its Limitations

Low-cost adsorbents, such as agricultural waste and industrial by-products, have gained attention as economical alternatives for heavy metal removal from water. A low-cost adsorbent is one that is derived from natural sources or waste, these materials are used as such or first undergo a minor and simple treatment process (Gupta et al., 2009).

Low-cost adsorbents such as activated carbon from coconut shells, rice husk ash, and zeolites exhibit promising adsorption capacities due to their abundance, renewability, and affordability, making them attractive in resource-limited settings (Li et al., 2023; Levesque et al., 2021). However, these materials face challenges. Their adsorption efficiency varies with factors like pH, temperature, and the targeted metal. Additionally, regeneration and reusability pose challenges, prompting research into more efficient chemical regeneration methods due to drawbacks associated with thermal regeneration (Alsawy et al., 2022). Additionally, the lack of standardised preparation methods and variability in the quality of raw materials can influence the overall performance of these adsorbents. Despite these limitations, ongoing research is focussing on optimising and enhancing the efficacy of low-cost adsorbents, recognising their potential to contribute to accessible and affordable water treatment solutions.

APPLICATIONS AND CASE STUDIES

Numerous publications have asserted the efficacy of biochar as a water treatment adsorbent, but a significant portion of these works relies on laboratory experiments. Consequently, there is a notable absence of studies examining its performance in large-scale applications. Critical considerations, including lifespan, regenerability, and mechanical properties such as strength and abrasion resistance, become pivotal factors when implementing biochar on a broader scale. In this section, two large scale processes that utilise biochar as an adsorbent will be discussed.

N-E-W Tech Water Treatment Process (University of Idaho)

Description

The N-E-W Tech water treatment process integrates reactive filtration with iron-functionalised biochar and ozone for catalytic oxidation water treatment, sterilising reclaimed water, and producing a value-added nutrient upcycled fertilizer product that addresses the need for carbon sequestration (Moller et al., 2016). This membrane-free approach addresses the negative aspects of several current alternative technologies to water reuse and recycling including their high energy inputs, removal but not destruction of contaminants and pathogens, high capital-maintenance-operations costs, as well as their poorer sustainability footprint. N-E-W Tech™ uses ozone, a chlorine disinfection alternative made from electricity and air, as an oxidant that is catalysed with iron salts to initiate aggressive oxidising chemical reactions that can destructively remove many, if not all, toxic organic compounds and pathogens into harmless carbonates, smaller non-toxic molecules, or to minute concentrations of no effect. Mineralised P and

N can be captured on Fe-modified biochar and recovered as a by-product enhanced efficiency fertilizer with economic value for direct field application in agriculture and horticulture.

Process

The mobile demonstration treatment plant, with a throughput of 4 m^3/h, employs ferric or ferrous chloride (4–25 mg/l) based on influent water quality. A concentrated slurry of 0.5-1% iron chloride (w/w) surface-modified biochar is introduced at 0.25-2 g/l. Ozone (9 mg/l) is added through a venturi. The process water undergoes mixing, enters a plug flow pipe reactor, and then proceeds to an up-flow sand bed reactive filter. A reject rate of 5-10% is determined by influent water quality and research goals. A second pass in filter 2 occurs without biochar addition. Reject water from filter 1 is directed to a 1 μm bag filter for nutrient upcycled biochar recovery (Fig. 4).

Figure 4. N-E-W tech process flow diagram (Moller et al., 2016)

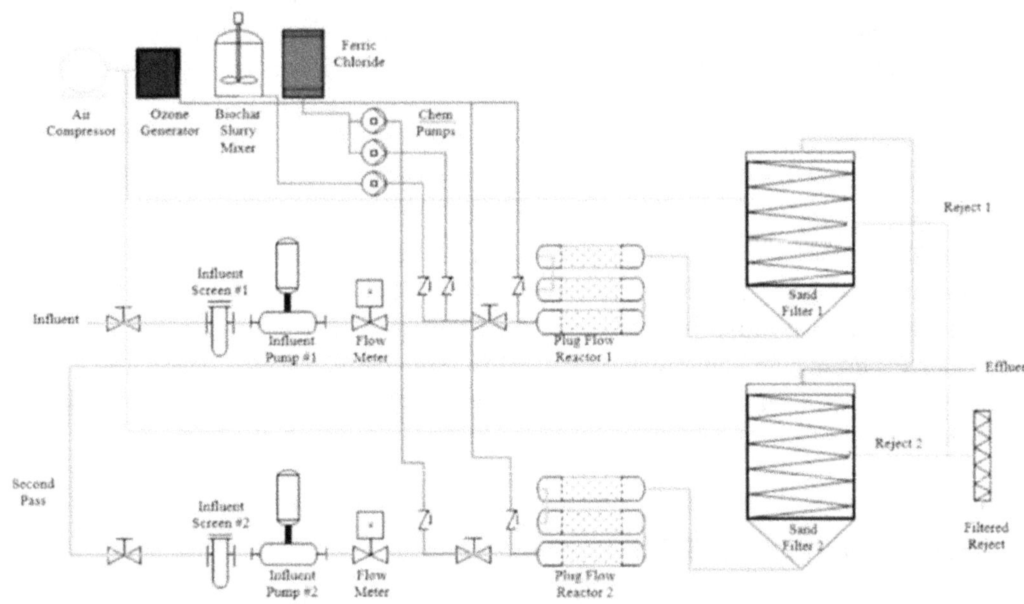

Table 3 shows the performance of the N-E-W Tech process when applied to municipality water. It is clear that the process exhibits state-of-the-art performance even though the total nitrogen removal efficiency is below the desired levels. The total nitrogen can be improved by focussing on biochar surface modification and process optimisation.

Table 3. Performance of N-E-W tech water treatment pilot plant utilising biochar as adsorbent

Parameter	Influent	Effluent	% Removal
E. coli count	2419	1	99.96
Total phosphorus (mg/l)	0.12	0.02	83.33
Ortho-phosphorus (mg/l)	0.06	0.002	96.67
Total nitrogen (mg/l)	30	9.9	67.00

Aqueous Solutions: 300 Litres per day Water Treatment System

Description

The Aqueous Solutions – 300 litres per day water treatment system is a simple and easy-to-operate process built from low-cost recycled materials mainly HDPE drums, PVC pipes and fittings. It is tailor-made for villages in developing countries. The system is designed to operate continuously otherwise dissolved oxygen will be depleted and impair water quality and aesthetics. Routine maintenance procedures such as backwashing to remove accumulated solids, filter harrowing and biochar replacement are carried out periodically.

Process

The process involves three stages: (1) pre-filtration with a 200 L HDPE drum filled with gravel layers, allowing sediment settling using gravity; (2) slow sand biofiltration with a 200 L HDPE drum filled with sand layers for physical straining and pathogenic cell removal; and (3) biochar adsorption column with a 200 L HDPE drum filled with biochar layers for removing heavy metals and organic chemical contaminats. The treated water is stored in a dedicated tank. Specific performance indicators were not provided by the system designers at the time of manuscript preparation.

CHALLENGES AND FUTURE OUTLOOK

Despite the thorough exploration of the potential benefits of utilising pyrolysis biochar derived from agricultural waste as an economical adsorbent for heavy metal removal, its use is not without drawbacks. Understanding the intricate relationship between thermal conversion conditions, biochar properties, and its behaviour towards water contaminants can facilitate its application for complex compounds. However, challenges associated with biochar use are noteworthy. Firstly, the variability in quality and properties of pyrolysis char depending on the feedstock used poses a challenge. Achieving consistent and reliable adsorption performance across diverse feedstocks necessitates either standardisation or customised approaches. Additionally, compliance with regulatory standards for the use of pyrolysis char in water treatment is indispensable, requiring navigation of evolving environmental regulations in different regions. Pyrolysis char faces competition from alternative water treatment technologies, necessitating a demonstration of its superiority in terms of efficiency, cost-effectiveness, and sustainability. Public acceptance hinges on perceptions of safety, effectiveness, and long-term impact, requiring clarification of misconceptions. Integration into existing water treatment infrastructure poses logistical challenges, demanding compatibility assessments. Proper disposal or regeneration of spent char requires eco-friendly solutions for enhanced sustainability. Scaling up from lab studies entails challenges in scalability, efficiency, and consistent performance, crucial for widespread adoption in water treatment.

Research in the field of using biochar as adsorbents is characterised by its dynamic and continually evolving nature, necessitating ongoing extensive efforts. Despite significant progress, a notable research gap exists, primarily due to the predominance of laboratory-scale experiments. Bridging this gap to facilitate the transition to commercial-scale applications is imperative for the practical implementation of biochar-based adsorption systems.

Recent research is increasingly focused on modifying biochar through functionalisation, introducing specific groups or nanoparticles to enhance its affinity for contaminants. This aims to improve both adsorption capacity and selectivity, optimising biochar's performance in water treatment. Another avenue involves integrating biochar into nanocomposites, exploring synergies with nanomaterials to create advanced adsorbents tailored for pollutant removal. This interdisciplinary approach holds promise for enhanced adsorption capabilities, addressing challenges associated with diverse contaminant profiles. As the field progresses, a shift from laboratory-scale investigations to larger-scale applications is crucial for validating the practical viability of biochar-based adsorption technologies. This transition will require collaborative efforts from researchers, industry stakeholders, and policymakers to bridge the existing gap and facilitate the widespread adoption of biochar as a versatile and effective adsorbent in real-world water treatment scenarios.

CONCLUSION

The chapter provides a comprehensive overview of the approach to utilising pyrolysis char as an adsorbent, with a primary emphasis on identifying optimal agricultural waste types for pyrolysis, detailing the pyrolysis process, and assessing the sustainability of such an initiative. It begins by outlining sources of heavy metal contamination, associated risks, and existing mitigation technologies for treating contaminated waters. Subsequently, the material characteristics of pyrolysis char relevant to contaminant adsorption are explored, accompanied by an examination of adsorption mechanisms and the factors influencing the adsorption process. The chapter conducted a comparative study, assessing various adsorbents alongside pyrolysis char to identify commonalities and strengths. This analysis lays the groundwork for potential composite adsorption materials, balancing cost and efficiency. Two case studies highlight pyrolysis char's effectiveness in heavy metal removal, with a comprehensive evaluation of performance metrics.

The overarching conclusion drawn from this chapter is the potential of biochar synthesised through the pyrolysis of agricultural waste for heavy metal removal via the adsorption process. The information presented contributes to a comprehensive understanding of the subject, laying the foundation for future research and practical applications in the field of water treatment.

REFERENCES

Ab Razak, N. H., Praveena, S. M., Aris, A. Z., & Hashim, Z. (2015). Drinking water studies: A review on heavy metal, application of biomaker and health risk assessment (a special focus in Malaysia). *Journal of Epidemiology and Global Health*, *5*(4), 297–310. doi:10.1016/j.jegh.2015.04.003 PMID:25944153

Abdullah, H., & Wu, H. (2009). Biochar as a fuel: 1. Properties and grindability of biochars produced from the pyrolysis of mallee woodunder slow-heating conditions. *Energy & Fuels*, *23*(8), 4174–4181. doi:10.1021/ef900494t

Adewuyi, A. (2020). Chemically Modified Biosorbents and Their Role in the Removal of Emerging Pharmaceutical Waste in the Water System. *Water (Basel)*, *12*(6), 1551. doi:10.3390/w12061551

Ahmad, M., Lee, S. S., Lee, S. E., Al-Wabel, M. I., Tsang, D. C. W., & Ok, Y. S. (2017). Biochar-induced changes in soil properties affected immobilization/mobilization of metals/metalloids in contaminated soils. *Journal of Soils and Sediments*, *17*(3), 717–730. doi:10.1007/s11368-015-1339-4

Ahmaruzzaman, M. (2010). A review on the utilization of fly ash. *Progress in Energy and Combustion Science*, *36*(3), 327–363. doi:10.1016/j.pecs.2009.11.003

Ahmed, M. B., Zhou, J. L., Ngo, H. H., & Guo, W. (2016). Insight into biochar properties and its cost analysis. *Biomass and Bioenergy*, *84*, 76–86. doi:10.1016/j.biombioe.2015.11.002

Ajala, E. O., Ighalo, J. O., Ajala, M. A., Adeniyi, A. G., & Ayanshola, A. M. (2021). Sugarcane bagasse: A biomass sufficiently applied for improving global energy, environment and economic sustainability. *Bioresources and Bioprocessing*, *8*(1), 1–25. doi:10.1186/s40643-021-00440-z

Al Arni, S. (2017). Comparison of slow and fast pyrolysis for converting biomass into fuel. *Renewable Energy*. Advance online publication. doi:10.1016/j.renene.2017.04.060

Al-Rumaihi, A., Shahbaz, M., Mckay, G., Mackey, H., & Al-Ansari, T. (2022). A review of pyrolysis technologies and feedstock: A blending approach for plastic and biomass towards optimum biochar yield. *Renewable & Sustainable Energy Reviews*, *167*, 112715. doi:10.1016/j.rser.2022.112715

Alsawy, T., Rashad, E., El-Qelish, M., & Mohammed, R. H. (2022). A comprehensive review on the chemical regeneration of biochar adsorbent for sustainable wastewater treatment. *NPJ Clean Water*, *5*(1), 29. doi:10.1038/s41545-022-00172-3

Batista, E. M. C. C., Shultz, J., Matos, T. T. S., Fornari, M. R., Ferreira, T. M., Szpoganicz, B., de Freitas, R. A., & Mangrich, A. S. (2018). Effect of surface and porosity of biochar on water holding capacity aiming indirectly at preservation of the Amazon biome. *Scientific Reports*, *81*(1), 1–9. doi:10.1038/s41598-018-28794-z PMID:30013110

Bavariani, M. Z., Ronaghi, A., & Ghasemi, R. (2019). Influence of pyrolysis temperatures on FTIR analysis, nutrient bioavailability, and agricultural use of poultry manure chars. *Communications in Soil Science and Plant Analysis*, *50*(4), 402–411. Advance online publication. doi:10.1080/00103624.2018.1563101

Bhurachevskaya, M., Minkina, T., & Bauer, T. (2023). Fabrication of biochar derived from different types of feedstocks as an efficient adsorbent for soil heavy metal removal. *Scientific Reports*, *13*(1), 2020. Advance online publication. doi:10.1038/s41598-023-27638-9 PMID:36737633

Botha, E., Smith, N., Hlabano-Moyo, B., & Bladergroen, B. J. (2021). The effect of slurry wet mixing time, thermal treatment and method of electrode preparation on membrane capacitive deionization performance. *Processes (Basel, Switzerland)*, *9*(1), 1. doi:10.3390/pr9010001

Bridgwater, A. V., Carson, P., & Coulson, M. (2007). A comparison of fast and slow pyrolysis liquids from mallee. *International Journal of Global Energy Issues*, *27*(2), 204–216. doi:10.1504/IJGEI.2007.013655

Brown, A. L., Dayton, D. C., Nimlos, M. R., & Daily, J. W. (2001). Characterization of biomass pyrolysis vapours with molecular beam, single photon ionization time-of-flight mass spectrometry. *Chemosphere*, *42*(5-7), 663–669. doi:10.1016/S0045-6535(00)00240-X PMID:11219692

Burachevskaya, M., Minkina, T., Bauer, T., Lobzenko, I., Fedorenko, A., Mazarji, M., Sushkova, S., Mandzhieva, S., Nazarenko, A., Butova, V., Wong, M. H., & Rajput, V. D. (2023). Fabrication of biochar derived from different types of feedstocks as an efficient adsorbent for soil heavy metal removal. *Scientific Reports*, *13*(1), 1–14. doi:10.1038/s41598-023-27638-9 PMID:36737633

Chaiwong, K., Kiatsiriroat, T., Vorayos, N., & Thararax, C. (2013). Study of bio-oil and bio-char production from algae by slow pyrolysis. *Biomass and Bioenergy*, *56*, 600–606. doi:10.1016/j.biombioe.2013.05.035

Chandel, A.K., da Silva, S.S., Carvalho, W., & Singh, O.V. (2011). Sugarcane bagasse and leaves: foreseeable biomass of biofuel and bio-products. *Journal of Chemical & Biotechnology*. doi:10.1002/jctb.2742

Chaturvedi, S., & Dave, P. N. (2012). Removal of iron for safe driking water. *Desalination*, *303*, 1–11. doi:10.1016/j.desal.2012.07.003

Chen, S., Qin, C., Wang, T., Chen, F., Li, X., Hou, H., & Zhou, M. (2019). Study on the adsorption of dyestuffs with different properties by sludge-rice husk biochar: Adsorption capacity, isotherm, kinetic, thermodynamics and mechanism. *Journal of Molecular Liquids*, *285*, 62–74. doi:10.1016/j.molliq.2019.04.035

Cheng, U., Hao, Z., Hao, C., Deng, Y., Li, X., Li, K., & Zhao, Y. (2019). A Review of Modification of Carbon Electrode Material in Capacitive Deionization. *RSC Advances*, *9*(42), 24401–24419. doi:10.1039/C9RA04426D PMID:35527893

Chiririwa, H., Matthews, T., Nyoni, B., Majoni, S., & Naidoo, B. (2017). Adorption of lead and copper by carbon black and sodium bentonite composite material: A study on adsorption isotherms and kinetics. *Asian Journal of Chemistry*, *29*(12), 2761–2766. doi:10.14233/ajchem.2017.20877

Cozmuta, L. M., Cozmuta, A. M., Peter, A., Nicula, C., Nsimba, E. B., & Tutu, H. (2012). The influence of pH on the adsorption of lead by Na clinoptilolite: Kinetic and equilibrium studies. *Water S.A.*, *38*(2), 269–278. doi:10.4314/wsa.v38i2.13

Das, T. K., & Poater, A. (2021). Review on the use of heavy metal deposits from water treatment waste towards catalytic chemical synthesis. *International Journal of Molecular Sciences*, *22*(24), 13383. doi:10.3390/ijms222413383 PMID:34948184

Demirbas, A. (2009). Biorefineries: Current activities and future developments. *Energy Conversion and Management, 50*(11), 2782–2801. doi:10.1016/j.enconman.2009.06.035

Dhyani, V., & Bhaskar, T. (2017). A comprehensive review on the pyrolysis of lignocellulosic biomass. *Renewable Energy, 129*, 695–716. doi:10.1016/j.renene.2017.04.035

Dilebo, W. B., Anchiso, M. D., Kidane, T. T., & Ayalew, M. E. (2023). Assessment of selected heavy metals concentration level of drinking water in Gazer town and selected Kebele, South Ari district, Southern Ethiopia. *International Journal of Analytical Chemistry, 2023*, 1–12. Advance online publication. doi:10.1155/2023/1524850 PMID:37180549

Dixit, S., & Hering, J. G. (2003). Comparison of arsenic (V) and arsenic (III) sorption onto iron oxide minerals: Implications for arsenic mobility. *Environmental Science & Technology, 37*(18), 4182–4189. doi:10.1021/es030309t PMID:14524451

Dube, D., Parekh, C. T., & Nyoni, B. (2016). Removal of chromium and nickel from electroplating wastewater using magnetite particulate adsorbent: (1) Effect of pH, contact time and dosage, (2) Adsorption isotherms and kinetics. *Modern Applied Science, 10*(7), 222–232. doi:10.5539/mas.v10n7p222

Dwivedi, S., & Dey, S. (2023). Review on biochar as an adsorbent material for removal of dyes from waterbodies. *International Journal of Environmental Science and Technology, 20*(8), 9335–9350. doi:10.1007/s13762-022-04364-9

El-Dessouky, H. T., & Ettouney, H. M. (2002). *Reverse osmosis*. Elservier Inc.

Ellens, C. J., & Brown, R. C. (2012). Optimisation of a free-fall reactor for the production of fast pyrolysis bio-oil. *Bioresource Technology, 103*(1), 374–380. doi:10.1016/j.biortech.2011.09.087 PMID:22036914

Elnour, A. Y., Alghyamah, A. A., Shaikh, H. M., Poulose, A. M., Al-Zahrani, S. M., Anis, A., & Al-Wabel, M. I. (2019). Effect of pyrolysis temperature on biochar microstructural evolution, physicochemical characteristics, and its influence on biochar/polypropylene composites. *Applied Sciences (Basel, Switzerland), 9*(6), 1149. doi:10.3390/app9061149

Faheem, J., Bao, H., Zheng, H., Tufail, S., Irshad, J., & Du, A. (2018). Adsorption-assisted decontamination of Hg (II) from aqueous solution by multi-functionalized corncob-derived biochar. *RSC Advances, 8*(67), 38425–38435. doi:10.1039/C8RA06622A PMID:35559052

Forgacs, Z., Massanyi, P., Lukac, N., & Somosy, Z. (2012). Reproductive toxicology of nickel – review. *Journal of Environmental Science and Health. Part A, Toxic/Hazardous Substances & Environmental Engineering, 47*(9), 1249–1260. doi:10.1080/10934529.2012.672114 PMID:22540648

Fouda-Mbanga, B. G., Seyisi, T., Nthwane, Y. B., Nyoni, B., & Tywabi-Ngeva, Z. (2023). A review on pollutants found in drinking water in Sub-Sahara African rural communities: Detection and potential low-cost remediation methods. *Industrial and Domestic Water Management, 3*(2), 67–89. doi:10.53623/idwm.v3i2.264

Freixo, R., Casanova, F., Ribeiro, A. B., Pereira, C. F., Costa, E. M., Pintado, M. E., & Ramos, O. L. (2023). Extraction methods and characterization of cellulose fractions from a sugarcane by-product for potential industry applications. *Industrial Crops and Products, 197*, 116615. doi:10.1016/j.indcrop.2023.116615

Gable, P., & Brown, R. C. (2016). Effect of biomass heating time on bio-oil yields in a free fall fast pyrolysis reactor. *Fuel*, *166*, 361–366. doi:10.1016/j.fuel.2015.10.073

Gonzalez, S., Lopez-Roldan, R., & Cortina, J. L. (2013). Presence of metals in drinking water distribution networks due to pipe material leaching: A review. *Toxicological and Environmental Chemistry*, *95*(6), 870–889. doi:10.1080/02772248.2013.840372

Goswami, L., Kushwaha, A., Kafle, S. R., & Kim, B. S. (2022). *Surface Modification of Biochar for Dye Removal from Wastewater*. Catalysis., doi:10.3390/catal12080817

Goyal, H. B., Seal, D., & Saxena, R. C. (2008). Bio-fuels from thermochemical conversion of renewable resources: A review. *Renewable & Sustainable Energy Reviews*, *12*(2), 504–517. doi:10.1016/j.rser.2006.07.014

Gupta, S., Mittal, Y., Panja, R., Prajapati, K. B., & Yadav, A. K. (2021). *Conventional wastewater treatment technologies*. Elservier Inc. doi:10.1016/B978-0-12-821009-3.00012-9

Gupta, V. K., Carrott, P. J. M., Carrott, M. M. L. R., & Suhas, R. (2009). Low-cost adsorbents: Growing approach to wastewater treatment – a Review. *Critical Reviews in Environmental Science and Technology*, *39*(10), 783–842. doi:10.1080/10643380801977610

Hadey, C., Allouch, M., Alami, N., Bouklifi, F., & Loulidi, I. (2022). Preparation and characterization of biochars obtained from biomasses for combustible briquette applications. *TheScientificWorldJournal*, *2022*, 1–13. Advance online publication. doi:10.1155/2022/2554475 PMID:36523325

Hanif, M. U., Capareda, S. C., Iqbal, H., Arazo, R. O., & Baig, M. A. (2016). Effects of Pyrolysis Temperature on Product Yields and Energy Recovery from Co-Feeding of Cotton Gin Trash, Cow Manure, and Microalgae: A Simulation Study. *PLoS One*, *11*(4), e0152230. Advance online publication. doi:10.1371/journal.pone.0152230 PMID:27043929

Horsfall, M. Jnr, & Spiff, M. I. (2005). Effects of temperature on the sorption of Pb^{2+} and Cd^{2+} from aqueous solution by *Caladium bicolor* (Wild Cocoyam) biomass. *Electronic Journal of Biotechnology*, *8*(2), 162–169. doi:10.2225/vol8-issue2-fulltext-4

Hualpa-Cutipa, E. H., Acosta, R. A. S., & Sangay-Tucto, S. (2022). *Recent trends for treatment of environmental contaminants in wastewater: An integrated valorization of industrial wastewater*. Elservier Inc.

Huang, D., Li, B., Ou, J., Xue, W., Li, J., Li, Z., Li, T., Chen, S., Deng, R., & Guo, X. (2020). Megamerger of biosorbents and catalytic technologies for the removal of heavy metals from wastewater: Preparation, final disposal, mechanism and influencing factors. *Journal of Environmental Management*, *261*, 109879. doi:10.1016/j.jenvman.2019.109879 PMID:32148248

Iwuozor, K. O., Emenike, E. C., Omonayin, E. O., Bamigbola, J. O., Ojo, H. T., Awoyale, A. A., Eletta, O. A. A., & Adeniyi, A. G. (2023). Unlocking the hidden value of pods: A review of thermochemical conversion processes for biochar production. *Bioresource Technology Reports*, *22*, 101488. Advance online publication. doi:10.1016/j.biteb.2023.101488

Izah, S. C., Chakrabarty, N., & Srivastav, A. L. (2016). A review on heavy metal concentration in potable water sources in Nigeria: Human health effects and mitigating measures. *Exposure and Health*, *8*(2), 285–304. Advance online publication. doi:10.1007/s12403-016-0195-9

Kajina, W., & Rousset, P. (2018). *Coupled effect of feedstock and pyrolysis temperature on biochar as soil amendment*. Agritrop.

Khamkeaw, A., Asavamongkolkul, T., Perngyai, T., Jongsomjit, B., & Phisalaphong, M. (2020). Interconnected micro, meso, and macro porous activated carbon from bacterial nanocellulose for superior adsorption properties and effective catalytic performance. *Molecules (Basel, Switzerland)*, *25*(18), 4063. doi:10.3390/molecules25184063 PMID:32899569

Kubier, A., Wilkin, R. T., & Pichler, T. (2019). Cadmium in soils and groundwater: A review. *Applied Geochemistry*, *108*, 1–16. doi:10.1016/j.apgeochem.2019.104388 PMID:32280158

Leng, L., Huang, H., Li, H., Li, J., & Zhou, W. (2019). Biochar stability assessment methods: A review. *The Science of the Total Environment*, *647*, 210–222. doi:10.1016/j.scitotenv.2018.07.402 PMID:30077850

Levesque, V., Oelbermann, M., & Ziadi, N. (2021). Biochar in temperate soils: Opportunities and challenges. *Canadian Journal of Soil Science*, *102*(1), 1–26. doi:10.1139/cjss-2021-0047

Li, J., Cao, L., Yuan, Y., Wang, R., Wen, Y., & Man, J. (2018). Comparative study for microcystin-LR sorption onto biochars produced from various plant- and animal-wastes at different pyrolysis temperatures: Influencing mechanisms of biochar properties. *Bioresource Technology*, *247*, 794–803. doi:10.1016/j.biortech.2017.09.120 PMID:30060415

Li, S., Xu, S., Liu, S., Yang, C., & Lu, Q. (2004). Fast pyrolysis of biomass in free-fall reactor for hydrogen-rich gas. *Fuel Processing Technology*, *85*(8-10), 1201–1211. doi:10.1016/j.fuproc.2003.11.043

Li, Y., Gupta, R., Zhang, Q., & You, S. (2023). Review of biochar production via crop residue pyrolysis: Development and perspectives. *Bioresource Technology*, *369*, 128423. doi:10.1016/j.biortech.2022.128423 PMID:36462767

Ling, E. J., Benham, B., & Forrester, S. (2011). *Virginia household water quality program: Heavy metals in household water*. Virginia Cooperative Extension.

Liu, R., Liu, G., Yousaf, B., Niu, Z., & Abbas, Q. (2022). Novel investigation of pyrolysis mechanisms and kinetics for functional groups in biomass matrix. *Renewable & Sustainable Energy Reviews*, *153*, 111761. doi:10.1016/j.rser.2021.111761

Malik, L. A., Bashir, A., Qureashi, A., & Pandith, A. H. (2019). Detection and removal of heavy metal ions: A review. *Environmental Chemistry Letters*, *17*(4), 1495–1521. doi:10.1007/s10311-019-00891-z

Manne, R., Kumaradoss, M. M., Iska, R. S. R., Devarajan, A., & Mekala, N. (2022). Water quality and risk assessment of copper content in drinking water stored in copper container. *Applied Water Science*, *12*(3), 27. doi:10.1007/s13201-021-01542-x

Mapombere, G., Nyoni, B., Sibali, L. L., Chiririwa, H., & Seodigeng, T. (2022). Adsorption of mercury by *Pterocarpus Anglolensis*: Study on adsorption isotherms and kinetics. *Iranian Journal of Chemistry & Chemical Enginerring*, *40*, 143–153.

Marshall, A. (2013). *Commercial Application of Pyrolysis Technology in Agriculture.* doi:10.13031/aim.20141909089

Moller, G., Baker, M., Dunkel, C., Shresta, D., & Strawn, D. (2016). N-E-W TechTM: Resource recovery in a novel functionalised biochar, catalytic oxidation, reactive filtration water treatment for N/P removal/recovery and pathogen/priority substance destruction. *10th European Waste Water Management Conference.*

Morgan, T. J., Turn, S. Q., George, A., & Aburto, J. (2015). Fast Pyrolysis Behaviour of Banagrass as a Function of Temperature and Volatiles Residence Time in a Fluidized Bed Reactor. *PLoS One, 10*(8), e0136511. Advance online publication. doi:10.1371/journal.pone.0136511 PMID:26308860

Muller, B. R. (2010). Effect of particle size and surface area on the adsorption of albumin-bonded bilirubin on activated carbon. *Carbon, 48*(12), 3607–3615. doi:10.1016/j.carbon.2010.06.011

Nanda, S., Mohanty, P., Pant, K. K., Naik, S., Kozinsk, J. A., & Dalai, A. K. (2013). Characterization of North American lignocellulosic biomass and biochars in terms of their candidacy for alternate renewable fuels. *BioEnergy Research, 6*(2), 663–677. doi:10.1007/s12155-012-9281-4

Nekouei, R. K., Pahlevani, F., Assefi, M., Maroufi, S., & Sahajwalla, V. (2019). Selective isolation of heavy metals from spent electronic waste solution by macroporous ion-exchange resins. *Journal of Hazardous Materials, 379*, 389–396. doi:10.1016/j.jhazmat.2019.03.013 PMID:30870643

Novak, J. M., Lima, I., & Xing, B. (2009). Characterization of designer biochar produced at different temperatures and their effects on a loamy sand. *Annalytical Environmental Science, 3*, 195–206.

Nyoni, B., Duma, S., Shabangu, S. V., & Hlangothi, S. P. (2020). Comparison of the slow pyrolysi behaviour and kinetics of coal, wood and algaeat high heating rates. *Natural Resources Research, 29*(6), 3943–3955. doi:10.1007/s11053-020-09687-3

Nyoni, B., & Hlangothi, S. P. (2023). Co-pyrolysis of low-grade bituminous coal and algal biomass in a rotary kiln: Effect of coal/algae ratio and kiln temperature on the yield and composition of the resultant oils. *Journal of Analytical and Applied Pyrolysis,* 171.

Nyoni, B., & Hlangothi, S. P. (2022). Intermediate pyrolysis of *Scenedesmus* microalgae in a rotary kiln pyrolyser. Effect of temperature on bio-oil yields and composition. *Biomass Conversion and Biorefinery.* Advance online publication. doi:10.1007/s13399-022-03378-0

Onay, O., & Kockar, O. M. (2003). Slow, fast and flash pyrolysis of rapeseed. *Renewable Energy, 28*(15), 2417–2433. doi:10.1016/S0960-1481(03)00137-X

Onay, O., & Kockar, O. M. (2004). Fixed-bed pyrolysis of rapeseed (Brassica napus L.). *Biomass and Bioenergy, 26*(3), 289–299. doi:10.1016/S0961-9534(03)00123-5

Pandey, S. (2022). Wood waste utilization and associated product development from under-utilized low-quality wood and its prospects in Nepal. *SN Applied Sciences, 4*(6), 168. doi:10.1007/s42452-022-05061-5

Pattanaik, L., Pattnaik, F., Saxena, D. K., & Naik, S. N. (2019). *Biofuels from agricultural wastes.* Elservier Inc. doi:10.1016/B978-0-12-815162-4.00005-7

Piersa, P., Unyay, H., Szufa, S., Lewandowska, W., Modrzewski, R., Ślężak, R., & Ledakowicz, S. (2022). An Extensive Review and Comparison of Modern Biomass Torrefaction Reactors vs. Biomass Pyrolysis—Part 1. *Energies*, *15*(6), 2227. doi:10.3390/en15062227

Pituello, C., Francioso, O., Simonetti, G., Pisi, A., Torreggiani, A., Berti, A., & Morari, F. (2015). Characterization of chemical–physical, structural and morphological properties of biochars from biowastes produced at different temperatures. *Journal of Soils and Sediments*, *15*(4), 792–804. doi:10.1007/s11368-014-0964-7

Rahimi, Z., Anand, A., & Gautam, S. (2022). An overview on thermochemical conversion and potential evaluation of biofuels derived from agricultural wastes. *Energy Nexus*, *7*, 100125. doi:10.1016/j.nexus.2022.100125

Raji, Z., Karim, A., Karam, A., & Khalloufi, S. (2023). Adsorption of Heavy Metals: Mechanisms, Kinetics, and Applications of Various Adsorbents in Wastewater Remediation – A Review. *Waste*, *1*(3), 775–805. doi:10.3390/waste1030046

Ramage, M. H., Burridge, H., Busse-Wicher, M., Fereday, G., Reynolds, T., Shah, D. U., Wu, G., Yu, L., Fleming, P., Densley-Tingley, D., Allwood, J., Dupree, P., Linden, P. F., & Scherman, O. (2017). The wood from trees: The use of timber in construction. *Renewable & Sustainable Energy Reviews*, *68*, 333–359. doi:10.1016/j.rser.2016.09.107

Raychaudhuri, S.S., Pramanick, P., Talukder, P., & Basak, A. (2021). *Polyamines, metallothioneins, and phytochelatins – Natural defence of plants to mitigate heavy metals.* doi:10.1016/B978-0-12-819487-4.00006-9

Raza, M., Inayat, A., Ahmed, A., Jamil, F., Ghenai, C., Naqvi, S. R., Shanableh, A., Ayoub, M., Waris, A., & Park, Y.-K. (2021). Progress of the Pyrolyzer Reactors and Advanced Technologies for Biomass Pyrolysis Processing. *Sustainability (Basel)*, *13*(19), 11061. doi:10.3390/su131911061

Sikder, S., & Joardar, J. C. (2019). Biochar production from poultry litter as management approach and effects on plant growth. *International Journal of Recycling of Organic Waste in Agriculture*, *8*(1), 47–58. doi:10.1007/s40093-018-0227-5

Singh, B. J., Chakraborty, A., & Sehgal, R. (2023). A Systematic Review of Industrial Wastewater Management: Evaluating Challenges and Enablers. *Journal of Environmental Management*, *348*, 119–230. doi:10.1016/j.jenvman.2023.119230 PMID:37832302

Singh, N., Gupta, V. K., Kumar, A., & Sharma, B. (2017). Synergistic effects of heavy metals and pesticides in living systems. *Frontiers in Chemistry*, *5*, 70. Advance online publication. doi:10.3389/fchem.2017.00070 PMID:29075624

Singh, R., Krishna, B. B., Mishra, G., Kumar, J., & Bhaskar, T. (2016). Strategies for selection of thermochemical processes for the valorisation of biomass. *Renewable Energy*, *98*, 226–237. doi:10.1016/j.renene.2016.03.023

Slezak, R., Unyay, H., Szufa, S., & Ledakowicz, S. (2023). An Extensive Review and Comparison of Modern Biomass Reactors Torrefaction vs. Biomass Pyrolizers—Part 2. *Energies*, *16*(5), 2212. Advance online publication. doi:10.3390/en16052212

Srivastava, V., Sarkar, A., Singh, S., Singh, P., de Araujo, A. S. F., & Singh, R. P. (2017). Agroecological responses to heavy metal pollution with special emphasis on soil health and plant perfomances. *Frontiers in Environmental Science*, *5*, 64. Advance online publication. doi:10.3389/fenvs.2017.00064

Srivatsav, P., Bhargav, B. S., Shanmugasundaram, V., Arun, J., Panchamoorthy, K., & Bhartnaga, A. (2020). Biochar as an eco-friendly and ecological adsorbent for the removal of colorants (dyes) from aqueous environment: A review. *Water (Basel)*, *12*(12), 3561. doi:10.3390/w12123561

Suman, S., & Gautam, S. (2017). Pyrolysis of coconut husk biomass: Analysis of its biochar properties. *Energy Sources. Part A, Recovery, Utilization, and Environmental Effects*, *39*(8), 761–767. doi:10.1080/15567036.2016.1263252

Tomczyk, A., Sokolowska, Z., & Boguta, P. (2020). Biochar Physicochemical properties: Pyrolysis temperature and feedstock kind effects. *Reviews in Environmental Science and Biotechnology*, *19*(1), 191–215. doi:10.1007/s11157-020-09523-3

United Nations. (2009). *Converting waste agricultural biomass into a resource-compendium of technologies*. UNEP.

Vasudevan, S., Jayaraj, J., Lakshmi, J., & Sozhan, G. (2009). Removal of iron from drinking water by electrocoagulation: Adsorption and kinetics studies. *Korean Journal of Chemical Engineering*, *26*(4), 1058–1064. doi:10.1007/s11814-009-0176-9

Wang, B., Lan, J., Bo, C., Gong, B., & Ou, J. (2023). Adsorption of heavy metal onto biomass-derived activated carbon [review]. *RSC Advances*, *13*(7), 4275–4302. doi:10.1039/D2RA07911A PMID:36760304

Wang, H., Yuan, X., Wu, Y., Huang, H., Zeng, G., Liu, Y., Wang, X., Lin, N., & Qi, Y. (2013). Adsorption characteristics and behaviours of graphene oxide for Zn (II) removal from aqueous solution. *Applied Surface Science*, *279*, 432–440. doi:10.1016/j.apsusc.2013.04.133

Yang, C., Liu, J., & Lu, S. (2021). Pyrolysis temperature affects pore characteristics of rice straw and canola stalk biochars and bio-char amended. *Geoderma*, *397*, 115097. doi:10.1016/j.geoderma.2021.115097

Yang, I., Kwon, D., Kim, M.-S., & Jung, J. C. (2018). A Comparative Study of Activated Carbon Aerogel and Commercial Activated Carbons as Electrode Materials for Organic Electric Double-Layer Capacitors. *Carbon*, *132*, 503–511. doi:10.1016/j.carbon.2018.02.076

Yin, H., He, B., Peng, H., Ye, J., Yang, F., & Zhang, N. (2008). Removal of Cr (IV) and Ni (II) from aqueous solution by fused yeast: Study of cations release and biosorption mechanism. *Journal of Hazardous Materials*, *158*(2-3), 568–576. doi:10.1016/j.jhazmat.2008.01.113 PMID:18346847

Yin, Q., Zhang, B., Wang, R., & Zhao, Z. (2017). Biochar as an adsorbent for inorganic nitrogen and phosphorus removal from water: A review. *Environmental Science and Pollution Research International*, *24*(34), 26297–26309. doi:10.1007/s11356-017-0338-y PMID:29039039

Yu, O. Y., Raichle, B., & Sink, S. (2013). Impact of biochar on the water holding capacity of loamy sand soil. *International Journal of Energy and Environmental Engineering*, *4*(1), 1–9. doi:10.1186/2251-6832-4-44

Yuan, J. H., Xu, R. K., & Zhang, H. (2011). The forms of alkalis in the biochar produced from crop residues at different temperatures. *Bioresource Technology, 102*(3), 3488–3497. doi:10.1016/j.biortech.2010.11.018 PMID:21112777

Zhao, R., Jiang, D., Coles, N., & Wu, J. (2015). Effects of biochar on the acidity of a loamy clay soil under different incubation conditions. *Journal of Soils and Sediments, 15*(9), 1919–1926. doi:10.1007/s11368-015-1143-1

Zhitkovich, A. (2011). Chromium in drinking water: Sources, metabolism, and cancer risks. *Chemical Research in Toxicology, 24*(10), 1617–1629. doi:10.1021/tx200251t PMID:21766833

KEY TERMS AND DEFINITIONS

Adsorption: A process that involves particles of a solute in solution adhering onto the surface of a solid material through attraction forces.

Agriculture: Any form of activity utilising land for producing crops and raising livestock for human use and sustenance.

Biochar: Biochar is a carbon-rich material produced through the pyrolysis of organic matter, and can be used to enhance soil fertility, sequester carbon, and improve environmental sustainability.

Biomass: Biomass is any organic material that originates from living organisms such plants and animals. Since living organisms such as plants convert sunlight, water and carbon dioxide into organic compounds via photosynthesis; biomass can be viewed as a store of solar energy.

Heavy Metals: Metals or metalloids with a high atomic mass and density that can be toxic to living organisms at elevated concentrations and pose environmental and health risks when present in soil, water, or air.

Pyrolysis: A process of decomposing a substance through subjecting it to high temperature in the absence of oxygen.

Waste: Unwanted material generated as by-products during a manufacturing process, often considered as unusable or no longer needed.

Chapter 12
Heavy Metal Bioaccumulation in Food Chains and Health Risks

Richa Saxena
https://orcid.org/0000-0002-9229-9235
Invertis University, India

Abhilekha Sharma
Noida International University, India

Vaishnavi Srivastava
Chhatrapati Shahu Ji Maharaj University, India

Dipti Bharti
Darbhanga College of Engineering, India

ABSTRACT

This chapter provides a comprehensive examination of the intricate interplay between heavy metal contamination, the food chain, and the associated health risks to human populations. The investigation encompasses a multifaceted exploration of the sources, pathways, and mechanisms involved in the bioaccumulation of heavy metals within the food web. The chapter initiates with a thorough analysis of the diverse sources of heavy metal contamination, ranging from industrial discharges to agricultural practices, elucidating the intricate routes through which these pollutants infiltrate ecosystems. Special emphasis is placed on understanding the dynamic interactions within soil-plant-animal systems that facilitate the uptake and transfer of heavy metals across trophic levels. This chapter amalgamates interdisciplinary insights from environmental science, toxicology, and public health, offering a valuable resource for researchers, policymakers, and practitioners engaged in the critical task of addressing heavy metal contamination in the food chain.

DOI: 10.4018/979-8-3693-1618-4.ch012

1. INTRODUCTION

In the modern era of industrialization, agriculture, and urbanization, the pervasive presence of heavy metal contamination in food chains has emerged as a critical global challenge with profound implications for both the environment and human health. Heavy metals, such as lead, cadmium, mercury, and arsenic, are persistent environmental pollutants released into ecosystems through anthropogenic activities (Ali et al., 2021; Nachana'a et al., 2019). The consequences of their infiltration into the intricate web of food chains extend far beyond immediate ecological disruptions, raising significant concerns about the safety and well-being of human populations that rely on these ecosystems for sustenance.

1.1 Background and Significance

Heavy metal contamination in food chains has emerged as a pressing environmental and public health concern, necessitating a comprehensive exploration of its origins and impacts (Kumar et al., 2019 ; Saxena et al.,2020) . The rapid industrialization, extensive agricultural practices, and urban development of recent decades have led to the release of significant amounts of heavy metals into the environment. These metals, including lead, cadmium, mercury, and arsenic, are persistent pollutants that, once introduced, can disrupt ecosystems and pose serious risks to human health.

The historical context of industrialization and its associated processes has substantially contributed to the widespread dispersion of heavy metals in air, water, and soil (Wu et al., 2016 ; Fang et al.,2019) . The combustion of fossil fuels, discharge of industrial effluents, and improper waste disposal have led to the accumulation of these contaminants in diverse ecological compartments. Understanding this historical background is crucial for delineating the pathways through which heavy metals enter the food chain, thereby enabling effective mitigation strategies (Bengtsson et al.,2018).

The significance of addressing heavy metal contamination is underscored by its potential to cause detrimental effects at multiple levels (Masri et al.,2021). Ecologically, these contaminants can disrupt the balance of ecosystems, affecting biodiversity, and impairing the vitality of soil and aquatic environments. Furthermore, heavy metals exhibit a remarkable capacity to bioaccumulate, progressively concentrating as they move up the trophic levels of the food chain. This bioaccumulation poses a direct threat to human populations consuming contaminated food, as these metals can elicit a range of adverse health effects, including neurotoxicity, developmental disorders, and carcinogenicity.

1.2 Scope of Heavy Metal Contamination in Food Chains

The scope of heavy metal contamination extends across various environmental matrices, influencing the entire food chain from soil to human consumers (Toussaint et al.,2019). Soil serves as the primary reservoir, where heavy metals initially accumulate through deposition and various human activities. From the soil, these contaminants are taken up by plants, entering the plant-animal food chain (Arslan et al.,2017). Aquatic ecosystems also play a crucial role, as waterborne heavy metals affect aquatic organisms, subsequently impacting organisms higher in the food web (Okereafor et al., 2020) .

The geographical and temporal scope of heavy metal contamination is diverse, with localized hotspots experiencing elevated concentrations due to specific industrial activities or historical practices (Alloway et al.,2013 ; Kaushal et al.,2021). Understanding this spatial variability is essential for targeted intervention and remediation efforts. Additionally, the temporal dynamics, influenced by seasonal variations and climate conditions, contribute to the complexity of heavy metal behavior in ecosystems.

2. SOURCES OF HEAVY METAL CONTAMINATION

Heavy metal contamination arises from various sources, and understanding these origins is essential for devising effective mitigation strategies (Sarwar et al.,2017). Here are some significant sources of heavy metal contamination:

2.1 Industrial Emissions and Discharges

Industrial activities stand as one of the primary contributors to heavy metal contamination in the environment (Błońska et al.,2016). The release of metals such as lead, mercury, and cadmium occurs through various processes, including manufacturing, mining, and energy production. Combustion of fossil fuels in power plants and industrial facilities releases metal-laden particulate matter into the air (Naghdi et al.,2023). These pollutants can settle onto soils and water bodies, initiating a cascade of environmental impacts. Additionally, industrial effluents containing heavy metals, often inadequately treated, find their way into rivers and other water sources, further exacerbating contamination (Chowdhary et al.,2023). Understanding the specific emissions from different industries and implementing effective control measures are crucial steps in mitigating industrial-related heavy metal contamination.

2.2 Agricultural Practices and Soil Amendments

Agriculture, while essential for food production, is a significant source of heavy metal introduction into the environment (Qin et al.,2021). The use of fertilizers, pesticides, and animal manure in farming can contribute to soil contamination. Some fertilizers may contain elevated levels of heavy metals, which, when applied excessively, accumulate in the soil (Atafar et al.,2010). Similarly, pesticides and herbicides can introduce metals into the ecosystem. Livestock farming, through the application of manure, can also be a source of heavy metal input. Understanding the intricacies of agricultural practices and adopting sustainable farming techniques are pivotal in minimizing heavy metal contamination in agricultural soils and the subsequent transfer into the food chain.

2.3 Urbanization and Anthropogenic Activities

The process of urbanization brings with it a suite of anthropogenic activities that contribute significantly to heavy metal contamination (Zhang et al.,2017). Urban areas are characterized by extensive infrastructure development, construction, and transportation networks. The wear and tear of vehicle components, such as brake pads, can release metal particles into the air. Construction activities disturb soil, potentially releasing metals that were previously sequestered. Stormwater runoff in urban areas can transport contaminants into water bodies (Kaushal et al.,2021). Furthermore, improper disposal of electronic waste (e-waste) and other products containing heavy metals adds to the complexity (Sovacool et al.,2019). Managing urbanization-induced heavy metal contamination requires a holistic approach that includes proper waste management, green infrastructure, and urban planning strategies aimed at reducing environmental impacts.

2.4 Agrochemicals

Agrochemicals, including fertilizers, pesticides, and herbicides, play a significant role in agricultural practices and can contribute to heavy metal contamination in the environment.

1. **Fertilizers**: While fertilizers are essential for enhancing soil fertility and crop productivity, certain types may contain heavy metals such as cadmium, lead, and arsenic as impurities. These metals can accumulate in soils over time, especially when fertilizers are applied in excessive amounts or without proper management practices. Additionally, phosphate fertilizers, in particular, can contain elevated levels of cadmium, which can leach into soil and water with repeated applications.
2. **Pesticides and Herbicides**: Many pesticides and herbicides contain heavy metal components or derivatives. For instance, copper-based fungicides and herbicides are commonly used in agriculture. While effective in pest and weed control, these products can introduce copper into the environment, potentially leading to soil and water contamination. Similarly, organomercury compounds were historically used as fungicides, contributing to mercury contamination in agricultural soils.
3. **Animal Manure**: Animal manure, often used as a natural fertilizer in agriculture, can also contain heavy metals. Livestock may accumulate metals from their feed or environment, and these metals can be excreted in their manure. When manure is applied to fields as fertilizer, it can introduce heavy metals into the soil, particularly if the animals were fed contaminated feed or raised in areas with high metal concentrations.

Mitigating heavy metal contamination from agrochemicals involves implementing best management practices in agriculture, such as:

- Proper selection and application of fertilizers, considering their heavy metal content and potential for leaching or runoff.
- Integrated Pest Management (IPM) strategies to minimize reliance on chemical pesticides and herbicides.
- Monitoring and testing of soils for heavy metal concentrations, especially in areas with intensive agricultural activities.
- Implementing practices to reduce nutrient runoff and erosion, such as cover cropping and conservation tillage.
- Proper management of animal manure, including composting and testing for heavy metal content before application to fields.

3. PATHWAYS OF HEAVY METAL ENTRY INTO THE FOOD CHAIN

The entry of heavy metals into the food chain involves complex mechanisms influenced by environmental interactions (Cao et al.,2021). Understanding these pathways is crucial for assessing the risks posed to ecosystems and human health.

3.1 Soil-Plant Uptake Mechanisms

Heavy metals enter the food chain through intricate soil-plant uptake mechanisms. In contaminated soils, plants absorb metals through their roots, a process influenced by factors such as soil pH, organic matter content, and the availability of specific metals (Bolan et al.,2014) . Essential nutrients like zinc and iron share similar uptake pathways with toxic heavy metals, leading to potential bioaccumulation in edible parts of plants. The translocation of metals within the plant and subsequent ingestion by herbivores or direct consumption by humans establishes a critical pathway for heavy metals to enter the terrestrial food chain.

3.2 Water Contamination and Aquatic Pathways

Water contamination is a major route through which heavy metals infiltrate the food chain, particularly in aquatic ecosystems. Industrial discharges, agricultural runoff, and atmospheric deposition contribute to the accumulation of metals in water bodies. Aquatic plants, filter-feeding organisms, and bottom-dwelling species can directly uptake heavy metals from water (Kahlon et al.,2018). Additionally, the ingestion of contaminated water by aquatic animals results in bioaccumulation, with predatory species at higher trophic levels accumulating higher concentrations. This aquatic pathway establishes a connection between freshwater and marine environments, impacting both aquatic organisms and the human consumers relying on aquatic food sources (Bakker et al.,2016).

3.3 Atmospheric Deposition and Bioaerosols

Atmospheric deposition, including the settling of particulate matter and gases, introduces heavy metals into ecosystems far from their original sources. Particles containing heavy metals can settle on soil surfaces, plants, and water bodies (Gupta et al.,2020). This atmospheric pathway is particularly relevant in regions impacted by industrial emissions, wildfires, or volcanic activities. Moreover, heavy metals can be associated with bioaerosols—small particles suspended in the air that may contain metals from sources such as dust, pollen, or combustion emissions. Deposition of these bioaerosols onto soil or water surfaces can contribute to the contamination of the food chain.

Understanding these pathways is essential for assessing the risk of heavy metal exposure to ecosystems and human populations. The complexity of these entry mechanisms emphasizes the need for interdisciplinary research and comprehensive monitoring programs to track the movement of heavy metals through different environmental compartments and trophic levels within the food chain (Szynkowska et al.,2018). Effective mitigation strategies should consider these pathways to minimize the impact of heavy metal contamination on both environmental health and human well-being.

Figure 1. Transmission of heavy metals through food chain

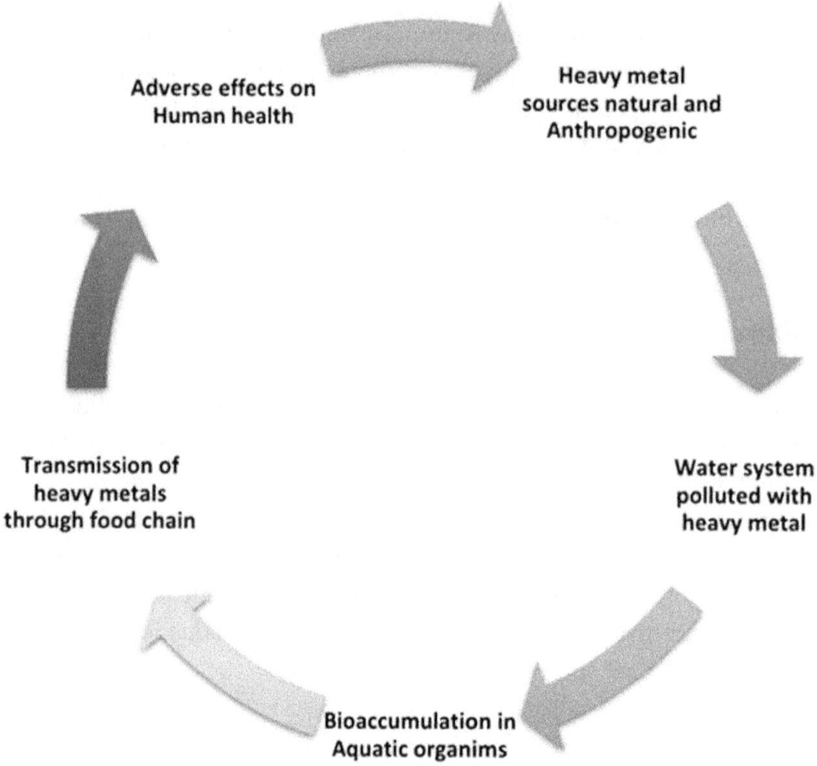

4. BIOACCUMULATION MECHANISMS

Bioaccumulation refers to the gradual buildup of substances, such as heavy metals, in living organisms over time (Sardar et al.,2013). This process involves intricate mechanisms that vary across ecosystems. Three primary bioaccumulation mechanisms are:

4.1 Uptake and Translocation in Plants

Bioaccumulation of heavy metals initiates with their uptake by plants from the soil. Roots absorb dissolved metal ions, and the uptake process is influenced by factors such as soil pH, metal speciation, and the presence of competing ions (Pang et al.,2023). Once inside the plant, metals may undergo complex transformations, including binding to proteins and translocation to different plant tissues. The translocation of metals from roots to shoots and, ultimately, to edible parts of the plant is a critical step in the bioaccumulation process (Gupta et al.,2019). Factors such as plant species, growth stage, and the chemical form of the metal influence the extent of translocation.

4.2 Trophic Transfer in Aquatic Ecosystems

In aquatic ecosystems, trophic transfer plays a pivotal role in the bioaccumulation of heavy metals. Phytoplankton, algae, and aquatic plants can directly uptake metals from water. Herbivorous organisms that consume these primary producers accumulate metals in their tissues. As predators consume herbivores, the concentration of metals increases further along the food chain (Gall et al.,2015). This biomagnification effect is particularly pronounced in aquatic systems, where top predators can accumulate high concentrations of heavy metals. The efficiency of trophic transfer is influenced by factors such as the species involved, the nutritional status of organisms, and the bioavailability of metals in the water.

4.3 Accumulation in Animal Tissues

Heavy metals further accumulate in animal tissues as a result of dietary intake and trophic interactions. Animals that feed on plants or prey with elevated metal concentrations can accumulate metals in their tissues (Zhang et al.,2021). The process is species-specific, with different organisms exhibiting varying levels of susceptibility to metal bioaccumulation. In some cases, certain tissues within an organism, such as the liver and kidneys, may preferentially accumulate higher concentrations of metals. The efficiency of metal elimination processes, such as excretion and detoxification, also plays a role in determining the overall metal burden in animal tissues. Human consumption of animals with accumulated heavy metals represents a significant pathway for human exposure and potential health risks.

4.4 Heavy Metal Effects on Animal Tissues and Metabolic Processes

1. **Lead (Pb)**:
    - **Affected Tissues**: Lead tends to accumulate primarily in bones, kidneys, liver, and brain tissues.
    - **Metabolic Processes**: Lead interferes with various enzymatic processes, disrupts calcium metabolism, and affects neurotransmitter function. It can also induce oxidative stress, leading to cellular damage and inflammation.
2. **Mercury (Hg)**:
    - **Affected Tissues**: Mercury accumulates in various tissues, including the brain, kidneys, liver, and muscles.
    - **Metabolic Processes**: Mercury disrupts the central nervous system, causing neurological and developmental disorders. It interferes with enzyme function, particularly those involved in sulfur metabolism. Methylmercury, a highly toxic form of mercury, can cross the blood-brain barrier and cause severe neurological damage.
3. **Cadmium (Cd)**:
    - **Affected Tissues**: Cadmium primarily accumulates in the kidneys, liver, and lungs.
    - **Metabolic Processes**: Cadmium disrupts calcium metabolism, leading to bone demineralization and renal dysfunction. It also induces oxidative stress and inflammation, contributing to tissue damage and increased risk of cancer, particularly in the lungs.
4. **Arsenic (As)**:
    - **Affected Tissues**: Arsenic accumulates in various tissues, including the skin, liver, kidneys, and lungs.

- **Metabolic Processes**: Arsenic interferes with cellular energy production, DNA repair mechanisms, and enzyme function. Chronic exposure to arsenic is associated with skin lesions, cardiovascular diseases, and various cancers, including skin, lung, and bladder cancer.

5. **Copper (Cu)**:
   - **Affected Tissues**: Copper primarily accumulates in the liver, brain, kidneys, and skeletal muscles.
   - **Metabolic Processes**: Copper plays a crucial role in enzymatic reactions, particularly those involved in energy metabolism and antioxidant defense. However, excessive copper accumulation can lead to oxidative stress and tissue damage, especially in the liver and brain. Disorders such as Wilson's disease result from impaired copper metabolism and excessive copper accumulation in tissues.
6. **Zinc (Zn)**:
   - **Affected Tissues**: Zinc is essential for various metabolic processes and is distributed throughout the body, with high concentrations in the muscles, bones, skin, and prostate gland.
   - **Metabolic Processes**: Zinc is a cofactor for numerous enzymes and plays a crucial role in immune function, wound healing, and DNA synthesis. However, excessive zinc intake can interfere with copper absorption and lead to gastrointestinal disturbances.

5. SPATIAL AND TEMPORAL DYNAMICS OF HEAVY METAL BIOACCUMULATION

The spatial and temporal dynamics of heavy metal bioaccumulation, focusing on seasonal variations and climate influences as well as regional disparities and hotspots (Dethier et al.,2019).

5.1 Seasonal Variations and Climate Influences

The bioaccumulation of heavy metals exhibits dynamic spatial and temporal patterns influenced by seasonal variations and climatic factors. Seasonal changes, such as variations in temperature, precipitation, and sunlight, can significantly impact the behavior of heavy metals in ecosystems. In colder seasons, biological activity may decrease, affecting the uptake and translocation of metals in plants and animals. Conversely, warmer seasons might enhance biological processes, leading to increased bioaccumulation (Wang et al.,2022).

Climate influences, including rainfall patterns, can affect the mobility of heavy metals in soil and their transport into water bodies. Heavy metals may be leached from soil during periods of intense rainfall, contributing to higher concentrations in surface waters (Bhateria et al.,2016). Conversely, arid conditions may result in the concentration of metals in specific areas due to reduced leaching.

Understanding the seasonal dynamics is crucial for assessing the risks of heavy metal exposure and designing effective monitoring programs. Additionally, climate change can alter precipitation patterns and temperatures, potentially influencing the bioaccumulation of heavy metals in unforeseen ways. Investigating these interactions is vital for predicting and mitigating the impacts of climate change on metal contamination.

5.2 Regional Disparities and Hotspots

Heavy metal bioaccumulation is not uniform across regions; rather, it exhibits spatial disparities influenced by various factors. Some areas may be more prone to metal contamination due to historical industrial activities, mining operations, or specific geological characteristics. Regions with elevated anthropogenic activities, such as urban centers and industrial zones, often experience higher concentrations of heavy metals in their ecosystems (Hou et al.,2017).

These localized areas with elevated metal concentrations are referred to as hotspots. Hotspots can arise from point-source pollution, such as industrial discharges or improper waste disposal (Wang et al.,2019). The accumulation of metals in these regions can have cascading effects on the surrounding environment, leading to potential risks for both ecological systems and human populations.

Identifying regional disparities and hotspots involves comprehensive environmental monitoring, soil and water analyses, and an understanding of the historical context of human activities. This knowledge is essential for targeted intervention strategies, regulatory measures, and remediation efforts aimed at mitigating the impact of heavy metal contamination in specific regions.

6. HEALTH RISKS ASSOCIATED WITH HEAVY METAL EXPOSURE

Heavy metal exposure poses significant health risks to human populations, with diverse metals contributing to a range of toxicological impacts. These health risks arise from chronic exposure through contaminated air, water, soil, and food (Mishra et al.,2019). The key aspects of the health risks associated with heavy metal exposure include:

6.1 Toxicological Impacts on Human Health

Heavy metal exposure poses significant health risks, as these elements can elicit toxicological effects on various physiological systems in humans. Different metals have distinct toxic profiles. For instance, lead exposure is associated with neurological and developmental effects, mercury can impact the nervous system, and cadmium has been linked to kidney damage and bone disorders (Engwa et al.,2019). Chronic exposure to heavy metals through the consumption of contaminated food, water, or air can lead to the accumulation of these substances in the body, potentially causing long-term health complications (Pandey et al.,2014). Understanding the toxicological impacts of specific metals is crucial for assessing health risks and developing guidelines to limit exposure.

6.2 Epidemiological Studies and Case Analyses

Epidemiological studies play a pivotal role in assessing the health risks associated with heavy metal exposure (Sabath et al.,2012). These studies analyze the relationships between exposure levels and health outcomes in human populations. By examining communities with known heavy metal contamination and comparing them to non-exposed populations, researchers can identify associations between exposure and health effects. Case analyses of specific incidents, such as industrial accidents or environmental disasters leading to heavy metal release, provide valuable insights into the immediate and long-term health consequences. Epidemiological evidence helps inform regulatory standards, public health interventions, and preventive measures to reduce heavy metal-related health risks.

6.3 Vulnerable Populations: Children, Pregnant Women, and Indigenous Communities

Certain demographic groups are particularly vulnerable to the health risks associated with heavy metal exposure.

- **Children:** Developing organs and systems make children more susceptible to the adverse effects of heavy metals. Lead exposure, for example, can lead to cognitive impairments and behavioral disorders in children (Zeng et al.,2016).
- **Pregnant Women:** Heavy metals can cross the placental barrier, potentially affecting fetal development. Exposure during pregnancy has been associated with adverse outcomes, including developmental abnormalities and preterm births.
- **Indigenous Communities:** Indigenous populations, often residing in areas with traditional resource use and facing unique cultural and socioeconomic factors, may be disproportionately affected by heavy metal contamination. This can result from factors such as mining activities on indigenous lands or the consumption of traditional foods from contaminated environments.

Understanding the vulnerabilities of these populations is critical for developing targeted public health strategies and interventions (Christmas et al.,2019). Efforts to mitigate heavy metal exposure must consider the specific needs and susceptibilities of these groups, emphasizing community engagement, cultural sensitivity, and equitable health interventions.

Table 1. Heavy metals and health implications - Sources, bioaccumulation, and risks

Heavy Metal	Sources	Bioaccumulation Pathways	Health Risks
Lead	- Industrial emissions and smelting	- Uptake by plants from contaminated soil	- Neurological damage
	- Old lead-based paint	- Accumulation in animal tissues	- Developmental issues in children
	- Leaded gasoline residues	- Biomagnification in aquatic food chains	- Cardiovascular effects
Mercury	- Industrial discharges	- Conversion to methylmercury in water	- Neurological damage and developmental issues
	- Artisanal gold mining	- Bioaccumulation in fish and seafood	- Cardiovascular effects
	- Coal-fired power plants	- Biomagnification in marine ecosystems	- Adverse effects on the immune system
Cadmium	- Industrial discharges	- Uptake by plants from contaminated soil	- Kidney damage and dysfunction
	- Phosphate fertilizers	- Accumulation in animal tissues	- Bone and lung diseases
	- Battery disposal	- Trophic transfer in aquatic ecosystems	- Increased cancer risk
Arsenic	-Agricultural pesticides	- Soil-plant uptake mechanisms	- Skin lesions and cancers
	- Mining activities	- Accumulation in rice and vegetables	- Cardiovascular effects and neurological disorders
	- Industrial effluents	- Trophic transfer in aquatic ecosystems	- Respiratory issues and gastrointestinal problems
Chromium	- Industrial discharges (chrome plating)	- Uptake by plants from contaminated soil	- Respiratory issues and lung cancer
	- Tannery waste	- Accumulation in animal tissues	- Skin irritation and allergies
	- Coal combustion residues	- Trophic transfer in aquatic ecosystems	- Gastrointestinal disorders and kidney damage
Copper	- Agricultural pesticides	- Uptake by plants from soil	- Gastrointestinal issues and liver damage
	- Mining activities	- Bioaccumulation in aquatic organisms	- Neurological disorders and headaches
	- Industrial effluents	- Biomagnification in marine ecosystems	- Hematological disorders
Nickel	- Industrial emissions and discharges	- Uptake by plants from contaminated soil	- Respiratory issues and lung cancer
	- Stainless steel production residues	- Accumulation in animal tissues	- Skin irritation and allergies
	- Nickel mining	- Trophic transfer in aquatic ecosystems	- Gastrointestinal disorders and developmental issues

7. CONTEMPORARY MONITORING TECHNIQUES AND REGULATORY FRAMEWORKS

Contemporary monitoring techniques and regulatory frameworks are essential components of environmental management, particularly in addressing heavy metal contamination. These elements work synergistically to assess, control, and mitigate the impact of heavy metals on ecosystems and human health (Mohammed et al.,2011).

7.1 Analytical Innovations in Heavy Metal Detection

Advancements in analytical techniques have significantly improved the precision and sensitivity of heavy metal detection, contributing to more effective environmental monitoring (Karimi et al.,2023). Some notable innovations include:

- **Atomic Absorption Spectrometry (AAS):** A classic technique, AAS measures the absorption of light by metal atoms and provides quantitative data on metal concentrations.
- **Inductively Coupled Plasma Mass Spectrometry (ICP-MS):** This technique allows for multi-elemental analysis with high sensitivity, making it invaluable for detecting trace levels of heavy metals in various environmental matrices.
- **X-ray Fluorescence (XRF):** XRF is a non-destructive technique that can analyze the elemental composition of samples, making it useful for on-site analysis of heavy metals in soil, sediments, and biological samples.
- **Biosensors:** Innovative biosensors use biological molecules to detect heavy metals selectively. These sensors are often portable and offer real-time monitoring capabilities.
- **Remote Sensing:** Satellite and airborne remote sensing technologies can provide valuable spatial information about environmental parameters, including heavy metal concentrations, over large areas.

These innovations in heavy metal detection empower environmental scientists and regulators to conduct more accurate and comprehensive monitoring, aiding in the identification of pollution sources and the assessment of environmental health (Nova et al.,2023).

7.2 Regulatory Approaches and Guidelines

Regulatory frameworks are essential for managing and mitigating heavy metal contamination (Yusuf, et al.,2022). Key aspects of regulatory approaches include:

- **Permissible Limits:** Governments and international bodies set permissible limits for heavy metal concentrations in air, water, soil, and food. These standards are based on scientific assessments of health risks.
- **Pollution Prevention:** Regulatory frameworks often emphasize pollution prevention measures for industries, prescribing technologies and practices to minimize heavy metal emissions and discharges.
- **Environmental Impact Assessments (EIA):** Before initiating projects with potential environmental impacts, regulatory agencies may require EIAs to evaluate and mitigate potential heavy metal contamination.
- **Monitoring and Reporting Requirements:** Regulators mandate routine monitoring by industries, laboratories, and environmental agencies. Non-compliance can result in penalties.

- **Public Disclosure:** Increasingly, regulatory frameworks incorporate provisions for public disclosure of environmental data, promoting transparency and community awareness.

7.3 Challenges and Opportunities in Implementation

Despite advancements, challenges persist in implementing effective monitoring and regulatory frameworks:

- **Resource Constraints:** Many regions, especially in developing countries, face resource constraints in terms of laboratory infrastructure, trained personnel, and financial resources for monitoring programs.
- **Emerging Contaminants:** Constantly evolving industrial processes introduce new heavy metals and contaminants, challenging regulators to keep pace with monitoring and regulation.
- **Global Cooperation:** Heavy metal pollution often transcends borders. Effective regulation requires international cooperation to address transboundary pollution and harmonize standards.
- **Community Engagement:** Ensuring the active involvement of local communities in monitoring and regulatory processes is crucial. Community-based monitoring programs can complement official efforts.
- **Technological Integration:** Integrating emerging technologies, such as artificial intelligence and big data analytics, into monitoring programs can enhance data analysis and interpretation.

8. MITIGATION STRATEGIES AND SUSTAINABLE PRACTICES

Mitigating heavy metal contamination requires a holistic approach that incorporates remediation techniques, sustainable practices, and community engagement (Song et al.,2018).

8.1 Remediation Techniques in Soil and Water

Remediation techniques play a crucial role in mitigating heavy metal contamination in soil and water:

- **Phytoremediation:** Certain plants, known as hyperaccumulators, can absorb and accumulate heavy metals in their tissues. Phytoremediation involves planting these species to extract and concentrate metals from the soil.
- **Bioremediation:** Microorganisms with the ability to transform or immobilize heavy metals are employed in bioremediation. This approach can enhance the natural processes of metal attenuation in soil and water.
- **Chemical Remediation:** Techniques such as soil amendments (e.g., liming) and chemical stabilizers are used to alter the chemical properties of the soil, reducing the mobility and bioavailability of heavy metals.
- **In-situ Immobilization:** This involves adding amendments to the contaminated site to reduce the mobility of metals in the soil, preventing their uptake by plants and minimizing leaching into groundwater.
- **Phytoextraction:** This involves the use of plants to extract metals from the soil, which are then harvested and properly disposed of, removing the contaminants from the site.

8.2 Sustainable Agriculture and Best Practices

Promoting sustainable agriculture practices can prevent and mitigate heavy metal contamination:

- **Organic Farming:** Organic farming practices, such as avoiding synthetic fertilizers and pesticides, reduce the introduction of heavy metals into the soil.
- **Crop Rotation and Cover Crops:** Alternating crops and using cover crops can enhance soil health and reduce the risk of heavy metal accumulation.
- **Proper Use of Fertilizers:** Careful application of fertilizers, avoiding excessive use, helps minimize the introduction of metal contaminants into agricultural soils.
- **Soil Conservation:** Implementing soil conservation practices, such as contour plowing and cover cropping, reduces erosion and the transport of heavy metals from agricultural fields to water bodies.

8.3 Community Engagement and Public Awareness

Community engagement and public awareness are integral components of successful heavy metal contamination mitigation:

- **Education and Outreach:** Public awareness campaigns and educational programs inform communities about the sources, risks, and mitigation strategies related to heavy metal contamination.
- **Community-Based Monitoring:** Involving local communities in monitoring programs empowers them to actively contribute to identifying contamination sources and implementing preventive measures.
- **Participatory Decision-Making:** Engaging communities in decision-making processes, especially in areas directly affected by contamination, fosters a sense of ownership and ensures that mitigation strategies align with local needs.
- **Advocacy and Policy Support:** Empowering communities to advocate for stronger environmental policies and regulations can lead to systemic changes and more effective mitigation efforts.

FUTURE DIRECTIONS AND RESEARCH NEEDS

As we navigate the future in addressing heavy metal contamination, key focus areas and research needs emerge for effective environmental management (Vijayaraghavan et al.,2015):

Emerging Threats and Unexplored Pathways

As we navigate the future of addressing heavy metal contamination, attention to emerging threats and unexplored pathways is critical:

- **Nanoparticles and Nano-Materials:** With the increasing use of nanoparticles in various industries, their potential environmental impact and contribution to heavy metal contamination need thorough investigation.

- **Emerging Contaminants:** Identifying and understanding new and emerging contaminants is essential, as industrial processes evolve, introducing novel pollutants that may pose risks to ecosystems and human health.
- **Climate Change Interactions:** The influence of climate change on heavy metal transport, fate, and bioavailability is an unexplored area. Changes in precipitation patterns, temperature, and extreme weather events can significantly impact the dynamics of heavy metals in the environment.
- **Urban Mining and E-Waste:** The rise of urban mining, involving the extraction of metals from electronic waste, brings new challenges. Understanding the environmental consequences of this practice is crucial for sustainable metal recovery.

Integrative Approaches for Comprehensive Solutions

To address heavy metal contamination comprehensively, integrative approaches are essential:

- **Interdisciplinary Research:** Collaborative efforts across disciplines, involving chemists, ecologists, geologists, engineers, and social scientists, are necessary to understand the complexities of heavy metal contamination fully.
- **Big Data and Technology Integration:** Leveraging big data analytics and advanced technologies, such as artificial intelligence and machine learning, can enhance the analysis and interpretation of large datasets, providing more insights into the spatial and temporal dynamics of heavy metal contamination.
- **Policy Integration:** Harmonizing environmental policies and regulations at local, national, and international levels ensures a cohesive approach to managing heavy metal contamination. This includes adopting consistent standards and fostering global cooperation to address transboundary pollution.
- **Community-Centric Approaches:** Engaging local communities in the development and implementation of mitigation strategies ensures that interventions are culturally sensitive, align with community needs, and promote long-term sustainability.
- **Long-Term Monitoring Programs:** Establishing robust and continuous monitoring programs provides valuable data for tracking the effectiveness of mitigation measures over time. Long-term studies are crucial for understanding the persistence and evolution of heavy metal contamination.

CONCLUSION

In the intricate tapestry of heavy metal contamination, this chapter has delved into multifaceted dimensions, ranging from sources and pathways to health risks and mitigation strategies. As we reflect on the current state of knowledge and anticipate future challenges, several key conclusions and considerations emerge. The dynamic nature of heavy metal contamination is apparent, shaped by diverse sources, environmental processes, and human activities. Understanding this complexity is paramount for effective management. The health risks associated with heavy metal exposure underscore the urgency of robust monitoring, regulatory frameworks, and mitigation strategies. Vulnerable populations, such as children and pregnant women, necessitate targeted interventions. The evolution of monitoring techniques, from traditional methods to cutting-edge technologies, empowers us to detect and respond to contamination more precisely. Integrating advancements like remote sensing and biosensors contributes to a comprehensive understanding of environmental health. Regulatory frameworks play a pivotal role in shaping preventive measures and enforcing compliance. Ongoing efforts to refine and integrate these frameworks are essential for global consistency and efficacy.

The journey toward mitigation involves not only technological solutions but also sustainable practices and community engagement. Embracing organic farming, implementing soil conservation, and involving communities in monitoring and decision-making processes are integral to long-term success. The horizon of heavy metal research extends to uncharted frontiers. Emerging threats, such as nanoparticles and urban mining, demand thorough exploration. Understanding the interplay between heavy metals and climate change adds a layer of complexity that requires sustained investigation. The path forward involves integration—integrating disciplines, technologies, policies, and community perspectives. Collaborative efforts between scientists, policymakers, industries, and local communities are indispensable for holistic solutions. The need for continuous monitoring and adaptive strategies cannot be overstated. As we navigate an ever-changing environmental landscape, long-term studies and the integration of evolving technologies are vital to staying ahead of emerging challenges. In conclusion, the chapter provides a snapshot of our current understanding of heavy metal contamination while emphasizing the need for continuous exploration and dynamic adaptation. Navigating the future of heavy metal management requires a commitment to innovation, collaboration, and a sustainable ethos—a collective endeavor to safeguard both our environment and the well-being of present and future generations.

REFERENCES

Ali, M. M., Hossain, D., Al-Imran, A., Khan, M. S., Begum, M., & Osman, M. H. (2021). Environmental pollution with heavy metals: A public health concern. *Heavy Metals-Their Environmental Impacts and Mitigation*, 771-783.

Alloway, B. J. (2013). Sources of heavy metals and metalloids in soils. *Heavy metals in soils: trace metals and metalloids in soils and their bioavailability*, 11-50.

Arslan, B., Djamgoz, M. B., & Akün, E. (2017). Arsenic: A review on exposure pathways, accumulation, mobility and transmission into the human food chain. *Reviews of Environmental Contamination and Toxicology*, *243*, 27–51. PMID:28005215

Atafar, Z., Mesdaghinia, A., Nouri, J., Homaee, M., Yunesian, M., Ahmadimoghaddam, M., & Mahvi, A. H. (2010). Effect of fertilizer application on soil heavy metal concentration. *Environmental Monitoring and Assessment*, *160*(1-4), 83–89. doi:10.1007/s10661-008-0659-x PMID:19058018

Bakker, E. S., Pagès, J. F., Arthur, R., & Alcoverro, T. (2016). Assessing the role of large herbivores in the structuring and functioning of freshwater and marine angiosperm ecosystems. *Ecography*, *39*(2), 162–179. doi:10.1111/ecog.01651

Bengtsson-Palme, J., Kristiansson, E., & Larsson, D. J. (2018). Environmental factors influencing the development and spread of antibiotic resistance. *FEMS Microbiology Reviews*, *42*(1), fux053. doi:10.1093/femsre/fux053 PMID:29069382

Bhateria, R., & Jain, D. (2016). Water quality assessment of lake water: A review. *Sustainable Water Resources Management, 2*(2), 161–173. doi:10.1007/s40899-015-0014-7

Błońska, E., Lasota, J., Szuszkiewicz, M., Łukasik, A., & Klamerus-Iwan, A. (2016). Assessment of forest soil contamination in Krakow surroundings in relation to the type of stand. *Environmental Earth Sciences, 75*(16), 1–15. doi:10.1007/s12665-016-6005-7

Bolan, N., Kunhikrishnan, A., Thangarajan, R., Kumpiene, J., Park, J., Makino, T., Kirkham, M. B., & Scheckel, K. (2014). Remediation of heavy metal (loid) s contaminated soils–to mobilize or to immobilize? *Journal of Hazardous Materials, 266*, 141–166. doi:10.1016/j.jhazmat.2013.12.018 PMID:24394669

Chowdhary, P., Bharagava, R. N., Mishra, S., & Khan, N. (2020). *Role of industries in water scarcity and its adverse effects on environment and human health. In Environmental Concerns and Sustainable Development* (Vol. 1). Air, Water and Energy Resources.

Christmas, H., & Srivastava, J. (2019). *Public health approaches in policing*. Public Health England and College of Policing.

Dethier, E. N., Sartain, S. L., & Lutz, D. A. (2019). Heightened levels and seasonal inversion of riverine suspended sediment in a tropical biodiversity hot spot due to artisanal gold mining. *Proceedings of the National Academy of Sciences of the United States of America, 116*(48), 23936–23941. doi:10.1073/pnas.1907842116 PMID:31712422

Engwa, G. A., Ferdinand, P. U., Nwalo, F. N., & Unachukwu, M. N. (2019). Mechanism and health effects of heavy metal toxicity in humans. *Poisoning in the modern world-new tricks for an old dog, 10*, 70-90.

Fang, T., Yang, K., Lu, W., Cui, K., Li, J., Liang, Y., Hou, G., Zhao, X., & Li, H. (2019). An overview of heavy metal pollution in Chaohu Lake, China: Enrichment, distribution, speciation, and associated risk under natural and anthropogenic changes. *Environmental Science and Pollution Research International, 26*(29), 29585–29596. doi:10.1007/s11356-019-06210-x PMID:31440974

Gall, J. E., Boyd, R. S., & Rajakaruna, N. (2015). Transfer of heavy metals through terrestrial food webs: A review. *Environmental Monitoring and Assessment, 187*(4), 1–21. doi:10.1007/s10661-015-4436-3 PMID:25800370

Gupta, N., Yadav, K. K., Kumar, V., Kumar, S., Chadd, R. P., & Kumar, A. (2019). Trace elements in soil-vegetables interface: Translocation, bioaccumulation, toxicity and amelioration-a review. *The Science of the Total Environment, 651*, 2927–2942. doi:10.1016/j.scitotenv.2018.10.047 PMID:30463144

Gupta, V. (2020). *Vehicle-generated heavy metal pollution in an urban environment and its distribution into various environmental components. In Environmental Concerns and Sustainable Development* (Vol. 1). Air, Water and Energy Resources.

Hou, D., O'Connor, D., Nathanail, P., Tian, L., & Ma, Y. (2017). Integrated GIS and multivariate statistical analysis for regional scale assessment of heavy metal soil contamination: A critical review. *Environmental Pollution, 231*, 1188–1200. doi:10.1016/j.envpol.2017.07.021 PMID:28939126

Kahlon, S. K., Sharma, G., Julka, J. M., Kumar, A., Sharma, S., & Stadler, F. J. (2018). Impact of heavy metals and nanoparticles on aquatic biota. *Environmental Chemistry Letters*, *16*(3), 919–946. doi:10.1007/s10311-018-0737-4

Karimi-Maleh, H., Darabi, R., Baghayeri, M., Karimi, F., Fu, L., Rouhi, J., Niculina, D. E., Gündüz, E. S., & Dragoi, E. N. (2023). Recent developments in carbon nanomaterials-based electrochemical sensors for methyl parathion detection. *Journal of Food Measurement and Characterization*, *17*(5), 5371–5389. doi:10.1007/s11694-023-02050-z

Kaushal, S. S., Likens, G. E., Pace, M. L., Reimer, J. E., Maas, C. M., Galella, J. G., Utz, R. M., Duan, S., Kryger, J. R., Yaculak, A. M., Boger, W. L., Bailey, N. W., Haq, S., Wood, K. L., Wessel, B. M., Park, C. E., Collison, D. C., Aisin, B. Y. I., Gedeon, T. M., ... Woglo, S. A. (2021). Freshwater salinization syndrome: From emerging global problem to managing risks. *Biogeochemistry*, *154*(2), 255–292. doi:10.1007/s10533-021-00784-w

Kaushal, S. S., Likens, G. E., Pace, M. L., Reimer, J. E., Maas, C. M., Galella, J. G., Utz, R. M., Duan, S., Kryger, J. R., Yaculak, A. M., Boger, W. L., Bailey, N. W., Haq, S., Wood, K. L., Wessel, B. M., Park, C. E., Collison, D. C., Aisin, B. Y. I., Gedeon, T. M., ... Woglo, S. A. (2021). Freshwater salinization syndrome: From emerging global problem to managing risks. *Biogeochemistry*, *154*(2), 255–292. doi:10.1007/s10533-021-00784-w

Kumar, S., Prasad, S., Yadav, K. K., Shrivastava, M., Gupta, N., Nagar, S., Bach, Q.-V., Kamyab, H., Khan, S. A., Yadav, S., & Malav, L. C. (2019). Hazardous heavy metals contamination of vegetables and food chain: Role of sustainable remediation approaches-A review. *Environmental Research*, *179*, 108792. doi:10.1016/j.envres.2019.108792 PMID:31610391

Masri, S., LeBrón, A. M., Logue, M. D., Valencia, E., Ruiz, A., Reyes, A., & Wu, J. (2021). Risk assessment of soil heavy metal contamination at the census tract level in the city of Santa Ana, CA: Implications for health and environmental justice. *Environmental Science. Processes & Impacts*, *23*(6), 812–830. doi:10.1039/D1EM00007A PMID:33954329

Mishra, S., Bharagava, R. N., More, N., Yadav, A., Zainith, S., Mani, S., & Chowdhary, P. (2019). Heavy metal contamination: an alarming threat to environment and human health. *Environmental biotechnology: For sustainable future*, 103-125.

Mohammed, A. S., Kapri, A., & Goel, R. (2011). Heavy metal pollution: source, impact, and remedies. *Biomanagement of metal-contaminated soils*, 1-28.

Nachana'a Timothy, E. T. W. (2019). Environmental pollution by heavy metal: An overview. *Chemistry (Basel, Switzerland)*, *3*(2), 72–82.

Naghdi, S., Mirmohammadi, M., Karimzadegan, H., & Ghodusi, J. (2023). Studying the health risk assessment and effect of heavy metals from automobile industries on the air quality of the 21st district of Tehran. *International Journal of Environmental Science and Technology*, *20*(10), 11255–11266. doi:10.1007/s13762-023-05156-5

Nova, K. (2023). AI-enabled water management systems: An analysis of system components and interdependencies for water conservation. *Eigenpub Review of Science and Technology*, *7*(1), 105–124.

Okereafor, U., Makhatha, M., Mekuto, L., Uche-Okereafor, N., Sebola, T., & Mavumengwana, V. (2020). Toxic metal implications on agricultural soils, plants, animals, aquatic life and human health. *International Journal of Environmental Research and Public Health, 17*(7), 2204. doi:10.3390/ijerph17072204 PMID:32218329

Pandey, G., & Madhuri, S. (2014). Heavy metals causing toxicity in animals and fishes. *Research Journal of Animal. Veterinary and Fishery Sciences, 2*(2), 17–23.

Pang, Y. L., Quek, Y. Y., Lim, S., & Shuit, S. H. (2023). Review on phytoremediation potential of floating aquatic plants for heavy metals: A promising approach. *Sustainability (Basel), 15*(2), 1290. doi:10.3390/su15021290

Qin, G., Niu, Z., Yu, J., Li, Z., Ma, J., & Xiang, P. (2021). Soil heavy metal pollution and food safety in China: Effects, sources and removing technology. *Chemosphere, 267*, 129205. doi:10.1016/j.chemosphere.2020.129205 PMID:33338709

Sabath, E., & Robles-Osorio, M. L. (2012). Renal health and the environment: Heavy metal nephrotoxicity. [English Edition]. *Nefrología : Publicación Oficial de la Sociedad Española Nefrologia, 32*(3), 279–286. PMID:22508139

Sardar, K., Ali, S., Hameed, S., Afzal, S., Fatima, S., Shakoor, M. B., ... Tauqeer, H. M. (2013). Heavy metals contamination and what are the impacts on living organisms. *Greener Journal of Environmental Management and Public Safety, 2*(4), 172-179.

Sarwar, N., Imran, M., Shaheen, M. R., Ishaque, W., Kamran, M. A., Matloob, A., Rehim, A., & Hussain, S. (2017). Phytoremediation strategies for soils contaminated with heavy metals: Modifications and future perspectives. *Chemosphere, 171*, 710–721. doi:10.1016/j.chemosphere.2016.12.116 PMID:28061428

Saxena, G., Purchase, D., Mulla, S. I., Saratale, G. D., & Bharagava, R. N. (2020). Phytoremediation of heavy metal-contaminated sites: Eco-environmental concerns, field studies, sustainability issues, and future prospects. *Reviews of Environmental Contamination and Toxicology, 249*, 71–131. PMID:30806802

Song, Y., Hou, D., Zhang, J., O'Connor, D., Li, G., Gu, Q., Li, S., & Liu, P. (2018). Environmental and socio-economic sustainability appraisal of contaminated land remediation strategies: A case study at a mega-site in China. *The Science of the Total Environment, 610*, 391–401. doi:10.1016/j.scitotenv.2017.08.016 PMID:28806555

Sovacool, B. K. (2019). Toxic transitions in the lifecycle externalities of a digital society: The complex afterlives of electronic waste in Ghana. *Resources Policy, 64*, 101459.

Szynkowska, M. I., Pawlaczyk, A., & Maćkiewicz, E. (2018). Bioaccumulation and biomagnification of trace elements in the environment. *Recent advances in trace elements*, 251-276.

Toussaint, B., Raffael, B., Angers-Loustau, A., Gilliland, D., Kestens, V., Petrillo, M., Rio-Echevarria, I. M., & Van den Eede, G. (2019). Review of micro-and nanoplastic contamination in the food chain. *Food Additives & Contaminants. Part A, Chemistry, Analysis, Control, Exposure & Risk Assessment, 36*(5), 639–673. doi:10.1080/19440049.2019.1583381 PMID:30985273

Vijayaraghavan, K., & Balasubramanian, R. (2015). Is biosorption suitable for decontamination of metal-bearing wastewaters? A critical review on the state-of-the-art of biosorption processes and future directions. *Journal of Environmental Management*, *160*, 283–296. doi:10.1016/j.jenvman.2015.06.030 PMID:26143501

Wang, D., Saleh, N. B., Byro, A., Zepp, R., Sahle-Demessie, E., Luxton, T. P., Ho, K. T., Burgess, R. M., Flury, M., White, J. C., & Su, C. (2022). Nano-enabled pesticides for sustainable agriculture and global food security. *Nature Nanotechnology*, *17*(4), 347–360. doi:10.1038/s41565-022-01082-8 PMID:35332293

Wang, X., Daigger, G., de Vries, W., Kroeze, C., Yang, M., Ren, N. Q., Liu, J., & Butler, D. (2019). Impact hotspots of reduced nutrient discharge shift across the globe with population and dietary changes. *Nature Communications*, *10*(1), 2627. doi:10.1038/s41467-019-10445-0 PMID:31201305

Wu, Q., Zhou, H., Tam, N. F., Tian, Y., Tan, Y., Zhou, S., Li, Q., Chen, Y., & Leung, J. Y. (2016). Contamination, toxicity and speciation of heavy metals in an industrialized urban river: Implications for the dispersal of heavy metals. *Marine Pollution Bulletin*, *104*(1-2), 153–161. doi:10.1016/j.marpolbul.2016.01.043 PMID:26856647

Yusuf, A., Sodiq, A., Giwa, A., Eke, J., Pikuda, O., Eniola, J. O., Ajiwokewu, B., Sambudi, N. S., & Bilad, M. R. (2022). Updated review on microplastics in water, their occurrence, detection, measurement, environmental pollution, and the need for regulatory standards. *Environmental Pollution*, *292*, 118421. doi:10.1016/j.envpol.2021.118421 PMID:34756874

Zeng, X., Xu, X., Boezen, H. M., & Huo, X. (2016). Children with health impairments by heavy metals in an e-waste recycling area. *Chemosphere*, *148*, 408–415. doi:10.1016/j.chemosphere.2015.10.078 PMID:26829309

Zhang, G., Bai, J., Xiao, R., Zhao, Q., Jia, J., Cui, B., & Liu, X. (2017). Heavy metal fractions and ecological risk assessment in sediments from urban, rural and reclamation-affected rivers of the Pearl River Estuary, China. *Chemosphere*, *184*, 278–288. doi:10.1016/j.chemosphere.2017.05.155 PMID:28601010

Zhang, H., Zhao, Y., Wang, Z., & Liu, Y. (2021). Distribution characteristics, bioaccumulation and trophic transfer of heavy metals in the food web of grassland ecosystems. *Chemosphere*, *278*, 130407. doi:10.1016/j.chemosphere.2021.130407 PMID:33823346

Chapter 13
Recovery of Heavy Metals by Biosorption and Regeneration of the Adsorbents

Jitu Saikia
https://orcid.org/0000-0002-3994-6990
Pandit Deendayal Upadhyaya Adarsha Mahavidyalaya, Behali, India

Rakesh Dutta
Pandit Deendayal Upadhyaya Adarsha Mahavidyalaya, Behali, India

ABSTRACT

In recent times, significant research is carried out regarding efficient and cost-effective removal of heavy metals from water systems with the help of adsorption. Out of different absorption processes, biosorption processes have some major advantages over the other available physical or chemical processes. They have higher applicability because of their low cost and eco-friendly nature. However, no sufficient attention was paid to the recycling of adsorbents and recovery of heavy metals from the absorbents. The adsorbents themselves become toxic after absorbing the heavy metal. So, it is important to dispose of the used absorbent or reuse the absorbent by recovering and recycling process. In this chapter, the authors focus on recycling the absorbent and recovering of heavy metal that are used in various biosorption processes. Various types of acids, alkalis, and chelating agents were used regenerating agents for the recycling and recovery process.

1. INTRODUCTION

The use of different heavy metals in various industries increased the risk on global public health and enhanced ecological problems round the globe. Along with natural sources, anthropogenic activities such as metal mining, industrial production and agricultural use contribute to environmental contamination of heavy metals. Other sources of pollution include metal corrosion, air deposition, soil erosion, leaching and metal evaporation from water resources. Industrial sources include paper processing factories,

DOI: 10.4018/979-8-3693-1618-4.ch013

plastics, microelectronics, textiles, wood preservation, and metal processing in refineries (Bashir et al. 2018; Chen et al. 2021).

Nowadays most of the surface water and groundwater sources are polluted with various essential and non-essential heavy metals from different source due to the industrialization and urbanization (Lata et al. 2015). Among different heavy metals, Copper (Cu), Cadmium (Cd), Nickel (Ni), Lead (Pb), Chromium (Cr), Mercury (Hg) and Zinc (Zn) are mainly responsible for the toxic effects among living organism (Tzou et al. 2007; O'Connell et al. 2008). Among the class of metalloids; boron, silicon, germanium, arsenic, antimony and tellurium are major heavy metal pollutant. However, among metalloids, arsenic toxicity is of major concern for the environment and is the most dominating toxic element found in water (Smedley and Kinniburg 2002).

Another highly toxic heavy metal which should be removed is mercury. It acts as neurotoxin that can leads to disfunction of central nervous system. High concentrations of mercury in our body can cause demage to pulmonary and kidney function, chest pain and dyspnoea (Tzou et al. 2007). Continuous exposure to lead metal can lead to anemia, diseases of the liver and kidneys, brain damage to living being and ultimately cause death (Smedley and Kinniburg 2002).

The presence of heavy metal in our ecosystem is common. However, presence of heavy metals in excess level in our eco-system leads to chronic poisoning. It affects the central nervous system, reduces energy level in body, alters blood composition of living beings and subsequently affects all vital organs like lungs, kidney, liver etc. of living organism (Baldwin and Marshal 1999). Heavy metals are non-biodegradable and it get accumulated in food chain of an ecosystem. Long-term exposure may lead to muscular and neurological disorder, including alzheimer's disease, parkinson's disease, muscular dystrophy and multiple sclerosis. Continuous exposure to higher concentration of heavy metals pollutant may even cause cancer to some people (Smedley and Kinniburg 2002).

Although heavy metals occurs naturally, does not cause environmental problems as it remains in immobilized form in sediments and as ores in nature. However, due to various anthropogenic activities like mining and industrial processes such as smelting, tannery, battery, from fly ash, electroplating, steel plant, distillery etc. causes increased deposition of heavy metals in soil and aquatic environment. Among toxic heavy metals, Arsenic, Cadmium, Lead, Mercury, Zinc and Chromium are responsible for most of the metal poisoning hazards faced by public and the environment. Arsenic is a ubiquitous element found in inorganic forms as trivalent arsenite and pentavalent arsenate and also in organic forms as monomethylarsonic acid (MMA), dimethylarsinic acid (DMA), and trimethylarsine oxide. Arsenic pollution occurs due to natural phenomena like volcanic eruptions and soil erosion, as well as human activities and industrial production of arsenic containg compounds used in agricultural applications, veterinary medicine, and medical treatment. Contamination with high levels of arsenic is of concern because it can cause various health effects, including cardiovascular and peripheral vascular diseases, developmental anomalies, neurologic and neurobehavioural disorders, diabetes, hearing loss, portal fibrosis, hematologic disorders, and carcinoma. The toxic effects to human health, sources and permissible limits of some heavy metals are listed in Table.1(Gouda et al. 2023, Hussain et al. 2021, Tripathi et al. 2023, Wierzba et al. 2022).

Table 1. Toxic effects of some heavy metals on human health (Fahimirad et al., 2017)

Toxic Heavy Metal	Major Sources	Effect on Human Health	Permissible Limit (ppm)
Arsenic	Pesticides, fungicides, metal smelting	Bronchitis, dermatitis	0.02
Cadmium	Welding, electroplating, pesticides, fertilizers, Ni-Cd batteries, burning of coal	Kidney damage, bronchitis, gastrointestinal disorder, bone marrow cancer	0.06
Lead	Paint, pesticides, smoking, automobile emission, minng, burning of coal	Liver, kidney, gastrointestinal damage, mental retardation in children	0.1
Mercury	Pesticides, batteries, paper industries	Damage to nervous system, protoplasm poisoning	0.01

Cadmium is widely distributed in the earth's crust at an average concentration of about 0.1 mg/kg, with the highest levels in sedimentary rocks and marine phosphates. Cadmium is frequently used in various industrial activities, including the production of alloys, pigments, and batteries. However, its commercial use has declined in developed countries due to environmental concerns. Human exposure to cadmium can be through inhalation or cigarette smoke, and ingestion of food. Exposure to cadmium can occur through working in primary metal industries, eating contaminated food, smoking cigarettes, and working in cadmium contaminated work places. Cadmium compounds are classified as human carcinogens by several regulatory agencies, with the International Agency for Research on Cancer and the U.S. National Toxicology Program concluding that there is adequate evidence that cadmium is a human carcinogen. Chromium enters environment from various natural and anthropogenic sources, with the largest release coming from industrial establishments. Industries with the largest contribution to chromium release include metal processing, tannery facilitics, chromate production, stainless steel welding, and ferrochrome and chrome pigment production (Tripathi et al. 2023, Wierzba et al. 2022).

Lead is used in various industrial, agricultural, and domestic applications, including lead-acid batteries, ammunitions, metal products, and devices to shield X-rays. However, the industrial use of lead has been significantly reduced from paints and ceramic products, caulking, and pipe solder. Human exposure to lead occurs mainly via inhalation of lead-contaminated dust particles or aerosols, and ingestion of lead-contaminated food, water, and paints. Adults absorb 35 to 50% of lead through drinking water, while children may have a higher absorption rate. The nervous system is the most vulnerable target of lead poisoning, with headache, poor attention span, irritability, loss of memory, and dullness being early symptoms. Lead exposure usually results from lead in deteriorating household paints, lead in the workplace, lead in crystals and ceramic containers that leaches into water and food, lead use in hobbies, and lead use in some traditional medicines and cosmetics. Lead exerts its toxic effect through biochemical processes that include lead's ability to inhibit or mimic the actions of calcium and interact with proteins. Within the skeleton, lead is incorporated into the mineral in place of calcium, binding to biological molecules and interfering with their functions through various mechanisms. Acute exposure to lead induces brain damage, kidney damage, and gastrointestinal diseases, while chronic exposure may cause adverse effects on the blood, central nervous system, blood pressure, kidneys, and vitamin D metabolism (Ordonez et al. 2023, Reis et al. 2020).

Mercury is found in nature in three forms: elemental, inorganic, and organic. It is a widespread environmental toxicant and pollutant that induces severe alterations in body tissues and causes adverse health effects. Both humans and animals are exposed to various chemical forms of mercury in the environment, including elemental mercury vapor (Hg^0), inorganic mercurous (Hg^{+1}), mercuric (Hg^{+2}), and organic mercury compounds. Humans are exposed to all forms of mercury through accidents, environmental pollution, food contamination, dental care, preventive medical practices, industrial and agricultural operations, and occupational operations. The major sources of chronic, low-level mercury exposure are dental amalgams and fish consumption. Mercury enters water as a natural process of off-gassing from the earth's crust and also through industrial pollution. Algae and bacteria methylate the mercury entering the waterways, and methyl mercury makes its way through the food chain into fish, shellfish, and eventually into humans (Mwandira et al. 2020, Olubode et al. 2022).

Various anthropogenic sources of heavy metal such as industrial effluents, waste water disposal (Tzou et al. 2007), fertilizers from agricultural flows (O'Connell et al. 2008)) and domestic sewage disposal to water flow carries heavy metals into the aquatic environment. Heavy metals not only pollute water, it is a widespread pollutant that pollutes the whole environment including air, soil and water. The availability of these metals in the environment is now a matter of high concern to the humanity. To overcome the health risk and to prevent environmental degradation, higher standards were imposed for pure drinking water and effluent discharge. To follow these strict standards properly, it is now a great challenge for the industries, drinking water supplier and researchers for better treatment of groundwater, surface water and industrial wastewater from heavy industries containing hazardous heavy metals (Tzou et al. 2007). Here, in this chapter we have mainly focus on the regeneration or recovery of different adsorbent materials after removal of heavy metal from the water sources which reduces the toxicity of the metal loaded adsorbents (Dusengemungu et al. 2020, Maisarah et al. 2021).

2. METHODS FOR REMOVAL OF HEAVY METAL

For the removal of heavy metal ions from the contaminant source of water such as surface water, groundwater or wastewater, several methods were employed such as physical, chemical or biological methods. The different methods which are used in this purpose are coagulation, filtration, flocculation, adsorption, reverse osmosis, activated sludge process, chemical precipitation, membrane separation process and bioremediation. Different conventional methods for treating effluents, such as precipitation, redox, membrane technologies and electrolysis are not cost effective and have problem of secondary waste product generation like sludge and other byproduct (Crini 2005). As a result, it creates another issue for the safe disposal of sludge and other byproducts (Mamatha et al. 2013). Heavy metals and metalloids such as arsenic removal techniques include coagulation, flocculation, precipitation, adsorption, ion exchange and membrane filtration. There are some other alternative methods such as ozone oxidation, bioremediation and electrochemical treatments which are used in the removal of arsenic (Ahmed 2005; Mandal and Sujuki 2002; Choong et al. 2007).

3. ADVANTAGES OF BIOSORPTION

Out of the various available techniques that are used for removal of heavy metal from contaminated water or from wastewater, adsorption or biosorption is considered as the most effective technique (Mohan and Pittman 2007). Biosorption methods is the most convenient, eco-friendly and cost effective method to separate heavy metals. It is an attractive method for this purpose because of its potential regeneration capacity and higher efficiency to remove metal ions. In contrary to the metal removal process, it is a major concern to recover the absorbent or recover the metal from the absorbent and recycle it for further processes (Maisarah et al. 2021).

The adsorption methods have various advantages over other above mention techniques, as it is easy to operate in field conditions, it has the potential capacity of regeneration and has a high efficiency of heavy metal ions removal (Ahmaruzzaman and Gupta 2011; Wu et al. 2013; Ranjan et al. 2009). Continuous research works were carried out for the improvements and development of effective and desirable adsorbents for achieving higher metal removal efficiency (Gupta and Nayak 2012). Different types of adsorbents have already been reported such as (1) activated carbon, (2) carbon nanotubes, (3) low-cost adsorbents and (4) bio-adsorbents that have been used for the purpose of heavy metal adsorption studies. Many researchers suggested that biosorption of heavy metals from waste water is relatively a better and efficient process that has been used as a promising and cost effective method in the removal of heavy metal ions.

Davis et al. (2003) has reported the removal of different heavy metal such as Cd (II), Cu(II), Zn(II), Pb(II), Cr(III) and Hg(II) by the use of the same adsorbent i.e. brown algae. Similarly, another researcher, Iqbal et al. (2002) reposted both the adsorption and desorption heavy metals such as Ni(II), Pb(II), Cd(II), Cu(II), Cr(III) and Zn(II) from waste-water by using bio-adsorbent i.e. petiolar felt sheath of palm-PFP (Dusengemungu et al. 2020).

4. DRAWBACKS OF BIOSORPTION

The different studies about biosorption reported that a particular adsorbent was not effective for removal of different types of water pollutants. Majority of the adsorbents is specific for some particular heavy metal removal. An adsorbent also has some other drawbacks such as it could not avail a good performance at commercial levels due to lack of suitable high adsorption capacity and availability to a large scale (Grassi et al. 2012). Moreover, very less attention has been paid by the researchers to the disposal or recycling or rageneration of used adsorbents containing adsorbed heavy metals. Gupta et al. (2000) reviewed and analyzed the efficiency of different bio-adsorbents used in biosorption of heavy metals and reported that biosorption process depends on the regeneration efficiency of bio-adsorbents after metal desorption.

The disposal of used adsorbents after recovery of contaminants such as adsorbed heavy metal will generate another issue i.e. it creates secondary pollution from the used adsorbents and the chemicals which are used to treat the adsorbents for metal recovery (Tzou et al. 2007). The metal-loaded adsorbents have very toxic effects on humans and environment. Therefore, the used adsorbents have to be released into the environment only after recovery of the heavy metals completely or only after the regeneration of adsorbent materials (Grassi et al. 2012).

5. TYPES OF HEAVY METAL POLLUTANTS

The heavy metals are generally categories in two types, essential and non-essential heavy metals. A trace amount of essential heavy metals is required by our human body (Shamim, 2018). The essential heavy metals are copper, zinc, iron, nickel etc. This essential heavy metals are responsible for majority of the biological functions in human body like oxidation and reduction processes, various transportation process etc., but its high concentration in the body is very much toxic to human health (Shamim, 2018; Iyengar, 1998). Essential heavy metals are also required by plants in small quantities. Deficiency of essential heavy metals in the body may cause many deficiency diseases or syndromes. However, non-essential heavy metals, as their name suggest, have no role in our body. They only induce toxic effects to the living being when present in the environment in very low concentration (Koller and Saleh, 2018).

Non-essential heavy metals are highly hazardous and toxic, and their low concentration in the body has very critical effects on the body. Some non-essential heavy metals like mercury may cause death to human or some heavy metal like arsenic may cause cancer. So, these heavy metals are categorized as carcinogenic (Ahalya et al., 2003). Furthermoe, heavy metals can also be categorized into different categories like macro nutrient elements, micro nutrient elements, highly toxic elements, precious elements and radioactive nuclides as shown in Figure.1 with some examples (Ahalya et al., 2003).

Figure 1. Classification of heavy metals with some examples

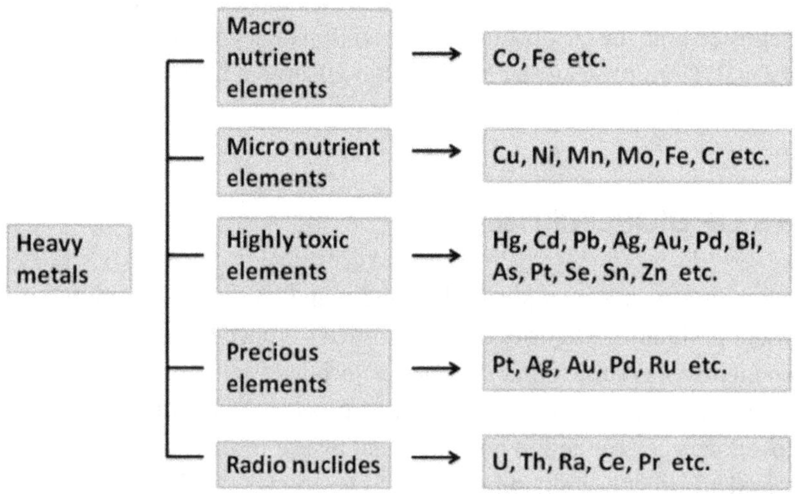

6. ADSORBENTS USED FOR BIOSORPTION

Biosorption of heavy metals can be categorized into two categories based on the bioadsorbent used, i.e.

(i) Organic non living materials used for heavy metal adsorption and
(ii) Organic living materials used for heavy metal adsorption

6.1. Organic Non Living Materials Used for Heavy Metal Adsorption

Different bioadsorbents are used for removal of different heavy metals. However, some of the bioadsorbents are specific for removal of some particular heavy metal from aqueous system. The cell walls of agricultural byproducts and waste materials such as fruit peels, seed hulls, corn cobs, include cellulosic material, which is known to include carboxylic or phenolic functional groups. The interaction of a metal ion with a functional group causes biosorption, which removes the metal ion from the medium based on the cation exchange between functional groups and metal ions (Abdi et al., 2015, Kelly-Vargas et al. 2012). The list of heavy metals along with their corresponding biosorbent is listed in the Table 2 as shown below.

Table 2. List of different bioadsorbents along with corresponding heavy metal

Bio-Adsorbent	Heavy Metal	Reference
Petiolar felt sheath palm	Ni(II), Pb(II), Cd(II), Cu(II), Cr(II), Zn(II)	Iqbal et al.(2002)
Spherical cellulose adsorbent	Cu(II)	Liu et al.(2002)
Coconut coir pith	As(V)	Anirudhan and Unnithan (2007)
Papaya wood	Cu(II), Cd(II), Zn(II)	Saeed et al.(2005)
Immobilized biomass (Garcinia cambogia)	As(III)	Kamala et al. (2005)
Natural zeolite	Fe(II), Cu(II), Zn(II), Mn(II)	Motsi et al.(2009)
Magnetic wheat straw	As(V), As(III)	Tian et al.(2011)
Inorganic solid waste materials (red mud)	Cd(II), Pb(II)	Zhou and Haynes (2011)
Orange peel powder with Fe3O4 (MNPOPP)	Cd(II)	Gupta and Nayak (2012)
Coffee waste	Cu(II), Cr(VI)	Kyzas (2012)
Red mud	As(V), As(III), Se(VI), Se(IV)	Zhou and Haynes (2012)

6.2. Living Organic Materials Used for Heavy Metal Adsorption

Based on their morphology, bacteria are categorized as rods, cocci, or spirillum. Functional groups like carboxyl, amino, phosphate, or sulfate are present in the slime layer to aid in the chelation of metals (Shamim, 2018). In general, the cell wall is in charge of the surface binding sites and binding strength for various metal ions based on various binding mechanisms. Bacteria are classified into two main groups: Gram positive and Gram negative. The cell wall of Gram-negative bacteria has an extra layer made of lipopolysaccharides and phospholipids. 10%–20% of the gram-negative cell wall is made up of peptidoglycan. Teichoic and teichuronic acids, as well as lipopolysaccharides, are the sources of the negative charge on Gram-negative bacteria. The ability to bind metal is also seen by extracellular poly-saccharides. The initial step in biosorption is when the metal ion comes into contact with the bacterial cell wall. The functional groups (amine, carboxyl, hydroxyl, phosphate, sulfate, etc.) on the cell wall are where the metal ions bind. The general process of metal uptake in bacteria entails the binding of metal ions to reactive groups on the cell wall, which is followed by the internalization of metal ions within the bacterial cell. Gram-positive bacteria are able to absorb more metal because they have glycoproteins. Because of phospholipids and LPS, Gram negative bacteria are seen to take in fewer metals (Gourdon et al. 1990). Table.3 lists the adsorption of heavy metals by living biosorbents.

Algae can include both macro and microalgae and are regarded as high-quality biosorbent materials due to their large size, strong potential for sorption, and lack of harmful chemical generation. They are primarily divided into three categories:

(i) Red algae,
(ii) Macroalgae (marine or brown algae), and
(iii) Microalgae (fresh water or green algae).

It is said that brown algae has a greater capacity for metal uptake among these three groups. The species of the algae, the ionic charge of the metal, and the chemical makeup of the metal ion solution are the factors that cause heavy metal ions to bind to the surface of the algae. Metal ion binding sites in the algal surface include carboxyl, imidazole, phosphate, sulfate, hydroxyl, and sulfhydryl groups. Algae absorb metals using a process akin to bacteria's, which involves attaching metal ions to surfaces. Abbas et al. state that there are two mechanisms at work in algal biosorption:

(i) The ion exchange approach, in which metal ions replace ions that are present on the algal surface (Ca, Mg, Na, and K), and
(ii) Complexation between functional groups and metal ions.

Algal cell walls are composed of polysaccharides (alginic acid, chitin, xylan, mannan), which contain functional groups (sulfate, hydroxyl, phosphate, imidazole, amino, and amine) that are known to function as metal binding sites. Ionic charge and covalent bonding are proposed theories about the metal binding mechanism. Whereas amino and carboxyl groups are engaged in covalent interaction between metal ions and functional groups, carboxyl and sulfate groups are involved in ionic bonding. Within the algal body, phytochelatins are generated in reaction to metal ions (Abbas et al., 2018).

Table 3. List of different bio-adsorbent along with corresponding heavy metal

Type of Microorganism	Microorganism	Heavy Metal	Reference
Bacteria	*Arthrobacter viscosus*	Pb	Hlihor et al. (2017)
	Pseudomonas aeruginosa RW9	Cr	Mat Arisah et al. (2021)
	Bacillus thuringiensis PW-05	Hg	Dash et al. (2014)
	Rhizopus stolonifer	Pb, Ni, Cd	Njokua et al. (2020)
	Bacillus subtilis	Pb, Cd	Abioye et al. (2018)
	Pseudomonas sp.	Pb, Cd	De et al. (2008)
	Streptomyces K11	Zn	Sedlakova-Kadukova et al. (2019)
	Streptomyces zinciresistens	Cd, Zn	Lin et al. (2012)
	Alcaligenes sp. MMA	Cr, Zn, Cd	Sodhi et al. (2020)
	Bacillus cereus RC-1	Zn, Cd, Pb	Huang et al. (2018)
Algae	Intact biomass (*Spirulina maxima*) Pretreated biomass (*Spirulina maxima*)	Pb	Gong et al. (2005)
	Marine algal biomass (*Cladophora fascicularis*)	Cd	Deng et al. (2008)
	Algal biomass (*Sargassum spp.*)	Pb	Jalali et al. (2002)
	Green algae (*Cladophora fascicularis*)	Pb	Deng et al. (2007)
	Spirogyra hyalina	As	Viraraghavan et al. (2011)
	Bifurcaria	Cd	Gina et al. (2002)
	Sargassum sp.	Hg, Cr	Subhashini et al. (2011)
	Pithophora spp.	Cr, Pb	Ramsenthil et al. (2010)
Yeast	Yeast biomass and waste yeast	Cd, Pb	Kordialik-Bogacka et al. (2011)
	Saccharomyces cerevisiae	Cd	Das et al. (2008)
	Saccharomyces cerevisiae	Cr, Hg	Anaemene et al. (2012)
	Mucor rouxii	Pb	Muraleedharan et al. (1991)
Mashroom	*Volvariella volvacea* (edible Mushroom) – mycelia, sporocarps	Cd, Pb, Cu, Cr	Purkayastha et al. (1991)
	Agaricus macrospores	Cd, Hg, Cu	García et al. (2005)

According to Das et al., Fungal cell walls have outstanding metal-binding characteristics because of their constituent parts. In addition to lipids, polysaccharides, and colors like melanin, the primary components of fungus cell walls include chitins, mannans, and glucans. The cell wall of fungi is 90% of polysaccharides. Functional groups such as carboxyl, phosphate, uranic acids, proteins, nitrogen-containing ligands, chitin, or chitosan are involved in metal binding. The ability of fungal cells to biosorb substances can be controlled using physical or chemical methods such as autoclaving, heating, dimethyl sulfoxide, laundry detergent, orthophosphoric acid, formaldehyde, gluteraldehyde, and NaOH. Wild mushrooms, also known as macrofungi, grows in a variety of habitats, including woods, contaminated soils, and bodies of water. Their sporocarps, mycelia, and fruiting bodies are used to absorb the metals (Abbas et al., 2018).

Yeast cells in their free state are not thought to be suitable candidates for biosorption. The problem of solid-liquid phase separation affects free cells. This issue appears to have less of an impact on flocculating cells. Increased surface to volume ratio for metal binding with metal binding sites can be obtained by pretreating yeast cells. In yeast cells, the pH values greater than 5 are reported to maximize the biosorption of metal.

7. MECHANISMS AND FACTORS AFFECTING BIOSORPTION EFFICIENCY

Optimizing the biosorption process requires an understanding of the underlying mechanisms involved. The process of heavy metal binding to bacterial cell wall can be

(i) Metabolism Dependent Biosorption

The Metabolism dependent biosorption is observed in living biological materials. It involves various mechanisms such as chelation, physical adsorption and adhesion to the surface. A single process like chelation or a combination of these processes occurs during this biosorption process.

(ii) Metabolism In-dependent Biosorption

This type of biosorption occurs mostly in the biomass consisting of dead cells. In this type of Biosorption adsorption process plays the key role which may be physical adsorption or ionic interaction depending on the nature of the heavy metal and the biomass used for biosorption process.

Factors such as the characteristics of metal ions nature of the biosorbent, pH, temperature and presence of other metal ion influence the efficiency of biosorption. Adjusting the pH of metal ion solutions can affect the electrostatic interactions between biomaterial surfaces and metal cations. pH affect the solubility of the metal ions and the binding sites of the biomass and the maximum adsorption takes place at the pH range of 2.5-6. The Biosorption efficiency is greatly dependent on the nature of the biosorbent used. It is observed that the biosorption ranges in different biosorbents. The efficiency of the biosorbents can be increased by a number of chemical treatments such as pretreatment with acid and alkali and physical treatments such as boiling, sonication, drying etc (Shamim, 2018).

8. RECOVERY OF HEAVY METAL AND REGENERATION OF THE BIOSORBENT

Recovery of the heavy metal and the regeneration of the adsorbent is very important from the point of view of the efficiency of the adsorbent and the process. On the other hand, release of metal loaded adsorbents into environment is not safe and should be released only after recovery of the metal from the spent adsorbents. The biggest challenge in this aspect is that the regenerated spent adsorbent losses its adsorption capacity and the desorption of the metals from the adsorbent surface is very difficult. The most efficient desorption agents used for the recovery of the metal from the spent adsorbent and the regeneration of the adsorbent are 30% H_2O_2 in 0.5 M HNO_3, 0.1 M HNO_3, 0.1 M $CaCl_2$, 0.3 M NaOH, NaCl, EDTA, Thermal desorption, $KMnO_4$, H_2SO_4, etc (Lata et al., 2015).

Srivastava and Goyal reported that desorption of metal ions in acidic media is comparatively rapid and higher than in basic and neutral media. Zhou and Haynes reported the possible mechanisms and conditions of desorption of metal ions by acids are as follows:

(1) Low pH,
(2) Strong competition of H^+ ions for adsorption sites and increases the desorption of metal ions into the solution,
(3) Acid reacts with residual alkalinity and lowers the desorption rates.

REFERENCES

Abbas, S. H., Ismail, I. M., Mostafa, T. M., & Sulaymon, A. H. (2014). Biosorption of heavy metals: A review. *Journal of Chemical Science and Technology.*, *3*(4), 74–102.

Abdi, O., & Kazemi, M. A. (2015). Review study of biosorption of heavy metals and compari- son between different biosorbents. *Journal of Materials & Environmental Sciences*, *6*(5), 1386–1399.

Abioye, O. P., Oyewole, O. A., Oyeleke, S. B., Adeyemi, M. O., & Orukotan, A. A. (2018). Biosorption of lead, chromium and cadmium in tannery effluent using indigenous microorganisms. *Brazilian Journal of Biological Sciences*, *5*(9), 25–32. doi:10.21472/bjbs.050903

Ahalya, N., Ramachandra, T. V., & Kanamadi, R. D. (2003). Bio sorption of heavy metals. *Research Journal of Chemistry and Environment*, *7*(4), 71–78.

Ahmaruzzaman, M., & Gupta, V. K. (2011). Rice husk and its ash as lowcost adsorbents in water and wastewater treatment. *Industrial & Engineering Chemistry Research*, *50*(24), 13589–13613. doi:10.1021/ie201477c

Ahmed, F. M. (2001). *An overview of arsenic removal technologies in Bangladesh and India. Technologies for arsenic removal from drinking water.* Bangladesh University of Engineering and Technology (BUET) and United Nations University (UNU). http://archive.unu.edu/env/Arsenic/Ahmed.pdf

Ahmed, M. F. (2005). *Arsenic mitigation technologies in south and east Asia. In: Towards a more effective operational response.* Technical Report No. 31303, Water and Sanitation Program- South Asia and the World Bank.

Anaemene, I. A. (2012). The use of Candida sp. in the biosorption of heavy metals from indus- trial effluent. *European Journal of Experimental Biology*, *2*(3), 484–488.

Anirudhan, T. S., & Unnithan, M. R. (2007). Arsenic (V) removal from aqueous solutions using an anion exchanger derived from coconut coir pith and its recovery. *Chemosphere*, *66*(1), 60–66. doi:10.1016/j.chemosphere.2006.05.031 PMID:16824580

Baldwin, D. R., & Marshal, W. J. (1999). Heavy metal poisoning and its laboratory investigation. *Annals of Clinical Biochemistry, 36*(3), 267–300. doi:10.1177/000456329903600301 PMID:10376071

Bashir, A., Malik, L. A., Ahad, S., Manzoor, T., Bhat, M. A., Dar, G. N., & Pandith, A. H. (2018). Removal of heavy metal ions from aqueous system by ion-exchange and biosorption methods. *Environmental Chemistry Letters, 17*(2), 729–754. doi:10.1007/s10311-018-00828-y

Chen, X., Tian, Z., Cheng, H., Xu, G., & Zhou, H. (2021). Adsorption process and mechanism of heavy metal ions by different components of cells, using yeast (Pichia pastoris) and Cu^{2+} as biosorption models. *RSC Advances, 11*(28), 17080–17091. doi:10.1039/D0RA09744F PMID:35479686

Choong, T. S. Y., Chuaha, T. G., Robiaha, Y., Koaya, F. L. G., & Azni, I. (2007). Arsenic toxicity, health hazards and removal techniques from water: An overview. *Desalination, 217*(1-3), 139–166. doi:10.1016/j.desal.2007.01.015

Crini, G. (2005). Recent developments in polysaccharide-based materials used as adsorbents in wastewater treatment. *Progress in Polymer Science, 30*(1), 38–70. doi:10.1016/j.progpolymsci.2004.11.002

Das, N., Vimala, R., & Karthika, P. (2008). Biosorption of heavy metals – An overview. *Indian Journal of Biotechnology, 7*, 159–169.

Dash, H. R., Mangwani, N., & Das, S. (2014). Characterization and potential application in mercury bioremediation of highly mercury- resistant marine bacterium Bacillus thuringiensis PW-05. *Environmental Science and Pollution Research International, 21*(4), 2642–2653. doi:10.1007/s11356-013-2206-8 PMID:24114385

Davis, T. A., Volesky, B., & Mucci, A. (2003). A review of the biochemistry of heavy metal biosorption by brown algae. *Water Research, 37*(18), 4311–4330. doi:10.1016/S0043-1354(03)00293-8 PMID:14511701

De, J., Ramaiah, N., & Vardanyan, L. (2008). Detoxification of toxic heavy metals by marine bacteria highly resistant to mercury. *Marine Biotechnology (New York, N.Y.), 10*(4), 471–477. doi:10.1007/s10126-008-9083-z PMID:18288535

Deng, L., Su, Y., Su, H., Wang, X., & Zhu, X. (2007). Sorption and desorption of lead(II) from wastewater by green algae Cladophora fascicularis. *Journal of Hazardous Materials, 143*(1-2), 220–225. doi:10.1016/j.jhazmat.2006.09.009 PMID:17049733

Deng, L., Zhu, X., Su, Y., Su, H., & Wang, X. (2008). Biosorption and desorption of Cd^{2+} from wastewater by dehydrated shreds of Cladophora fascicularis. *Chinese Journal of Oceanology and Limnology, 26*(1), 45–49. doi:10.1007/s00343-008-0045-0

Dusengemungu, L., Kasali, G., Gwanama, C., & Ouma, K. O. (2020). Recent Advances in Biosorption of Copper and Cobalt by Filamentous Fungi. *Frontiers in Microbiology, 11*, 582016. doi:10.3389/fmicb.2020.582016 PMID:33408701

Fahimirad, S., Hatami, M. (2017). Heavy metal-mediated changes in growth and phytochemicals of edible and medicinal plants. *Medicinal plants and environmental challenges*, 189-214. doi:10.1007/978-3-319-68717-9_11

García, M. A., Alonso, J., & Melgar, M. J. (2005). Agaricus macrosporus as a potential bioremediation agent for substrates contaminated with heavy metals. *Journal of Chemical Technology and Biotechnology*, *80*(3), 325–330. doi:10.1002/jctb.1203

Gina, K. Y., Tanga, Y. Z., & Aziz, M. A. (2002). Derivation and application of a new model for heavy metal biosorption by algae. *Water Research*, *36*(5), 1313–1323. doi:10.1016/S0043-1354(01)00332-3 PMID:11902786

Gong, R., Ding, Y., Liu, H., Chen, Q., & Liu, Z. (2005). Lead biosorption and desorption by intact and pretreated Spirulina maxima biomass. *Chemosphere*, *58*(1), 125–130. doi:10.1016/j.chemosphere.2004.08.055 PMID:15522341

Gouda, S. A., & Taha, A. (2023). Biosorption of Heavy Metals as a New Alternative Method for Wastewater Treatment: A Review. *Egyptian Journal of Aquatic Biology & Fisheries*, *27*(2), 135–153. doi:10.21608/ejabf.2023.291671

Gourdon, R., Bhende, S., Rus, E., & Sofer, S. S. (1990). Comparison of cadmium biosorption by gram-positive and gram-negative bacteria from activated sludge. *Biotechnology Letters*, *12*(11), 839–842. doi:10.1007/BF01022606

Gupta, R., Khan, P. A. S., Saxena, R. K., & Mohapatra, H. (2000). Microbial biosorbents: Meeting challenges of heavy metal pollution in aqueous solutions. *Current Science*, *78*(8), 25.

Gupta, V. K., & Nayak, A. (2012). Cadmium removal and recovery from aqueous solutions by novel adsorbents prepared from orange peel and Fe_2O_3 nanoparticles. *Chemical Engineering Journal*, *180*, 81–90. doi:10.1016/j.cej.2011.11.006

Hlihor, R. M., Ros¸ca, M., Tavares, T., & Gavrilescu, M. (2017). The role of Arthrobacter viscosus in the removal of Pb(II) from aqueous solutions. *Water Science and Technology*, *76*(7), 1726–1738. doi:10.2166/wst.2017.360 PMID:28991789

Huang, F., Wang, Z. H., Cai, Y. X., Chen, S. H., Tian, J. H., & Cai, K. Z. (2018). Heavy metal bioaccumulation and cation release by growing Bacillus cereus RC-1 under culture conditions. *Ecotoxicology and Environmental Safety*, *157*, 216–226. doi:10.1016/j.ecoenv.2018.03.077 PMID:29625395

Hussain, S. A., Hasan, N. K., & Al-Abodi, E. E. (2021). Biosorption to removing heavy metals from wastewater. *The International Conference of Chemistry, 1853*. 10.1088/1742-6596/1853/1/012012

Iqbal, M., Saeed, A., & Akhtar, N. (2002). Petiolar felt sheath of palm: A new adsorbent for the removal of heavy metals from contaminated water. *Bioresource Technology*, *81*(2), 151–153. doi:10.1016/S0960-8524(01)00126-2 PMID:11762907

Iyengar, G. V. (1998). Re-evaluation of the trace element content in Reference man. *Radiation Physics and Chemistry*, *51*(4-6), 545–560. doi:10.1016/S0969-806X(97)00202-8

Jalali, R., Ghafourian, H., Asef, Y., Davarpanah, S. J., & Sepeh, S. (2002). Removal and recovery of lead using nonliving biomass of marine algae. *Journal of Hazardous Materials*, *92*(3), 253–262. doi:10.1016/S0304-3894(02)00021-3 PMID:12031610

Kamala, C. T., Chu, K. H., & Char, N. S. (2005). Removal of arsenic(III) from aqueous solutions using fresh and immobilized plant biomass. *Water Research*, *39*(13), 2815–2826. doi:10.1016/j.watres.2005.04.059 PMID:15993920

Kelly-Vargas, K., Cerro-Lopez, M., Reyna-Tellez, S., Bandala, E. R., & Sanchez-Salas, J. L. (2012). Biosorption of heavy metals in polluted water, using different waste fruit cortex. *Physics and Chemistry of the Earth Parts A/B/C*, *37-39*, 26–29. doi:10.1016/j.pce.2011.03.006

Koller, M., & Saleh, H. M. (2018). Introductory chapter: introducing heavy metals. Tech Open. doi:10.5772/intechopen.74783

Kordialik-Bogacka, E. (2011). Cadmium and lead recovery from yeast biomass. *Central European Journal of Chemistry*, *9*, 320–325.

Kyzas, G. Z. (2012). Commercial coffee wastes as materials for adsorption of heavy metals from aqueous solutions. *Materials (Basel)*, *5*(10), 1826–1840. doi:10.3390/ma5101826

Lata, S., Singh, P. K., & Samadder, S. R. (2015). Regeneration of adsorbents and recovery of heavy metals: A review. *International Journal of Environmental Science and Technology*, *12*(4), 1461–1478. doi:10.1007/s13762-014-0714-9

Lin, Y., Wang, X., Wang, B., Mohamad, O., & Wei, G. (2012). Bioaccumulation characterization of zinc and cadmium by Streptomyces zinciresistens, a novel actinomycete. *Ecotoxicology and Environmental Safety*, *77*, 7–17. doi:10.1016/j.ecoenv.2011.09.016 PMID:22001317

Liu, M., Deng, Y., Zhan, H., & Zhang, X. (2002). Adsorption and desorption of copper(II) from solutions on new spherical cellulose adsorbent. *Journal of Applied Polymer Science*, *84*(3), 478–485. doi:10.1002/app.10114

Maisarah, A., Adityosulindro, S., & Wulandari, D. (2021). Utilization of wild algae biomass as biosorbent for removal of heavy metal Zinc (Zn^{2+}) from aqueous solution. *IOP Conference Series. Earth and Environmental Science*, *824*(1), 012017. doi:10.1088/1755-1315/824/1/012017

Mamatha, M., Aravinda, H. B., Puttaiah, E. T., & Manjappa, S. (2013). Adsorbent from Pongamia Pinnata tree bark for zinc adsorption. *Int. J. Sci. Eng. Technol.*, *2*(1), 22–25.

Mandal, B. K., & Sujuki, K. J. (2002). Arsenic round the world: A review. *Talanta*, *58*(1), 201–235. doi:10.1016/S0039-9140(02)00268-0 PMID:18968746

Mat Arisah, F., Amir, A. F., Ramli, N., Ariffin, H., Maeda, T., Hassan, M. A., & Mohd Yusoff, M. Z. (2021). Bacterial Resistance against heavy metals in Pseudomonas aeruginosa RW9 involving hexavalent chromium removal. *Sustainability (Basel)*, *13*(17), 9797. doi:10.3390/su13179797

Mohan, D., & Pittman, J. C. U. (2007). Arsenic removal from water/wastewater using adsorbents-A critical review. *International Journal of Mineral Processing*, *92*, 42–48. PMID:17324507

Motsi, T., Rowson, N. A., & Simmons, M. J. H. (2009). Adsorption of heavy metals from acid mine drainage by natural zeolite. *International Journal of Mineral Processing*, *92*(1-2), 42–48. doi:10.1016/j.minpro.2009.02.005

Muraleedharan, T. R., Iyengar, L., & Venkobachar, C. (1991). Biosorption: An attractive alternative for metal removal and recovery. *Current Science*, *61*(6), 379–385.

Mwandira, W., Nakashima, K., Kawasaki, S., Arabelo, A., Banda, K., Nyambe, I., Chirwa, M., Ito, M., Sato, T., Igarashi, T., Nakata, H., Nakayama, S., & Ishizuka, M. (2020). Biosorption of Pb (II) and Zn (II) from aqueous solution by *Oceanobacillus profundus* isolated from an abandoned mine. *Scientific Reports*, *10*(1), 21189. doi:10.1038/s41598-020-78187-4 PMID:33273589

Njokua, K. L. (2020). Microbial Remediation of Heavy Metals Contaminated Media by Bacillus megaterium and Rhizopus stolonifer. *Scientific African*, *10*, e00545. doi:10.1016/j.sciaf.2020.e00545

O'Connell, D. W., Birkinshawb, C., & O'Dwyer, T. F. (2008). Heavy metal adsorbents prepared from the modification of cellulose: A review. *Bioresource Technology*, *99*(15), 6709–6724. doi:10.1016/j.biortech.2008.01.036 PMID:18334292

Olubode, T. P., Amusat, A. I., Olawale, B. R., & Adekola, F. F. (2022). Biosorption of heavy metal using bacterial isolatesfrom e-waste soil. *African Journal of Microbiological Research*, *16*(7), 268–272.

Ordonez, J. I., Cortes, S., Maluenda, P., & Soto, I. (2023). Biosorption of Heavy Metals with Algae: Critical Review of Its Application in Real Effluents. *Sustainability (Basel)*, *15*(6), 5521. doi:10.3390/su15065521

Purkayastha, R. P., & Mitra, A. K. (1992). Metal uptake by mycelia during submerged growth and by sporocarps of an edible fungus Volvariella volvacea. *Indian Journal of Experimental Biology*, *30*, 1184–1187. PMID:1294483

Ramsenthil, R., & Meyyappan, R. (2010). Single and multi-component biosorption of copper and zinc ions using microalgal resin. *International Journal of Environmental Sciences and Development*, *1*(4), 298–301. doi:10.7763/IJESD.2010.V1.58

Ranjan, D., Talat, M., & Hasan, S. H. (2009). Biosorption of arsenic from aqueous solution using agricultural residue 'rice polish. *Journal of Hazardous Materials*, *166*(2-3), 1050–1059. doi:10.1016/j.jhazmat.2008.12.013 PMID:19131161

Reis, P. C., Reis, M. A. M., & Freitas, F. (2020). Biosorption of Heavy Metals by the Bacterial Exopolysaccharide FucoPol. *Applied Sciences (Basel, Switzerland)*, *10*(19), 6708. doi:10.3390/app10196708

Saeed, A., Akhterb, M. W., & Iqbal, M. (2005). Removal and recovery of heavy metals from aqueous solution using papaya wood as a new biosorbent. *Separation and Purification Technology*, *45*(1), 25–31. doi:10.1016/j.seppur.2005.02.004

Sedlakova-Kadukova, J., Kopcakova, A., Gresakova, L., Godany, A., & Pristas, P. (2019). Bioaccumulation and biosorption of zinc by a novel Streptomyces K11 strain isolated from highly alkaline aluminium brown mud disposal site. *Ecotoxicology and Environmental Safety*, *167*, 204–211. doi:10.1016/j.ecoenv.2018.09.123 PMID:30340085

Shamim, S. (2018). Biosorption of heavy metals. Tech Open. doi:10.5772/intechopen.72099

Smedley, P. L., & Kinniburg, D. G. (2002). A review of the source, behavior and distribution of arsenic in natural waters. *Applied Geochemistry*, *17*(5), 517–568. doi:10.1016/S0883-2927(02)00018-5

Sodhi, K. K., Kumar, M., & Singh, D. K. (2020). Multi-metal resistance and potential of Alcaligenes sp. MMA for the removal of heavy metals. *SN Applied Sciences*, *2*(11), 1885. doi:10.1007/s42452-020-03583-4

Srivastava, S., & Goyal, P. (2010). Reusability of biomaterial: a cost effective approach. *Environ Sci Eng*, 93–96.

Subhashini, S., Kaliappan, S., & Velan, M. (2011). Removal of heavy metal from aqueous solu- tion using Schizosaccharomyces pombe in free and alginate immobilized cells. In *2nd International Conference on Environmental Science and Technology IPCBEE.* Vol. 6. Singapore: IACSIT Press.

Tian, Y., Wu, M., Lin, X., Huang, P., & Huang, Y. (2011). Synthesis of magnetic wheat straw for arsenic adsorption. *Journal of Hazardous Materials*, *193*, 10–16. doi:10.1016/j.jhazmat.2011.04.093 PMID:21862216

Tripathi, M., Singh, P., Singh, R., Bala, S., Pathak, N., Singh, S., Chauhan, R. S., & Singh, P. K. (2023). Microbial biosorbent for remediation of dyes and heavy metals pollution: A green strategy for sustainable environment. *Frontiers in Microbiology*, *14*, 1168954. doi:10.3389/fmicb.2023.1168954 PMID:37077243

Tzou, Y., Wang, S., Hsu, L., Chan, R., & Lin, C. (2007). Deintercalation of Li/Al LDH and its application to recover adsorbed chromate from used adsorbent. *Applied Clay Science*, *37*(1-2), 107–114. doi:10.1016/j.clay.2006.10.007

Viraraghavan, T., & Srinivasan, A. (2011). Fungal Biosorption and biosorbents. In Microbial Biosorption of Metals. Springer. doi:10.1007/978-94-007-0443-5_6

Wierzba, S., Makuchowska-Fryc, J., Kłos, A., Ziembik, Z., & Ochędzan-Siodłak, W. (2022). Role of calcium carbonate in the process of heavy metal biosorption from solutions: Synergy of metal removal mechanisms. *Scientific Reports*, *12*(1), 17668. doi:10.1038/s41598-022-22603-4 PMID:36271239

Wu, Y., Wen, Y., Zhou, J., Cao, J., Jin, Y., & Wu, Y. (2013). Comparative and competitive adsorption of Cr(VI), As(III), and Ni(II) onto coconut charcoal. *Environmental Science and Pollution Research International*, *20*(4), 2210–2219. doi:10.1007/s11356-012-1066-y PMID:22791350

Zhou, Y., & Haynes, R. J. (2011). A comparison of inorganic solid wastes as adsorbents of heavy metal cations in aqueous solution and their capacity for desorption and regeneration. *Water, Air, and Soil Pollution*, *218*(1-4), 457–470. doi:10.1007/s11270-010-0659-7

Zhou, Y., & Haynes, R. J. (2012). A Comparison of Water Treatment Sludge and Red Mud as Adsorbents of As and Se in Aqueous Solution and Their Capacity for Desorption and Regeneration. *Water, Air, and Soil Pollution*, *223*(9), 5563–5573. doi:10.1007/s11270-012-1296-0

Zhou, Y., & Hynes, R. J. (2011). A comparison of inorganic solid wastes as adsorbents of heavy metal cations in aqueous solution and their capacity for desorption and regeneration. *Water, Air, and Soil Pollution*, *218*(1-4), 457–470. doi:10.1007/s11270-010-0659-7

Compilation of References

. Rafeeq, Hamza, Nadia Afsheen, Sadia Rafique, Arooj Arshad, Maham Intisar, Asim Hussain, Muhammad Rajendran, P., J. Muthukrishnan, and P. Gunasekaran. "Microbes in heavy metal remediation." (2003).

Ab Razak, N. H., Praveena, S. M., Aris, A. Z., & Hashim, Z. (2015). Drinking water studies: A review on heavy metal, application of biomaker and health risk assessment (a special focus in Malaysia). *Journal of Epidemiology and Global Health*, *5*(4), 297–310. doi:10.1016/j.jegh.2015.04.003 PMID:25944153

Abbas, Ismail, Mostafa, & Abbas. (2014). Sulaymon Biosorption of Heavy Metals: A Review Salman. *Journal of Chemical Science and Technology*, *3*, 74–102.

Abbas, M., & Nussrat, T. (n.d.). Statistical analysis of experimental data for adsorption process of cadmium by watermelon rinds in continuous packed bed column. *Innovation Journal of Creative Change, 13*(3), 124-1121.

Abbas, A., Al-Amer, A. M., Laoui, T., Al-Marri, M. J., Nasser, M. S., Khraisheh, M., & Atieh, M. A. (2016). Heavy metal removal from aqueous solution by advanced carbon nanotubes:critical review of adsorption applications. *Separation and Purification Technology*, *157*, 141–161. doi:10.1016/j.seppur.2015.11.039

Abbas, S. H., Ismail, I. M., Mostafa, T. M., & Sulaymon, A. H. (2014). Biosorption of heavy metals: A review. *Journal of Chemical Science and Technology.*, *3*(4), 74–102.

Abdel-Aty, M., Ammar, S., Abdel Ghafar, H., & Ali, K. (2013). Biosorption of cadmium and lead from aqueous solution by fresh water alga *Anabaena sphaerica* biomass. *Journal of Advanced Research*, *4*(4), 367–374. doi:10.1016/j.jare.2012.07.004

Abdi, O., & Kazemi, M. (2015). A review study of biosorption of heavy metals and comparison between different biosorbents. *Journal of Materials & Environmental Sciences*, *6*(5), 1386–1399.

Abdi, O., & Kazemi, M. A. (2015). Review study of biosorption of heavy metals and compari- son between different biosorbents. *Journal of Materials & Environmental Sciences*, *6*(5), 1386–1399.

Abdullah, H., & Wu, H. (2009). Biochar as a fuel: 1. Properties and grindability of biochars produced from the pyrolysis of mallee woodunder slow-heating conditions. *Energy & Fuels*, *23*(8), 4174–4181. doi:10.1021/ef900494t

Abigail, M. E. A., Samuel, M. S., & Chidambaram, R. (2015). Hexavalent chromium biosorption studies using *Penicillium griseofulvum* MSR1 a novel isolate from tannery effluent site: Box–Behnken optimization, equilibrium, kinetics and thermodynamic studies. *Journal of the Taiwan Institute of Chemical Engineers*, *49*, 156–164. doi:10.1016/j.jtice.2014.11.026

Abioye, O. P., Oyewole, O. A., Oyeleke, S. B., Adeyemi, M. O., & Orukotan, A. A. (2018). Biosorption of lead, chromium and cadmium in tannery effluent using indigenous microorganisms. *Brazilian Journal of Biological Sciences*, *5*(9), 25–32. doi:10.21472/bjbs.050903

Acharya, J., Kumar, U., & Rafi, P. M. (2018). Removal of Heavy Metal Ions from Wastewater by Chemically Modified Agricultural Waste Material as Potential Adsorbent-A Review. *International Journal of Current Engineering and Technology*, *8*(3). Advance online publication. doi:10.14741/ijcet/v.8.3.6

Acosta, J. A., Faz, A., Martínez-Martínez, S., Zornoza, R., Carmona, D. M., & Kabas, S. (2011). Multivariate statistical and GIS-based approach to evaluate heavy metals behavior in mine sites for future reclamation. *Journal of Geochemical Exploration*, *109*(1-3), 8–17. doi:10.1016/j.gexplo.2011.01.004

Adam, O. (2016). Removal of Resorcinol from Aqueous Solution by Activated Carbon: Isotherms, Thermodynamics and Kinetics. *American Chemical Science Journal*, *16*(1), 1–13. doi:10.9734/ACSJ/2016/27637

Addo, M. A., Darko, E. O., Gordon, C., Nyarko, B. J. B., Gbadago, J. K., Nyarko, E., Affum, H. A., & Botwe, B. O. (2012). Evaluation of heavy metals contamination of soil and vegetation in the vicinity of a cement factory in the Volta region, Ghana. *International Journal of Science and Technology*, *2*, 40–50.

Adegoke, K. A., Adesina, O. O., Okon-Akan, O. A., Adegoke, O. R., Olabintan, A. B., Ajala, O. A., Olagoke, H., Maxakato, N. W., & Bello, O. S. (2022). Sawdust-biomass based materials for sequestration of organic and inorganic pollutants and potential for engineering applications. *Current Research in Green and Sustainable Chemistry*, *5*, 100274. doi:10.1016/j.crgsc.2022.100274

Adegoke, K. A., Akinnawo, S. O., Adebusuyi, T. A., Ajala, O. A., Adegoke, R. O., Maxakato, N. W., & Bello, O. S. (2023). Modified biomass adsorbents for removal of organic pollutants: A review of batch and optimization studies. *International Journal of Environmental Science and Technology*, *20*(10), 11615–11644. doi:10.1007/s13762-023-04872-2

Adesuyi, A. A., Njoku, K. L., & Akinola, M. O. (2015). Assessment of Heavy Metals Pollution in Soils and Vegetation around Selected Industries in Lagos State, Nigeria. *Journal of Geoscience and Environment Protection*, *3*(7), 11–19. doi:10.4236/gep.2015.37002

Adewuyi, A. (2020). Chemically Modified Biosorbents and Their Role in the Removal of Emerging Pharmaceutical Waste in the Water System. *Water (Basel)*, *12*(6), 1551. doi:10.3390/w12061551

Adriano, D. C., Knox, A., Geebelen, W., Vangronsveld, J., & Mench, M. 2000. Univ of Georgia. Unpublished data.

Agboola, O. D., & Benson, N. U. (2021). Physisorption and Chemisorption Mechanisms Influencing Micro (Nano) Plastics-Organic Chemical Contaminants Interactions: A Review. *Frontiers in Environmental Science*, *9*, 678574. Advance online publication. doi:10.3389/fenvs.2021.678574

Agency for Toxic Substances and Disease Registry (ATSDR). (2000). *Toxicological Profile for Arsenic TP-92/09. Georgia*. Center for Disease Control.

Agency for Toxic Substances and Disease Registry. (1999). *Toxicological profile for mercury*. Academic Press.

Agency for Toxic Substances and Disease Registry. (2007a). *Toxicological profile for arsenic*. Academic Press.

Agency for Toxic Substances and Disease Registry. (2007b). *Toxicological profile for lead*. Academic Press.

Agency for Toxic Substances and Disease Registry. (2012). Toxicological profile for cadmium. Academic Press.

Aghili, S., & Golzary, A. (2023). Greening the earth, healing the soil: A comprehensive life cycle assessment of phytoremediation for heavy metal contamination. *Environmental Technology & Innovation*, *103241*, 103241. Advance online publication. doi:10.1016/j.eti.2023.103241

Ahalya, N., Kanamadi, R. D., & Ramachandra, T. V. (2006). Biosorption of iron(III) from aqueous solutions using the husk of Cicer arientinum. *Indian Journal of Chemical Technology*, *13*(2), 122–127.

Ahalya, N., Ramachandra, T. V., & Kanamadi, R. D. (2003). Bio sorption of heavy metals. *Research Journal of Chemistry and Environment*, *7*(4), 71–78.

Ahalya, N., Ramachandra, T. V., & Kanamadi, R. D. (2003). Biosorption of Heavy Metals. *Research Journal of Chemistry and Environment*, *7*, 71–79.

Ahemad, M., & Mulugeta, K. (2013). Recent Trends in Microbial Biosorption of Heavy Metals: A Review. *Biochemistry and Molecular Biology (New York, N.Y.)*, *1*(1), 19–26. doi:10.12966/bmb.06.02.2013

Ahluwalia, S. S., & Goyal, D. (2007). Microbial and plant derived biomass for removal of heavy metals from wastewater. *Bioresource Technology*, *98*(12), 2243–2257. doi:10.1016/j.biortech.2005.12.006 PMID:16427277

Ahmad, F. A. (2023). The use of agro-waste-based adsorbents as sustainable, renewable, and low-cost alternatives for the removal of ibuprofen and carbamazepine from water. *Heliyon*, *9*(6), e16449. doi:10.1016/j.heliyon.2023.e16449 PMID:37292321

Ahmad, F., Haydar, S., Bhatti, A., & Bari, J. (2014). Application of artificial neural network for the prediction of biosorption capacity of immobilized *Bacillus subtilis* for the removal of cadmium ions from aqueous solution. *Biochemical Engineering Journal*, *84*, 83–90. doi:10.1016/j.bej.2014.01.004

Ahmad, F., & Zaidi, S. (2021). Potential Use of Agro/Food Wastes as Biosorbents in the Removal of Heavy Metals. In *Emerging Contaminants*. IntechOpen. doi:10.5772/intechopen.94175

Ahmadi, H., Hafiz, S. S., Sharifi, H., Rene, N. N., Habibi, S. S., & Hussain, S. (2022). Low cost biosorbent (Melon Peel) for effective removal of Cu (II), Cd (II), and Pb (II) ions from aqueous solution. *Case Studies in Chemical and Environmental Engineering*, *6*, 100242. doi:10.1016/j.cscee.2022.100242

Ahmad, M., Lee, S. S., Lee, S. E., Al-Wabel, M. I., Tsang, D. C. W., & Ok, Y. S. (2017). Biochar-induced changes in soil properties affected immobilization/mobilization of metals/metalloids in contaminated soils. *Journal of Soils and Sediments*, *17*(3), 717–730. doi:10.1007/s11368-015-1339-4

Ahmaruzzaman, M. (2010). A review on the utilization of fly ash. *Progress in Energy and Combustion Science*, *36*(3), 327–363. doi:10.1016/j.pecs.2009.11.003

Ahmaruzzaman, M., & Gupta, V. K. (2011). Rice husk and its ash as lowcost adsorbents in water and wastewater treatment. *Industrial & Engineering Chemistry Research*, *50*(24), 13589–13613. doi:10.1021/ie201477c

Ahmed, F. M. (2001). *An overview of arsenic removal technologies in Bangladesh and India. Technologies for arsenic removal from drinking water.* Bangladesh University of Engineering and Technology (BUET) and United Nations University (UNU). http://archive.unu.edu/env/Arsenic/Ahmed.pdf

Ahmed, M. F. (2005). *Arsenic mitigation technologies in south and east Asia. In: Towards a more effective operational response.* Technical Report No. 31303, Water and Sanitation Program- South Asia and the World Bank.

Ahmed, M. B., Zhou, J. L., Ngo, H. H., & Guo, W. (2016). Insight into biochar properties and its cost analysis. *Biomass and Bioenergy*, *84*, 76–86. doi:10.1016/j.biombioe.2015.11.002

Ahmed, S. F. P., Kumar, P. S., Rozbu, M. R., Chowdhury, A. T., Nuzhat, S., Rafa, N., Mahlia, T. M. I., Ong, H. C., & Mofijur, M. (2022). Heavy metal toxicity, sources, and remediation techniques for contaminated water and soil. *Environmental Technology & Innovation*, *25*, 102114. doi:10.1016/j.eti.2021.102114

Ahn, C. K., Park, D., Woo, S. H., & Park, J. M. (2009). Removal of cationic heavy metal from aqueous solution by activated carbon impregnated with anionic surfactants. *Journal of Hazardous Materials*, *164*(2-3), 1130–1136. doi:10.1016/j.jhazmat.2008.09.036 PMID:19022570

Ajala, E. O., Ighalo, J. O., Ajala, M. A., Adeniyi, A. G., & Ayanshola, A. M. (2021). Sugarcane bagasse: A biomass sufficiently applied for improving global energy, environment and economic sustainability. *Bioresources and Bioprocessing*, *8*(1), 1–25. doi:10.1186/s40643-021-00440-z

Akar, T., Akar, T., Kaynak, Z., Anilan, B., Cabuk, A., Tabak, O., Demir, T., & Gedikbey, T. (2009). Removal of copper(II) ions from synthetic solution and real wastewater by the combined action of dried *Trametes versicolor* cells and montmorillonite. *Hydrometallurgy*, *97*(1-2), 98–104. doi:10.1016/j.hydromet.2009.01.009

Akbari, M., Hallajisani, A., Keshtkar, R., Shahbeig, H., & Ali, S. (2015). Equilibrium and kinetic study and modelling of Cu(II) and Co(II) synergistic biosorption from Cu(II)-Co(II) single and binary mixtures on brown algae *C. indica*. *Journal of Environmental Chemical Engineering*, *3*(1), 140–149. doi:10.1016/j.jece.2014.11.004

Åkesson, A., Bjellerup, P., Lundh, T., Lidfeldt, J., Nerbrand, C., Samsioe, G., ... Järup, L. (2006). Cadmium-induced effects on bone in a population-based study of women. *Environmental Health Perspectives*, *114*(6), 830–834. doi:10.1289/ehp.8763 PMID:16759980

Akhtar, K., Khalid, M., Akhtar, W., & Ghauri, M. (2009). Removal and recovery of uranium from aqueous solutions by Ca-alginate immobilized *Trichoderma harzianum*. *Bioresource Technology*, *100*(20), 4551–4558. doi:10.1016/j.biortech.2009.03.073

Akinhanmi, T. F., Ofudje, E. A., Adeogun, A. I., Aina, P., & Joseph, I. M. (2020). Orange peel as low-cost adsorbent in the elimination of Cd(II) ion: Kinetics, isotherm, thermodynamic and optimization evaluations. *Bioresources and Bioprocessing*, *7*(1), 34. doi:10.1186/s40643-020-00320-y

Akoto, O., Ephraim, J. H., & Darko, G. (2008). Heavy Metals Pollution in Surface Soils in the Vicinity of Abundant Railway Servicing Workshop in Kumasi, Ghana. *International Journal of Environmental Research, 2*(4), 359-364.

Akpomie, K. G., & Conradie, J. (2020). Biogenic and chemically synthesized Solanum tuberosum peel–silver nanoparticle hybrid for the ultrasonic aided adsorption of bromophenol blue dye. *Scientific Reports, 10*(1), 1–18. doi:10.1038/s41598-020-74254-y

Aksu, Z. (2001). Equilibrium and kinetic modelling of cadmium(II) biosorption by C. vulgaris in a batch system: Effect of temperature. *Separation and Purification Technology*, *21*(3), 285–294. doi:10.1016/S1383-5866(00)00212-4

Aksu, Z. (2005). Application of biosorption for the removal of organic pollutants: A review. *Process Biochemistry (Barking, London, England)*, *40*(3–4), 997–1026. doi:10.1016/j.procbio.2004.04.008

Akthar, M. N., Sastry, K. S., & Mohan, P. M. (1996). Mechanism of metal ion biosorption by fungal biomass. *Biometals*, *9*(1), 21–28. doi:10.1007/BF00188086

Al Arni, S. (2017). Comparison of slow and fast pyrolysis for converting biomass into fuel. *Renewable Energy*. Advance online publication. doi:10.1016/j.renene.2017.04.060

Alalwan, H. A., Kadhom, M. A., & Alminshid, A. H. (2020). Removal of heavy metals fromwastewater using agricultural byproducts. *Journal of Water Supply: Research & Technology - Aqua*, *69*(2), 99–112. doi:10.2166/aqua.2020.133

Albadarin, A. B., Mangwandi, C., Walker, G. M., Allen, S. J., Ahmad, M. N., & Kharaisheh, M. (2013). Influence of solution chemistry on Cr (VI) reduction and complexation onto date pits/tea waste biomaterials. *Journal of Environmental Management*, *114*, 190–201. doi:10.1016/j.jenvman.2012.09.017 PMID:23134975

Albatrni, H., Qiblawey, H., & Al-Marri, M. J. (2022). Walnut shell based adsorbents: A review study on preparation, mechanism, and application. *Journal of Water Process Engineering*, *45*, 102527. doi:10.1016/j.jwpe.2021.102527

Al-Enezi, G., Hamoda, M. F., & Fawzi, N. (2004). Ion exchange extraction of heavy metals from wastewater sludges. *Journal of Environmental Science and Health. Part A, Toxic/Hazardous Substances & Environmental Engineering*, *39*(2), 455–464. doi:10.1081/ESE-120027536 PMID:15027828

Alengebawy, A., Abdelkhalek, S. T., Qureshi, S. R., & Wang, M. Q. (2021). Heavy metals and pesticides toxicity in agricultural soil and plants: Ecological risks and human health implications. *Toxics*, *9*(3), 42. doi:10.3390/toxics9030042 PMID:33668829

Al-Homaidan, A., Alabdullatif, A., Al-Hazzani, A., Al-Ghanayem, A., & Alabbad, F. (2015). Adsorptive removal of cadmium ions by *Spirulina platensis* dry biomass. *Saudi Journal of Biological Sciences*, *22*(6), 795–800. doi:10.1016/j.sjbs.2015.06.010

Ali, M. M., Hossain, D., Al-Imran, A., Khan, M. S., Begum, M., & Osman, M. H. (2021). Environmental pollution with heavy metals: A public health concern. *Heavy Metals-Their Environmental Impacts and Mitigation*, 771-783.

Ali, A., Sihabudeen, M., & Hussain, Z. (2016). Biosorption of heavy metals by *Pseudomonas* bacteria. *International Research Journal of Engineering and Technology*, *3*(8), 1446–1450.

Ali, H., Khan, E., & Ilahi, I. (2019). Environmental chemistry and ecotoxicology of hazardous heavy metals: Environmental persistence, toxicity, and bioaccumulation. *Journal of Chemistry*, *2019*, 1–14. Advance online publication. doi:10.1155/2019/6730305

Ali, I. H., & Alrafai, H. A. (2016). Kinetic, isotherm and thermodynamic studies on biosorption of chromium(VI) by using activated carbon from leaves of *Ficus nitida*. *Chemistry Central Journal*, *10*(1), 1–7. doi:10.1186/s13065-016-0180-1 PMID:27252777

Ali, S., Abbas, Z., Rizwan, M., Zaheer, I. E., Yavaş, İ., Ünay, A., Abdel-DAIM, M., Bin-Jumah, M., Hasanuzzaman, M., & Kalderis, D. (1927). Application of floating aquatic plants in phytoremediation of heavy metals polluted water: A review. *Sustainability (Basel)*, *12*(5), 1927. Advance online publication. doi:10.3390/su12051927

Alloway, B. J. (2013). Sources of heavy metals and metalloids in soils. *Heavy metals in soils: trace metals and metalloids in soils and their bioavailability*, 11-50.

Alloway, B. J. (1995). Soil Processes and Behavior of Heavy Metals. In B. J. Alloway (Ed.), *Heavy Metals in Soils* (p. 368). Blackie Academic and Professional Publication. doi:10.1007/978-94-011-1344-1

Al-Qahtani, K. (2016). Water purification using different waste fruit cortexes for removal of heavy metals. *Journal of Taibah University for Science : JTUSCI*, *10*(5), 700–708. doi:10.1016/j.jtusci.2015.09.001

Al-Rashdi, B. A. M., Johnson, D. J., & Hilal, N. (2013). Removal of heavy metal ions bynanofiltration. *Desalination*, *315*, 2–17. doi:10.1016/j.desal.2012.05.022

Al-Rumaihi, A., Shahbaz, M., Mckay, G., Mackey, H., & Al-Ansari, T. (2022). A review of pyrolysis technologies and feedstock: A blending approach for plastic and biomass towards optimum biochar yield. *Renewable & Sustainable Energy Reviews*, *167*, 112715. doi:10.1016/j.rser.2022.112715

Alsawy, T., Rashad, E., El-Qelish, M., & Mohammed, R. H. (2022). A comprehensive review on the chemical regeneration of biochar adsorbent for sustainable wastewater treatment. *NPJ Clean Water*, *5*(1), 29. doi:10.1038/s41545-022-00172-3

Altowayti, W., Othman, N., Al-Gheethi, A., Dzahir, N., Asharuddin, S., Alshalif, A., Nasser, I., Tajarudin, H., & AL-Towayti, F. A. H. (2021). Adsorption of Zn^{2+} from synthetic wastewater using dried watermelon rind (D-WMR): An overview of nonlinear and linear regression and error analysis. *Molecules (Basel, Switzerland)*, *26*(20), 6176. doi:10.3390/molecules26206176

Amalina, F., Razak, A. S. A., Krishnan, S., Zularisam, A. W., & Nasrullah, M. (2022). A comprehensive assessment of the method for producing biochar, its characterization, stability, and potential applications in regenerative economic sustainability – A review. *Cleaner Materials*, *3*, 100045. doi:10.1016/j.clema.2022.100045

Amini, M., Younesi, H., Bahramifar, N., Lorestani, A. A. Z., Ghorbani, F., Daneshi, A., & Sharifzadeh, M. (2008). Application of response surface methodology for optimization of lead biosorption in an aqueous solution by *Aspergillus niger. Journal of Hazardous Materials*, *154*(1–3), 694–702. doi:10.1016/j.jhazmat.2007.10.114 PMID:18068898

Anaemene, I. A. (2012). The use of Candida sp. in the biosorption of heavy metals from indus- trial effluent. *European Journal of Experimental Biology*, *2*(3), 484–488.

Anastopoulos, I., & Kyzas, G. Z. (2015). Progress in batch bio-sorption of heavy metals onto algae. *Journal of Molecular Liquids*, *209*, 77–86. doi:10.1016/j.molliq.2015.05.023

Angulo-Bejarano, P. I., Puente-Rivera, J., & Cruz-Ortega, R. (2021). Metal and metalloid toxicity in plants: An overview on molecular aspects. *Plants*, *10*(4), 635. doi:10.3390/plants10040635 PMID:33801570

Anirudhan, T. S., & Unnithan, M. R. (2007). Arsenic (V) removal from aqueous solutions using an anion exchanger derived from coconut coir pith and its recovery. *Chemosphere*, *66*(1), 60–66. doi:10.1016/j.chemosphere.2006.05.031 PMID:16824580

Ansari, M. I., Masood, F., & Malik, A. (2011). Bacterial Biosorption: A Technique for Remediation of Heavy Metals. *Microbes and Microbial Technology*, 283-319.

Antony, J. (2003). A systematic methodology for Design of Experiments. In J. Antony (Ed.), *Design of Experiments for Engineers and Scientists* (pp. 29–43). Butterworth-Heinemann. doi:10.1016/B978-075064709-0/50005-3

Apostol, L. C., Ghinea, C., Alves, M., & Gavrilescu, M. (2016). Removal of Erythrosine B dye from water effluents using crop waste pumpkin seed hulls as adsorbent. *Desalination and Water Treatment*, *57*(47), 22585–22608. doi:10.1080/19443994.2015.1132477

Arias, J. Z., Reuter, T., Sabir, A., & Gilroyed, B. H. (2018). Ambient alkaline hydrolysis and anaerobic digestion as a mortality management strategy for whole poultry carcasses. *Waste Management (New York, N.Y.)*, *81*, 71–77. doi:10.1016/j.wasman.2018.09.049 PMID:30527045

Arief, V. O., Trilestari, K., Sunarso, J., Indraswati, N., & Ismadji, S. (2008). Recent progress on Biosorption of Heavy Metals from Liquids Using Low Cost Biosorbents: Characterization, Biosorption Parameters and Mechanical Studies. *Clean (Weinheim)*, *36*(12), 937–962. doi:10.1002/clen.200800167

Armstrong, R. (2023). Towards the microbial home: An overview of developments in next-generation sustainable architecture. *Microbial Biotechnology*, *16*(6), 1112–1130. doi:10.1111/1751-7915.14256 PMID:37070748

Arslan, B., Djamgoz, M. B., & Akün, E. (2017). Arsenic: A review on exposure pathways, accumulation, mobility and transmission into the human food chain. *Reviews of Environmental Contamination and Toxicology*, *243*, 27–51. PMID:28005215

Atafar, Z., Mesdaghinia, A., Nouri, J., Homaee, M., Yunesian, M., Ahmadimoghaddam, M., & Mahvi, A. H. (2010). Effect of fertilizer application on soil heavy metal concentration. *Environmental Monitoring and Assessment*, *160*(1-4), 83–89. doi:10.1007/s10661-008-0659-x PMID:19058018

Ayawei, N., Ebelegi, A. N., & Wankasi, D. (2017). Modelling and interpretation of adsorption isotherms. *Journal of Chemistry*, *2017*, 1–11. doi:10.1155/2017/3039817

Aydar, A. Y. (2018). Utilization of Response Surface Methodology in Optimization of Extraction of Plant Materials. In V. Silva (Ed.), *Statistical Approaches with Emphasis on Design of Experiments Applied to Chemical Processes*. IntechOpen. doi:10.5772/intechopen.73690

Ayele, A., Haile, S., Alemu, D., & Kamara, M. (2021). Comparative Utilization of Dead and Live Fungal Biomass for the Removal of Heavy Metal: A Concise Review. *TheScientificWorldJournal*, *2021*, 10. doi:10.1155/2021/5588111 PMID:33927581

Ayers, P. W. (2005). An elementary deviation of the hard/soft-acid/base principle. *The Journal of Chemical Physics*, *122*(14), 141102. Advance online publication. doi:10.1063/1.1897374 PMID:15847508

Ayub, ur Rehman, Umar, Zuhra, & Shabaan. (2022). Role of legumes in phytoremediation of heavy metals. In Advances in Legumes for Sustainable Intensification (pp. 345-360). Academic Press.

Azad, M. A. K., Amin, L., & Sidik, N. M. (2014). Genetically engineered organisms for bioremediation of pollutants in contaminated sites. *Chinese Science Bulletin*, *59*(8), 703–714. doi:10.1007/s11434-013-0058-8

Azhar, W., & Iram, S. (2014). kinetics, equilibrium and thermodynamics studies on biosorption of heavy metals by fungal biomass. *International Journal of Engineering Sciences & Emerging Technologies*, *7*(3), 693–700. doi:10.1007/s11356-022-22860-w

Bădescu, I. S., Bulgariu, D., Ahmad, I., & Bulgariu, L. (2018). Valorisation possibilities of exhausted biosorbents loaded with metal ions – A review. *Journal of Environmental Management*, *224*, 288–297. doi:10.1016/j.jenvman.2018.07.066 PMID:30055461

Bakan, B., Bernet, N., Bouchez, T., Boutrou, R., Choubert, J. M., Dabert, P., Duquennoi, C., Ferraro, V., García-Bernet, D., Gillot, S., Mery, J., Rémond, C., Steyer, J.-P., Trably, E., & Tremier, A. (2022). Circular Economy Applied to Organic Residues and Wastewater: Research Challenges. *Waste and Biomass Valorization*, *13*(2), 1267–1276. doi:10.1007/s12649-021-01549-0

Bakker, E. S., Pagès, J. F., Arthur, R., & Alcoverro, T. (2016). Assessing the role of large herbivores in the structuring and functioning of freshwater and marine angiosperm ecosystems. *Ecography*, *39*(2), 162–179. doi:10.1111/ecog.01651

Balali-Mood, M., Naseri, K., Tahergorabi, Z., Khazdair, M. R., & Sadeghi, M. (2021). Toxic mechanisms of five heavy metals: Mercury, lead, chromium, cadmium, and arsenic. *Frontiers in Pharmacology*, *227*, 643972. Advance online publication. doi:10.3389/fphar.2021.643972 PMID:33927623

Bala, S., Garg, D., Sridhar, K., Inbaraj, B. S., Singh, R., Kamma, S., Tripathi, M., & Sharma, M. (2023). Transformation of Agro-Waste into Value-Added Bioproducts and Bioactive Compounds: Micro/Nano Formulations and Application in the Agri-Food-Pharma Sector. *Bioengineering (Basel, Switzerland)*, *10*(2), 152. doi:10.3390/bioengineering10020152 PMID:36829646

Baldwin, D. R., & Marshal, W. J. (1999). Heavy metal poisoning and its laboratory investigation. *Annals of Clinical Biochemistry*, *36*(3), 267–300. doi:10.1177/000456329903600301 PMID:10376071

Balíková, K., Farkas, B., Matúš, P., & Urík, M. (2022). Prospects of Biogenic Xanthan and Gellan in Removal of Heavy Metals from Contaminated Waters. *Polymers*, *14*(23), 5326. Advance online publication. doi:10.3390/polym14235326

Balkis, N. (Ed.). (2012). *Water Pollution*. InTech. doi:10.5772/1418

Barakat, M. A. (2011). New trends in removing heavy metals from industrial wastewater. *Arabian Journal of Chemistry*, *4*(4), 361–377. doi:10.1016/j.arabjc.2010.07.019

Barros, M. S. D., Arroyo, P. A., & Silva, E. A. (2013). General Aspects of Aqueous Sorption Process in Fixed Beds. In H. Nakajima (Ed.), *Mass Transfer - Advances in Sustainable Energy and Environment Oriented Numerical Modeling.* IntechOpen. doi:10.5772/51954

Baselt, R. C., & Cravey, R. H. (1995). *Disposition of Toxic Drugs and Chemicals in Man* (4th ed.). Year Book Medical Publishers.

Bashir, A., Malik, L. A., Ahad, S., Manzoor, T., Bhat, M. A., Dar, G. N., & Pandith, A. H. (2018). Removal of heavy metal ions from aqueous system by ion-exchange and biosorption methods. *Environmental Chemistry Letters, 17*(2), 729–754. doi:10.1007/s10311-018-00828-y

Basu, A., Ali, S. S., Hossain, S. K. S., & Asif, M. (2022). A Review of the Dynamic Mathematical Modeling of Heavy Metal Removal with the Biosorption Process. *Processes (Basel, Switzerland), 10*(6), 1154. doi:10.3390/pr10061154

Batista, E. M. C. C., Shultz, J., Matos, T. T. S., Fornari, M. R., Ferreira, T. M., Szpoganicz, B., de Freitas, R. A., & Mangrich, A. S. (2018). Effect of surface and porosity of biochar on water holding capacity aiming indirectly at preservation of the Amazon biome. *Scientific Reports, 81*(1), 1–9. doi:10.1038/s41598-018-28794-z PMID:30013110

Batool, F., Akbar, J., Iqbal, S., Noreen, S., & Bukhari, S. N. A. (2018). Study of isothermal, kinetic, and thermodynamic parameters for adsorption of cadmium: An overview of linear and nonlinear approach and error analysis. *Bioinorganic Chemistry and Applications, 2018*, e3463724. doi:10.1155/2018/3463724 PMID:30065756

Battarbee, R. W., Anderson, N. J., Appleby, P. G., Flower, R. J., Fritz, S. C., Haworth, E. Y., ... Stevenson, A. C. (1988). *Lake acidification in the United Kingdom 1800-1986.* Academic Press.

Bavariani, M. Z., Ronaghi, A., & Ghasemi, R. (2019). Influence of pyrolysis temperatures on FTIR analysis, nutrient bioavailability, and agricultural use of poultry manure chars. *Communications in Soil Science and Plant Analysis, 50*(4), 402–411. Advance online publication. doi:10.1080/00103624.2018.1563101

Becker, M., Ahn, K., Bacher, M., Xu, C., Sundberg, A., Willför, S., Rosenau, T., & Potthast, A. (2021). Comparative hydrolysis analysis of cellulose samples and aspects of its application in conservation science. *Cellulose (London, England), 28*(13), 8719–8734. doi:10.1007/s10570-021-04048-6 PMID:34316103

Beetul, K., Gopeechund, A., Kaullysing, D., Moorgawa, S. M., Puchooa, D., & Bhagooli, R. (2016). Challenges and opportunities in the present era of marine algal applications. *Algae-Organisms for Imminent Biotechnology*, 40.

Begum, S. A. S., Tharakeswar, Y., Kalyan, Y., & Naidu, G. R. (2015). Biosorption of Cd (II), Cr (VI) & Pb (II) from aqueous solution using *Mirabilis jalapa* as adsorbent. *Journal of Encapsulation and Adsorption Sciences, 5*(2), 2. doi:10.4236/jeas.2015.52007

Behloul, M., Lounici, H., Abdi, N., Drouiche, N., & Mameri, N. (2017). Adsorption study of metribuzin pesticide on fungus Pleurotus mutilus. *International Biodeterioration & Biodegradation, 119*, 687–695. doi:10.1016/j.ibiod.2016.07.005

Bengtsson-Palme, J., Kristiansson, E., & Larsson, D. J. (2018). Environmental factors influencing the development and spread of antibiotic resistance. *FEMS Microbiology Reviews, 42*(1), fux053. doi:10.1093/femsre/fux053 PMID:29069382

Beni, A. A., & Esmaeili, A. (2020). Biosorption, an efficient method for removing heavy metals from industrial effluents: A Review. *Environmental Technology & Innovation, 17*, 100503. doi:10.1016/j.eti.2019.100503

Benila, M., & Sumithra, P. (2017). Optimization of Chromium Biosorption by Fungal Adsorbent, *Trichoderma sp.* BSCR02 and its Desorption Studies. *Journal of Biosciences, 24*(2), 65–71. https://doi.org/.hjb.2017.08.005 doi:10.1016/j

Bhakta, J. N., Ohnishi, K., Munekage, Y., Iwasaki, K., & Wei, M. Q. (2012). Characterization of lactic acid bacteria-based probiotics as potential heavy metal sorbents. *Journal of Applied Microbiology, 112*(6), 1193–1206. doi:10.1111/j.1365-2672.2012.05284.x PMID:22404232

Bharat, S., & Manchanda, D. (2017). Efficient bio-adsorbents for removal of heavy metals from water: A review. *International Journal of Chemical Studies, 5*(4), 1691–1694.

Bhateria, R., & Dhaka, R. (2019). Optimization and statistical modelling of cadmium biosorption process in aqueous medium by *Aspergillus niger* using response surface methodology and principal component analysis. *Ecological Engineering, 135*, 127–138. doi:10.1016/j.ecoleng.2019.05.010

Bhateria, R., & Jain, D. (2016). Water quality assessment of lake water: A review. *Sustainable Water Resources Management, 2*(2), 161–173. doi:10.1007/s40899-015-0014-7

Bhatia, L., Jha, H., Sarkar, T., & Sarangi, P. K. (2023). Food Waste Utilization for Reducing Carbon Footprints towards Sustainable and Cleaner Environment: A Review. *International Journal of Environmental Research and Public Health, 20*(3), 2318. doi:10.3390/ijerph20032318 PMID:36767685

Bhurachevskaya, M., Minkina, T., & Bauer, T. (2023). Fabrication of biochar derived from different types of feedstocks as an efficient adsorbent for soil heavy metal removal. *Scientific Reports, 13*(1), 2020. Advance online publication. doi:10.1038/s41598-023-27638-9 PMID:36737633

Bilal, M., Hafiz, M. N., Tahir, R., Ali, R., Faran, N., & Juan Eduardo, S. H. (2018). Bio sorption interplay between marine algae and potentially toxic elements-a review. *Marine Drugs, 16*(2), 1–16. doi:10.3390/md16020065 PMID:29463058

Bilal, M., Ihsanullah, I., Younas, M., & Shah, M. U. H. (2021). Recent advances in applications of low-cost adsorbents for the removal of heavy metals from water: A critical review. *Separation and Purification Technology, 278*, 119510. doi:10.1016/j.seppur.2021.119510

Bilos, C., Colomobo, J. C., Skorupka, C. N., & Rodriguez Presa, M. J. (2001). Source distribution and variability of airborne trace metals in La Plata city area, Argentina. *Environmental Pollution, 111*(1), 149–158. doi:10.1016/S0269-7491(99)00328-0 PMID:11202709

Bingol, D., Tekin, N., & Alkan, M. (2010). Brilliant Yellow dye adsorption onto sepiolite using a full factorial design. *Applied Clay Science, 50*(3), 315–321. doi:10.1016/j.clay.2010.08.015

Biswas, G., Kumari, M., Adhikari, K., & Dutta, S. (2017). Application of response surface methodology for optimization of biosorption of fluoride from groundwater using *Shorea robusta* flower petal. *Applied Water Science, 7*(8), 4673–4690. doi:10.1007/s13201-017-0630-5

Biswas, S., Bal, M., Behera, S., Sen, T., & Meikap, B. (2019). Process optimization study of Zn^{2+} adsorption on biochar-alginate composite adsorbent by response surface methodology (RSM). *Water (Basel), 11*(2), 325. doi:10.3390/w11020325

Bittihn, P., Didovyk, A., Tsimring, L. S., & Hasty, J. (2020). Genetically engineered control of phenotypic structure in microbial colonies. *Nature Microbiology, 5*(5), 697–705. doi:10.1038/s41564-020-0686-0 PMID:32284568

Blaga, A. C., Tanasă, A. M., Cimpoesu, R., Tataru-Farmus, R.-E., & Suteu, D. (2022). Biosorbents Based on Biopolymers from Natural Sources and Food Waste to Retain the Methylene Blue Dye from the Aqueous Medium. *Polymers, 14*(13), 2728. doi:10.3390/polym14132728 PMID:35808773

Blazquez, G., Hernainz, F., Calero, M., Martın-Lara, M.A., & Tenorio, G. (2009). The effect of pH on the biosorption of Cr (III) and Cr (VI) with olive stone. *Chemical Engineering Journal, 148*(2-3), 473–479. https://doi.org/.2008.09.026 doi:10.1016/j.cej

Błońska, E., Lasota, J., Szuszkiewicz, M., Łukasik, A., & Klamerus-Iwan, A. (2016). Assessment of forest soil contamination in Krakow surroundings in relation to the type of stand. *Environmental Earth Sciences*, *75*(16), 1–15. doi:10.1007/s12665-016-6005-7

Bobaker, A. M., Alakili, I., Sarmani, S. B., Al-Ansari, N., & Yaseen, Z. M. (2019). Determination and assessment of the toxic heavy metal elements abstracted from the traditional plant cosmetics and medical remedies: Case study of Libya. *International Journal of Environmental Research and Public Health*, *16*(11), 1957. doi:10.3390/ijerph16111957 PMID:31159472

Bolan, N., Kunhikrishnan, A., Thangarajan, R., Kumpiene, J., Park, J., Makino, T., Kirkham, M. B., & Scheckel, K. (2014). Remediation of heavy metal (loid) s contaminated soils–to mobilize or to immobilize? *Journal of Hazardous Materials*, *266*, 141–166. doi:10.1016/j.jhazmat.2013.12.018 PMID:24394669

Boshagh, F., & Rostami, K. (2020). A review of application of experimental design techniques related to dark fermentative hydrogen production. *Journal of Renewable Energy and Environment*, *7*(2), 27–42.

Botha, E., Smith, N., Hlabano-Moyo, B., & Bladergroen, B. J. (2021). The effect of slurry wet mixing time, thermal treatment and method of electrode preparation on membrane capacitive deionization performance. *Processes (Basel, Switzerland)*, *9*(1), 1. doi:10.3390/pr9010001

Brady, D., Stoll, A. D., Starke, L., & Duncan, J. R. (1994). Chemical and enzymatic extraction of Heavy metal binding polymers from isolated cell wall of Saccharomyces cerevisiae. *Biotechnology and Bioengineering*, *44*(3), 297–302. doi:10.1002/bit.260440307 PMID:18618746

Brathwaite, R. L., & Rabone, S. D. C. (1995). Heavy metals sulphide deposits and geochemical surveys for heavy metals in New Zealand. *Journal of the Royal Society of New Zealand*, *15*(4), 363–370. doi:10.1080/03036758.1985.10421713

Bridgwater, A. V., Carson, P., & Coulson, M. (2007). A comparison of fast and slow pyrolysis liquids from mallee. *International Journal of Global Energy Issues*, *27*(2), 204–216. doi:10.1504/IJGEI.2007.013655

Briffa, J., Sinagra, E., & Blundell, R. (2020). Heavy metal pollution in the environment and their toxicological effects on humans. *Heliyon*, *6*(9), e04691. Advance online publication. doi:10.1016/j.heliyon.2020.e04691 PMID:32964150

Brown, M. J., & Margolis, S. (2012). Lead in drinking water and human blood lead levels in the United States. In The Morbidity and Mortality Weekly Report (MMWR). Washington, DC: Center for Disease Control and Prevention (CDC).

Brown, A. L., Dayton, D. C., Nimlos, M. R., & Daily, J. W. (2001). Characterization of biomass pyrolysis vapours with molecular beam, single photon ionization time-of-flight mass spectrometry. *Chemosphere*, *42*(5-7), 663–669. doi:10.1016/S0045-6535(00)00240-X PMID:11219692

Bulgariu, L., Bulgariu, D., & Rusu, C. (2015). Marine Algae Biomass for Removal of Heavy Metal Ions. In S.-K. Kim (Ed.), *Springer Handbook of Marine Biotechnology* (pp. 611–648). Springer. doi:10.1007/978-3-642-53971-8_25

Burakov, A. E., Galunin, E. V., Burakova, I. V., Kucherova, A. E., Agarwal, S., Tkachev, A. G., & Gupta, V. K. (2018). Adsorption of heavy metals on conventional and nanostructured materials for wastewater treatment purposes: A review. *Ecotoxicology and Environmental Safety*, *148*, 702–712. doi:10.1016/j.ecoenv.2017.11.034 PMID:29174989

Calderón, O. A. R., Abdeldayem, O. M., Pugazhendhi, A., & Rene, E. R. (2020). Current Updates and Perspectives of Biosorption Technology: An Alternative for the Removal of Heavy Metals from Wastewater. *Current Pollution Reports*.

Calero, M., Perez, A., Blazquez, G., Ronda, A., & Martin-Lara, M. (2013). Characterization of chemically modified biosorbents from olive tree pruning for the biosorption of lead. *Ecological Engineering*, *58*, 344–354. doi:10.1016/j.ecoleng.2013.07.012

Carolin, C. F., Kumar, P. S., Saravanan, A., Joshiba, G. J., & Naushad, M. (2017). Efficient techniques for the removal of toxic heavy metals from aquatic environment: A review. *Journal of Environmental Chemical Engineering*, *5*(3), 2782–2799. doi:10.1016/j.jece.2017.05.029

Carpi, A. (Ed.). (2011). *Progress in molecular and environmental bioengineering: from analysis and modeling to technology applications*. BoD–Books on Demand.

Carris, L. M., Little, C. R., & Stiles, C. M. (2012). Introduction to fungi. *The Plant Health Instructor*, 48.

Carrizales, L., Razo, I., Tellez-Hernandez, J. I., Torres-Nerio, R., Torres, A., Batres, L. E., Cubillas, A.-C., & Diaz-Barriga, F. (2006). Exposure to arsenic and lead of children living near a Cu-smelter in San Luis Potosi, Mexico: Importance of soil contamination for exposure of children. *Environmental Research*, *101*(1), 1–10. doi:10.1016/j.envres.2005.07.010 PMID:16171795

Cases & de Lorenzo. (2005). *Genetically modified organisms for the environment: stories of success and failure and what we have learned from them.* Academic Press.

Centers for Disease Control (CDC). (1991). *Preventing Lead Poisoning in Young children: A statement by the Centers for Disease Control*. CDC.

Centers for Disease Control and Prevention (CDC). (2001). Managing Elevated Blood Lead Levels Among Young Children: Recommendations From the Advisory Committee on Childhood Lead Poisoning Prevention. CDC.

Chaiwong, K., Kiatsiriroat, T., Vorayos, N., & Thararax, C. (2013). Study of bio-oil and bio-char production from algae by slow pyrolysis. *Biomass and Bioenergy*, *56*, 600–606. doi:10.1016/j.biombioe.2013.05.035

Chakraborty, S., Chowdhury, S., & Das Saha, P. (2012). Insight into biosorption equilibrium, kinetics and thermodynamics of crystal violet onto *Ananas comosus* (pineapple) leaf powder. *Applied Water Science*, *2*(2), 135–141. doi:10.1007/s13201-012-0030-9

Chandel, A.K., da Silva, S.S., Carvalho, W., & Singh, O.V. (2011). Sugarcane bagasse and leaves: foreseeable biomass of biofuel and bio-products. *Journal of Chemical & Biotechnology*. doi:10.1002/jctb.2742

Chang, C., & Li, R. (2019). Agricultural waste. *Water Environment Research*, *91*(10), 1150–1167. doi:10.1002/wer.1211 PMID:31433884

Chang, J. S., & Hong, J. (1994). Biosorption of mercury by the inactivated cells of Pseudomonas aeruinosa. *Biotechnology and Bioengineering*, *44*(8), 999–1006. doi:10.1002/bit.260440817 PMID:18618919

Chaturvedi, S., & Dave, P. N. (2012). Removal of iron for safe driking water. *Desalination*, *303*, 1–11. doi:10.1016/j.desal.2012.07.003

Cheng, J., Gao, J., Zhang, J., Yuan, W., Yan, S., Zhou, J., Zhao, J., & Feng, S. (2021). Optimization of hexavalent chromium biosorption by *Shewanella putrefaciens* using the Box-Behnken Design. *Water, Air, and Soil Pollution*, *232*(3), 92. doi:10.1007/s11270-020-04947-7

Cheng, U., Hao, Z., Hao, C., Deng, Y., Li, X., Li, K., & Zhao, Y. (2019). A Review of Modification of Carbon Electrode Material in Capacitive Deionization. *RSC Advances*, *9*(42), 24401–24419. doi:10.1039/C9RA04426D PMID:35527893

Chen, S., Qin, C., Wang, T., Chen, F., Li, X., Hou, H., & Zhou, M. (2019). Study on the adsorption of dyestuffs with different properties by sludge-rice husk biochar: Adsorption capacity, isotherm, kinetic, thermodynamics and mechanism. *Journal of Molecular Liquids*, *285*, 62–74. doi:10.1016/j.molliq.2019.04.035

Chen, X., Tian, Z., Cheng, H., Xu, G., & Zhou, H. (2021). Assorption process and mechanism of heavy metal ions by di ff erent components of cells, using yeast (Pichia pastoris) and Cu^{2+} as biosorption models. *RSC Advances*, *28*(28), 17080–17091. doi:10.1039/D0RA09744F PMID:35479686

Chen, Y., Song, R., Yang, R. F., Wu, N., & Li, J. (2014). A novel dopamine D3 receptor antagonist YQA14 inhibits methamphetamine self-administration and relapse to drug-seeking behaviour in rats. *European Journal of Pharmacology*, *743*, 126–132. doi:10.1016/j.ejphar.2014.09.026 PMID:25261038

Chen, Z., Ahmed, I., Osman, D., Rooney, W., Oh, W., & Yap, P. (2023). Remediation of Heavy Metals in Polluted Water by Immobilized Algae: Current Applications and Future Perspectives. *Sustainability (Basel)*, *15*(6), 5128. doi:10.3390/su15065128

Chheang, L., Thongkon, N., Sriwiriyarat, T., & Thanasupsin, S. P. (2021). Heavy metal contamination and human health implications in the Chan Thnal reservoir, Cambodia. *Sustainability (Basel)*, *13*(24), 13538. doi:10.3390/su132413538

Chibuike, G. U., & Obiora, S. C. (2014). Heavy Metal Polluted Soils: Effect on Plants and Bioremediation Methods. *Applied and Environmental Soil Science*, *2014*, 752708. Advance online publication. doi:10.1155/2014/752708

Chiririwa, H., Matthews, T., Nyoni, B., Majoni, S., & Naidoo, B. (2017). Adorption of lead and copper by carbon black and sodium bentonite composite material: A study on adsorption isotherms and kinetics. *Asian Journal of Chemistry*, *29*(12), 2761–2766. doi:10.14233/ajchem.2017.20877

Chojnacka, K. (2006). Biosorption of Cr(III) by wheat straw and grass: A systematic characteristics of new biosorbents. *Polish Journal of Environmental Studies*, *15*(6), 845–852.

Chojnacka, K., Chojnacka, A., & Gorecka, H. (2005). Biosorption of Cr(III), Cd(II) and Cu(II) ions by blue-green algae *Spirulina Sp. Chemosphere*, *59*, 75–84. doi:10.1016/j.chemosphere.2004.10.005 PMID:15698647

Choong, G., Liu, Y., & Templeton, D. M. (2014). Interplay of calcium and cadmium in mediating cadmium toxicity. *Chemico-Biological Interactions*, *211*, 54–65. doi:10.1016/j.cbi.2014.01.007 PMID:24463198

Choong, T. S. Y., Chuaha, T. G., Robiaha, Y., Koaya, F. L. G., & Azni, I. (2007). Arsenic toxicity, health hazards and removal techniques from water: An overview. *Desalination*, *217*(1-3), 139–166. doi:10.1016/j.desal.2007.01.015

Cho, W., Kang, S. B., Kim, S., Yun, Y., & Won, S. (2016). Reusable polyethylenimine-coated polysulfone/bacterial biomass composite fibre biosorbent for recovery of Pd(II) from acidic solutions. *Chemical Engineering Journal*, *302*, 545–551. doi:10.1016/j.cej.2016.05.091

Chowdhary, P., Bharagava, R. N., Mishra, S., & Khan, N. (2020). *Role of industries in water scarcity and its adverse effects on environment and human health. In Environmental Concerns and Sustainable Development* (Vol. 1). Air, Water and Energy Resources.

Christmas, H., & Srivastava, J. (2019). *Public health approaches in policing*. Public Health England and College of Policing.

Ciobanu, A.-A., Vasile, G., & Bulgariu, L. (2023). Biosorption of Cu(II) ions on algae biomass and derived biochar. *Environmental Engineering and Management Journal*, *22*(5), 899–907. doi:10.30638/eemj.2023.072

Cozma, P., Hlihor, R.-M., Rosca, M., Simion, I.-M., Apostol, M., & Gavrilescu, M. (2021). Progress in microbial bioremediation of industrial effluents using different bioreactors design: An Overview. *2021 International Conference on E-Health and Bioengineering (EHB)*, 1–4. 10.1109/EHB52898.2021.9657749

Cozmuta, L. M., Cozmuta, A. M., Peter, A., Nicula, C., Nsimba, E. B., & Tutu, H. (2012). The influence of pH on the adsorption of lead by Na clinoptilolite: Kinetic and equilibrium studies. *Water S.A.*, *38*(2), 269–278. doi:10.4314/wsa.v38i2.13

Crini, G. (2005). Recent developments in polysaccharide-based materials used as adsorbents in wastewater treatment. *Progress in Polymer Science*, *30*(1), 38–70. doi:10.1016/j.progpolymsci.2004.11.002

Cui, Y., Zhu, Y., Zhai, R., Huang, Y., Qiu, Y., & Liang, J. (2005). Exposure to metal mixtures and human health impacts in a contaminated area in Nanning, China. *Environment International*, *31*(6), 784–790. doi:10.1016/j.envint.2005.05.025 PMID:15979144

Dai, Y., Sun, Q., Wang, W., Lu, L., Liu, M., Li, J., Yang, S., Sun, Y., Zhang, K., Xu, J., Zheng, W., Hu, Z., Yang, Y., Gao, Y., Chen, Y., Zhang, X., Gao, F., & Zhang, Y. (2018). Utilizations of agricultural waste as adsorbent for the removal of contaminants: A review. *Chemosphere*, *211*, 235–253. doi:10.1016/j.chemosphere.2018.06.179 PMID:30077103

Daoud, J. I. (2017). Multicollinearity and regression analysis. *Journal of Physics: Conference Series*, *949*, 012009. doi:10.1088/1742-6596/949/1/012009

Dash, H. R., Mangwani, N., & Das, S. (2014). Characterization and potential application in mercury bioremediation of highly mercury- resistant marine bacterium Bacillus thuringiensis PW-05. *Environmental Science and Pollution Research International*, *21*(4), 2642–2653. doi:10.1007/s11356-013-2206-8 PMID:24114385

Das, N., Vimala, R., & Karthika, P. (2008). Biosorption of heavy metals - An overview. *Indian Journal of Biotechnology*, *7*(2), 159–169.

Das, N., Vimala, R., & Karthika, P. (2008). Biosorption of heavy metals – An overview. *Indian Journal of Biotechnology*, *7*, 159–169.

Das, T. K., & Poater, A. (2021). Review on the use of heavy metal deposits from water treatment waste towards catalytic chemical synthesis. *International Journal of Molecular Sciences*, *22*(24), 13383. doi:10.3390/ijms222413383 PMID:34948184

Daurov, D., Zhambakin, K., & Shamekova, M. (2023). Phytoremediation as a way to clean technogenically polluted areas of Kazakhstan. *Brazilian Journal of Biology*, *83*, e271684. doi:10.1590/1519-6984.271684 PMID:37222372

Davison, J. (2005). Risk mitigation of genetically modified bacteria and plants designed for bioremediation. *Journal of Industrial Microbiology & Biotechnology*, *32*(11-12), 639–650. doi:10.1007/s10295-005-0242-1 PMID:15973534

Davis, T. A., Volesky, B., & Mucci, A. (2003). A review of the biochemistry of heavy metal biosorption by brown algae. *Water Research*, *37*(18), 4311–4330. doi:10.1016/S0043-1354(03)00293-8 PMID:14511701

Davis, T. A., Volesky, B., & Vieira, R. (2000). Sargassum seaweed as biosorbent for heavy metals. *Water Research*, *34*(17), 4270–4278. doi:10.1016/S0043-1354(00)00177-9

De Flora, S., Bagnasco, M., Serra, D., & Zanacchi, P. (1990). Genotoxicity of chromium compounds: A review. *Mutation Research*, *238*(2), 99–172. doi:10.1016/0165-1110(90)90007-X PMID:2407950

De Gisi, S., Lofrano, G., Grassi, M., & Notarnicola, M. (2016). Characteristics and adsorption capacities of low-cost sorbents for wastewater treatment: A review. *Sustainable Materials and Technologies*, *9*, 10–40. doi:10.1016/j.susmat.2016.06.002

de Oliveira, L. G., de Paiva, A. P., Balestrassi, P. P., Ferreira, J. R., da Costa, S. C., & da Silva Campos, P. H. (2019). Response surface methodology for advanced manufacturing technology optimization: Theoretical fundamentals, practical guidelines, and survey literature review. *International Journal of Advanced Manufacturing Technology*, *104*(5), 1785–1837. doi:10.1007/s00170-019-03809-9

Deckers, M., De Loose, M., Papazova, N., Deforce, D., Fraiture, M.-A., & Nancy, H. C. (2022). Roosens. "First monitoring for unauthorized genetically modified bacteria in food enzymes from the food market.". *Food Control*, *135*, 108665. doi:10.1016/j.foodcont.2021.108665

Deering, T. B., Dickson, E. R., Fleming, C. R., Geall, M. G., McCall, J. T., & Baggenstoss, A. H. (1977). Effect of d-penicillamine on copper retention in patients with primary biliary cirrhosis. *Gastroenterology*, *72*(6), 1208–1212. doi:10.1016/S0016-5085(77)80012-7 PMID:870369

De, J., Ramaiah, N., & Vardanyan, L. (2008). Detoxification of toxic heavy metals by marine bacteria highly resistant to mercury. *Marine Biotechnology (New York, N.Y.)*, *10*(4), 471–477. doi:10.1007/s10126-008-9083-z PMID:18288535

Demiral, H., & Gündüzoğlu, G. (2010). Removal of nitrate from aqueous solutions by activated carbon prepared from sugar beet bagasse. *Bioresource Technology*, *101*(6), 1675–1680. doi:10.1016/j.biortech.2009.09.087 PMID:19854640

Demirbas, A. (2009). Biorefineries: Current activities and future developments. *Energy Conversion and Management*, *50*(11), 2782–2801. doi:10.1016/j.enconman.2009.06.035

Deng, L., Su, Y., Su, H., Wang, X., & Zhu, X. (2007). Sorption and desorption of lead(II) from wastewater by green algae Cladophora fascicularis. *Journal of Hazardous Materials*, *143*(1-2), 220–225. doi:10.1016/j.jhazmat.2006.09.009 PMID:17049733

Deng, L., Zhu, X., Su, Y., Su, H., & Wang, X. (2008). Biosorption and desorption of Cd^{2+} from wastewater by dehydrated shreds of Cladophora fascicularis. *Chinese Journal of Oceanology and Limnology*, *26*(1), 45–49. doi:10.1007/s00343-008-0045-0

Desta, M. B. (2013). Batch sorption experiments: Langmuir and Freundlich isotherm studies for the adsorption of textile metal ions onto teff straw (*Eragrostis tef*) agricultural waste. *Journal of Thermodynamics*, *2013*, e375830. doi:10.1155/2013/375830

Dethier, E. N., Sartain, S. L., & Lutz, D. A. (2019). Heightened levels and seasonal inversion of riverine suspended sediment in a tropical biodiversity hot spot due to artisanal gold mining. *Proceedings of the National Academy of Sciences of the United States of America*, *116*(48), 23936–23941. doi:10.1073/pnas.1907842116 PMID:31712422

Dey, S., Haripavan, N., Basha, S. R., & Babu, G. V. (2021). Removal of ammonia and nitrates from contaminated water by using solid waste bio-adsorbents. *Current Research in Chemical Biology*, *1*, 100005. doi:10.1016/j.crchbi.2021.100005

Dhankhar, R., & Hooda, A. (2011). Fungal biosorption – an alternative to meet the challenges of heavy metal pollution in aqueous solutions. *Environmental Technology*, *32*(5), 467–491. doi:10.1080/09593330.2011.572922 PMID:21877528

Dhyani, V., & Bhaskar, T. (2017). A comprehensive review on the pyrolysis of lignocellulosic biomass. *Renewable Energy*, *129*, 695–716. doi:10.1016/j.renene.2017.04.035

Diale, P. P., Hildebrandt, D., Glasser, D., Matambo, T. S., & Makgato, S. S. (2023). An analysis of the processes, kinetics and equilibrium of iron's biosorption on immobilized green microalgae. *South African Journal of Chemical Engineering*, *45*, 210–220. doi:10.1016/j.sajce.2023.05.007

Díaz, M. J., Ruiz-Montoya, M., Palma, A., & de-Paz, M.-V. (2021). Thermogravimetry Applicability in Compost and Composting Research: A Review. *Applied Sciences (Basel, Switzerland)*, *11*(4), 1692. doi:10.3390/app11041692

Dilebo, W. B., Anchiso, M. D., Kidane, T. T., & Ayalew, M. E. (2023). Assessment of selected heavy metals concentration level of drinking water in Gazer town and selected Kebele, South Ari district, Southern Ethiopia. *International Journal of Analytical Chemistry*, *2023*, 1–12. Advance online publication. doi:10.1155/2023/1524850 PMID:37180549

Ding, C., Zhang, T., Wang, X., Zhou, F., Yang, Y., & Yin, Y. (2013). Effects of soil type and genotype on lead concentration in rootstalk vegetables and the selection of cultivars for food safety. *Journal of Environmental Management*, *122*, 8–14. doi:10.1016/j.jenvman.2013.02.026 PMID:23542226

Ding, X., Tan, X., Hu, N., Li, Y., Wang, Y.-D., & Tan, Y. (2012). Removal and recovery of uranium(VI) from aqueous solutions by immobilized *Aspergillus niger* powder beads. *Bioprocess and Biosystems Engineering*, *35*(9), 1567–1576. doi:10.1007/s00449-012-0747-8

Ding, Y., Jing, D., Gong, H., Zhou, I., & Yang, X. (2012). Biosorption of aquatic Cd(II) by unmodified rice straw. *Bioresource Technology*, *114*, 20–25. doi:10.1016/j.biortech.2012.01.110 PMID:22445266

Dixit, S., & Hering, J. G. (2003). Comparison of arsenic (V) and arsenic (III) sorption onto iron oxide minerals: Implications for arsenic mobility. *Environmental Science & Technology*, *37*(18), 4182–4189. doi:10.1021/es030309t PMID:14524451

Domard, A. (1987). pH and c.d. measurements on a fully deacetylated chitosan: Application to CuII—polymer interactions. *International Journal of Biological Macromolecules*, *9*(2), 98–104. doi:10.1016/0141-8130(87)90033-X

Domingo, J. L. (1989). Cobalt in the environment and its toxicological implications. *Reviews of Environmental Contamination and Toxicology*, *108*, 105–132. doi:10.1007/978-1-4613-8850-0_3 PMID:2646660

Dopp, E., Hartmann, L. M., Florea, A. M., Rettenmier, A. W., & Hirner, A. V. (2004). Environmental distribution, analysis, and toxicity of organometal (loid) compounds. *Critical Reviews in Toxicology*, *34*(3), 301–333. doi:10.1080/10408440490270160 PMID:15239389

Doyle, R. J., Matthews, T. H., & Streips, U. N. (1980). Chemical basis for selectivity of metal ions by the Bacillus subtilis cell wall. *Journal of Bacteriology*, *143*(1), 471–480. doi:10.1128/jb.143.1.471-480.1980 PMID:6772632

Dua, M., Singh, A., Sethunathan, N., & Johri, A. (2002). Biotechnology and bioremediation: Successes and limitations. *Applied Microbiology and Biotechnology*, *59*(2-3), 143–152. doi:10.1007/s00253-002-1024-6 PMID:12111139

Dube, D., Parekh, C. T., & Nyoni, B. (2016). Removal of chromium and nickel from electroplating wastewater using magnetite particulate adsorbent: (1) Effect of pH, contact time and dosage, (2) Adsorption isotherms and kinetics. *Modern Applied Science*, *10*(7), 222–232. doi:10.5539/mas.v10n7p222

Dudka, S., & Adriano, D. C. (1997). Environmental impacts of metal ore mining and processing: A review. *Journal of Environmental Quality*, *26*(3), 590–602. doi:10.2134/jeq1997.00472425002600030003x

Duque-Acevedo, M., Belmonte-Ureña, L. J., Cortés-García, F. J., & Camacho-Ferre, F. (2020). Agricultural waste: Review of the evolution, approaches and perspectives on alternative uses. *Global Ecology and Conservation*, *22*, e00902. doi:10.1016/j.gecco.2020.e00902

Duque, E., Udaondo, Z., Molina, L., de la Torre, J., Godoy, P., & Ramos, J. L. (2022). Providing octane degradation capability to Pseudomonas putida KT2440 through the horizontal acquisition of oct genes located on an integrative and conjugative element. *Environmental Microbiology Reports*, *14*(6), 934–946. doi:10.1111/1758-2229.13097 PMID:35651318

Dusengemungu, L., Kasali, G., Gwanama, C., & Ouma, K. O. (2020). Recent Advances in Biosorption of Copper and Cobalt by Filamentous Fungi. *Frontiers in Microbiology*, *11*, 582016. doi:10.3389/fmicb.2020.582016 PMID:33408701

Dwivedi, S., & Dey, S. (2023). Review on biochar as an adsorbent material for removal of dyes from waterbodies. *International Journal of Environmental Science and Technology*, *20*(8), 9335–9350. doi:10.1007/s13762-022-04364-9

Ebelegi, A. N., Ayawei, N., & Wankasi, D. (2020). Interpretation of adsorption thermodynamics and kinetics. *Open Journal of Physical Chemistry*, *10*(3), 166–182. doi:10.4236/ojpc.2020.103010

Ebelegi, A. N., Elijah, N. T., Godwin, J., Ebelegi, A. N., Elijah, N. T., & Godwin, J. (2023). Evaluation of Physicochemical Parameters of Biosorbents Produced from Groundnut Hull Using Microwave Assisted Irradiation Method. *Open Journal of Physical Chemistry*, *13*(1), 1–12. doi:10.4236/ojpc.2023.131001

Eccles, H. (1995). Removal of heavy metals from effluent streams - Why select a biological process? *International Biodeterioration & Biodegradation*, *35*(1–3), 5–16. doi:10.1016/0964-8305(95)00044-6

Eivazihollagh, A., Svanedal, I., Edlund, H., & Norgren, M. (2019). On chelating surfactants: Molecular perspectives and application prospects. *Journal of Molecular Liquids*, *278*, 688–705. doi:10.1016/j.molliq.2019.01.076

El Mourabit, S. (2016). *Optimization of Concrete Beam Bridges: Development of Software for Design Automation and Cost Optimization* [Independent thesis Advanced level, KTH, School of Architecture and the Built Environment (ABE), Civil and Architectural Engineering, Structural Engineering and Bridges]. https://urn.kb.se/resolve?urn=urn:nbn:se:kth:diva-189273

El-Dessouky, H. T., & Ettouney, H. M. (2002). *Reverse osmosis*. Elservier Inc.

Elgallal, M., Fletcher, L., & Evans, B. (2016). Assessment of potential risks associated with chemicals in wastewater used for irrigation in arid and semiarid zones: A review. *Agricultural Water Management*, *177*, 419–431. doi:10.1016/j.agwat.2016.08.027

Elgarahy, A. M., Elwakeel, K. Z., Mohammad, S. H., & Elshoubaky, G. A. (2021). A critical review of biosorption of dyes, heavy metals and metalloids from wastewater as an efficient and green process. *Cleaner Engineering and Technology*, *4*, 100209. doi:10.1016/j.clet.2021.100209

El-Kady, A. A., & Abdel-Wahhab, M. A. (2018). Occurrence of trace metals in foodstuffs and their health impact. *Trends in Food Science & Technology*, *75*, 36–45. doi:10.1016/j.tifs.2018.03.001

Ellens, C. J., & Brown, R. C. (2012). Optimisation of a free-fall reactor for the production of fast pyrolysis bio-oil. *Bioresource Technology*, *103*(1), 374–380. doi:10.1016/j.biortech.2011.09.087 PMID:22036914

Elnabi, M. K. A. (2023). Toxicity of Heavy Metals and Recent Advances in Their Removal: A Review. *Toxics*, *11*(7), 580. doi:10.3390/toxics11070580 PMID:37505546

Elnour, A. Y., Alghyamah, A. A., Shaikh, H. M., Poulose, A. M., Al-Zahrani, S. M., Anis, A., & Al-Wabel, M. I. (2019). Effect of pyrolysis temperature on biochar microstructural evolution, physicochemical characteristics, and its influence on biochar/polypropylene composites. *Applied Sciences (Basel, Switzerland)*, *9*(6), 1149. doi:10.3390/app9061149

Eman, Z. G. (2012). Production and characteristics of a heavy metals removing bioflocculant produced by Pseudomonas aeruginosa. *Polish Journal of Microbiology*, *61*(4), 281–289. doi:10.33073/pjm-2012-038 PMID:23484411

Engwa, G. A., Ferdinand, P. U., Nwalo, F. N., & Unachukwu, M. N. (2019). Mechanism and health effects of heavy metal toxicity in humans. *Poisoning in the modern world-new tricks for an old dog*, *10*, 70-90.

Engwa, A. G., Ihekwoaba, C. J., Ilo, U. S., Unaegbu, M., Ayuk, L. E., & Osuji, A. G. (2015). Determination of some soft drink constituents and contamination by some heavy metals in Nigeria. *Toxicology Reports*, *2*, 384–390. doi:10.1016/j.toxrep.2015.01.014 PMID:28962372

Espinoza-Sánchez, M. A., Arévalo-Niño, K., Quintero-Zapata, I., Castro-González, I., & Almaguer-Cantú, V. (2019). Cr(VI) adsorption from aqueous solution by fungal bioremediation based using *Rhizopus sp. Journal of Environmental Management*, *251*, 109595. doi:10.1016/j.jenvman.2019.109595 PMID:31561145

Fadel, M., Hassanein, N. M., Elshafei, M. M., Mostafa, A. H., Ahmed, M. A., & Khater, H. M. (2017). Biosorption of manganese from groundwater by biomass of Saccharomyces cerevisiae. *HBRC Journal*, *13*(1), 106–113. doi:10.1016/j.hbrcj.2014.12.006

Faheem, J., Bao, H., Zheng, H., Tufail, S., Irshad, J., & Du, A. (2018). Adsorption-assisted decontamination of Hg (II) from aqueous solution by multi-functionalized corncob-derived biochar. *RSC Advances*, *8*(67), 38425–38435. doi:10.1039/C8RA06622A PMID:35559052

Fahimirad, S., Hatami, M. (2017). Heavy metal-mediated changes in growth and phytochemicals of edible and medicinal plants. *Medicinal plants and environmental challenges*, 189-214. doi:10.1007/978-3-319-68717-9_11

Fang, T., Yang, K., Lu, W., Cui, K., Li, J., Liang, Y., Hou, G., Zhao, X., & Li, H. (2019). An overview of heavy metal pollution in Chaohu Lake, China: Enrichment, distribution, speciation, and associated risk under natural and anthropogenic changes. *Environmental Science and Pollution Research International*, *26*(29), 29585–29596. doi:10.1007/s11356-019-06210-x PMID:31440974

Fan, J.-X., Niu, M.-T., Qin, Y.-T., Sun, Y.-X., & Zhang, X.-Z. (2022). Progress of engineered bacteria for tumor therapy. *Advanced Drug Delivery Reviews*, *185*, 114296. doi:10.1016/j.addr.2022.114296 PMID:35439571

Fan, Y., Li, H., Xue, Z., Zhang, Q., & Cheng, F. (2017). Accumulation characteristics and potential risk of heavy metals in soil-vegetable system under greenhouse cultivation condition in Northern China. *Ecological Engineering*, *102*, 367–373. doi:10.1016/j.ecoleng.2017.02.032

Farhan, N., & Khadom, A. (2015). Biosorption of heavy metals from aqueous solutions by *Saccharomyces Cerevisiae. International Journal of Industrial Chemistry*, *6*(2), 119–130. doi:10.1007/s40090-015-0038-8

Fato, F. P., Li, D. W., Zhao, L. J., Qiu, K., & Long, Y. T. (2019). Simultaneous Removal of Multiple Heavy Metal Ions from River Water Using Ultrafine Mesoporous Magnetite Nanoparticles. *ACS Omega*, *4*(4), 7543–7549. doi:10.1021/acsomega.9b00731

Fernandez-Lopez, A., Angosto, M., & Aviles, D. (2014). Biosorption of hexavalent chromium from aqueous medium with opuntia biomass. *TheScientificWorldJournal*, *4*, 1–8. doi:10.1155/2014/670249

Fertu, D. I., Bulgariu, L., & Gavrilescu, M. (2022). Modeling and optimization of heavy metals biosorption by low-cost sorbents using response surface methodology. *Processes (Basel, Switzerland)*, *10*(3), 523. doi:10.3390/pr10030523

Filote, C., Roşca, M., Hlihor, R. M., Cozma, P., Simion, I. M., Apostol, M., & Gavrilescu, M. (2021). Sustainable application of biosorption and bioaccumulation of persistent pollutants in wastewater treatment: Current practice. *Processes (Basel, Switzerland)*, *9*(10), 1696. doi:10.3390/pr9101696

Filote, C., Roşca, M., Simion, I. M., & Hlihor, R. M. (2022). Continuous systems bioremediation of wastewaters loaded with heavy metals using microorganisms. *Processes (Basel, Switzerland)*, *10*(9), 1758. doi:10.3390/pr10091758

Florea, A. M., & Büsselberg, D. (2006). Occurrence, use and potential toxic effects of metals and metal compounds. *Biometals*, *19*(4), 419–427. doi:10.1007/s10534-005-4451-x PMID:16841251

Forgacs, Z., Massanyi, P., Lukac, N., & Somosy, Z. (2012). Reproductive toxicology of nickel – review. *Journal of Environmental Science and Health. Part A, Toxic/Hazardous Substances & Environmental Engineering*, *47*(9), 1249–1260. doi:10.1080/10934529.2012.672114 PMID:22540648

Fouda-Mbanga, B. G., Seyisi, T., Nthwane, Y. B., Nyoni, B., & Tywabi-Ngeva, Z. (2023). A review on pollutants found in drinking water in Sub-Sahara African rural communities: Detection and potential low-cost remediation methods. *Industrial and Domestic Water Management*, *3*(2), 67–89. doi:10.53623/idwm.v3i2.264

Franca, F. C. S. S., & Albuquerque, A. M. A. (2017). Heavy metals deposited in the culture of lettuce (Lactuca sativa L.) by the influence of vehicular traffic in Pernambuco, Brazil. *Food Chemistry*, *215*, 171–176. doi:10.1016/j.foodchem.2016.07.168 PMID:27542464

Freixo, R., Casanova, F., Ribeiro, A. B., Pereira, C. F., Costa, E. M., Pintado, M. E., & Ramos, O. L. (2023). Extraction methods and characterization of cellulose fractions from a sugarcane by-product for potential industry applications. *Industrial Crops and Products*, *197*, 116615. doi:10.1016/j.indcrop.2023.116615

French, K. E., Zhou, Z., & Terry, N. (2020). French, Katherine E., Zhongrui Zhou, and Norman Terry. "Horizontal 'gene drives' harness indigenous bacteria for bioremediation.". *Scientific Reports*, *10*(1), 15091. doi:10.1038/s41598-020-72138-9 PMID:32934307

Fu, F., & Wang, Q. (2011). Removal of heavy metal ions from wastewaters: A review. *Journal of Environmental Management*, *92*(3), 407–418. doi:10.1016/j.jenvman.2010.11.011 PMID:21138785

Fu, J., Zhang, A., Wang, T., Qu, G., Shao, J., Yuan, B., Wang, Y., & Jiang, G. (2013). Influence of e-waste dismantling and its regulations: Temporal trend, spatial distribution of heavy metals in rice grains, and its potential health risk. *Environmental Science & Technology*, *47*(13), 7437–7445. doi:10.1021/es304903b PMID:23725497

Fu, Z., & Xi, S. (2020). The effects of heavy metals on human metabolism. *Toxicology Mechanisms and Methods*, *30*(3), 167–176. doi:10.1080/15376516.2019.1701594 PMID:31818169

Gable, P., & Brown, R. C. (2016). Effect of biomass heating time on bio-oil yields in a free fall fast pyrolysis reactor. *Fuel*, *166*, 361–366. doi:10.1016/j.fuel.2015.10.073

Gadd, G. (2009). Biosorption: Critical review of scientific rationale, environmental importance and significance for pollution treatment. *Journal of Chemical Technology and Biotechnology*, *84*(1), 13–28. doi:10.1002/jctb.1999

Gadd, G. (2009). Heavy Metal Pollutants: Environmental and Biotechnological Aspects. In *Encyclopedia of Microbiology* (pp. 321–334). Elsevier. doi:10.1016/B978-012373944-5.00149-8

Gadd, G. M. (1990). Fungi and yeasts for metal binding. In H. L. Ehrlich & C. L. Brierley (Eds.), *Microbial Mineral Recovery* (pp. 249–275). McGraw-Hill.

Gadd, G. M. (2004). Micobial Influence on Metal Mobilility and Application for Bioremediation. *Geoderma*, *122*(2-4), 109–119. doi:10.1016/j.geoderma.2004.01.002

Gadd, G. M., & White, C. (1992). Removal of thorium from stimulated acid process streams by fungal biomass: Potential for thorium desorption and reuse of biomass and desorbent. *Journal of Chemical Technology and Biotechnology*, *55*(1), 39–44. doi:10.1002/jctb.280550107

Gallagher, C. M., Chen, J. J., & Kovach, J. S. (2011). The relationship between body iron stores and blood and urine cadmium concentrations in US never-smoking, non-pregnant women aged 20–49 years. *Environmental Research*, *111*(5), 702–707. doi:10.1016/j.envres.2011.03.007 PMID:21507392

Gall, J. E., Boyd, R. S., & Rajakaruna, N. (2015). Transfer of heavy metals through terrestrial food webs: A review. *Environmental Monitoring and Assessment*, *187*(4), 201. doi:10.1007/s10661-015-4436-3 PMID:25800370

Gao, C., Xiao, W., Ji, G., Zhang, Y., Cao, Y., & Han, L. (2017). Regularity and mechanism of wheat straw properties change in ball milling process at cellular scale. *Bioresource Technology*, *241*, 214–219. doi:10.1016/j.biortech.2017.04.115 PMID:28570886

Gao, J.-J., Wang, B., Li, Z.-J., Xu, J., Fu, X.-Y., Han, H.-J., Wang, L.-J., Zhang, W.-H., Deng, Y.-D., Wang, Y., Gong, Z.-H., Tian, Y.-S., Peng, R.-H., & Yao, Q.-H. (2022). Metabolic engineering of Oryza sativa for complete biodegradation of thiocyanate. *The Science of the Total Environment*, *820*, 153283. doi:10.1016/j.scitotenv.2022.153283 PMID:35066037

Garbarino, J. R., Hayes, H., & Roth, D. (1995). Contaminants in the Mississippi River US Geological survey circular. Academic Press.

Garbisu, C., & Alkorta, I. (1999). Utilization of genetically engineered microorganisms (GEMs) for bioremediation. *Journal of Chemical Technology and Biotechnology, 74*(7), 599–606. doi:10.1002/(SICI)1097-4660(199907)74:7<599::AID-JCTB82>3.0.CO;2-G

Garcia Mendieta, A., Olguin, M. T., & Solache Rios, M. (2012). Biosorption properties of green tomato husk for iron, manganese and iron-manganese from aquoes solution. *Desalination, 284*, 167-174.

García, M. A., Alonso, J., & Melgar, M. J. (2005). Agaricus macrosporus as a potential bioremediation agent for substrates contaminated with heavy metals. *Journal of Chemical Technology and Biotechnology, 80*(3), 325–330. doi:10.1002/jctb.1203

Gaspareto, T. C., Scapini, T., Venturin, B., Tápparo, D. C., Camargo, A. F., Di Luccio, M., Valerio, A., Favretto, R., Antes, F. G., Steinmetz, R. L. R., Treichel, H., & Kunz, A. (2020). Pre-treatment Strategies for Value Addition in Poultry Litter. *Frontiers in Bioengineering and Biotechnology, 8*, 477. Advance online publication. doi:10.3389/fbioe.2020.00477 PMID:32523940

Gautam, P. K., Gautam, R. K., Chattopadhyaya, M. C., Banerjee, S., Chattopadhyaya, M. C., & Pandey, J. D. (2016). *Heavy metals in the environment: Fate, transport, toxicity and remediation technologies. Thermodynamic profiling of pollutants*. Academic Press.

Gavrilescu, M. (2004). Removal of heavy metals from the environment by biosorption. *Engineering in Life Sciences, 4*(3), 219–232. doi:10.1002/elsc.200420026

Gebreslassie, Y. T. (2020). Equilibrium, kinetics, and thermodynamic studies of malachite green adsorption onto fig (*Ficus cartia*) leaves. *Journal of Analytical Methods in Chemistry, 7384675*, 1–11. Advance online publication. doi:10.1155/2020/7384675 PMID:32190403

Gelais, A. T. S., & Costa-Picrce, B. A. (2016). Mercury concentrations in Northwest Atlantic winter-caught, male spiny dogfish (*Squalus acanthias*): A geographic mercury comparison and risk-reward framework for human consumption. *Marine Pollution Bulletin, 102*(1), 199–205. doi:10.1016/j.marpolbul.2015.12.009 PMID:26707980

Genchi, G., Carocci, A., Lauria, G., Sinicropi, M. S., & Catalano, A. (2020). Nickel: Human health and environmental toxicology. *International Journal of Environmental Research and Public Health, 17*(3), 679. doi:10.3390/ijerph17030679 PMID:31973020

Georganas, A., Giamouri, E., Pappas, A. C., Papadomichelakis, G., Galliou, F., Manios, T., Tsiplakou, E., Fegeros, K., & Zervas, G. (2020). Bioactive Compounds in Food Waste: A Review on the Transformation of Food Waste to Animal Feed. *Foods, 9*(3), 291. doi:10.3390/foods9030291 PMID:32150822

Gerardo, B., Cabral Pinto, M., Nogueira, J., Pinto, P., Almeida, A., Pinto, E., Marinho-Reis, P., Diniz, L., Moreira, P. I., Simões, M. R., & Freitas, S. (2020). Associations between trace elements and cognitive decline: An exploratory 5-year follow-up study of an elderly cohort. *International Journal of Environmental Research and Public Health, 17*(17), 6051. doi:10.3390/ijerph17176051 PMID:32825289

Ghannam, H. E., El Haddad, E. S. E., & Talab, A. S. (2015). Bioaccumulation of heavy metals in tilapia fish organs. *Journal of Biodiversity and Environmental Sciences, 7*(2), 88–99.

Gheorghe, S., Stoica, C., Vasile, G. G., Nita-Lazar, M., Stanescu, E., & Lucaciu, I. E. (2017). Metals toxic effects in aquatic ecosystems: modulators of water quality. *Water Quality, 87*, 59-89. doi:10.5772/65744

Ghimire, K. N., Inoue, K., Ohto, K., & Hayashida, T. (2008). Adsorption study of metal ions onto crosslinked seaweed Laminaria japonica. *Bioresource Technology, 99*(1), 32–37. doi:10.1016/j.biortech.2006.11.057 PMID:17267212

Ghoneim, M., El-Desoky, S., El-Moselhy, M., Amer, A., Abou El-Naga, H., Mohamedein, I., & Al-Prol, E. (2014). Removal of cadmium from aqueous solution using marine green algae, *Ulva lactuca*. *Egyptian Journal of Aquatic Research, 40*(3), 235–242. doi:10.1016/j.ejar.2014.08.005

Ghoniem, A. A., El-Naggar, N. E.-A., Saber, W. I. A., El-Hersh, M. S., & El-khateeb, A. Y. (2020). Statistical modeling-approach for optimization of Cu^{2+} biosorption by *Azotobacter nigricans* NEWG-1; characterization and application of immobilized cells for metal removal. *Scientific Reports, 10*(1), 9491. doi:10.1038/s41598-020-66101-x PMID:32528020

Giese, E., Silva, D., Costa, A., Almeida, S., & Dussan, K. (2020). Immobilized microbial nanoparticles for biosorption. *Critical Reviews in Biotechnology, 40*(5), 653–666. doi:10.1080/07388551.2020.1751583

Gina, K. Y., Tanga, Y. Z., & Aziz, M. A. (2002). Derivation and application of a new model for heavy metal biosorption by algae. *Water Research, 36*(5), 1313–1323. doi:10.1016/S0043-1354(01)00332-3 PMID:11902786

Girish, C. (2018). Various Impregnation Methods Used for the Surface Modification of the Adsorbent: A Review. *International Journal of Engineering & Technology, 7*(4.7), 330. doi:10.14419/ijet.v7i4.7.20571

Gómez, X., Cuetos, M. J., García, A. I., & Morán, A. (2007). An evaluation of stability by thermogravimetric analysis of digestate obtained from different biowastes. *Journal of Hazardous Materials, 149*(1), 97–105. doi:10.1016/j.jhazmat.2007.03.049 PMID:17448599

Gong, R., Ding, Y., Liu, H., Chen, Q., & Liu, Z. (2005). Lead biosorption and desorption by intact and pretreated Spirulina maxima biomass. *Chemosphere, 58*(1), 125–130. doi:10.1016/j.chemosphere.2004.08.055 PMID:15522341

Gonzalez, S., Lopez-Roldan, R., & Cortina, J. L. (2013). Presence of metals in drinking water distribution networks due to pipe material leaching: A review. *Toxicological and Environmental Chemistry, 95*(6), 870–889. doi:10.1080/02772248.2013.840372

Goswami, L., Kushwaha, A., Kafle, S. R., & Kim, B. S. (2022). *Surface Modification of Biochar for Dye Removal from Wastewater*. Catalysis., doi:10.3390/catal12080817

Gouda, S. A., & Taha, A. (2023). Biosorption of Heavy Metals as a New Alternative Method for Wastewater Treatment: A Review. *Egyptian Journal of Aquatic Biology & Fisheries, 27*(2), 135–153. doi:10.21608/ejabf.2023.291671

Gourdon, R., Bhende, S., Rus, E., & Sofer, S. S. (1990). Comparison of cadmium biosorption by gram- positive and gram-negative bacteria from activated sludge. *Biotechnology Letters, 12*(11), 839–842. doi:10.1007/BF01022606

Goyal, H. B., Seal, D., & Saxena, R. C. (2008). Bio-fuels from thermochemical conversion of renewable resources: A review. *Renewable & Sustainable Energy Reviews, 12*(2), 504–517. doi:10.1016/j.rser.2006.07.014

Gray, C. W., McLaren, R. G., & Roberts, A. H. C. (2003). Atmospheric accessions of heavy metals to some New Zealand pastoral soils. *The Science of the Total Environment, 305*(1–3), 105–115. doi:10.1016/S0048-9697(02)00404-7 PMID:12670761

Greeshma, K., Kim, H. S., & Ramanan, R. (2022). The emerging potential of natural and synthetic algae-based microbiomes for heavy metal removal and recovery from wastewaters. *Environmental Research, 215*, 14238. doi:10.1016/j.envres.2022.114238 PMID:36108721

Groen, J. C., Maldonado, L., Moulijn, J. A., & Pérez-Ramírez, J. (2006). *On the role of iron in preparation of mesoporous Fe-MFI zeolites via desilication.* doi:10.1016/S0167-2991(06)80916-X

Gross, J. B. Jr, Ludwig, J., Wiesner, R. H., McCall, J. T., LaRusso, N. F., & Ivanova, A. (1985). Abnormalities in tests of copper metabolism in primary sclerosing cholangitis. *Gastroenterology, 89*(2), 272–278. doi:10.1016/0016-5085(85)90326-9 PMID:4007418

Guertin, J. (2005). Toxicity and health effects of chromium (all oxidation states). In J. Guertin, J. A. Jacobs, & C. P. Avakian (Eds.), *Chromium (VI) Handbook* (pp. 216–234). CRC Press.

Günther, V., Lindert, U., & Schaffner, W. (2012). The taste of heavy metals: Gene regulation by MTF-1. *Biochimica et Biophysica Acta, 1823*(9), 1416–1425. doi:10.1016/j.bbamcr.2012.01.005 PMID:22289350

Guo, J., Zheng, X. D., Chen, Q. B., Zhang, L., & Xu, X. P. (2012). Biosorption of Cd(II) from aqueous solution by *Pseudomonas plecoglossicida. Current Microbiology, 65*(4), 350–355. doi:10.1007/s00284-012-0164-x PMID:22706778

Gupta, N., Yadav, K. K., Kumar, V., Kumar, S., Chadd, R. P., & Kumar, A. (2019). Trace elements in soil-vegetables interface: Translocation, bioaccumulation, toxicity and amelioration-a review. *The Science of the Total Environment, 651*, 2927–2942. doi:10.1016/j.scitotenv.2018.10.047 PMID:30463144

Gupta, R., Khan, P. A. S., Saxena, R. K., & Mohapatra, H. (2000). Microbial biosorbents: Meeting challenges of heavy metal pollution in aqueous solutions. *Current Science, 78*(8), 25.

Gupta, S., Mittal, Y., Panja, R., Prajapati, K. B., & Yadav, A. K. (2021). *Conventional wastewater treatment technologies*. Elservier Inc. doi:10.1016/B978-0-12-821009-3.00012-9

Gupta, V. (2020). *Vehicle-generated heavy metal pollution in an urban environment and its distribution into various environmental components. In Environmental Concerns and Sustainable Development* (Vol. 1). Air, Water and Energy Resources.

Gupta, V. K., Carrott, P. J. M., Carrott, M. M. L. R., & Suhas, R. (2009). Low-cost adsorbents: Growing approach to wastewater treatment – a Review. *Critical Reviews in Environmental Science and Technology, 39*(10), 783–842. doi:10.1080/10643380801977610

Gupta, V. K., & Nayak, A. (2012). Cadmium removal and recovery from aqueous solutions by novel adsorbents prepared from orange peel and Fe_2O_3 nanoparticles. *Chemical Engineering Journal, 180*, 81–90. doi:10.1016/j.cej.2011.11.006

Gupta, V., Nayak, A., & Agarwal, S. (2015). Bio-adsorbents for remediation of heavy metals: Current status and their future prospects. *Environmental Engineering Research, 20*(1), 1–8. doi:10.4491/eer.2015.018

Gu, S., & Lan, C. Q. (2024). Mechanism of heavy metal ion biosorption by microalgal cells: A mathematic approach. *Journal of Hazardous Materials, 463*, 132875. doi:10.1016/j.jhazmat.2023.132875 PMID:37918069

Gworek, B., Dmuchowski, W., & Baczewska-Dąbrowska, A. H. (2020). Mercury in the terrestrial environment: A review. *Environmental Sciences Europe, 32*(1), 128. Advance online publication. doi:10.1186/s12302-020-00401-x

Hadey, C., Allouch, M., Alami, N., Bouklifi, F., & Loulidi, I. (2022). Preparation and characterization of biochars obtained from biomasses for combustible briquette applications. *TheScientificWorldJournal, 2022*, 1–13. Advance online publication. doi:10.1155/2022/2554475 PMID:36523325

Hajahmadi, Z., Younesi, H., Bahramifar, N., Khakpour, H., & Pirzadeh, K. (2015). Multicomponent isotherm for biosorption of Zn(II), Co(II) and Cd(II) from ternary mixture onto pretreated dried *Aspergillus niger* biomass. *Water Resources and Industry, 11*, 71–80. doi:10.1016/j.wri.2015.07.003

Halim, M., Conte, P., & Piccolo, A. (2002). Potential Availability of Heavy Metals to Phytoextraction from Contaminated Soils Induced by Exogenous Humid Substances. *Chemosphere, 52*(1), 265–275. doi:10.1016/S0045-6535(03)00185-1 PMID:12729711

Halttunen, T., Salminen, S., & Tahvonen, R. (2007). Rapid Removal of Lead and Cadmium from Water by Specific Lactic Acid Bacteria. *International Journal of Food Microbiology, 114*(1), 30–35. doi:10.1016/j.ijfoodmicro.2006.10.040 PMID:17184867

Hamelink, J. L., Landrum, P. F., Harold, B. L., & William, B. H. (Eds.). (1994). *Bioavailability: Physical, Chemical, and Biological Interactions*. CRC Press Inc.

Hammaini, A., Gonzalez, F., Ballester, A., Blazquez, M. L., & Munoz, J. A. (2003). Simultaneous uptake of metals by activated sludge. *Minerals Engineering*, *16*(8), 723–729. doi:10.1016/S0892-6875(03)00166-3

Hamza, S. M., Ahmed, H. F., & Mohammmad, E. A. M. (2010). Optimization of cadmium, zinc and copper biosorption in an aquoes solution by *Saccharomyces cerevisiae. The Journal of American Science*, *6*, 597–604.

Hanif, M. (2011). Biosorption of Cr (III) and Cr (VI) by Newly Isolated White Rot Fungi: Batch and Column Studies. *Asian Journal of Chemistry*, *23*(8), 3375–3383.

Hanif, M. U., Capareda, S. C., Iqbal, H., Arazo, R. O., & Baig, M. A. (2016). Effects of Pyrolysis Temperature on Product Yields and Energy Recovery from Co-Feeding of Cotton Gin Trash, Cow Manure, and Microalgae: A Simulation Study. *PLoS One*, *11*(4), e0152230. Advance online publication. doi:10.1371/journal.pone.0152230 PMID:27043929

Han, X., Wong, Y. S., & Tam, N. F. Y. (2006). surface complexation mechanism and modelling in Cr(III) biosorption by a microalgal isolate, *Chloerlla miniata. Journal of Colloid and Interface Science*, *303*(2), 365–371. doi:10.1016/j.jcis.2006.08.028 PMID:16962604

Harada, M. (1995). Minamata disease: Methylmercury poisoning in Japan caused by environmental pollution. *Critical Reviews in Toxicology*, *25*(1), 1–24. doi:10.3109/10408449509089885 PMID:7734058

Harsha Vardhan, K., Kumar, P. S., & Panda, R. C. (2019). A review on heavy metal pollution, toxicity and remedial measures: Current trends and future perspectives. *Journal of Molecular Liquids*, *290*, 111197. doi:10.1016/j.molliq.2019.111197

Harshala, K., & Wagh, N. D. (2022). Use of Agricultural Waste-Based Biosorbents for the Removal of Heavy Metals from Aqueous Solution: A Review. *Nature Environment and Pollution Technology*, *21*(3), 1003–1014. doi:10.46488/NEPT.2022.v21i03.007

Hartmann, A. J., & Lamb, R. N. (1997). X-ray photoemission spectroscopy of thin films. *Current Opinion in Solid State and Materials Science*, *2*(5), 511–516. doi:10.1016/S1359-0286(97)80038-1

Hasan, H., & Srivastava, P. (2009). Batch and continuous biosorption of Cu(II) by immobilized biomass of *Arthrobacter sp. Journal of Environmental Management*, *90*(11), 3313–3332. doi:10.1016/j.jenvman.2009.05.005

Hasan, M., Al-Tameemi, I., & Abbas, M. (2021). Orange peels as a sustainable material for treating water polluted with antimony. *Journal of Ecological Engineering*, *22*(2), 25–35. doi:10.12911/22998993/130632

Hasan, S. H., Srivastava, P., & Talat, M. (2009). Biosorption of Pb(II) from water using biomass of *Aeromonas hydrophila*: Central composite design for optimization of process variables. *Journal of Hazardous Materials*, *168*(2), 1155–1162. doi:10.1016/j.jhazmat.2009.02.142 PMID:19345004

Hassan, S. H., Awad, Y. M., Kabir, M. H., Oh, S. E., & Joo, J. H. (2010). Bacterial Biosorption of heavy metals. *Biotechnology: Cracking New Pastures,* 79-110.

Hayat, M. T., Nauman, M., Nazir, N., Ali, S., & Bangash, N. (2018). Environmental hazards of cadmium: past, present, and future. *Cadmium Toxic. Toler. Plants Physiol. Remediat.*, 163–183. doi:10.1016/B978-0-12-814864-8.00007-3

Haynes, W. M. (2011). *CRC handbook of chemistry and physics* (W. M. Haynes, Ed.; 92nd ed.). CRC Press.

Helmenstine, A. M. (2018). *Heavy metal definition and list*. https://www.thoughtco.com

He, M., Wang, N., Long, X., Zhang, C., Ma, C., Zhong, Q., Wang, A., Wang, Y., Pervaiz, A., & Shan, J. (2019). Antimony speciation in the environment: Recent advances in understanding the biogeochemical processes and ecological effects. *Journal of Environmental Sciences (China)*, *75*, 14–39. doi:10.1016/j.jes.2018.05.023 PMID:30473279

Hemathilake, D. M. K. S., & Gunathilake, D. M. C. C. (2022). Agricultural productivity and food supply to meet increased demands. In *Future Foods* (pp. 539–553). Elsevier. doi:10.1016/B978-0-323-91001-9.00016-5

He, Q., Bai, Y., Lu, Y., Cui, B., Huang, Z., Yang, Q., Jiang, D., & Shao, D. (2022). Isolation and characterization of cellulose nanocrystals from Chinese medicine residues. *Biomass Conversion and Biorefinery*. Advance online publication. doi:10.1007/s13399-022-03380-6 PMID:36259074

Herawati, N., Suzuki, S., Hayashi, K., Rivai, I. F., & Koyama, H. (2000). Cadmium, copper, and zinc levels in rice and soil of Japan, Indonesia, and China by soil type. *Bulletin of Environmental Contamination and Toxicology*, *64*, 33–39. doi:10.1007/s001289910006 PMID:10606690

He, Z. L., Yang, X. E., & Stoffella, P. J. (2005). Trace elements in agroecosystems and impacts on the environment. *Journal of Trace Elements in Medicine and Biology*, *19*(2-3), 125–140. doi:10.1016/j.jtemb.2005.02.010 PMID:16325528

Hlihor, R. M., Diaconu, M., Fertu, D., Chelaru, C., Sandu, I., & Tavares, T. (2013). Bioremediation of Cr(VI) Polluted wastewaters by sorption on heat inactivated *Saccharomyces cerevisiae* biomass. *International Journal of Environmental Research*, *7*(3), 581–594. doi:10.22059/ijer.2013.638

Hlihor, R. M., Diaconu, M., Leon, F., Curteanu, S., Tavares, T., & Gavrilescu, M. (2015). Experimental analysis and mathematical prediction of Cd(II) removal by biosorption using support vector machines and genetic algorithms. *New Biotechnology*, *32*(3), 358–368. doi:10.1016/j.nbt.2014.08.003 PMID:25224921

Hlihor, R. M., Roşca, M., Drăgoi, E. N., Simion, I. M., Favier, L., & Gavrilescu, M. (2023). New insights into the application of fungal biomass for chromium(VI) bioremoval from aqueous solutions using Design of Experiments and Differential Evolution based neural network approaches. *Chemical Engineering Research & Design*, *190*, 233–254. doi:10.1016/j.cherd.2022.12.024

Hlihor, R. M., Roşca, M., Tavares, T., & Gavrilescu, M. (2017). The role of *Arthrobacter viscosus* in the removal of Pb(II) from aqueous solutions. *Water Science and Technology*, *76*(7), 1726–1738. doi:10.2166/wst.2017.360 PMID:28991789

Hlihor, R.-M., Cozma, P., Ghinea, C., Pavel, V. L., Robu, B. M., & Gavrilescu, M. (2010). Thermodynamic study of Cd(II) sorption on soil. *Scientific Papers Series A. Agronomy*, *54*(2), 92–95. http://www.uaiasi.ro/revagrois/PDF/2010_2_94.pdf

Hoang, M. T., Pham, T. D., Pham, T. T., Nguyen, M. K., Nu, D. T. T., Nguyen, T. H., Bartling, S., & Van der Bruggen, B. (2021). Esterification of sugarcane bagasse by citric acid for Pb2+ adsorption: Effect of different chemical pretreatment methods. *Environmental Science and Pollution Research International*, *28*(10), 11869–11881. doi:10.1007/s11356-020-07623-9 PMID:31953762

Holliday, M. C., Parsons, D. R., & Zein, S. H. (2022). Agricultural Pea Waste as a Low-Cost Pollutant Biosorbent for Methylene Blue Removal: Adsorption Kinetics, Isotherm And Thermodynamic Studies. *Biomass Conversion and Biorefinery*. Advance online publication. doi:10.1007/s13399-022-02865-8

Hong, M., Yu, L., Wang, Y., Zhang, J., Chen, Z., Dong, L., Zan, Q., & Li, R. (2019). Heavy metal adsorption with zeolites: The role of hierarchical pore architecture. *Chemical Engineering Journal*, *359*, 363–372. doi:10.1016/j.cej.2018.11.087

Horsfall, M. Jnr, & Spiff, M. I. (2005). Effects of temperature on the sorption of Pb^{2+} and Cd^{2+} from aqueous solution by *Caladium bicolor* (Wild Cocoyam) biomass. *Electronic Journal of Biotechnology*, *8*(2), 162–169. doi:10.2225/vol8-issue2-fulltext-4

Hou, D., O'Connor, D., Nathanail, P., Tian, L., & Ma, Y. (2017). Integrated GIS and multivariate statistical analysis for regional scale assessment of heavy metal soil contamination: A critical review. *Environmental Pollution*, *231*, 1188–1200. doi:10.1016/j.envpol.2017.07.021 PMID:28939126

Hualpa-Cutipa, E. H., Acosta, R. A. S., & Sangay-Tucto, S. (2022). *Recent trends for treatment of environmental contaminants in wastewater: An integrated valorization of industrial wastewater*. Elservier Inc.

Huang, D., Li, B., Ou, J., Xue, W., Li, J., Li, Z., Li, T., Chen, S., Deng, R., & Guo, X. (2020). Megamerger of biosorbents and catalytic technologies for the removal of heavy metals from wastewater: Preparation, final disposal, mechanism and influencing factors. *Journal of Environmental Management*, *261*, 109879. doi:10.1016/j.jenvman.2019.109879

Huang, F., Wang, Z. H., Cai, Y. X., Chen, S. H., Tian, J. H., & Cai, K. Z. (2018). Heavy metal bioaccumulation and cation release by growing Bacillus cereus RC-1 under culture conditions. *Ecotoxicology and Environmental Safety*, *157*, 216–226. doi:10.1016/j.ecoenv.2018.03.077 PMID:29625395

Huang, L., Wang, Q., Zhou, Q., Ma, L., Wu, Y., Liu, Q., Wang, S., & Feng, Y. (2020). Cadmium uptake from soil and transport by leafy vegetables: A meta-analysis. *Environmental Pollution*, *264*, 114677. doi:10.1016/j.envpol.2020.114677 PMID:32388299

Huang, S. S., Liao, Q. L., Hua, M., Wu, X. M., Bi, K. S., Yan, C. Y., Chen, B., & Zhang, X. Y. (2007). Survey of Heavy Metal Pollution and Assessment of Agricultural Soil in Yangzhong District, Jiangsu Province, China. *Chemosphere*, *67*(11), 2148–2155. doi:10.1016/j.chemosphere.2006.12.043 PMID:17275882

Hubbe, M., Azizian, S., & Douven, S. (2019). Implications of apparent pseudo-second-order adsorption kinetics onto cellulosic materials: A review. *BioResources*, *14*(3), 7582–7626. doi:10.15376/biores.14.3.Hubbe

Hughes, M. F. (2002). Arsenic toxicity and potential mechanisms of action. *Toxicology Letters*, *133*(1), 1–16. doi:10.1016/S0378-4274(02)00084-X PMID:12076506

Hu, J., Wu, F., Wu, S., Sun, X., Lin, X., & Wong, M. H. (2013). Phytoavailability and phytovariety codetermine the bioaccumulation risk of heavy metal from soils, focusing on Cd-contaminated vegetable farms around the Pearl River Delta, China. *Ecotoxicology and Environmental Safety*, *91*, 18–24. doi:10.1016/j.ecoenv.2013.01.001 PMID:23399100

Hussain, S. A., Hasan, N. K., & Al-Abodi, E. E. (2021). Biosorption to removing heavy metals from wastewater. *The International Conference of Chemistry, 1853*. 10.1088/1742-6596/1853/1/012012

Hutton, M., & Symon, C. (1986). The quantities of cadmium, lead, mercury and arsenic entering the UK environment from human activities. *The Science of the Total Environment*, *57*, 129–150. doi:10.1016/0048-9697(86)90018-5 PMID:3810138

Hu, X., Cao, J., Yang, H., Li, D., Qiao, Y., Zhao, J., Zhang, Z., & Huang, L. (2020). Pb^{2+} biosorption from aqueous solutions by live and dead biosorbents of the hydrocarbon-degrading strain *Rhodococcus* sp. HX-2. *PLoS One*, *15*(1), e0226557. doi:10.1371/journal.pone.0226557 PMID:31995615

Ibrahim, U. B., Yusuf, I., Saleh, A., Fardami, A. Y., Yahaya, H. I., Jodi, A. M., Sanusi, Z. M., Kawo, A. H., Nataala, M. K., & Yahaya, S. (2023). Biomass assessment and optimization of *Alcaligenes faecalis* isolated from some Nigerian mining sites for heavy metal uptake using response surface methodology model. *UMYU Scientifica*, *2*(3), 128–141. doi:10.56919/usci.2323.019

Igiri, B., Okoduwa, S., Idoko, G., Akabuogu, E., Adeyi, A., & Ejiogu, I. (2018). Toxicity and bioremediation of heavy metals contaminated ecosystem from tannery wastewater: A Review. *Journal of Toxicology*, *27*, 2568038. doi:10.1155/2018/2568038

Iqbal, M., & Saeed, A. (2007). Production of an immobilized hybrid biosorbent for the sorption of Ni(II) from aqueous solution. *Process Biochemistry (Barking, London, England)*, *41*(2), 48–57. doi:10.1016/j.procbio.2006.07.022

Iqbal, M., Saeed, A., & Akhtar, N. (2002). Petiolar felt sheath of palm: A new adsorbent for the removal of heavy metals from contaminated water. *Bioresource Technology*, *81*(2), 151–153. doi:10.1016/S0960-8524(01)00126-2 PMID:11762907

Ivanova, D., Kadukova, J., Kavulicova, J., & Horvathova, H. (2012). Determination of the Functional Groups in Algae *Parachlorella Kessleri* by Potentiometric Titrations. *Nova Biotechnologica et Chimica*, *11*(2), 93–99. doi:10.2478/v10296-012-0010-3

Iwuozor, K. O., Emenike, E. C., Omonayin, E. O., Bamigbola, J. O., Ojo, H. T., Awoyale, A. A., Eletta, O. A. A., & Adeniyi, A. G. (2023). Unlocking the hidden value of pods: A review of thermochemical conversion processes for biochar production. *Bioresource Technology Reports*, *22*, 101488. Advance online publication. doi:10.1016/j.biteb.2023.101488

Iyengar, G. V. (1998). Re-evaluation of the trace element content in Reference man. *Radiation Physics and Chemistry*, *51*(4-6), 545–560. doi:10.1016/S0969-806X(97)00202-8

Iyengar, V., & Nair, P. (2000). Global outlook on nutrition and the environment: Meeting the challenges of the next millennium. *The Science of the Total Environment*, *249*(1-3), 331–346. doi:10.1016/S0048-9697(99)00529-X PMID:10813462

Izah, S. C., Chakrabarty, N., & Srivastav, A. L. (2016). A review on heavy metal concentration in potable water sources in Nigeria: Human health effects and mitigating measures. *Exposure and Health*, *8*(2), 285–304. Advance online publication. doi:10.1007/s12403-016-0195-9

Jaishankar, M., Tseten, T., Anbalagan, N., Mathew, B. B., & Beeregowda, K. N. (2014). Toxicity, mechanism, and health effects of some heavy metals. *Interdisciplinary Toxicology*, *7*(2), 60–72. doi:10.2478/intox-2014-0009 PMID:26109881

Jalali, R., Ghafourian, H., Asef, Y., Davarpanah, S. J., & Sepeh, S. (2002). Removal and recovery of lead using nonliving biomass of marine algae. *Journal of Hazardous Materials*, *92*(3), 253–262. doi:10.1016/S0304-3894(02)00021-3 PMID:12031610

James, G., Witten, D., Hastie, T., & Tibshirani, R. (2013). *An Introduction to Statistical Learning* (Vol. 103). Springer. doi:10.1007/978-1-4614-7138-7

Jana, A., Bhattacharya, P., Sarkar, S., Majumdar, S., & Ghosh, S. (2016). An ecofriendly approach towards remediation of high lead containing toxic industrial effluent by a combined biosorption and microfiltration process: A total reuse prospect. *Desalination and Water Treatment*, *57*(12), 5498–5513. doi:10.1080/19443994.2015.1004596

Jankovic, A., Chaudhary, G., & Goia, F. (2021). Designing the design of experiments (DOE) – An investigation on the influence of different factorial designs on the characterization of complex systems. *Energy and Building*, *250*, 111298. doi:10.1016/j.enbuild.2021.111298

Javanbakht, V., Alavi, S. A., & Zilouei, H. (2014). Mechanisms of heavy metal removal using microorganisms as biosorbent. *Water Science and Technology*, *69*(9), 1775–1787. doi:10.2166/wst.2013.718 PMID:24804650

Javier, I., Ordóñez, I., Cortés, S., Maluenda, P., & Soto, I. (2023). Biosorption of Heavy Metals with Algae: Critical Review of Its Application in Real Effluents. *Sustainability (Basel)*, *15*(6), 5521. doi:10.3390/su15065521

Jawed, A., Saxena, V., & Pandey, L. M. (2020). Engineered nanomaterials and their surface functionalization for the removal of heavy metals: A review. *Journal of Water Process Engineering*, *33*, 101009. doi:10.1016/j.jwpe.2019.101009

Jayathilakan, K., Sultana, K., Radhakrishna, K., & Bawa, A. S. (2012). Utilization of byproducts and waste materials from meat, poultry and fish processing industries: A review. *Journal of Food Science and Technology*, *49*(3), 278–293. doi:10.1007/s13197-011-0290-7 PMID:23729848

Jiang, L., Zhou, W., Liu, D., Liu, T., & Wang, Z. (2017). Biosorption isotherm study of Cd^{2+}, Pb^{2+} and Zn^{2+} biosorption onto marine bacterium *Pseudoalteromonas* sp. SCSE709-6 in multiple systems. *Journal of Molecular Liquids, 247*, 230–237. doi:10.1016/j.molliq.2017.09.117

Jiménez-Cedillo, M. J., Olguín, M. T., Fall, C., & Colin-Cruz, A. (2013). As (III) and As (V)sorption on iron-modified non-pyrolyzed and pyrolyzed biomass from Petroselinum crispum(parsley). *Journal of Environmental Management, 117*, 242–252. doi:10.1016/j.jenvman.2012.12.023 PMID:23376307

Jinsong, H. J., Pau, J., & Chen, A. (2014). Comprehensive review on biosorption of heavy metals by algal biomass: Materials, performances, chemistry, and modelling simulation tools. *Bioresource Technology, 160*, 67–78. doi:10.1016/j.biortech.2014.01.068 PMID:24630371

Jin, Y., Wang, X., Zang, T., Hu, Y., Hu, X., Ren, G., Xu, X., & Qu, J. (2016). Biosorption of lead(II) by *Arthrobacter* sp. 25: Process optimization and mechanism. *Journal of Microbiology and Biotechnology, 26*(8), 1428–1438. doi:10.4014/jmb.1603.03074 PMID:27197671

Juela, D., Vera, M., Cruzat, C., Alvarez, X., & Vanegas, E. (2021). Adsorption properties of sugarcane bagasse and corn cob for the sulfamethoxazole removal in a fixed-bed column. *Sustainable Environment Research, 31*(1), 27. doi:10.1186/s42834-021-00102-x

Junaid, M., Hashmi, M. Z., Malik, R. N., & Pei, D. S. (2016). Toxicity and oxidative stress induced by chromium in workers exposed from different occupational settings around the globe: A review. *Environmental Science and Pollution Research International, 23*(20), 20151–20167. doi:10.1007/s11356-016-7463-x PMID:27562808

Kahlon, S. K., Sharma, G., Julka, J. M., Kumar, A., Sharma, S., & Stadler, F. J. (2018). Impact of heavy metals and nanoparticles on aquatic biota. *Environmental Chemistry Letters, 16*(3), 919–946. doi:10.1007/s10311-018-0737-4

Kajina, W., & Rousset, P. (2018). *Coupled effect of feedstock and pyrolysis temperature on biochar as soil amendment.* Agritrop.

Kajitvichyanukul, P., Ananpattarachai, J., & Pongpom, S. (2005). Sol-gel preparation andproperties study of TiO2 thin film for photocatalytic reduction of chromium (VI) in thephotocatalysis process. *Science and Technology of Advanced Materials, 6*(3–4), 352–358. doi:10.1016/j.stam.2005.02.014

Kajjumba, G. W., Emik, S., Öngen, A., Kurtulus Özcan, H., & Aydın, S. (2019). Modelling of Adsorption Kinetic Processes - Errors, Theory and Application. In S. Edebali (Ed.), *Advanced Sorption Process Applications*. IntechOpen. doi:10.5772/intechopen.80495

Kalita, N., & Baruah, P. P. (2023). Cyanobacteria as a potent platform for heavy metals biosorption: Uptake, responses and removal mechanisms. *Journal of Hazardous Materials Advances, 11*, 100349. doi:10.1016/j.hazadv.2023.100349

Kamala, C. T., Chu, K. H., & Char, N. S. (2005). Removal of arsenic(III) from aqueous solutions using fresh and immobilized plant biomass. *Water Research, 39*(13), 2815–2826. doi:10.1016/j.watres.2005.04.059 PMID:15993920

Kanamarlapudi, R. K., Chintalpudi, V. K., & Muddada, S. (2018). Application of bio sorption for removal of heavy metals from waste water. In Tech Open, 70-116.

Kanamarlapudi, S. L. R. K., Chintalpudi, V. K., & Muddaada, S. (2018). Application of Biosorption for Removal of Heavy Metals from Waste Water. School of Environmental Sciences, 69-116.

Kanamarlapudi, S. L. R. K., Kumar Chintalpudi, V., Muddada, S., Kanamarlapudi, S. L. R. K., Kumar Chintalpudi, V., & Muddada, S. (2018). *Application of Biosorption for Removal of Heavy Metals from Wastewater.* Biosorption. doi:10.5772/intechopen.77315

Kapahi, M., & Sachdeva, S. (2019). Bioremediation options for heavy metal pollution. *Journal of Health & Pollution, 9*(24), 191203. doi:10.5696/2156-9614-9.24.191203 PMID:31893164

Kapoor, A., & Viraraghavan, T. (1995). Fungal biosorption — an alternative treatment option for heavy metal bearing wastewaters: A review. *Bioresource Technology, 53*, 195–206.

Kapoor, A., & Viraraghavan, T. (1997). Fungi as biosorbents. In J. Wase & C. Forster (Eds.), *Biosorbents for metal ions, 67–86.*

Karić, N., Maia, A. S., Teodorović, A., Atanasova, N., Langergraber, G., Crini, G., Ribeiro, A. R. L., & Đolić, M. (2022). Bio-waste valorisation: Agricultural wastes as biosorbents for removal of (in)organic pollutants in wastewater treatment. *Chemical Engineering Journal Advances, 9*, 100239. doi:10.1016/j.ceja.2021.100239

Karim, A., Raji, Z., Karam, A., & Khalloufi, S. (2023). Valorization of Fibrous Plant-Based Food Waste as Biosorbents for Remediation of Heavy Metals from Wastewater—A Review. *Molecules (Basel, Switzerland), 28*(10), 4205. doi:10.3390/molecules28104205 PMID:37241944

Karimi-Maleh, H., Darabi, R., Baghayeri, M., Karimi, F., Fu, L., Rouhi, J., Niculina, D. E., Gündüz, E. S., & Dragoi, E. N. (2023). Recent developments in carbon nanomaterials-based electrochemical sensors for methyl parathion detection. *Journal of Food Measurement and Characterization, 17*(5), 5371–5389. doi:10.1007/s11694-023-02050-z

Kaushal, S. S., Likens, G. E., Pace, M. L., Reimer, J. E., Maas, C. M., Galella, J. G., Utz, R. M., Duan, S., Kryger, J. R., Yaculak, A. M., Boger, W. L., Bailey, N. W., Haq, S., Wood, K. L., Wessel, B. M., Park, C. E., Collison, D. C., Aisin, B. Y. I., Gedeon, T. M., ... Woglo, S. A. (2021). Freshwater salinization syndrome: From emerging global problem to managing risks. *Biogeochemistry, 154*(2), 255–292. doi:10.1007/s10533-021-00784-w

Kaye, G. W. C., & Laby, T. H. (1986). *Tables of physical and chemical constants* (15th ed.). John Wiley and Sons Ltd.

Kazantzis, G. (2000). Thallium in the environment and health effects. *Environmental Geochemistry and Health, 22*(4), 275–280. doi:10.1023/A:1006791514080

Kelly-Vargas, K., Cerro-Lopez, M., Reyna-Tellez, S., Bandala, E. R., & Sanchez-Salas, J. L. (2012). Biosorption of heavy metals in polluted water, using different waste fruit cortex. *Physics and Chemistry of the Earth Parts A/B/C, 37-39*, 26–29. doi:10.1016/j.pce.2011.03.006

Keskin Gündoğdu, T., Deniz, İ., Çalışkan, G., Şahin, E. S., & Azbar, N. (2016). Experimental design methods for bioengineering applications. *Critical Reviews in Biotechnology, 36*(2), 368–388. doi:10.3109/07388551.2014.973014 PMID:25373790

Keskinan, O., Goksu, M. Z. L., Yuceer, A., Basibuyuk, M., & Forster, C. (2003). Heavy metal adsorption characteristics of a submerged aqueous plant (Myriophyllum spicatum). *Process Biochemistry (Barking, London, England), 39*(2), 179–183. doi:10.1016/S0032-9592(03)00045-1

Khamkeaw, A., Asavamongkolkul, T., Perngyai, T., Jongsomjit, B., & Phisalaphong, M. (2020). Interconnected micro, meso, and macro porous activated carbon from bacterial nanocellulose for superior adsorption properties and effective catalytic performance. *Molecules (Basel, Switzerland), 25*(18), 4063. doi:10.3390/molecules25184063 PMID:32899569

Khan, S., Rehman, S., Khan, A. Z., Khan, M. A., & Shah, M. T. (2010). Soil and vegetables enrichment with heavy metals from geological sources in Gilgit, northern Pakistan. *Ecotoxicology and Environmental Safety, 73*(7), 1820–1827. doi:10.1016/j.ecoenv.2010.08.016 PMID:20810165

Khan, S., Reid, B. J., Li, G., & Zhu, Y. G. (2014). Application of biochar to soil reduces cancer risk via rice consumption: A case study in Miaoqian village, Longyan, China. *Environment International*, *68*, 154–161. doi:10.1016/j.envint.2014.03.017 PMID:24727070

Khosravi, J., & Alamdari, A. (2009). Copper removal from oil-field brine by coprecipitation. *Journal of Hazardous Materials*, *166*(2–3), 695–700. doi:10.1016/j.jhazmat.2008.11.079 PMID:19157701

Kildegaard, G., Balbi, M. D. P., Salierno, G., Cassanello, M., De Blasio, C., & Galvagno, M. (2022). A Cleaner Delignification of Urban Leaf Waste Biomass for Bioethanol Production, Optimised by Experimental Design. *Processes*, *10*(5), 943. doi:10.3390/pr10050943

Kim, I., Lee, M., & Wang, S. (2014). Heavy metal removal in groundwater originating from acid mine drainage using dead *Bacillus drentensis* sp. immobilized in polysulfone polymer. *Journal of Environmental Management*, *146*, 568–574. doi:10.1016/j.jenvman.2014.05.042

Kim, S. Y., Jin, M. R., Chung, C. H., Yun, Y.-S., Jahng, K. Y., & Yu, K.-Y. (2015). Biosorption of cationic basic dye and cadmium by the novel biosorbent *Bacillus catenulatus* JB-022 strain. *Journal of Bioscience and Bioengineering*, *119*(4), 433–439. doi:10.1016/j.jbiosc.2014.09.022 PMID:25454694

Kim, Y., & Park, S. (2014). Iron deficiency increases blood concentrations of neurotoxic metals in children. *Korean Journal of Pediatrics*, *57*(8), 345–350. doi:10.3345/kjp.2014.57.8.345 PMID:25210521

Kinoshita, H., Ohtake, F., Ariga, Y., & Kimura, K. (2016). Comparison and Characterization of Biosorption by Weissellaviridescens MYU 205 of Periodic Group 12 Metal Ions. *Animal Science Journal*, *87*(2), 271–276. doi:10.1111/asj.12425 PMID:26223952

Kiran, I., Akar, T., & Tunali, S. (2005). Biosorption of Pb (II) and Cu (II) from aqueous solution by pretreated biomass of Neurospora crassa. *Process Biochemistry (Barking, London, England)*, *40*(11), 3550–3558. doi:10.1016/j.procbio.2005.03.051

Kishor, B. (2014). Removal of Heavy Metals from Wastewater using Iron Plaque of Typha-latifolia & Bacterial Precipitated Fe (OH) 3 Rich Soil. *International Journal of Engineering Research & Technology (Ahmedabad)*, *3*(5), 26–31.

Kiyono, M., & Pan-Hou, H. (2006). Genetic engineering of bacteria for environmental remediation of mercury. *Journal of Health Science*, *52*(3), 199–204. doi:10.1248/jhs.52.199

Koelmel, J., Prasad, M. N. V., Velvizhi, G., Butti, S. K., & Venkata, M. S. (2016). Metalliferous waste in India and knowledge explosion in Metal Recovery Techniques and processes for the prevention of pollution. *Environ. Mat. Was.*, *15*, 339–390. doi:10.1016/B978-0-12-803837-6.00015-9

Koller, M., & Saleh, H. M. (2018). Introductory chapter: introducing heavy metals. In Tech Open, 1-11.

Koller, M., & Saleh, H. M. (2018). Introductory chapter: introducing heavy metals. Tech Open. doi:10.5772/intechopen.74783

Konig-Peter, A., Kilar, F., & Pernyeszi, T. (2019). Copper(II) biosorption characteristics of lyophilized and thermally treated *Pseudomonas* cells. *Environmental Engineering and Management Journal*, *18*(2), 455–464. doi:10.30638/eemj.2019.043

Kordialik-Bogacka, E. (2011). Cadmium and lead recovery from yeast biomass. *Central European Journal of Chemistry*, *9*, 320–325.

Koul, B., Yakoob, M., & Shah, M. P. (2022). Agricultural waste management strategies for environmental sustainability. *Environmental Research*, *206*, 112285. doi:10.1016/j.envres.2021.112285 PMID:34710442

Kratochvil, D., & Volesky, B. (1998). Advances in the biosorption of heavy metals. *Trends in Biotechnology*, *16*(7), 291–300. doi:10.1016/S0167-7799(98)01218-9

Kravanja, K. A., Finšgar, M., Knez, Ž., & Knez Marevci, M. (2022). Supercritical Fluid Technologies for the Incorporation of Synthetic and Natural Active Compounds into Materials for Drug Formulation and Delivery. *Pharmaceutics*, *14*(8), 1670. Advance online publication. doi:10.3390/pharmaceutics14081670

Kubier, A., Wilkin, R. T., & Pichler, T. (2019). Cadmium in soils and groundwater: A review. *Applied Geochemistry*, *108*, 1–16. doi:10.1016/j.apgeochem.2019.104388 PMID:32280158

Kulka, M. (2016). A review of paraoxonase 1 properties and diagnostic applications. *Polish Journal of Veterinary Sciences*, *19*(1), 225–232. doi:10.1515/pjvs-2016-0028 PMID:27096809

Kulkarni, R., Shetty, K. V., & Srinikethan, G. (2014). Cadmium (II) and nickel (II) biosorption by *Bacillus laterosporus* (MTCC 1628). *Journal of the Taiwan Institute of Chemical Engineers*, *45*(4), 1628–1635. doi:10.1016/j.jtice.2013.11.006

Kumar, H., Bhardwaj, K., Sharma, R., Nepovimova, E., Kuča, K., Dhanjal, D. S., Verma, R., Bhardwaj, P., Sharma, S., & Kumar, D. (2020). Fruit and Vegetable Peels: Utilization of High Value Horticultural Waste in Novel Industrial Applications. *Molecules (Basel, Switzerland)*, *25*(12), 2812. doi:10.3390/molecules25122812 PMID:32570836

Kumar, R. R., Park, B. J., & Cho, J. Y. (2013). Application and environmental risks of livestock manure. *Journal of the Korean Society for Applied Biological Chemistry*, *56*(5), 497–503. doi:10.1007/s13765-013-3184-8

Kumar, R. S., Singh, D., Bose, S. K., & Trivedi, P. K. (2020). Biodegradation of environmental pollutant through pathways engineering and genetically modified organisms approaches. In *Microorganisms for Sustainable Environment and Health* (pp. 137–165). Elsevier. doi:10.1016/B978-0-12-819001-2.00007-3

Kumar, R., Bhatia, D., Singh, R., Rani, S., & Bishnoi, N. (2011). Sorption of heavy metals from electroplating effluent using immobilized biomass *Trichoderma viride* in a continuous packed-bed column. *International Biodeterioration & Biodegradation*, *65*(8), 1133–1139. doi:10.1016/j.ibiod.2011.09.003

Kumar, S., & Asif, P. (2019). Equilibrium and Kinetic Studies of Biosorptive Removal of 2,4,6-Trichlorophenol from Aqueous Solutions Using Untreated Agro-Waste Pine Cone Biomass. *Processes (Basel, Switzerland)*, *7*(10), 757. doi:10.3390/pr7100757

Kumar, S., Prasad, S., Yadav, K. K., Shrivastava, M., Gupta, N., Nagar, S., Bach, Q.-V., Kamyab, H., Khan, S. A., Yadav, S., & Malav, L. C. (2019). Hazardous heavy metals contamination of vegetables and food chain: Role of sustainable remediation approaches-A review. *Environmental Research*, *179*, 108792. doi:10.1016/j.envres.2019.108792 PMID:31610391

Kurniawan, T. A., Chan, G. Y., Lo, W. H., & Babel, S. (2006). Physico-chemical treatmenttechniques for wastewater laden with heavy metals. *Chemical Engineering Journal*, *118*(1–2), 83–98. doi:10.1016/j.cej.2006.01.015

Kuyucak, N., & Volesky, B. (1988). Biosorbents for recovery of metals from industrial solutions. *Biotechnology Letters*, *10*(2), 137–142. doi:10.1007/BF01024641

Kwikima, M. M., Mateso, S., & Chebude, Y. (2021). Potentials of agricultural wastes as the ultimate alternative adsorbent for cadmium removal from wastewater. A review. *Scientific African*, *13*, e00934. doi:10.1016/j.sciaf.2021.e00934

Kyzas, G. Z. (2012). Commercial coffee wastes as materials for adsorption of heavy metals from aqueous solutions. *Materials (Basel)*, *5*(10), 1826–1840. doi:10.3390/ma5101826

Lai, H., Ding, X., Cui, M., Zheng, J., Chen, Z., Pei, J., & Zhang, J. (2023). Mechanisms and influencing factors of biomineralization based heavy metal remediation: A review. *Biogeotechnics*, *1*(3), 100039. doi:10.1016/j.bgtech.2023.100039

Lanati, A., Poli, C., Imberti, M., Menegon, A., & Grohovaz, F. (2015). A Design of Experiment Approach to Optimize an Image Analysis Protocol for Drug Screening. In V. Zazzu, M. B. Ferraro, & M. R. Guarracino (Eds.), *Mathematical Models in Biology: Bringing Mathematics to Life* (pp. 65–84). Springer International Publishing. doi:10.1007/978-3-319-23497-7_5

Landrigan, P. J. (1982). Occupational and community exposures to toxic metals: Lead, cadmium, mercury and arsenic. *The Western Journal of Medicine*, *137*(6), 531. PMID:7164433

Lanphear, B. P., Matte, T. D., Rogers, J., Clickner, R. P., Dietz, B., Bornschein, R. L., Succop, P., Mahaffey, K. R., Dixon, S., Galke, W., Rabinowitz, M., Farfel, M., Rohde, C., Schwartz, J., Ashley, P., & Jacobs, D. E. (1998). The contribution of lead-contaminated house dust and residential soil to children's blood lead levels. A pooled analysis of 12 epidemiologic studies. *Environmental Research*, *79*(1), 51–68. doi:10.1006/enrs.1998.3859 PMID:9756680

Lara-Torres, S., Figueiredo, D., Paz, S., Gutiérrez, A. J., Rubio, C., González-Weller, D., Revert, C., & Hardisson, A. (2021). Determination and risk assessment of toxic metals in lipsticks from Europe and China. *Journal of Trace Elements in Medicine and Biology*, *67*, 126792. doi:10.1016/j.jtemb.2021.126792 PMID:34022566

Largitte, L., & Pasquier, R. (2016). A review of the kinetics adsorption models and their application to the adsorption of lead by an activated carbon. *Chemical Engineering Research & Design*, *109*, 495–504. doi:10.1016/j.cherd.2016.02.006

Lata, S., Singh, P. K., & Samadder, S. R. (2015). Regeneration of adsorbents and recovery of heavy metals: A review. *International Journal of Environmental Science and Technology*, *12*(4), 1461–1478. doi:10.1007/s13762-014-0714-9

Lau, T. C., Ang, P. O., & Wong, P. K. (2003). Development of seaweed biomass as a biosorbent for metal ions. *Water Science and Technology*, *47*(10), 49–54. doi:10.2166/wst.2003.0536 PMID:12862216

Lazli, W., & Hellal, A. (2020). Biosorption of phenol by modified dead leaves of *Posidonia oceanica* immobilized in calcium alginate beads: Optimal experimental parameters using central composite design. *Water S.A.*, *46*(2 April). Advance online publication. doi:10.17159//wsa/2020.v46.i2.8248

Lee, K.-M., & Gilmore, D. F. (2006). Statistical experimental design for bioprocess modeling and optimization analysis. *Applied Biochemistry and Biotechnology*, *135*(2), 101–115. doi:10.1385/ABAB:135:2:101 PMID:17159235

Lemke, M., & DeSalle, R. (2023). The next generation of microbial ecology and its importance in environmental sustainability. *Microbial Ecology*, *85*(3), 781–795. doi:10.1007/s00248-023-02185-y PMID:36826587

Leng, L., Huang, H., Li, H., Li, J., & Zhou, W. (2019). Biochar stability assessment methods: A review. *The Science of the Total Environment*, *647*, 210–222. doi:10.1016/j.scitotenv.2018.07.402 PMID:30077850

Lenntech, K. (2004). *Water treatment and air purification*. Rotter Dam Seweg.

Lentini, P., Zanoli, L., Granata, A., Signorelli, S. S., Castellino, P., & Dell'Aquila, R. (2017). Kidney and heavy metals – the role of environmental exposure. *Molecular Medicine Reports*, *15*(5), 3413–3419. doi:10.3892/mmr.2017.6389 PMID:28339049

Levesque, V., Oelbermann, M., & Ziadi, N. (2021). Biochar in temperate soils: Opportunities and challenges. *Canadian Journal of Soil Science*, *102*(1), 1–26. doi:10.1139/cjss-2021-0047

Li, H., Fagerberg, B., Sallsten, G., Borné, Y., Hedblad, B., Engström, G., Barregard, L., & Andersson, E. M. (2019). Smoking-induced risk of future cardiovascular disease is partly mediated by cadmium in tobacco: Malmö Diet and Cancer Cohort study. *Environmental Health*, *18*(1), 56. doi:10.1186/s12940-019-0495-1 PMID:31200698

Li, H., Li, Z., Liu, T., Xiao, X., Peng, Z., & Deng, L. (2008). A novel technology for biosorption and recovery hexavalent chromium in waste water by bio-functional magnetic beads. *Bioresource Technology*, *99*(14), 6271–6279. doi:10.1016/j.biortech.2007.12.002

Li, J., Cao, L., Yuan, Y., Wang, R., Wen, Y., & Man, J. (2018). Comparative study for microcystin-LR sorption onto biochars produced from various plant- and animal-wastes at different pyrolysis temperatures: Influencing mechanisms of biochar properties. *Bioresource Technology*, *247*, 794–803. doi:10.1016/j.biortech.2017.09.120 PMID:30060415

Li, M., Pi, H., Yang, Z., Reiter, R. J., Xu, S., Chen, X., ... Zhang, Y. (2016). Melatonin antagonizes cadmium-induced neurotoxicity by activating the transcription factor EB-dependent autophagy–lysosome machinery in mouse neuroblastoma cells. *Journal of Pineal Research*, *61*(3), 353–369. doi:10.1111/jpi.12353 PMID:27396692

Limcharoensuk, T., Sooksawat, N., Sumarnrote, A., Awutpet, T., Kruatrachue, M., Pokethitiyook, P., & Auesukaree, C. (2015). Bioaccumulation and biosorption of Cd^{2+} and Zn^{2+} by bacteria isolated from a zinc mine in Thailand. *Ecotoxicology and Environmental Safety*, *122*, 322–330. doi:10.1016/j.ecoenv.2015.08.013 PMID:26300116

Lin, C., Lu, W., & Wu, Y. (2005). Agricultural soils irrigated with acidic mine water: Acidity, heavy metals, and crop contamination. *Soil Research (Collingwood, Vic.)*, *43*(7), 819–826. doi:10.1071/SR04148

Ling, E. J., Benham, B., & Forrester, S. (2011). *Virginia household water quality program: Heavy metals in household water*. Virginia Cooperative Extension.

Lin, Y., Wang, X., Wang, B., Mohamad, O., & Wei, G. (2012). Bioaccumulation characterization of zinc and cadmium by Streptomyces zinciresistens, a novel actinomycete. *Ecotoxicology and Environmental Safety*, *77*, 7–17. doi:10.1016/j.ecoenv.2011.09.016 PMID:22001317

Lin, Z., Li, J., Luan, Y., & Dai, W. (2020). Application of algae for heavy metal adsorption: A 20-year meta-analysis. *Ecotoxicology and Environmental Safety*, *190*, 110089. doi:10.1016/j.ecoenv.2019.110089 PMID:31896472

Li, R., Wu, H., Ding, J., Gan, L., & Li, Y. (2017). Mercury pollution in vegetables, grains and soils from areas surrounding coal-fired power plants. *Scientific Reports*, *7*(1), 46545. doi:10.1038/srep46545 PMID:28484233

Li, S., Xu, S., Liu, S., Yang, C., & Lu, Q. (2004). Fast pyrolysis of biomass in free-fall reactor for hydrogen-rich gas. *Fuel Processing Technology*, *85*(8-10), 1201–1211. doi:10.1016/j.fuproc.2003.11.043

Litvak, P., Slavkovich, V., Liu, X., Popovac, D., Preteni, E., Capuni-Paracka, S., ... Graziano, J. (1998). Hyperproduction of erythropoietin in nonanemic lead-exposed children. *Environmental Health Perspectives*, *106*(6), 361–364. doi:10.1289/ehp.98106361 PMID:9618353

Liu, C., Ngo, H. H., & Guo, W. (2012). Watermelon rind: Agro waste or superior biosorbent. *Applied Biochemistry and Biotechnology*, *167*(6), 1699–1715. doi:10.1007/s12010-011-9521-7 PMID:22222432

Liu, G., Dai, Z., Liu, X., Dahlgren, R. A., & Xu, J. (2022). Modification of agricultural wastes to improve sorption capacities for pollutant removal from water – a review. *Carbon Research*, *1*(1), 24. doi:10.1007/s44246-022-00025-1

Liu, J., Zhou, X., Wang, T., Fan, L., Liu, S., Wu, N., Xu, A., Qian, X., Li, Z., Jiang, M., Zhou, J., & Dong, W. (2022). Construction and comparison of synthetic microbial consortium system (SMCs) by non-living or living materials immobilization and application in acetochlor degradation. *Journal of Hazardous Materials*, *438*, 129460. doi:10.1016/j.jhazmat.2022.129460 PMID:35803189

Liu, L., Bilal, M., Duan, X., & Hafiz, M. N. (2019). Iqbal. "Mitigation of environmental pollution by genetically engineered bacteria—Current challenges and future perspectives.". *The Science of the Total Environment*, *667*, 444–454. doi:10.1016/j.scitotenv.2019.02.390 PMID:30833243

Liu, L., Liu, J., Liu, X., Dai, C., Zhang, Z., Song, W., & Chu, Y. (2019). Kinetic and equilibrium of U(VI) biosorption onto the resistant bacterium *Bacillus amyloliquefaciens*. *Journal of Environmental Radioactivity*, *203*, 117–124. doi:10.1016/j.jenvrad.2019.03.008

Liu, L., Xia, M., Hao, J., Xu, H., & Song, W. (2021). Biosorption of Pb(II) by the resistant *Enterobacter* sp.: Investigated by kinetics, equilibrium and thermodynamics. *Archives of Environmental Protection*, *47*(3), 28–36. doi:10.24425/aep.2021.138461

Liu, M., Deng, Y., Zhan, H., & Zhang, X. (2002). Adsorption and desorption of copper(II) from solutions on new spherical cellulose adsorbent. *Journal of Applied Polymer Science*, *84*(3), 478–485. doi:10.1002/app.10114

Liu, P., Zhao, H., Wang, L., Liu, Z., Wei, J., Wang, Y., Jiang, L., Dong, L., & Zhang, Y. (2011). Analysis of heavy metal sources for vegetable soils from Shandong Province, China. *Agricultural Sciences in China*, *10*(1), 109–119. doi:10.1016/S1671-2927(11)60313-1

Liu, R., Guan, Y., Chen, L., & Lian, B. (2018). Adsorption and desorption characteristics of Cd2+ and Pb2+ by micro and nano-sized biogenic CaCO3. *Frontiers in Microbiology*, *9*(JAN), 294049. doi:10.3389/fmicb.2018.00041

Liu, R., Liu, G., Yousaf, B., Niu, Z., & Abbas, Q. (2022). Novel investigation of pyrolysis mechanisms and kinetics for functional groups in biomass matrix. *Renewable & Sustainable Energy Reviews*, *153*, 111761. doi:10.1016/j.rser.2021.111761

Liu, Y., & Liu, Y. J. (2008). Biosorption isotherms, kinetics and thermodynamics. *Separation and Purification Technology*, *61*(3), 229–242. doi:10.1016/j.seppur.2007.10.002

Liu, Y., & Shen, L. (2008). A general rate law equation for biosorption. *Biochemical Engineering Journal*, *38*(3), 390–394. doi:10.1016/j.bej.2007.08.003

Liu, Z., Xu, Z., Xu, L., Buyong, F., Chay, T. C., Li, Z., Cai, Y., Hu, B., Zhu, Y., & Wang, X. (2022). Modified biochar: Synthesis and mechanism for removal of environmental heavy metals. *Carbon Research*, *1*(1), 1–21. doi:10.1007/s44246-022-00007-3

Li, Y., Gao, L., Lu, Z., Wang, Y., Wang, Y., & Wan, S. (2020). Enhanced Removal of Heavy Metals from Water by Hydrous Ferric Oxide-Modified Biochar. *ACS Omega*, *5*(44), 28702–28711. doi:10.1021/acsomega.0c03893

Li, Y., Gupta, R., Zhang, Q., & You, S. (2023). Review of biochar production via crop residue pyrolysis: Development and perspectives. *Bioresource Technology*, *369*, 128423. doi:10.1016/j.biortech.2022.128423 PMID:36462767

Li, Y., Li, L., & Yu, J. (2017). Applications of Zeolites in Sustainable Chemistry. *Chem*, *3*(6), 928–949. doi:10.1016/j.chempr.2017.10.009

Li, Z., Zheng, Z., Li, H., Xu, D., Li, X., Xiang, L., & Tu, S. (2023). Review on Rice Husk Biochar as an Adsorbent for Soil and Water Remediation. *Plants*, *12*(7), 1524. doi:10.3390/plants12071524 PMID:37050150

Lloyd, J. R., Lovley, D. R., & Macaskie, L. E. (2003). Biotechnological application of metal-reducing microorganisms. *Advances in Applied Microbiology*, *53*, 85–128. doi:10.1016/S0065-2164(03)53003-9 PMID:14696317

Loh, N., Loh, H.-P., Wang, L.K., & Wang, M.-H.-S. (2016). Health effects and control of toxic lead in the environment. *Natural Resources and Control Processes*, 233–284. doi:10.1007/978-3-319-26800-2_5

Loranger, S., & Zayed, J. (1995). Environmental and occupational exposure to manganese: A multimedia assessment. *International Archives of Occupational and Environmental Health*, *67*(2), 101–110. doi:10.1007/BF00572233 PMID:7545647

Lu, H., Shi, X., Costa, M., & Huang, C. (2005). Carcinogenic effect of nickel compounds. *Molecular and Cellular Biochemistry*, *279*(1-2), 45–67. doi:10.1007/s11010-005-8215-2 PMID:16283514

Luo, X., Huo, X., Zhang, Y., Cheng, Z., Chen, S., & Xu, X. (2021). Increased intestinal permeability with elevated peripheral blood endotoxin and inflammatory indices for e-waste lead exposure in children. *Chemosphere*, *279*, 130862. doi:10.1016/j.chemosphere.2021.130862 PMID:34134434

Mackova, Macek, & Kotrba. (2011). *Microbial Biosorption of Metals.* Springer.

Madeła, M., & Skuza, M. (2021). Towards a Circular Economy: Analysis of the Use of Biowaste as Biosorbent for the Removal of Heavy Metals. *Energies*, *14*(17), 5427. doi:10.3390/en14175427

Mahalakshmi, M. (2012). Characteristic levels of heavy metals in canned tuna fish. *Journal of Toxicology and Environmental Health Sciences*, *4*(2), 43–45. doi:10.5897/JTEHS11.079

Mahato, N., Sharma, K., Sinha, M., Baral, E. R., Koteswararao, R., Dhyani, A., Hwan Cho, M., & Cho, S. (2020). Biosorbents, industrially important chemicals and novel materials from citrus processing waste as a sustainable and renewable bioresource: A review. *Journal of Advanced Research*, *23*, 61–82. doi:10.1016/j.jare.2020.01.007 PMID:32082624

Mahmoud, E., Abdou, H., Mohamed, S., & Osman, M. (2016). Engineered *Staphylococcus aureus* via immobilization on magnetic Fe3O4-phthalate nanoparticles for biosorption of divalent ions from aqueous solutions. *Journal of Environmental Chemical Engineering*, *4*(4), 3810–3824. doi:10.1016/j.jece.2016.08.022

Mahmoud, E., Yakout, A., Abdel-Aal, H., & Osman, M. (2011). Enhanced biosorptive removal of cadmium from aqueous solutions by silicon dioxide nano-powder, heat inactivated and immobilized *Aspergillus ustus. Desalination*, *279*(1-3), 291–297. doi:10.1016/j.desal.2011.06.023

Mahmoud, E., Yakout, A., Hany, A., & Osman, M. (2015). Speciation and selective biosorption of Cr(III) and Cr(VI) using nanosilica immobilized-fungi biosorbents. *Journal of Environmental Engineering*, *141*(4), 1–9. doi:10.1061/(ASCE)EE.1943-7870.0000899

Maisarah, A., Adityosulindro, S., & Wulandari, D. (2021). Utilization of wild algae biomass as biosorbent for removal of heavy metal Zinc (Zn^{2+}) from aqueous solution. *IOP Conference Series. Earth and Environmental Science*, *824*(1), 012017. doi:10.1088/1755-1315/824/1/012017

Maiti, S., Prasad, B., & Minocha, A. K. (2022). Optimization of Pb(II) removal from aqueous solutions by fly ash using Box-Behnken design. *Environmental Engineering and Management Journal*, *21*(1), 125–135. doi:10.30638/eemj.2022.012

Malik, L. A., Bashir, A., Qureashi, A., & Pandith, A. H. (2019). Detection and removal of heavy metal ions: A review. *Environmental Chemistry Letters*, *17*(4), 1495–1521. doi:10.1007/s10311-019-00891-z

Mamatha, M., Aravinda, H. B., Puttaiah, E. T., & Manjappa, S. (2013). Adsorbent from Pongamia Pinnata tree bark for zinc adsorption. *Int. J. Sci. Eng. Technol.*, *2*(1), 22–25.

Mañay, N., Cousillas, A. Z., Alvarez, C., & Heller, T. (2008). Lead contamination in Uruguay: The "La Teja" neighborhood case. *Reviews of Environmental Contamination and Toxicology*, *195*, 93–115. PMID:18418955

Mandal, A., Singh, N., & Nain, L. (2017). Agro-waste biosorbents: Effect of physico-chemical properties on atrazine and imidacloprid sorption. *Journal of Environmental Science and Health. Part B, Pesticides, Food Contaminants, and Agricultural Wastes*, *52*(9), 671–682. doi:10.1080/03601234.2017.1331677 PMID:28679066

Mandal, B. K., & Sujuki, K. J. (2002). Arsenic round the world: A review. *Talanta*, *58*(1), 201–235. doi:10.1016/S0039-9140(02)00268-0 PMID:18968746

Manne, R., Kumaradoss, M. M., Iska, R. S. R., Devarajan, A., & Mekala, N. (2022). Water quality and risk assessment of copper content in drinking water stored in copper container. *Applied Water Science*, *12*(3), 27. doi:10.1007/s13201-021-01542-x

Mansour, S. A., Belal, M. H., Abou-Arab, A. A. K., & Gad, M. F. (2009). Monitoring of pesticides and heavy metals in cucumber fruits produced from different farming systems. *Chemosphere*, *75*(5), 601–609. doi:10.1016/j.chemosphere.2009.01.058 PMID:19237184

Manzoor, J., Sharma, M., & Wani, K. A. (2018). Heavy metals in vegetables and their impact on the nutrient quality of vegetables: A review. *Journal of Plant Nutrition*, *41*(13), 1744–1763. doi:10.1080/01904167.2018.1462382

Mao, J., Kim, S., Wu, X. H., Kwak, I., Zhou, T., & Yun, Y. (2015). A sustainable cationic chitosan/*E. coli* fibre biosorbent for Pt(IV) removal and recovery in batch and column systems. *Separation and Purification Technology*, *143*, 32–39. doi:10.1016/j.seppur.2015.01.023

Mapombere, G., Nyoni, B., Sibali, L. L., Chiririwa, H., & Seodigeng, T. (2022). Adsorption of mercury by *Pterocarpus Anglolensis*: Study on adsorption isotherms and kinetics. *Iranian Journal of Chemistry & Chemical Enginerring*, *40*, 143–153.

Marín-Rangel, V. M., Cortés-Martínez, R., Villanueva, R. A., Garnica-Romo, M. G., & Martínez-Flores, H. E. (2012). As (V) biosorption in an aqueous solution using chemically treated lemon (Citrus aurantifolia Swingle) residues. *Journal of Food Science*, *77*(1), T10–T14. doi:10.1111/j.1750-3841.2011.02466.x PMID:22122309

Marshall, A. (2013). *Commercial Application of Pyrolysis Technology in Agriculture.* doi:10.13031/aim.20141909089

Martin, S., & Griswold, W. (2009). Human health effects of heavy metals. *Environmental Science and Technology Briefs for Citizens*, *15*, 1–6.

Martorell, I., Perelló, G., Martí-cid, R., Llobet, J.M., Castell, V., Domingo, J.L. (2011). *Human exposure to arsenic, cadmium, mercury, and lead from foods in Catalonia, Spain: Temporal Trend.* Academic Press.

Masindi, V., & Muedi, K. L. (2018). Environmental contamination by heavy metals. In *Heavy Metals*. InTech. doi:10.5772/intechopen.76082

Masri, S., LeBrón, A. M., Logue, M. D., Valencia, E., Ruiz, A., Reyes, A., & Wu, J. (2021). Risk assessment of soil heavy metal contamination at the census tract level in the city of Santa Ana, CA: Implications for health and environmental justice. *Environmental Science. Processes & Impacts*, *23*(6), 812–830. doi:10.1039/D1EM00007A PMID:33954329

Mat Arisah, F., Amir, A. F., Ramli, N., Ariffin, H., Maeda, T., Hassan, M. A., & Mohd Yusoff, M. Z. (2021). Bacterial Resistance against heavy metals in Pseudomonas aeruginosa RW9 involving hexavalent chromium removal. *Sustainability (Basel)*, *13*(17), 9797. doi:10.3390/su13179797

Mathew, B. B., & Krishnamurthy, N. B. (2018). Screening and identification of bacteria isolated from industrial area groundwater to study lead sorption: Kinetics and statistical optimization of biosorption parameters. *Groundwater for Sustainable Development*, *7*, 313–327. doi:10.1016/j.gsd.2018.07.007

Ma, Y., Chen, H., Lan, C., & Ren, J. (2018). Help, hope and hype: Ethical considerations of human microbiome research and applications. *Protein & Cell*, *9*(5), 404–415. doi:10.1007/s13238-018-0537-4 PMID:29675808

Mead, M. N. (2010). Cadmium confusion: Do consumers need protection? *Environmental Health Perspectives*, *118*(12), 528–534. doi:10.1289/ehp.118-a528 PMID:21123140

Meier, L. (2014). *Statistical and Numerical Methods for Chemical Engineers.* https://people.math.ethz.ch/~muellepa/2016Stat/Script/cheming-e-part-I.pdf

Meltzer, H. M., Brantsaeter, A. L., Borch-Iohnsen, B., Ellingsen, D. G., Alexander, J., Thomassen, Y., Stigum, H., & Ydersbond, T. A. (2010). Low iron stores are related to higher blood concentrations of manganese, cobalt, and cadmium in non-smoking, Norwegian women in the HUNT 2 study. *Environmental Research*, *110*(5), 497–504. doi:10.1016/j.envres.2010.03.006 PMID:20381026

Mhlanga, S. D., Mamba, B. B., Krause, R. W., & Malefetse, T. J. (2007). Removal of organic contaminants from water using nanosponge cyclodextrin polyurethanes. Journal of Chemical Technology & Biotechnology: International Research in Process. *Environmental & Clean Technology*, *82*(4), 382–388.

Michalak, I., & Chojnacka, K. (2010). Interactions of metal cations with anionic groups on the cell wall of the macroalga Vaucheria sp. *Engineering in Life Sciences*, *10*(3), 209–217. doi:10.1002/elsc.200900039

Michalak, I., Chojnacka, K., & Krowiak, A. W. (2013). State of the art for the bio sorption process-A review. *Applied Biochemistry and Biotechnology*, *170*(6), 1389–16. doi:10.1007/s12010-013-0269-0 PMID:23666641

Mishra, S., Bharagava, R. N., More, N., Yadav, A., Zainith, S., Mani, S., & Chowdhary, P. (2019). Heavy metal contamination: an alarming threat to environment and human health. *Environmental biotechnology: For sustainable future*, 103-125.

Mishra, A., Dubey, A., & Shinghal, S. (2015). Biosorption of chromium(VI) from aqueous solutions using waste plant biomass. *International Journal of Environmental Science and Technology*, *12*(4), 1415–1426. doi:10.1007/s13762-014-0516-0

Mishra, S., Singh, G., Gupta, A., & Tiwari, R. K. (2023). Heavy Metal/Metalloid Contamination: Their Sources in Environment and Accumulation in Food Chain. In R. P. Singh, P. Singh, & A. Srivastava (Eds.), *Heavy Metal Toxicity: Environmental Concerns, Remediation and Opportunities*. Springer. doi:10.1007/978-981-99-0397-9_2

Mishra, V., Balomajumder, C., & Agarwal, V. K. (2010). Zn (II) ion biosorption onto the surfaceof eucalyptus leaf biomass: Isotherm, kinetic, and mechanistic modeling. *Clean (Weinheim)*, *38*(11), 1062–1073. doi:10.1002/clen.201000030

Misra, D. N. (1973). Jovanovich adsorption isotherm for heterogeneous surfaces. *Journal of Colloid and Interface Science*, *43*(1), 85–88. doi:10.1016/0021-9797(73)90350-0

Mitra, S., Chakraborty, A. J., Tareq, A. M., Emran, T. B., Nainu, F., Khusro, A., Idris, A. M., Khandaker, M. U., Osman, H., Alhumaydhi, F. A., & Simal-Gandara, J. (2022). Impact of heavy metals on the environment and human health: Novel therapeutic insights to counter the toxicity. *Journal of King Saud University. Science*, *34*(3), 101865. doi:10.1016/j.jksus.2022.101865

Moghaddam, R., Fatemi, S., & Keshtkar, A. (2013). Adsorption of lead (Pb^{2+}) and uranium (UO_2^{2+}) cations by brown algae; experimental and thermodynamic modelling. *Chemical Engineering Journal*, *231*, 294–303. doi:10.1016/j.cej.2013.07.037

Mohammadi, A., Mahmoudnia, F., Mohammadi, A., & Mahmoudnia, F. (2023). *Biological Treatment of Heavy Metals with Algae*. Heavy Metals - Recent Advances., doi:10.5772/intechopen.110301

Mohammed, A. S., Kapri, A., & Goel, R. (2011). Heavy metal pollution: source, impact, and remedies. *Biomanagement of metal-contaminated soils*, 1-28.

Mohan, V. (2018). Across India, high levels of toxins in groundwater. *The Times of India*.

Mohan, D., & Pittman, J. C. U. (2007). Arsenic removal from water/wastewater using adsorbents-A critical review. *International Journal of Mineral Processing*, *92*, 42–48. PMID:17324507

Mohapatra, R. K., Parhi, P. K., Pandey, S., Bindhani, B. K., Thatoi, H., & Panda, C. R. (2019). Active and passive biosorption of Pb(II)using live and dead biomass of marine bacterium *Bacillus xiamenensis* PbRPSD202: Kinetics and isotherm studies. *Journal of Environmental Management, 247*, 121–134. doi:10.1016/j.jenvman.2019.06.073 PMID:31238200

Mohmand, A. Q. K., Kousar, M. W., Zafar, H., Bukhari, K. T., & Khan, M. Z. (2011). Medical importance of fungi with special emphasis on mushrooms. *Isra Medical Journal, 3*, 1–44.

Mokarram, M., Saber, A., & Sheykhi, V. (2020). Effects of heavy metal contamination on river water quality due to release of industrial effluents. *Journal of Cleaner Production, 277*, 123380. doi:10.1016/j.jclepro.2020.123380

Moller, G., Baker, M., Dunkel, C., Shresta, D., & Strawn, D. (2016). N-E-W TechTM: Resource recovery in a novel functionalised biochar, catalytic oxidation, reactive filtration water treatment for N/P removal/recovery and pathogen/priority substance destruction. *10th European Waste Water Management Conference.*

Mondal, N. K., Samanta, A., Dutta, S., & Chattoraj, S. (2017). Optimization of Cr(VI) biosorption onto *Aspergillus niger* using 3-level Box-Behnken design: Equilibrium, kinetic, thermodynamic and regeneration studies. *Journal of Genetic Engineering and Biotechnology, 15*(1), 151–160. doi:10.1016/j.jgeb.2017.01.006 PMID:30647651

Montgomery, D. C. (2013). *Design and Analysis of Experiments* (8th ed.). John Wiley & Sons, Inc.

Moore, T. A., Black, A., Centeno, J. A., Harding, J. S., & Trumm, D. A. (Eds.). (2005). *Metal Contaminants in New Zealand* (pp. 195–219). Resolutionz Press.

Morgan, T. J., Turn, S. Q., George, A., & Aburto, J. (2015). Fast Pyrolysis Behaviour of Banagrass as a Function of Temperature and Volatiles Residence Time in a Fluidized Bed Reactor. *PLoS One, 10*(8), e0136511. Advance online publication. doi:10.1371/journal.pone.0136511 PMID:26308860

Morosanu, I., Teodosiu, C., Paduraru, C., Ibanescu, D., & Tofan, L. (2017). Biosorption of lead ions from aqueous effluents by rapeseed biomass. *New Biotechnology, 39*, 110–124. doi:10.1016/j.nbt.2016.08.002

Mosa, K., Saadoun, I., Kumar, K., Helmy, M., & Dhankher, O. (2016). Potential biotechnological strategies for the cleanup of heavy metals and metalloids. *Frontiers in Plant Science, 7*, 303. doi:10.3389/fpls.2016.00303

Mosbah, R., & Sahmoune M. N. (2013). Biosorption of heavy metals by Streptomyces species - an overview. *Central European Journal of Chemistry, 11*, 1412-1422.

Motsi, T., Rowson, N. A., & Simmons, M. J. H. (2009). Adsorption of heavy metals from acid mine drainage by natural zeolite. *International Journal of Mineral Processing, 92*(1-2), 42–48. doi:10.1016/j.minpro.2009.02.005

Muhammad, S., Abdul Khalil, H. P. S., Abd Hamid, S., Albadn, Y. M., Suriani, A. B., Kamaruzzaman, S., Mohamed, A., Allaq, A. A., & Yahya, E. B. (2022). Insights into Agricultural-Waste-Based Nano-Activated Carbon Fabrication and Modifications for Wastewater Treatment Application. *Agriculture, 12*(10), 1737. doi:10.3390/agriculture12101737

Muller, B. R. (2010). Effect of particle size and surface area on the adsorption of albumin-bonded bilirubin on activated carbon. *Carbon, 48*(12), 3607–3615. doi:10.1016/j.carbon.2010.06.011

Muraleedharan, T. R., Iyengar, L., & Venkobachar, C. (1991). Biosorption: An attractive alternative for metal removal and recovery. *Current Science, 61*(6), 379–385.

Mustapha, M. U., & Halimmon, N. (2015). Microorganisms and bio sorption of heavy metals in the environment: A review paper. *Journal of Microbial & Biochemical Technology, 7*(5), 253–256. doi:10.4172/1948-5948.1000219

Mutanda, I., Sun, J., Jiang, J., & Zhu, D. (2022). Bacterial membrane transporter systems for aromatic compounds: Regulation, engineering, and biotechnological applications. *Biotechnology Advances*, *59*, 107952. doi:10.1016/j.biotechadv.2022.107952 PMID:35398204

Mwandira, W., Nakashima, K., Kawasaki, S., Arabelo, A., Banda, K., Nyambe, I., Chirwa, M., Ito, M., Sato, T., Igarashi, T., Nakata, H., Nakayama, S., & Ishizuka, M. (2020). Biosorption of Pb (II) and Zn (II) from aqueous solution by *Oceanobacillus profundus* isolated from an abandoned mine. *Scientific Reports*, *10*(1), 21189. doi:10.1038/s41598-020-78187-4 PMID:33273589

Nabulo, G., Black, C. R., & Young, S. D. (2011). Trace metal uptake by tropical vegetables grown on soil amended with urban sewage sludge. *Environmental Pollution*, *159*(2), 368–376. doi:10.1016/j.envpol.2010.11.007 PMID:21129831

Nachana'a Timothy, E. T. W. (2019). Environmental pollution by heavy metal: An overview. *Chemistry (Basel, Switzerland)*, *3*(2), 72–82.

Naghdi, S., Mirmohammadi, M., Karimzadegan, H., & Ghodusi, J. (2023). Studying the health risk assessment and effect of heavy metals from automobile industries on the air quality of the 21st district of Tehran. *International Journal of Environmental Science and Technology*, *20*(10), 11255–11266. doi:10.1007/s13762-023-05156-5

Nanda, S., Mohanty, P., Pant, K. K., Naik, S., Kozinsk, J. A., & Dalai, A. K. (2013). Characterization of North American lignocellulosic biomass and biochars in terms of their candidacy for alternate renewable fuels. *BioEnergy Research*, *6*(2), 663–677. doi:10.1007/s12155-012-9281-4

Ncibi, M. C. (2008). Applicability of some statistical tools to predict optimum adsorption isotherm after linear and non-linear regression analysis. *Journal of Hazardous Materials*, *153*(1), 207–212. doi:10.1016/j.jhazmat.2007.08.038 PMID:17900804

Neikov, O. D. (2009). Mechanical Crushing and Grinding. In *Handbook of Non Ferrous Metal Powders* (pp. 47–62). Elsevier. doi:10.1016/B978-1-85617-422-0.00002-1

Nekouei, R. K., Pahlevani, F., Assefi, M., Maroufi, S., & Sahajwalla, V. (2019). Selective isolation of heavy metals from spent electronic waste solution by macroporous ion-exchange resins. *Journal of Hazardous Materials*, *379*, 389–396. doi:10.1016/j.jhazmat.2019.03.013 PMID:30870643

Nestle, N. (2002). NMR studies on heavy metal immobilization in biosorbents and mineral matrices. *Reviews in Environmental Science and Biotechnology*, *1*(3), 215–225. doi:10.1023/A:1021255727484

Nestle, N. F. E. I., & Kimmich, R. (2000). NMR imaging of heavy metal absorption in alginate, immobilized cells, and kombu algal biosorbents. *Biotechnology and Bioengineering*, *51*(5), 538–543. doi:10.1002/(SICI)1097-0290(19960905)51:5<538::AID-BIT5>3.0.CO;2-D PMID:18629817

Ngeno, E. C., Mbuci, K. E., Necibi, M. C., Shikuku, V. O., Olisah, C., Ongulu, R., Matovu, H., Ssebugere, P., Abushaban, A., & Sillanpää, M. (2022). Sustainable re-utilization of waste materials as adsorbents for water and wastewater treatment in Africa: Recent studies, research gaps, and way forward for emerging economies. *Environmental Advances*, *9*, 100282. doi:10.1016/j.envadv.2022.100282

Nieboe, E., & Richardson, D. H. S. (1980). The replacement of the no-descript term 'heavy metals' by a biologically and chemically significant classification of metal ions. *Environmental Pollution. Series B. Chemical and Physical*, *1*(1), 3–26. doi:10.1016/0143-148X(80)90017-8

Nishijo, M., Nakagawa, H., Suwazono, Y., Nogawa, K., & Kido, T. (2017). Causes of death in patients with Itai-itai disease suffering from severe chronic cadmium poisoning: A nested case–control analysis of a follow-up study in Japan. *BMJ Open*, *7*(7), e015694. doi:10.1136/bmjopen-2016-015694 PMID:28710217

Njokua, K. L. (2020). Microbial Remediation of Heavy Metals Contaminated Media by Bacillus megaterium and Rhizopus stolonifer. *Scientific African*, *10*, e00545. doi:10.1016/j.sciaf.2020.e00545

Nkwunonwo, U. C., Odika, P. O., & Onyia, N. I. (2020). A review of the health implications of heavy metals in food chain in Nigeria. *TheScientificWorldJournal*, *2020*, 1–11. Advance online publication. doi:10.1155/2020/6594109 PMID:32351345

Noli, F., Kapashi, E., & Kapnisti, M. (2019). Biosorption of uranium and cadmium using sorbents based on Aloe vera wastes. *Journal of Environmental Chemical Engineering*, *7*(2), 102985. doi:10.1016/j.jece.2019.102985

Noor, A., & Khan, S. A. (2023). Agricultural Wastes as Renewable Biomass to Remediate Water Pollution. *Sustainability (Basel)*, *15*(5), 4246. doi:10.3390/su15054246

Norseth, T. (1981). The carcinogenicity of chromium. *Environmental Health Perspectives*, *40*, 121–130. doi:10.1289/ehp.8140121 PMID:7023928

Norton, L., Kangaratnam, B., & McKenzie, T. (2004). Bio sorption of zinc from aqueous solutions using bio solids. *Advances in Environmental Research*, *8*(3-4), 629–635. doi:10.1016/S1093-0191(03)00035-2

Nourmoradi, H., Foroghi, M., Farhadkhani, M., & Vahid Dastjerdi, M. (2013). Assessment of lead and cadmium levels in frequently used cosmetic products in Iran. *Journal of Environmental and Public Health*, *2013*, 962727. doi:10.1155/2013/962727 PMID:24174937

Nova, K. (2023). AI-enabled water management systems: An analysis of system components and interdependencies for water conservation. *Eigenpub Review of Science and Technology*, *7*(1), 105–124.

Novak, J. M., Lima, I., & Xing, B. (2009). Characterization of designer biochar produced at different temperatures and their effects on a loamy sand. *Annalytical Environmental Science*, *3*, 195–206.

Nyika, J., & Dinka, M. (2023a). *Global industrial impacts of heavy metal pollution in sub-Saharan Africa*. IGI Global. doi:10.4018/978-1-6684-7116-6

Nyika, J., & Dinka, M. (2023b). Factors and mechanisms regulating heavy metal phycoremediation in polluted water. *Discover Water*, *3*(1), 14. doi:10.1007/s43832-023-00041-1

Nyika, J., & Dinka, M. (2023c). (in press). The potential of reusing fruit bio-adsorbents for water purification: A mini-review. *Materials Today: Proceedings*. Advance online publication. doi:10.1016/j.matpr.2023.07.285

Nyika, J., & Dinka, M. (2023d). The application of microorganism-derived biosorbents in the removal of heavy metals and dyes. In K. Ukhurebor, U. Aigble, & R. Onyancha (Eds.), *Adsorption applications for environmental sustainability*. IOP Publishing Limited. doi:10.1088/978-0-7503-5598-8ch8

Nyoni, B., Duma, S., Shabangu, S. V., & Hlangothi, S. P. (2020). Comparison of the slow pyrolysi behaviour and kinetics of coal, wood and algaeat high heating rates. *Natural Resources Research*, *29*(6), 3943–3955. doi:10.1007/s11053-020-09687-3

Nyoni, B., & Hlangothi, S. P. (2022). Intermediate pyrolysis of *Scenedesmus* microalgae in a rotary kiln pyrolyser. Effect of temperature on bio-oil yields and composition. *Biomass Conversion and Biorefinery*. Advance online publication. doi:10.1007/s13399-022-03378-0

Nyoni, B., & Hlangothi, S. P. (2023). Co-pyrolysis of low-grade bituminous coal and algal biomass in a rotary kiln: Effect of coal/algae ratio and kiln temperature on the yield and composition of the resultant oils. *Journal of Analytical and Applied Pyrolysis,* 171.

O'Connell, D. W., Birkinshawb, C., & O'Dwyer, T. F. (2008). Heavy metal adsorbents prepared from the modification of cellulose: A review. *Bioresource Technology*, *99*(15), 6709–6724. doi:10.1016/j.biortech.2008.01.036 PMID:18334292

O'Neal, S. L., & Zheng, W. (2015). Manganese toxicity upon overexposure: A decade in review. *Current Environmental Health Reports*, *2*(3), 315–328. doi:10.1007/s40572-015-0056-x PMID:26231508

Obaideen, K., Shehata, N., Sayed, E. T., Abdelkareem, M. A., Mahmoud, M. S., & Olabi, A. G. (2022). The role of wastewater treatment in achieving sustainable development goals (SDGs) and sustainability guideline. *Energy Nexus*, *7*, 100112. doi:10.1016/j.nexus.2022.100112

Obiora, S. C., Chukwu, A., & Davies, T. C. (2016). Heavy metals and health risk assessment of arable soils and food crops around Pb–Zn mining localities in Enyigba, southeastern Nigeria. *Journal of African Earth Sciences*, *116*, 182–189. doi:10.1016/j.jafrearsci.2015.12.025

Obradovic, B. (2020). Guidelines for general adsorption kinetics modeling. *Hemijska Industrija*, *74*(1), 65–70. doi:10.2298/HEMIND200201006O

Offiong, N.-A. O., Edet, J. B., Shaibu, S. E., Akan, N. E., Atakpa, E. O., Sanganyado, E., Okop, I. J., Benson, N. U., & Okoh, A. (2023). Metagenomics: An emerging tool for the chemistry of environmental remediation. *Frontiers in Environmental Chemistry*, *4*, 7. doi:10.3389/fenvc.2023.1052697

Ojedokun, A. T., & Bello, O. S. (2016). Sequestering heavy metals from wastewater using cowdung. *Water Resources and Industry*, *13*, 7–13. doi:10.1016/j.wri.2016.02.002

Ojuederie, O., & Babalola, O. (2017). Microbial and Plant-Assisted Bioremediation of Heavy Metal Polluted Environments: A Review. *International Journal of Environmental Research and Public Health*, *14*(12), 1504. doi:10.3390/ijerph14121504 PMID:29207531

Okereafor, U., Makhatha, M., Mekuto, L., Uche-Okereafor, N., Sebola, T., & Mavumengwana, V. (2020). Toxic metal implications on agricultural soils, plants, animals, aquatic life and human health. *International Journal of Environmental Research and Public Health*, *17*(7), 2204. doi:10.3390/ijerph17072204 PMID:32218329

Okunola, O. J., Uzairu, A., & Ndukwe, G. (2007). Levels of trace metals in soil and vegetation along major and minor roads in metropolitan city of Kaduna, Nigeria. Academic Press.

Olubode, T. P., Amusat, A. I., Olawale, B. R., & Adekola, F. F. (2022). Biosorption of heavy metal using bacterial isolatesfrom e-waste soil. *African Journal of Microbiological Research*, *16*(7), 268–272.

Onay, O., & Kockar, O. M. (2003). Slow, fast and flash pyrolysis of rapeseed. *Renewable Energy*, *28*(15), 2417–2433. doi:10.1016/S0960-1481(03)00137-X

Onay, O., & Kockar, O. M. (2004). Fixed-bed pyrolysis of rapeseed (Brassica napus L.). *Biomass and Bioenergy*, *26*(3), 289–299. doi:10.1016/S0961-9534(03)00123-5

Osman, A. I., El-Monaem, E. M. A., Elgarahy, A. M., Aniagor, C. O., Hosny, M., Farghali, M., Rashad, E., Ejimofor, M. I., López-Maldonado, E. A., Ihara, I., Yap, P. S., Rooney, D. W., & Eltaweil, A. S. (2023). Methods to prepare biosorbents and magnetic sorbents for water treatment: a review. *Environmental Chemistry Letters, 21*(4), 2337–2398. doi:10.1007/s10311-023-01603-4

Othman, Y. A., Al-Assaf, A., Tadros, M. J., & Albalawneh, A. (2021). Heavy metals and microbes accumulation in soil and food crops irrigated with wastewater and the potential human health risk: A metadata analysis. *Water (Basel)*, *13*(23), 3405. doi:10.3390/w13233405

Pagnanelli, F. (2011). Equilibrium, Kinetic and Dynamic Modelling of Biosorption Processes. In P. Kotrba, M. Mackova, & T. Macek (Eds.), *Microbial Biosorption of Metals* (pp. 59–120). Springer Netherlands. doi:10.1007/978-94-007-0443-5_4

Pagnanelli, F., Toro, L., & Veglio, F. (2002). Oilve mill solid residues as heavy metal sorbent material: A preliminary study. *Waste Management (New York, N.Y.)*, *22*(8), 901–907. doi:10.1016/S0956-053X(02)00086-7 PMID:12423052

Paknikar, K. M., Palnitkar, U. S., & Puranik, P. R. (1993). In A. E. Torma, M. L. Apel, & C. L. Brierely (Eds.), *Biohydrometallurgical Technologies, V2* (pp. 229–236). The Minerals, Metals & Materials Society.

Palmer, P. B., & O'Connell, D. G. (2009). Regression analysis for prediction: Understanding the process. *Cardiopulmonary Physical Therapy Journal*, *20*(3), 23–26. doi:10.1097/01823246-200920030-00004 PMID:20467520

Pandey, G., & Madhuri, S. (2014). Heavy metals causing toxicity in animals and fishes. *Research Journal of Animal. Veterinary and Fishery Sciences*, *2*(2), 17–23.

Pandey, S. (2022). Wood waste utilization and associated product development from under-utilized low-quality wood and its prospects in Nepal. *SN Applied Sciences*, *4*(6), 168. doi:10.1007/s42452-022-05061-5

Pang, Y. L., Quek, Y. Y., Lim, S., & Shuit, S. H. (2023). Review on phytoremediation potential of floating aquatic plants for heavy metals: A promising approach. *Sustainability (Basel)*, *15*(2), 1290. doi:10.3390/su15021290

Parmar, P., Shukla, A., Goswami, D., Patel, B., & Saraf, M. (2020). Optimization of cadmium and lead biosorption onto marine *Vibrio alginolyticus* PBR1 employing a Box-Behnken design. *Chemical Engineering Journal Advances*, *4*, 100043. doi:10.1016/j.ceja.2020.100043

Particle Size Reduction and Enlargement. (2019). *Coulson and Richardson's Chemical Engineering*. Elsevier. doi:10.1016/B978-0-08-101098-3.00006-8

Patel, H., Shakhreliya, S., Maurya, R., Pandey, V. C., Gohil, N., Bhattacharjee, G., Alzahrani, K. J., & Singh, V. (2022). CRISPR-assisted strategies for futuristic phytoremediation. In *Assisted phytoremediation* (pp. 203–220). Elsevier. doi:10.1016/B978-0-12-822893-7.00006-9

Pathak, P., Mandavgane, S., & Kulkarni, B. (2015). Fruit peel waste as a novel cost bioadsorbent. *Reviews in Chemical Engineering*, *31*(4), 361–381. doi:10.1515/revce-2014-0041

Patrick, L. (2003). Toxic Metals and Antioxidants: Part II. The Role of Antioxidants in Arsenic and Cadmium Toxicity. *Alternative Medicine Review*, *8*(2). PMID:12777158

Patrón-Prado, M., Acosta-Vargas, B., Serviere-Zaragoza, E., & Méndez-Rodríguez, L. C. (2010). Copper and cadmium biosorption by dried seaweed sargassum sinicola in saline wastewater. *Water, Air, and Soil Pollution*, *210*(1-4), 197–202. doi:10.1007/s11270-009-0241-3

Pattanaik, L., Pattnaik, F., Saxena, D. K., & Naik, S. N. (2019). *Biofuels from agricultural wastes*. Elservier Inc. doi:10.1016/B978-0-12-815162-4.00005-7

Paul, D., & Sinha, S. N. (2013). Bacteria showing phosphate solubilizing efficiency in river sediment. *Electronic Journal of Biosciences*, *1*(1), 1–5.

Pedersen, A. J. (2003). Characterization and electrolytic treatment of wood combustion fly ash for the removal of cadmium. *Biomass and Bioenergy*, *25*(4), 447–458. doi:10.1016/S0961-9534(03)00051-5

Peixoto, R. S., Voolstra, C. R., Sweet, M., Duarte, C. M., Carvalho, S., Villela, H., Lunshof, J. E., Gram, L., Woodhams, D. C., Walter, J., Roik, A., Hentschel, U., Thurber, R. V., Daisley, B., Ushijima, B., Daffonchio, D., Costa, R., Keller-Costa, T., Bowman, J. S., ... Berg, G. (2022). Harnessing the microbiome to prevent global biodiversity loss. *Nature Microbiology*, *7*(11), 1726–1735. doi:10.1038/s41564-022-01173-1 PMID:35864220

Peng, X., Jiang, Y., Chen, Z., Osman, A. I., Farghali, M., Rooney, D. W., & Yap, P.-S. (2023). Recycling municipal, agricultural and industrial waste into energy, fertilizers, food and construction materials, and economic feasibility: A review. *Environmental Chemistry Letters*, *21*(2), 765–801. doi:10.1007/s10311-022-01551-5

Perić, J., Trgo, M., & Vukojević Medvidović, N. (2004). Removal of zinc, copper and lead by naturalzeolite - a comparison of adsorption isotherms. *Water Research*, *38*(7), 1893–1899. doi:10.1016/j.watres.2003.12.035 PMID:15026244

Perpetuo, E. A., Souza, C. B. & Nascimento, C. A. O. (2014). Engineering Bacteria for Bioremediation. *Progress in Molecular and Environmental Engineering - From Analysis and Modelling to Technology Applications,* 605-632.

Perumal, K., Antony, J., & Muthuramalingam, S. (2021). Heavy metal pollutants and their spatial distribution in surface sediments from Thondi coast, Palk Bay, South India. *Environmental Sciences Europe*, *33*(2), 63–83. doi:10.1186/s12302-021-00501-2

Perussello, C. A., Zhang, Z., Marzocchella, A., & Tiwari, B. K. (2017). Valorization of Apple Pomace by Extraction of Valuable Compounds. *Comprehensive Reviews in Food Science and Food Safety*, *16*(5), 776–796. doi:10.1111/1541-4337.12290 PMID:33371603

Peter, A. L. J., & Viraraghavan, T. (2005). Thallium: A review of public health and environmental concerns. *Environment International*, *31*(4), 493–501. doi:10.1016/j.envint.2004.09.003 PMID:15788190

Pham, V. H. T., Kim, J., Chang, S., & Chung, W. (2022). Bacterial Biosorbents, an Efficient Heavy Metals Green Clean-Up Strategy: Prospects, Challenges, and Opportunities. *Microorganisms*, *10*(3), 610. doi:10.3390/microorganisms10030610 PMID:35336185

Pieper, D. H., & Reineke, W. (2000). Engineering bacteria for bioremediation. *Current Opinion in Biotechnology*, *11*(3), 262–270. doi:10.1016/S0958-1669(00)00094-X PMID:10851148

Piersa, P., Unyay, H., Szufa, S., Lewandowska, W., Modrzewski, R., Ślężak, R., & Ledakowicz, S. (2022). An Extensive Review and Comparison of Modern Biomass Torrefaction Reactors vs. Biomass Pyrolysis—Part 1. *Energies*, *15*(6), 2227. doi:10.3390/en15062227

Pituello, C., Francioso, O., Simonetti, G., Pisi, A., Torreggiani, A., Berti, A., & Morari, F. (2015). Characterization of chemical–physical, structural and morphological properties of biochars from biowastes produced at different temperatures. *Journal of Soils and Sediments*, *15*(4), 792–804. doi:10.1007/s11368-014-0964-7

Ponnusami, V., Krithika, V., Madhuram, R., & Srivastava, S. N. (2007). Biosorption of reactive dye using acid-treated rice husk: Factorial design analysis. *Journal of Hazardous Materials*, *142*(1), 397–403. doi:10.1016/j.jhazmat.2006.08.040 PMID:17011118

Priya, A. K., Gnanasekaran, L., Dutta, K., Rajendran, S., Balakrishnan, D., & Soto-Moscoso, M. (2022). Biosorption of heavy metals by microorganisms: Evaluation of different underlying mechanisms. *Chemosphere*, *307*, 135957. doi:10.1016/j.chemosphere.2022.135957 PMID:35985378

Priya, A. K., Muruganandam, M., Ali, S. S., & Kornaros, M. (2023). Clean-Up of Heavy Metals from Contaminated Soil by Phytoremediation: A Multidisciplinary and Eco-Friendly Approach. *Toxics*, *11*(5), 422. doi:10.3390/toxics11050422 PMID:37235237

Priya, A., Dutta, K., & Daverey, A. (2022). A comprehensive biotechnological and molecular insight into plastic degradation by microbial community. *Journal of Chemical Technology and Biotechnology*, *97*(2), 381–390. doi:10.1002/jctb.6675

Priyadarshanee, M., & Das, S. (2020). Biosorption and removal of toxic heavy metals by metal tolerating bacteria for bioremediation of metal contamination: A comprehensive review. *Journal of Environmental Chemical Engineering*, *9*(1), 104686. doi:10.1016/j.jece.2020.104686

Puranik, P. R., & Paknikar, K. M. (1997). Biosorption of lead and zinc from solutions using Streptoverticillium cinnamoneum waste biomass. *Journal of Biotechnology*, *55*(2), 113–124. doi:10.1016/S0168-1656(97)00067-9 PMID:9232033

Purkayastha, R. P., & Mitra, A. K. (1992). Metal uptake by mycelia during submerged growth and by sporocarps of an edible fungus Volvariella volvacea. *Indian Journal of Experimental Biology*, *30*, 1184–1187. PMID:1294483

Qin, G., Niu, Z., Yu, J., Li, Z., Ma, J., & Xiang, P. (2021). Soil heavy metal pollution and food safety in China: Effects, sources and removing technology. *Chemosphere*, *267*, 129205. doi:10.1016/j.chemosphere.2020.129205 PMID:33338709

Rafeeq, H., Afsheen, N., Rafique, S., Arshad, A., Intisar, M., Hussain, A., Bilal, M., & Hafiz, M. N. (2023). Iqbal. "Genetically engineered microorganisms for environmental remediation.". *Chemosphere*, *310*, 136751. doi:10.1016/j.chemosphere.2022.136751 PMID:36209847

Rahimi, Z., Anand, A., & Gautam, S. (2022). An overview on thermochemical conversion and potential evaluation of biofuels derived from agricultural wastes. *Energy Nexus*, *7*, 100125. doi:10.1016/j.nexus.2022.100125

Rai, P. K. (2012). An eco-sustainable green approach for heavy metals management: Two case studies of developing industrial region. *Environmental Monitoring and Assessment*, *184*(1), 421–448. doi:10.1007/s10661-011-1978-x PMID:21465134

Rai, P. K. (2018). Heavy metal phyto-technologies from Ramsar wetland plants: Green approach. *Chemistry and Ecology*, *34*(8), 786–796. doi:10.1080/02757540.2018.1501476

Rai, P. K., Lee, S. S., Zhang, M., Tsang, Y. F., & Kim, K. H. (2019). Heavy metals in food crops: Health risks, fate, mechanisms, and management. *Environment International*, *125*, 365–385. doi:10.1016/j.envint.2019.01.067 PMID:30743144

Raize, O., Argaman, Y., & Yannai, S. (2004). Mechanisms of biosorption of different heavy metals by brown marine macroalgae. *Biotechnology and Bioengineering*, *87*(4), 451–458. doi:10.1002/bit.20136 PMID:15286981

Rajewski, J., & Dobrzyńska-Inger, A. (2021). Application of Response Surface Methodology (RSM) for the optimization of chromium(III) synergistic extraction by supported liquid membrane. *Membranes*, *11*(11), 854. doi:10.3390/membranes11110854 PMID:34832083

Raji, Z., Karim, A., Karam, A., & Khalloufi, S. (2023). Adsorption of Heavy Metals: Mechanisms, Kinetics, and Applications of Various Adsorbents in Wastewater Remediation -. *Review*, 775–805.

Raji, Z., Karim, A., Karam, A., & Khalloufi, S. (2023). Adsorption of Heavy Metals: Mechanisms, Kinetics, and Applications of Various Adsorbents in Wastewater Remediation—A Review. *Waste*, *1*(3), 775–805. doi:10.3390/waste1030046

Ramage, M. H., Burridge, H., Busse-Wicher, M., Fereday, G., Reynolds, T., Shah, D. U., Wu, G., Yu, L., Fleming, P., Densley-Tingley, D., Allwood, J., Dupree, P., Linden, P. F., & Scherman, O. (2017). The wood from trees: The use of timber in construction. *Renewable & Sustainable Energy Reviews*, *68*, 333–359. doi:10.1016/j.rser.2016.09.107

Ramrakhiani, L., Majumder, R., & Khowala, S. (2011). Removal of hexavalent chromium by heat inactivated fungal biomass of *Termitomyces clypeatus*. *Chemical Engineering Journal*, *171*(3), 1060–1068. doi:10.1016/j.cej.2011.05.002

Ramsenthil, R., & Meyyappan, R. (2010). Single and multi-component biosorption of copper and zinc ions using microalgal resin. *International Journal of Environmental Sciences and Development, 1*(4), 298–301. doi:10.7763/IJESD.2010.V1.58

Rangabhashiyam, S., Anu, N., Giri Nandagopal, M. S., & Selvaraju, N. (2014). Relevance of isotherm models in biosorption of pollutants by agricultural byproducts. *Journal of Environmental Chemical Engineering, 2*(1), 398–414. doi:10.1016/j.jece.2014.01.014

Rangabhashiyam, S., & Balasubramanian, P. (2019). The potential of lignocellulosic biomass precursors for biochar production: Performance, mechanism and wastewater application—A review. *Industrial Crops and Products, 128*, 405–423. doi:10.1016/j.indcrop.2018.11.041

Ranjan, D., Talat, M., & Hasan, S. H. (2009). Biosorption of arsenic from aqueous solution using agricultural residue 'rice polish. *Journal of Hazardous Materials, 166*(2-3), 1050–1059. doi:10.1016/j.jhazmat.2008.12.013 PMID:19131161

Rao, J. S., & Kumar, B. (2012). 3D Blade root shape optimization. In *10th International Conference on Vibrations in Rotating Machinery* (pp. 173–188). Woodhead Publishing. 10.1533/9780857094537.4.173

Rashid, A., Bhatti, H. N., Iqbal, M., & Noreen, S. (2016). Fungal biomass composite with bentonite efficiency for nickel and zinc adsorption: A mechanistic study. *Ecological Engineering, 91*, 459–471. doi:10.1016/j.ecoleng.2016.03.014

Raut, N. A., Kokare, D. M., Randive, K. R., Bhanvase, B. A., & Dhoble, S. J. (2023). Introduction: fundamentals of waste removal technologies. In 360-Degree Waste Management, Volume 1 (pp. 1–16). Elsevier. doi:10.1016/B978-0-323-90760-6.00009-6

Rawat, M., Ramanathan, A. L., & Kuriakose, T. (2013). *Characterisation of municipal solid waste compost (MSWC) from selected Indian cities—a case study for its sustainable utilisation.* Academic Press.

Raychaudhuri, S.S., Pramanick, P., Talukder, P., & Basak, A. (2021). *Polyamines, metallothioneins, and phytochelatins – Natural defence of plants to mitigate heavy metals.* doi:10.1016/B978-0-12-819487-4.00006-9

Raza, M., Inayat, A., Ahmed, A., Jamil, F., Ghenai, C., Naqvi, S. R., Shanableh, A., Ayoub, M., Waris, A., & Park, Y.-K. (2021). Progress of the Pyrolyzer Reactors and Advanced Technologies for Biomass Pyrolysis Processing. *Sustainability (Basel), 13*(19), 11061. doi:10.3390/su131911061

Razzak, S. A., Faruque, M. O., Alsheikh, Z., Alsheikhmohamad, L., Alkuroud, D., Alfayez, A., Hossain, S. M. Z., & Hossain, M. M. (2022). A comprehensive review on conventional and biological-driven heavy metals removal from industrial wastewater. *Environmental Advances, 7*, 100168. doi:10.1016/j.envadv.2022.100168

Reddy, A., Lakshmipathy, R., & Sarada, N. C. (2014). Application of *Citrullus lanatus* rind as biosorbent for removal of trivalent chromium from aqueous solution. *Alexandria Engineering Journal, 53*(4), 969–975. doi:10.1016/j.aej.2014.07.006

Reddy, D. H. K., Lee, S. M., & Seshaiah, K. (2012). Biosorption of toxic heavy metal ions from water environment using honeycomb biomass-an industrial waste material. *Water, Air, and Soil Pollution, 223*(9), 5967–5982. doi:10.1007/s11270-012-1332-0

Redha, A. A., & Redha, A. A. (2020). Removal of heavy metals from aqueous media by biosorption Removal of heavy metals from aqueous media by biosorption. *Arab Journal of Basic and Applied Sciences, 27*(1), 183–193. doi:10.1080/25765299.2020.1756177

Reis, P. C., Reis, M. A. M., & Freitas, F. (2020). Biosorption of Heavy Metals by the Bacterial Exopolysaccharide FucoPol. *Applied Sciences (Basel, Switzerland), 10*(19), 6708. doi:10.3390/app10196708

Resources, N.National Academies of Sciences, Engineering, and Medicine. (2016). Regulation of current and future genetically engineered crops. In *Genetically engineered crops: Experiences and prospects.* National Academies Press.

Rezaei, H. (2016). Biosorption of chromium by using *Spirulina sp. Arabian Journal of Chemistry*, *9*(6), 846–853. doi:10.1016/j.arabjc.2013.11.008

Rice, K. M., Walker, E. M., Wu, M., Gillette, C., & Blough, E. R. (2014). Environmental mercury and its toxic effects. *Journal of Preventive Medicine and Public Health*, *47*(2), 74–83. doi:10.3961/jpmph.2014.47.2.74 PMID:24744824

Rokonuzzaman, M. D., Li, W. C., Wu, C., & Ye, Z. H. (2022). Human health impact due to arsenic contaminated rice and vegetables consumption in naturally arsenic endemic regions. *Environmental Pollution*, *308*, 119712. doi:10.1016/j.envpol.2022.119712 PMID:35798190

Rosca, M., Hlihor, R.-M., Cozma, P., Comăniţă, E.-D., Simion, I. M., & Gavrilescu, M. (2015). Potential of biosorption and bioaccumulation processes for heavy metals removal in bioreactors. *2015 E-Health and Bioengineering Conference (EHB)*, 1–4. doi:10.1109/EHB.2015.7391487

Roşca, M., Diaconu, M., Hlihor, R.-M., Cozma, P., & Gavrilescu, M. (2020). Prediction of equillibrium sorption isotherm for cadmium biosorption by microorganisms: comparison of linear and nonlinear methods. *2020 International Conference on E-Health and Bioengineering (EHB)*, 1–4. 10.1109/EHB50910.2020.9280180

Rosca, M., Hlihor, R. M., Cozma, P., Simion, I. M., Filote, C., Grecu, C., Stoleru, V., & Gavrilescu, M. (2021). Scaling-up strategies of heavy metals microbial bioremediation. *2021 International Conference on E-Health and Bioengineering (EHB)*, 1–4. 10.1109/EHB52898.2021.9657641

Roşca, M., Hlihor, R.-M., Cozma, P., Drăgoi, E. N., Diaconu, M., Silva, B., Tavares, T., & Gavrilescu, M. (2018). Comparison of *Rhodotorula* sp. and *Bacillus megaterium* in the removal of cadmium ions from liquid effluents. *Green Processing and Synthesis*, *7*(1), 74–88. doi:10.1515/gps-2016-0218

Rosca, M., Hlihor, R.-M., & Gavrilescu, M. (2019). Bioremediation of Persistent Toxic Substances: From Conventional to New Approaches in Using Microorganisms and Plants. In P. K. Arora (Ed.), *Microbial Technology for the Welfare of Society* (Vol. 17, pp. 289–312). Springer Singapore. doi:10.1007/978-981-13-8844-6_14

Roşca, M., Silva, B., Tavares, T., & Gavrilescu, M. (2023). Biosorption of hexavalent chromium by *Bacillus megaterium* and *Rhodotorula* sp. inactivated biomass. *Processes (Basel, Switzerland)*, *11*(1), 1. Advance online publication. doi:10.3390/pr11010179

Roy, P., Mohanty, A. K., Dick, P., & Misra, M. (2023). A Review on the Challenges and Choices for Food Waste Valorization: Environmental and Economic Impacts. *ACS Environmental Au*, *2*(2), 58–75. doi:10.1021/acsenvironau.2c00050 PMID:36941850

Saadi, R., Saadi, Z., Fazaeli, R., & Fard, N. E. (2015). Monolayer and multilayer adsorption isotherm models for sorption from aqueous media. *Korean Journal of Chemical Engineering*, *32*(5), 787–799. doi:10.1007/s11814-015-0053-7

Saady, N. M. C., Rezaeitavabe, F., & Ruiz Espinoza, J. E. (2021). Chemical Methods for Hydrolyzing Dairy Manure Fiber: A Concise Review. *Energies*, *14*(19), 6159. doi:10.3390/en14196159

Sabath, E., & Robles-Osorio, M. L. (2012). Renal health and the environment: Heavy metal nephrotoxicity. [English Edition]. *Nefrología : Publicación Oficial de la Sociedad Española Nefrologia*, *32*(3), 279–286. PMID:22508139

Sadh, P. K., Duhan, S., & Duhan, J. S. (2018). Agro-industrial wastes and their utilization using solid state fermentation: A review. *Bioresources and Bioprocessing*, *5*(1), 1. doi:10.1186/s40643-017-0187-z

Sadhukhan, B., Mondal, N. K., & Chattoraj, S. (2016). Optimisation using central composite design (CCD) and the desirability function for sorption of methylene blue from aqueous solution onto *Lemna major. Karbala International Journal of Modern Science*, *2*(3), 145–155. doi:10.1016/j.kijoms.2016.03.005

Saeed, A., Akhterb, M. W., & Iqbal, M. (2005). Removal and recovery of heavy metals from aqueous solution using papaya wood as a new biosorbent. *Separation and Purification Technology*, *45*(1), 25–31. doi:10.1016/j.seppur.2005.02.004

Sağ, Y. (2001). biosorption of heavy metals by fungal biomass and modelling of fungal biosorption: A review. *Seperation and Purification Methods*, *30*, 1-48.

Sahmoune, M. N. (2018). Performance of *Streptomyces rimosus* biomass in biosorption of heavy metals from aqueous solutions. *Microchemical Journal*, *141*, 87–95. doi:10.1016/j.microc.2018.05.009

Salonen, J. T., Malin, R., Tuomainen, T. P., Nyyssönen, K., Lakka, T. A., Lehtimäki, T., & Flather, M. D. (1999). Polymorphism in high density lipoprotein paraoxonase gene and risk of acute myocardial infarction in men: Prospective nested case-control study. *British Medical Journal*, *319*(7208), 487–488. doi:10.1136/bmj.319.7208.487 PMID:10454401

Sánchez-Castro, I., Molina, L., Prieto-Fernández, M. Á., & Segura, A. (2023). Past, present and future trends in the remediation of heavy-metal contaminated soil-Remediation techniques applied in real soil-contamination events. *Heliyon*, *9*(6), e16692. Advance online publication. doi:10.1016/j.heliyon.2023.e16692 PMID:37484356

Sanfeliu, C., Sebastia, J., Cristofol, R., & Rodriquez-Farre, E. (2003). Neurotoxicity of organomercurial compounds. *Neurotoxicity Research*, *5*(4), 283–305. doi:10.1007/BF03033386 PMID:12835120

Sao, K. (2014). A review on heavy metals uptake by plants through biosorption. *International Proceedings of Economics Development and Research*, *75*(17), 78–83. doi:10.7763/IPEDR

Sapwarobol, S., Saphyakhajorn, W., & Astina, J. (2021). Biological Functions and Activities of Rice Bran as a Functional Ingredient: A Review. *Nutrition and Metabolic Insights*, *14*, 117863882110585. doi:10.1177/11786388211058559 PMID:34898989

Sardar, K., Ali, S., Hameed, S., Afzal, S., Fatima, S., Shakoor, M. B., ... Tauqeer, H. M. (2013). Heavy metals contamination and what are the impacts on living organisms. *Greener Journal of Environmental Management and Public Safety*, *2*(4), 172-179.

Sarkar, S., Skalicky, M., Hossain, A., Brestic, M., Saha, S., Garai, S., Ray, K., & Brahmachari, K. (2020). Management of Crop Residues for Improving Input Use Efficiency and Agricultural Sustainability. *Sustainability (Basel)*, *12*(23), 9808. doi:10.3390/su12239808

Sarker, A., Al Masud, M. A., Deepo, D. M., Das, K., Nandi, R., Ansary, M. W. R., Islam, A. R. M. T., & Islam, T. (2023). Biological and green remediation of heavy metal contaminated water and soils: A state-of-the-art review. *Chemosphere*, *138861*. Advance online publication. doi:10.1016/j.chemosphere.2023.138861 PMID:37150456

Sarojini, G., Babu, S. V., Rajamohan, N., Kumar, P. S., & Rajasimman, M. (2021). Surface modified polymer-magnetic-algae nanocomposite for the removal of chromium- equilibrium and mechanism studies. *Environmental Research*, *201*, 111626. doi:10.1016/j.envres.2021.111626 PMID:34217718

Sarret, G., Manceau, A., Spadini, L., Roux, J. C., Hazemann, J. L., Soldo, Y., Berard, L. E., & Menthonnex, J. J. (1998). Structural determination of Zn and Pb binding sites in Penicillium chrysogenum cell walls by EXAFS spectroscopy. *Environmental Science & Technology*, *32*(11), 1648–1655. doi:10.1021/es9709684

Sarwar, N., Imran, M., Shaheen, M. R., Ishaque, W., Kamran, M. A., Matloob, A., Rehim, A., & Hussain, S. (2017). Phytoremediation strategies for soils contaminated with heavy metals: Modifications and future perspectives. *Chemosphere*, *171*, 710–721. doi:10.1016/j.chemosphere.2016.12.116 PMID:28061428

Saxena, A., Bhardwaj, M., Allen, T., Kumar, S., & Sahney, R. (2017). Adsorption of heavy metals from wastewater using agricultural–industrial wastes as biosorbents. *Water Science*, *31*(2), 189–197. doi:10.1016/j.wsj.2017.09.002

Saxena, G., Purchase, D., Mulla, S. I., Saratale, G. D., & Bharagava, R. N. (2020). Phytoremediation of heavy metal-contaminated sites: Eco-environmental concerns, field studies, sustainability issues, and future prospects. *Reviews of Environmental Contamination and Toxicology*, *249*, 71–131. PMID:30806802

Sayler, G. S., & Ripp, S. (2000). Field applications of genetically modified bacteria for bioremediation processes. *Current Opinion in Biotechnology*, *11*, 286–289. doi:10.1016/S0958-1669(00)00097-5 PMID:10851144

Schaefer, H. R., Dennis, S., & Fitzpatrick, S. (2020). Cadmium: Mitigation strategies to reduce dietary exposure. *Journal of Food Science*, *85*(2), 260–267. doi:10.1111/1750-3841.14997 PMID:31957884

Schut, S., Zauner, S., Hampel, G., König, H., & Claus, H. (2011). Biosorption of copper by wine-relevant lactobacilli. *International Journal of Food Microbiology*, *145*(1), 126–131. doi:10.1016/j.ijfoodmicro.2010.11.039 PMID:21195499

Schutte, R., Nawrot, T. S., Richart, T., Thijs, L., Vanderschueren, D., Kuznetsova, T., Van Hecke, E., Roels, H. A., & Staessen, J. A. (2008). Bone resorption and environmental exposure to cadmium in women: A population study. *Environmental Health Perspectives*, *116*(6), 777–783. doi:10.1289/ehp.11167 PMID:18560534

Schwartz, G. G., Il'yasova, D., & Ivanova, A. (2003). Urinary cadmium, impaired fasting glucose, and diabetes in the NHANES III. *Diabetes Care*, *26*(2), 468–470. doi:10.2337/diacare.26.2.468 PMID:12547882

Sedlakova-Kadukova, J., Kopcakova, A., Gresakova, L., Godany, A., & Pristas, P. (2019). Bioaccumulation and biosorption of zinc by a novel *Streptomyces* K11 strain isolated from highly alkaline aluminium brown mud disposal site. *Ecotoxicology and Environmental Safety*, *167*, 204–211. doi:10.1016/j.ecoenv.2018.09.123 PMID:30340085

Selatnia, A., Boukazoula, A., Kechid, N., Bakhti, M. Z., & Chergui, A. (2004). Biosorption of Fe3+ from aqueous solution by a bacterial dead Streptomyces rimosus biomass. *Process Biochemistry (Barking, London, England)*, *39*(11), 1643–1651. doi:10.1016/S0032-9592(03)00305-4

Sen, T. K. (2023). Agricultural Solid Wastes Based Adsorbent Materials in the Remediation of Heavy Metal Ions from Water and Wastewater by Adsorption: A Review. *Molecules (Basel, Switzerland)*, *28*(14), 5575. doi:10.3390/molecules28145575 PMID:37513447

Shaheen, N., Irfan, N., Khan, I. N., Islam, S., Islam, M., & Ahmed, M. (2016). Presence of heavy metals in fruits and vegetables: Health risk implications in Bangladesh. *Chemosphere*, *152*, 431–438. doi:10.1016/j.chemosphere.2016.02.060 PMID:27003365

Shamim, S. (2018). Biosorption of heavy metals. In Tech Open, 21-49.

Shamim, S. (2018). Biosorption of Heavy Metals. In J. Derco & B. Vrana (Eds.), *Biosorption*. IntechOpen. doi:10.5772/intechopen.72099

Sharma, B., & Shukla, P. (2022). Futuristic avenues of metabolic engineering techniques in bioremediation. *Biotechnology and Applied Biochemistry*, *69*(1), 51–60. doi:10.1002/bab.2080 PMID:33242354

Sharma, H., Rawal, N., & Mathew, B. B. (2015). The characteristics, toxicity and effects of cadmium. *International Journal of Nanotechnology and Nanoscience*, *3*, 1–9.

Shartooh, S. M. (2013). Lettuce Leaves as Biosorbent Material to Remove Heavy Metal Ions from Industrial Wastewater. *Journal of Baghdad for Science*, *11*(3), 1164–1170.

Sheng, P. X., Ting, Y. P., Chen, J. P., & Hong, L. (2004). Sorption of lead, copper, cadmium, zinc, and nickel by marine algal biomass: Characterization of biosorptive capacity and investigation of mechanisms. *Journal of Colloid and Interface Science*, *275*(1), 131–141. doi:10.1016/j.jcis.2004.01.036

Shkliarenko, Y., Halysh, V., & Nesterenko, A. (2023). Adsorptive Performance of Walnut Shells Modified with Urea and Surfactant for Cationic Dye Removal. *Water (Basel)*, *15*(8), 1536. doi:10.3390/w15081536

Shrivastava, S., & Gupta, M. (2021). Application of Box-Behnken model to study the biosorption of chromium Cr(VI) ion by *Aspergillus niger* isolated from polluted site. *International Journal of Current Microbiology and Applied Sciences*, *10*(7), 414–424. doi:10.20546/ijcmas.2021.1007.045

Shukla, K. P., Singh, N. K., & Sharma, S. (2010). Bioremediation: Developments, current practices and perspectives. *Genetic Engineering and Biotechnology Journal*, *3*, 1–20.

Sibi, G. (2016). Biosorption of chromium from electroplating and galvanizing industrial effluents under extreme conditions using *Chlorella vulgaris*. *Green Energy & Environment*, *1*(2), 172–177. doi:10.1016/j.gee.2016.08.002

Siegel, S. M. M., Galun, M., & Siegel, B. Z. (1990). Filamentous fungi as metal biosorbents: A review. *Water, Air, and Soil Pollution*, *53*(3-4), 335–344. doi:10.1007/BF00170747

Sikder, S., & Joardar, J. C. (2019). Biochar production from poultry litter as management approach and effects on plant growth. *International Journal of Recycling of Organic Waste in Agriculture*, *8*(1), 47–58. doi:10.1007/s40093-018-0227-5

Silva, J. A. (2023). Wastewater treatment and reuse for sustainable water resources management: A systematic literature review. *Sustainability (Basel)*, *15*(14), 10940. doi:10.3390/su151410940

Singh, A., Sharma, A., Verma, R. K., Chopade, R. L., Pandit, P. P., Nagar, V., Aseri, V., Choudhary, S. K., Awasthi, G., Awasthi, K. K., Sankhla, M. S., Singh, A., Sharma, A., Verma, R. K., Chopade, R. L., Pandit, P. P., Nagar, V., Aseri, V., Choudhary, S. K., … Sankhla, M. S. (2022). Heavy Metal Contamination of Water and Their Toxic Effect on Living Organisms. *The Toxicity of Environmental Pollutants*. doi:10.5772/intechopen.105075

Singh, A., Ahmed, M., Saini, P., & Kaur, D. (2023). *Biosorption: Heavy Metals Removal through Biofilms and Microbial Polysaccharides. Food and Industrial Bioprocessing*. New India Publishing Agency.

Singh, A., & Saini, P. (2020). Biosorption: A Novel Biotechnological Application for Removal of Hazardous Pollutants. In *Innovations in Food Technology: Current Perspectives and Future Goals*. Springer. doi:10.1007/978-981-15-6121-4_24

Singh, A., Saini, P., Srivastava, U., Ahmed, M., & Iqbal, U. (2021). Cadmium Toxicology and its Prevention through Lactic Acid Bacteria. *Journal of Food and Agriculture Research*, *1*(2), 221–248.

Singh, A., Saini, P., Srivastava, U., & Kushwaha, R. (2021). Biosorption of Cadmium and Lead through Lactobacillus reuteri and Lactobacillus plantarum and its Exopolysaccharides: A Comparative Study. *International Journal of Current Research and Review*, *13*(15), 30–35. doi:10.52403/ijrr.20210506

Singh, B. J., Chakraborty, A., & Sehgal, R. (2023). A Systematic Review of Industrial Wastewater Management: Evaluating Challenges and Enablers. *Journal of Environmental Management*, *348*, 119–230. doi:10.1016/j.jenvman.2023.119230 PMID:37832302

Singh, J. S., Abhilash, P. C., Singh, H. B., Singh, R. P., & Singh, D. P. (2011). Genetically engineered bacteria: An emerging tool for environmental remediation and future research perspectives. *Gene*, *480*(1-2), 1–9. doi:10.1016/j.gene.2011.03.001 PMID:21402131

Singh, N., Gupta, V. K., Kumar, A., & Sharma, B. (2017). Synergistic effects of heavy metals and pesticides in living systems. *Frontiers in Chemistry*, *5*, 70. Advance online publication. doi:10.3389/fchem.2017.00070 PMID:29075624

Singh, R., Chadetrik, R., Kumar, R., Bishnoi, K., Bhatia, D., Kumar, A., Bishnoi, N. R., & Singh, N. (2010). Biosorption optimization of lead(II), cadmium(II) and copper(II) using response surface methodology and applicability in isotherms and thermodynamics modeling. *Journal of Hazardous Materials*, *174*(1), 623–634. doi:10.1016/j.jhazmat.2009.09.097 PMID:19836883

Singh, R., Gautam, N., Mishra, A., & Gupta, R. (2011). Heavy metals and living systems: An overview. *Indian Journal of Pharmacology*, *43*(3), 246–253. doi:10.4103/0253-7613.81505 PMID:21713085

Singh, R., Krishna, B. B., Mishra, G., Kumar, J., & Bhaskar, T. (2016). Strategies for selection of thermo-chemical processes for the valorisation of biomass. *Renewable Energy*, *98*, 226–237. doi:10.1016/j.renene.2016.03.023

Sinha, S. N., & Paul, D. (2012). Detoxification of heavy metals by biosurfactants. *Bulletin of Environmental and Scientific Research*, *1*(3-4), 1–3.

Sinha, S. N., & Paul, D. (2014). Heavy metal tolerance and accumulation by bacterial strains isolated from wastewater. *Journal of Chemical, Biological and Physical Sciences*, *4*, 812–817.

Sinha, S. N., Paul, D., & Biswas, K. (2016). Effects of Moringa oleifera Lam. and Azadirachta indica A. Juss. leaf extract in treatment of tannery effluent. *Our Nature*, *14*(1), 47–53. doi:10.3126/on.v14i1.16440

Slezak, R., Unyay, H., Szufa, S., & Ledakowicz, S. (2023). An Extensive Review and Comparison of Modern Biomass Reactors Torrefaction vs. Biomass Pyrolizers—Part 2. *Energies*, *16*(5), 2212. Advance online publication. doi:10.3390/en16052212

Smedley, P. L., & Kinniburg, D. G. (2002). A review of the source, behavior and distribution of arsenic in natural waters. *Applied Geochemistry*, *17*(5), 517–568. doi:10.1016/S0883-2927(02)00018-5

Sodhi, K. K., Kumar, M., & Singh, D. K. (2020). Multi-metal resistance and potential of Alcaligenes sp. MMA for the removal of heavy metals. *SN Applied Sciences*, *2*(11), 1885. doi:10.1007/s42452-020-03583-4

Song, Y., Hou, D., Zhang, J., O'Connor, D., Li, G., Gu, Q., Li, S., & Liu, P. (2018). Environmental and socio-economic sustainability appraisal of contaminated land remediation strategies: A case study at a mega-site in China. *The Science of the Total Environment*, *610*, 391–401. doi:10.1016/j.scitotenv.2017.08.016 PMID:28806555

Sovacool, B. K. (2019). Toxic transitions in the lifecycle externalities of a digital society: The complex afterlives of electronic waste in Ghana. *Resources Policy, 64*, 101459.

Spungen, J. H. (2019). Children's exposures to lead and cadmium: FDA total diet study 2014–16. *Food Additives & Contaminants. Part A, Chemistry, Analysis, Control, Exposure & Risk Assessment*, *36*(6), 893–903. doi:10.1080/19440049.2019.1595170 PMID:30985263

Srivastava, S., & Goyal, P. (2010). Reusability of biomaterial: a cost effective approach. *Environ Sci Eng*, 93–96.

Srivastava, P., & Kowshik, M. (2013). Mechanisms of metal resistance and homeostasis in haloarchaea. *Archaea (Vancouver, B.C.)*, *2013*, 732864. doi:10.1155/2013/732864 PMID:23533331

Srivastava, V., Sarkar, A., Singh, S., Singh, P., De Araujo, A. S., & Singh, R. P. (2017). Agroecological responses of heavy metal pollution with special emphasis on soil health and plant performances. *Frontiers in Environmental Science*, *5*, 64. doi:10.3389/fenvs.2017.00064

Srivatsav, P., Bhargav, B. S., Shanmugasundaram, V., Arun, J., Panchamoorthy, K., & Bhartnaga, A. (2020). Biochar as an eco-friendly and ecological adsorbent for the removal of colorants (dyes) from aqueous environment: A review. *Water (Basel)*, *12*(12), 3561. doi:10.3390/w12123561

Stasinos, S., & Zabetakis, I. (2013). The uptake of nickel and chromium from irrigation water by potatoes, carrots, and onions. *Ecotoxicology and Environmental Safety*, *91*, 122–128. doi:10.1016/j.ecoenv.2013.01.023 PMID:23433838

Stumm, W., & Morgan, J. J. (1996). *Aquatic chemistry*. Wiley.

Subhashini, S., Kaliappan, S., & Velan, M. (2011). Removal of heavy metal from aqueous solu- tion using Schizosaccharomyces pombe in free and alginate immobilized cells. In *2nd International Conference on Environmental Science and Technology IPCBEE*. Vol. 6. Singapore: IACSIT Press.

Sud, D., Mahajan, G., & Kaur, M. P. (2008). Agricultural waste material as potential adsorbent for sequestering heavy metal ions from aqueous solutions - a review. *Bioresource Technology*, *99*(14), 6017–6027. doi:10.1016/j.biortech.2007.11.064 PMID:18280151

Suman, S., & Gautam, S. (2017). Pyrolysis of coconut husk biomass: Analysis of its biochar properties. *Energy Sources. Part A, Recovery, Utilization, and Environmental Effects*, *39*(8), 761–767. doi:10.1080/15567036.2016.1263252

Sundar, S., & Chakravarty, J. (2010). Antimony toxicity. *International Journal of Environmental Research and Public Health*, *7*(12), 4267–4277. doi:10.3390/ijerph7124267 PMID:21318007

Sun, F., Wu, F., Liao, H., & Xing, B. (2011). Biosorption of antimony(V) by freshwater cyanobacter *Mycrocystis* biomass. *Chemical Engineering Journal*, *171*, 1082–1090. doi:10.1016/j.cej.2011.05.004

Swiatek, M. Z., & Krzywonos, M. (2014). Potentials of biosorption and bioaccumulation processes for heavy metal removal. *Polish Journal of Environmental Studies*, *23*(2), 551–561.

Szynkowska, M. I., Pawlaczyk, A., & Maćkiewicz, E. (2018). Bioaccumulation and biomagnification of trace elements in the environment. *Recent advances in trace elements*, 251-276.

Tafakori, V., Zadmard, R., Tabandeh, F., Amoozegar, M. A., & Ahmadian, G. (2017). Equilibrium isotherm, kinetic modeling, optimization, and characterization studies of cadmium adsorption by surface-engineered *Escherichia coli*. *Iranian Biomedical Journal*, *21*(6), 380. doi:10.18869/acadpub.ibj.21.6.380 PMID:28555492

Taka, A. L., & Pillay, K. (2017). Nanosponge cyclodextrin polyurethanes andtheir modification with nanomaterials for the removal of pollutants from wastewater: A review. *Carbohydrate Polymers*, *159*, 94–107. doi:10.1016/j.carbpol.2016.12.027 PMID:28038758

Tan, S., & Ting, Y. (2012). Efficacy and reusability of alginate-immobilized live and heat-inactivated *Trichoderma asperellum* cells for Cu (II) removal from aqueous solution. *Bioresource Technology*, *123*, 290–295. doi:10.1016/j.biortech.2012.07.082

Tan, Y. H., Davis, J. A., Fujikawa, K., Ganesh, N. V., Demchenko, A. V., & Stine, K. J. (2012). Surface area and pore size characteristics of nanoporous gold subjected to thermal, mechanical, or surface modification studied using gas adsorption isotherms, cyclic voltammetry, thermogravimetric analysis, and scanning electron microscopy. *Journal of Materials Chemistry*, *22*(14), 6733. doi:10.1039/c2jm16633j PMID:22822294

Tasar, S., Kaya, F., & Ozer, A. (2014). Biosorption of lead (II) ions from aqueous solution by peanut shells: Equilibrium, thermodynamic and kinetic studies. *Journal of Environmental Chemical Engineering*, *2*(2), 1018–1026. doi:10.1016/j.jece.2014.03.015

Tchounwou, P. B., Yedjou, C. G., Patlolla, A. K., & Sutton, D. J. (2012). Heavy metal toxicity and the environment. *Molecular, clinical and environmental toxicology: volume 3: environmental toxicology*, 133-164.

Tchounwou, P. B., Yedjou, C. G., Patlolla, A. K., & Sutton, D. J. (2012). Heavy metal toxicity and the environment. *Molecular, Clinical and Environmental Toxicology: volume 3: environmental toxicology*, 133-164. doi:10.1007/978-3-7643-8340-4_6

Tchounwou, P. B., Ayensu, W. K., Ninashvilli, N., & Sutton, D. (2003). Environmental exposures to mercury and its toxicopathologic implications for public health. *Environmental Toxicology*, *18*(3), 149–175. doi:10.1002/tox.10116 PMID:12740802

Tchounwou, P. B., Centeno, J. A., & Patlolla, A. K. (2004). Arsenic toxicity, mutagenesis and carcinogenesis – a health risk assessment and management approach. *Molecular and Cellular Biochemistry*, *255*(1/2), 47–55. doi:10.1023/B:MCBI.0000007260.32981.b9 PMID:14971645

Tchounwou, P. B., Patlolla, A. K., & Centeno, J. A. (2003). Carcinogenic and systemic health effects associated with arsenic exposure-a critical review. *Toxicologic Pathology*, *31*(6), 575–588. doi:10.1080/714044691 PMID:14585726

Tchounwou, P. B., Wilson, B., & Ishaque, A. (1999). Important considerations in the development of public health advisories for arsenic and arsenic-containing compounds in drinking water. *Reviews on Environmental Health*, *14*(4), 211–229. doi:10.1515/REVEH.1999.14.4.211 PMID:10746734

Téllez, A. A. (2022). Prediction of the impact induced by Cd in binary interactions with other divalent metals on wild-type and Cd-resistant strains of Dictyosphaerium chlorelloides. *Environmental Science and Pollution Research International*, 1–11.

Temel, F. A., Avci, E., & Turan, N. G. (2022). Investigation of copper(II), zinc(II) and lead(II) removal onto expanded perlite by adsorption from the wastes of metal casting industry: Statistical modeling and optimization. *Environmental Engineering and Management Journal*, *21*(5), 757–767. doi:10.30638/eemj.2022.070

Teo, J., Goh, K., Ahuja, A., Ng, H., & Poon, W. (1997). Intracranial vascular calcifications, glioblastoma multiforme, and lead poisoning. *AJNR. American Journal of Neuroradiology*, *18*, 576–579. PMID:9090426

Thurmer, K., Williams, E., & Reutt-Robey, J. (2002). Autocatalytic oxidation of lead crystallite surfaces. *Science*, *297*(5589), 2033–2035. doi:10.1126/science.297.5589.2033 PMID:12242437

Tian, Y., Wu, M., Lin, X., Huang, P., & Huang, Y. (2011). Synthesis of magnetic wheat straw for arsenic adsorption. *Journal of Hazardous Materials*, *193*, 10–16. doi:10.1016/j.jhazmat.2011.04.093 PMID:21862216

Timková, I., Sedláková-Kaduková, J., & Pristaš, P. (2018). Biosorption and Bioaccumulation Abilities of Actinomycetes/Streptomycetes Isolated from Metal Contaminated Sites. *Separations*, *5*(4), 54. doi:10.3390/separations5040054

Tinega, J., Omosa, I., & Letema, S. (2023). Potential removal of chromium (VI) ions by macadamia nutshell from steel industry wastewater. *Environmental Engineering and Management Journal*, *22*(3), 411–427. doi:10.30638/eemj.2023.032

Tiquia-Arashiro, S., Li, X., Pokhrel, K., Kassem, A., Abbas, L., Coutinho, O., Kasperek, D., Najaf, H., & Opara, S. (2023). Applications of Fourier Transform-Infrared spectroscopy in microbial cell biology and environmental microbiology: Advances, challenges, and future perspectives. *Frontiers in Microbiology*, *14*, 1304081. Advance online publication. doi:10.3389/fmicb.2023.1304081

Tiwari, S., Bhardwaj, U., & Dhingra, N. (2022). Towards Further Understanding the Mechanism of Biosorption Through Various Biosorbents. *International Journal of Pharmaceutical Sciences and Research*, *13*(4), 1441–1455. doi:10.13040/IJPSR.0975-8232.13(4).1441-55

Tobin, J. M., Cooper, D. G., & Neufeld, R. J. (1984). Uptake of metal ions by Rhizopus Arrhizus Biomass. *Applied and Environmental Microbiology*, *74*(4), 821–824. doi:10.1128/aem.47.4.821-824.1984 PMID:16346521

Tomczyk, A., Sokolowska, Z., & Boguta, P. (2020). Biochar Physicochemical properties: Pyrolysis temperature and feedstock kind effects. *Reviews in Environmental Science and Biotechnology*, *19*(1), 191–215. doi:10.1007/s11157-020-09523-3

Torkia, Y. B., Bouaziz, N., Al-Muhtaseb, S. A., & Lamine, A. B. (2014). Adsorption energy and pore-size distributions of activated carbons calculated using Hill's model. *Adsorption Science and Technology*, *32*(7), 571–590. doi:10.1260/0263-6174.32.7.571

Toussaint, B., Raffael, B., Angers-Loustau, A., Gilliland, D., Kestens, V., Petrillo, M., Rio-Echevarria, I. M., & Van den Eede, G. (2019). Review of micro-and nanoplastic contamination in the food chain. *Food Additives & Contaminants. Part A, Chemistry, Analysis, Control, Exposure & Risk Assessment*, *36*(5), 639–673. doi:10.1080/19440049.2019.1583381 PMID:30985273

Tran, H. N., Nguyen, H. C., Woo, S. H., Nguyen, T. V., Vigneswaran, S., Hosseini-Bandegharaei, A., Rinklebe, J., Kumar Sarmah, A., Ivanets, A., Dotto, G. L., Bui, T. T., Juang, R. S., & Chao, H. P. (2019). Removal of various contaminants from water by renewable lignocellulose-derived biosorbents: A comprehensive and critical review. *Critical Reviews in Environmental Science and Technology*, *49*(23), 2155–2219. doi:10.1080/10643389.2019.1607442

Tripathi, M., Singh, P., Singh, R., Bala, S., Pathak, N., Singh, S., Chauhan, R. S., & Singh, P. K. (2023). Microbial biosorbent for remediation of dyes and heavy metals pollution: A green strategy for sustainable environment. *Frontiers in Microbiology*, *14*, 1168954. doi:10.3389/fmicb.2023.1168954 PMID:37077243

Tsezos, M., Remoundaki, E., & Hatzikioseyian, A. (2014). Biosorption - Principles and Applications for Metal Immobilization from waste-water streams. *Clean Production and Nano Technologies*, 23–33.

Tsezos, M., & Volesky, B. (1982). The mechanism of Uranium Biosorption by Rhizopus arrhizus. *Biotechnology and Bioengineering*, *24*(2), 385–401. doi:10.1002/bit.260240211 PMID:18546310

Turer, D. G., & Maynard, B. J. (2003). Heavy metal contamination in highway soils. Comparison of Corpus Christi, Texas and Cincinnati, Ohio shows organic matter is key to mobility. *Clean Technologies and Environmental Policy*, *4*(4), 235–245. doi:10.1007/s10098-002-0159-6

Turkdogan, M. K., Fevzi, K., Kazim, K., Ilyas, T., & Ismail, U. (2003). Heavy metals in soil, vegetables, and fruits in the endemic upper gastrointestinal cancer region of Turkey. *Environmental Toxicology and Pharmacology*, *13*(3), 175–179. doi:10.1016/S1382-6689(02)00156-4 PMID:21782652

Turner, N. H. (1975). Kinetics of chemisorption: An examination of the Elovich equation. *Journal of Catalysis*, *36*(3), 262–265. doi:10.1016/0021-9517(75)90035-4

Tzou, Y., Wang, S., Hsu, L., Chan, R., & Lin, C. (2007). Deintercalation of Li/Al LDH and its application to recover adsorbed chromate from used adsorbent. *Applied Clay Science*, *37*(1-2), 107–114. doi:10.1016/j.clay.2006.10.007

Uddin, M. M., Zakeel, M. C. M., Zavahir, J. S., Marikar, F. M., & Jahan, I. (2021). Heavy metal accumulation in rice and aquatic plants used as human food: A general review. *Toxics*, *9*(12), 360. doi:10.3390/toxics9120360 PMID:34941794

Umas, B., & Roslan, H. A. (2016). Mechanism of Biosorption of Pb (II) and Cu (II) ions using Dead Biomass of Fusarium equiseti strain UMAS and Penicillium citrinum. *Journal of Biochemistry. Microbiology*, *4*(2), 1–6.

Uriaegbu, M., Engwa, G. A., Abaa, Q. D., Aliozo, S. O., Ayuk, E. L., Osuji, G. A., & Onwurah, E. I. (2016). Heavy metal, nutrient, and antioxidant status of selected fruit samples sold in Enugu, Nigeria. *International Journal of Food Contamination*, *3*(7), 1–8. doi:10.1186/s40550-016-0031-9

United Nations. (2009). *Converting waste agricultural biomass into a resource-compendium of technologies*. UNEP.

United States Environmental Protection Agency. (2002). *Lead Compounds.* Technology Transfer Network- Air Toxics Website. Available: https://www.epa.gov/cgi-bin/epaprintonly.cgi

Urgun-Demirtas, M., Stark, B., & Pagilla, K. (2006). Use of genetically engineered microorganisms (GEMs) for the bioremediation of contaminants. *Critical Reviews in Biotechnology*, *26*(3), 145–164. doi:10.1080/07388550600842794 PMID:16923532

Uysal, O. G., Demir, A., & Gören, N. (2022). Investigation of cobalt(II) adsorption from aqueous solution using *Genista albida* as a low-cost adsorbent: Optimization based upon response surface methodology. *Environmental Engineering and Management Journal*, *21*(4), 637–650. doi:10.30638/eemj.2022.059

Vaishaly, A., Blessy, M. B., & Krishnamurthy, N. (2015). Health effects caused by metal contaminated ground water. *International Journal of Advances in Scientific Research*, *1*(2), 60–64. doi:10.7439/ijasr.v1i2.1798

Valls, M., & Lorenzo, V. (2022). Exploiting the Genetic and Biochemical Capacities of Bacteria for the Remediation of Heavy Metal Pollution. *FEMS Microbiology Reviews*, *26*, 327–338. doi:10.1016/S0168-6445(02)00114-6 PMID:12413663

van der Wal, A., Norde, W., Zehnder, A. J. B., & Lyklema, J. (1997). Determination of the total charge in the cell walls of gram-positive bacteria. *Colloids and Surfaces. B, Biointerfaces*, *9*(1-2), 81–100. doi:10.1016/S0927-7765(96)01340-9

Vasudevan, S., Jayaraj, J., Lakshmi, J., & Sozhan, G. (2009). Removal of iron from drinking water by electrocoagulation: Adsorption and kinetics studies. *Korean Journal of Chemical Engineering*, *26*(4), 1058–1064. doi:10.1007/s11814-009-0176-9

Veglio', F., & Beolchini, F. (1997). Removal of metals by biosorption: A review. *Hydrometallurgy*, *44*(3), 301–316. doi:10.1016/S0304-386X(96)00059-X

Velkova, Z., Kirova, G., Stoytcheva, M., Kostadinova, S., Todorova, K., & Gochev, V. (2018). Immobilized microbial biosorbents for heavy metal removal. *Engineering in Life Sciences*, *18*(12), 871–881. doi:10.1002/elsc.201800017

Venkata, A., Kumar, A., Darwish, N. A., & Hilal, N. A. (2009). Study of various parameters in the biosorption of heavy metals on activated sludge. *World Applied Sciences Journal*, *5*, 32–40.

Venkobachar, C. (1990). Metal Removal by Waste Biomass to Upgrade Wastewater Treatment Plants. *Water Science and Technology*, *22*(7-8), 319–320. doi:10.2166/wst.1990.0277

Verkleji, J. A. S. (1993). The effects of heavy metals stress on higher plants and their use as biomonitors. In B. Markert (Ed.), *Plants as Bioindicators: Indicators of Heavy Metals in the Terrestrial Environment* (pp. 415–424). VCH.

Verma, A., Shalu, Singh, A., Bishnoi, N. R., & Gupta, A. (2013). Biosorption of Cu(II) using free and immobilized biomass of *Penicillium citrinum. Ecological Engineering*, *61*, 486–49. doi:10.1016/j.ecoleng.2013.10.008

Veza, I., Spraggon, M., Fattah, I. M. R., & Idris, M. (2023). Response surface methodology (RSM) for optimizing engine performance and emissions fueled with biofuel: Review of RSM for sustainability energy transition. *Results in Engineering*, *18*, 101213. doi:10.1016/j.rineng.2023.101213

Vijayakumar, S., Sasikala, M., & Ramesh, R. (2012). Lead poisoning—An overview. *International Journal of Phramcology and Toxicology*, *2*, 70–82.

Vijayaraghavan, K., & Balasubramanian, R. (2015). Is biosorption suitable for decontamination of metal-bearing wastewaters? A critical review on the state-of-the-art of biosorption processes and future directions. *Journal of Environmental Management*, *160*, 283–296. doi:10.1016/j.jenvman.2015.06.030 PMID:26143501

Vijayaraghavan, K., & Yun, Y. S. (2007). Utilization of fermentation waste (Corynebacterium glutamicum) for biosorption of Reactive Black 5 from aqueous solution. *Journal of Hazardous Materials, 141*(1), 45–52. doi:10.1016/j.jhazmat.2006.06.081 PMID:16879915

Vijayaraghavan, K., & Yun, Y. S. (2008). Bacterial biosorbents and biosorption. *Biotechnology Advances, 26*(3), 266–291. doi:10.1016/j.biotechadv.2008.02.002 PMID:18353595

Villarín, M. C., & Merel, S. (2020). Paradigm shifts and current challenges in wastewater management. *Journal of Hazardous Materials, 390*, 122139. doi:10.1016/j.jhazmat.2020.122139 PMID:32007860

Viraraghavan, T., & Srinivasan, A. (2011). Fungal Biosorption and biosorbents. In Microbial Biosorption of Metals. Springer. doi:10.1007/978-94-007-0443-5_6

Vishan, I., Saha, B., Sivaprakasam, S., & Kalamdhad, A. (2019). Evaluation of Cd(II) biosorption in aqueous solution by using lyophilized biomass of novel bacterial strain *Bacillus badius* AK: Biosorption kinetics, thermodynamics and mechanism. *Environmental Technology & Innovation, 14*, 100323. doi:10.1016/j.eti.2019.100323

Volesky, B. (2007). Biosorption and Me. *Water Research, 41*(18), 4017–4029. doi:10.1016/j.watres.2007.05.062 PMID:17632204

Volesky, B., & Holant, Z. R. (1995). Biosorption of Heavy Metals. *Biotechnology Progress, 11*(3), 235–250. doi:10.1021/bp00033a001 PMID:7619394

Volesky, B., & Tsezos, M. (1981). Biosorption of uranium and thorium. *Biotechnology and Bioengineering, 23*(3), 583–604. doi:10.1002/bit.260230309

Vuković, G. D., Marinković, A. D., Čolić, M., Ristić, M. D., Aleksić, R., Perić-Grujić, A. A., & Uskoković, P. S. (2010). Removal of cadmium from aqueous solutions by oxidized and ethylenediamine-functionalized multi-walled carbon nanotubes. *Chemical Engineering Journal, 157*(1), 238–248. doi:10.1016/j.cej.2009.11.026

Waalkes, M. P., & Rehm, S. (1992). Carcinogenicity of oral cadmium in the male Wistar (WF/NCr) rat: Effect of chronic dietary zinc deficiency*1. *Fundamental and Applied Toxicology, 19*(4), 512–520. doi:10.1016/0272-0590(92)90089-Z PMID:1426709

Walker, C. H., Sibly, R. M., Hopkin, S. P., & D.B.P., T. (2012). *Principles of Ecotoxicology* (4th ed.). CRC Press.

Wang, Y., Pan, C., Chu, W., Vipin, A., Sun, L. (2019). Environmental remediation applications of carbon nanotubes and graphene oxide: Adsorption and catalysis. *Nanomaterials, 9*(439).

Wang, B., Lan, J., Bo, C., Gong, B., & Ou, J. (2023). Adsorption of heavy metal onto biomass-derived activated carbon [review]. *RSC Advances, 13*(7), 4275–4302. doi:10.1039/D2RA07911A PMID:36760304

Wang, C., Wang, Z., Wang, X., Li, N., Tao, J., Zheng, W., Yan, B., Cui, X., Cheng, Z., & Chen, G. (2022). A Review on the Hydrothermal Treatment of Food Waste: Processing and Applications. *Processes (Basel, Switzerland), 10*(11), 2439. doi:10.3390/pr10112439

Wang, D., Saleh, N. B., Byro, A., Zepp, R., Sahle-Demessie, E., Luxton, T. P., Ho, K. T., Burgess, R. M., Flury, M., White, J. C., & Su, C. (2022). Nano-enabled pesticides for sustainable agriculture and global food security. *Nature Nanotechnology, 17*(4), 347–360. doi:10.1038/s41565-022-01082-8 PMID:35332293

Wang, G., Zhang, S., Yao, P., Chen, Y., Xu, X., Li, T., & Gong, G. (2018). Removal of Pb(II) from aqueous solutions by *Phytolacca americana L.* biomass as a low cost biosorbent. *Arabian Journal of Chemistry, 11*(1), 99–110. doi:10.1016/j.arabjc.2015.06.011

Wang, H., Mi, X., Li, Y., & Zhan, S. (2020). 3D graphene-based macrostructures for water treatment. *Advanced Materials, 32*(3), 1806843. doi:10.1002/adma.201806843 PMID:31074916

Wang, H., Yuan, X., Wu, Y., Huang, H., Zeng, G., Liu, Y., Wang, X., Lin, N., & Qi, Y. (2013). Adsorption characteristics and behaviours of graphene oxide for Zn (II) removal from aqueous solution. *Applied Surface Science, 279*, 432–440. doi:10.1016/j.apsusc.2013.04.133

Wang, J., & Chen, C. (2006). Biosorption of heavy metals by Saccharomyces cerevisiae: A review. *Biotechnology Advances, 24*(5), 427–445. doi:10.1016/j.biotechadv.2006.03.001 PMID:16737792

Wang, J., & Chen, C. (2009). Biosorbents for heavy metals removal and their future. *Biotechnology Advances, 27*(2), 195–226. doi:10.1016/j.biotechadv.2008.11.002 PMID:19103274

Wang, J., & Guo, X. (2020a). Adsorption kinetic models: Physical meanings, applications, and solving methods. *Journal of Hazardous Materials, 390*, 122156. doi:10.1016/j.jhazmat.2020.122156 PMID:32006847

Wang, J., & Guo, X. (2020b). Adsorption isotherm models: Classification, physical meaning, application and solving method. *Chemosphere, 258*, 127279. doi:10.1016/j.chemosphere.2020.127279 PMID:32947678

Wang, X. F., Xing, M. L., Shen, Y., Zhu, X., & Xu, L. H. (2006). Oral administration of Cr (VI) induced oxidative stress, DNA damage and apoptotic cell death in mice. *Toxicology, 228*(1), 16–23. doi:10.1016/j.tox.2006.08.005 PMID:16979809

Wang, X., Daigger, G., de Vries, W., Kroeze, C., Yang, M., Ren, N. Q., Liu, J., & Butler, D. (2019). Impact hotspots of reduced nutrient discharge shift across the globe with population and dietary changes. *Nature Communications, 10*(1), 2627. doi:10.1038/s41467-019-10445-0 PMID:31201305

Wani, A. L., Ara, A., & Usmani, J. A. (2015). Lead toxicity: A review. *Interdisciplinary Toxicology, 8*(2), 55–64. doi:10.1515/intox-2015-0009 PMID:27486361

Waqas, M., Li, G., Khan, S., Shamshad, I., Reid, B. J., Qamar, Z., & Chao, C. (2015). Application of sewage sludge and sewage sludge biochar to reduce polycyclic aromatic hydrocarbons (PAH) and potentially toxic elements (PTE) accumulation in tomato. *Environmental Science and Pollution Research International, 22*(16), 12114–12123. doi:10.1007/s11356-015-4432-8 PMID:25877899

Wase, J., & Forster, C. (1997). *Biosorbents for Metal Ions*. Taylor & Francis.

Wei, W., Wang, Q., Li, A., Yang, J., Ma, F., Pi, S., & Wu, D. (2016). Biosorption of Pb (II) from aqueous solution by extracellular polymeric substances extracted from *Klebsiella sp.* J1: Adsorption behaviour and mechanism assessment. *Scientific Reports, 6*(1), 31575. doi:10.1038/srep31575

Wierzba, S. (2015). Biosorption of lead(II), zinc(II) and nickel(II) from industrial wastewater by *Stenotrophomonas maltophilia* and *Bacillus subtilis*. *Polish Journal of Chemical Technology, 17*(1), 79–87. doi:10.1515/pjct-2015-0012

Wierzba, S., Makuchowska-Fryc, J., Kłos, A., Ziembik, Z., & Ochędzan-Siodłak, W. (2022). Role of calcium carbonate in the process of heavy metal biosorption from solutions: Synergy of metal removal mechanisms. *Scientific Reports, 12*(1), 17668. doi:10.1038/s41598-022-22603-4 PMID:36271239

Witek-Krowiak, A., & Harikishore Kumar Reddy, D. (2013). Removal of microelemental Cr(III) and Cu(II) by using soybean meal waste - Unusual isotherms and insights of binding mechanisms. *Bioresource Technology, 127*, 350–357. doi:10.1016/j.biortech.2012.09.072 PMID:23138058

Witek-Krowiak, A., Szafran, R. G., & Modelski, S. (2011). Biosorption of heavy metals from aqueous solutions onto peanut shell as a low-cost biosorbent. *Desalination, 265*(1–3), 126–134. doi:10.1016/j.desal.2010.07.042

Witkowska, D., Słowik, J., & Chilicka, K. (2021). Heavy metals and human health: Possible exposure pathways and the competition for protein binding sites. *Molecules (Basel, Switzerland)*, *26*(19), 6060. doi:10.3390/molecules26196060 PMID:34641604

Woldetsadik, D., Drechsel, P., Keraita, B., Itanna, F., & Gebrekidan, H. (2017). Heavy metal accumulation and health risk assessment in wastewater-irrigated urban vegetable farming sites of Addis Ababa, Ethiopia. *International Journal of Food Contamination*, *4*(1), 9. doi:10.1186/s40550-017-0053-y

Wood, T. K. (2008). Molecular approaches in bioremediation. *Current Opinion in Biotechnology*, *19*(6), 572–578. doi:10.1016/j.copbio.2008.10.003 PMID:19000765

World Health Organization. (2000). *Air quality guidelines for Europe* (2nd ed.). World Health Organization. https://www.who.int/en/newsroom/fact-sheets/detail/lead-poisoning-and-health

Wu, C., Li, F., Yi, S., & Ge, F. (2021). Genetically engineered microbial remediation of soils co-contaminated by heavy metals and polycyclic aromatic hydrocarbons: Advances and ecological risk assessment. *Journal of Environmental Management*, *296*, 113185. doi:10.1016/j.jenvman.2021.113185

Wu, Q., Zhou, H., Tam, N. F., Tian, Y., Tan, Y., Zhou, S., Li, Q., Chen, Y., & Leung, J. Y. (2016). Contamination, toxicity and speciation of heavy metals in an industrialized urban river: Implications for the dispersal of heavy metals. *Marine Pollution Bulletin*, *104*(1-2), 153–161. doi:10.1016/j.marpolbul.2016.01.043 PMID:26856647

Wu, Y., Wen, Y., Zhou, J., Cao, J., Jin, Y., & Wu, Y. (2013). Comparative and competitive adsorption of Cr(VI), As(III), and Ni(II) onto coconut charcoal. *Environmental Science and Pollution Research International*, *20*(4), 2210–2219. doi:10.1007/s11356-012-1066-y PMID:22791350

Wyroba, E., Suski, S., Miller, K., & Bartosiewicz, R. (2015). Biomedical and agricultural applications of energy dispersive X-ray spectroscopy in electron microscopy. *Cellular & Molecular Biology Letters*, *20*(3). Advance online publication. doi:10.1515/cmble-2015-0028 PMID:26208393

Xie, S., Tran, H.-T., Pu, M., & Zhang, T. (2023). Transformation characteristics of organic matter and phosphorus in composting processes of agricultural organic waste: Research trends. *Materials Science for Energy Technologies*, *6*, 331–342. doi:10.1016/j.mset.2023.02.006

Xin, Z., Wenchao, Z., Zhenguang, Y., Yiguo, H., Zhengtao, L., Xianliang, Y., Xiaonan, W., Tingting, L., & Liming, Z. (2015). Species sensitivity analysis of heavy metals to freshwater organisms. *Ecotoxicology (London, England)*, *24*(7–8), 1621–1631. doi:10.1007/s10646-015-1500-2 PMID:26104218

Xiong, T., Dumat, C., Dappe, V., Vezin, H., Schreck, E., Shahid, M., Pierart, A., & Sobanska, S. (2017). Copper oxide nanoparticle foliar uptake, phytotoxicity, and consequences for sustainable urban agriculture. *Environmental Science & Technology*, *51*(9), 5242–5251. doi:10.1021/acs.est.6b05546 PMID:28383257

Xu, J., Song, C., Zhang, Q., Pan, H., Liang, Y., Fan, X., & Li, Y. (2011). Characterization of metal removal of immobilized *Bacillus strain CR-7* biomass from aqueous solutions. *Journal of Hazardous Materials*, *187*(1-3), 450–458. doi:10.1016/j.jhazmat.2011.01.047

Xu, L., Lu, A., Wang, J., Ma, Z., Pan, L., Feng, X., & Luan, Y. (2015). Accumulation status, sources and phytoavailability of metals in greenhouse vegetable production systems in Beijing, China. *Ecotoxicology and Environmental Safety*, *122*, 214–220. doi:10.1016/j.ecoenv.2015.07.025 PMID:26264401

Xu, R., Li, Q., Yang, Y., Jin, S., Liao, L., Wu, Z., Yin, Z., Xu, B., Nan, X., He, Y., Zhu, B., & Jiang, T. (2022). Removal of heavy metal(loid)s from aqueous solution by biogenic FeS–kaolin composite: Behaviors and mechanisms. *Chemosphere*, *299*, 134382. doi:10.1016/j.chemosphere.2022.134382

Yaashikaa, P. R., Kumar, P. S., Varjani, S., & Saravanan, A. (2020). A critical review on the biochar production techniques, characterization, stability and applications for circular bioeconomy. *Biotechnology Reports (Amsterdam, Netherlands), 28*, e00570. doi:10.1016/j.btre.2020.e00570

Yan, G., & Viraraghavan, T. (2008). Mechanism of Biosorption of Heavy Metals by *Mucor rouxii. Engineering in Life Sciences, 4*(4), 363–371. doi:10.1002/elsc.200820237

Yang, C., Liu, J., & Lu, S. (2021). Pyrolysis temperature affects pore characteristics of rice straw and canola stalk biochars and bio-char amended. *Geoderma, 397*, 115097. doi:10.1016/j.geoderma.2021.115097

Yang, I., Kwon, D., Kim, M.-S., & Jung, J. C. (2018). A Comparative Study of Activated Carbon Aerogel and Commercial Activated Carbons as Electrode Materials for Organic Electric Double-Layer Capacitors. *Carbon, 132*, 503–511. doi:10.1016/j.carbon.2018.02.076

Yang, L., & Chen, J. P. (2008). Biosorption of hexavalent chromium onto raw and chemically modified Sargassum sp. *Bioresource Technology, 99*(2), 297–307. doi:10.1016/j.biortech.2006.12.021 PMID:17336517

Yang, X., Wan, Y., Zheng, Y., He, F., Yu, Z., Huang, J., Wang, H., Ok, Y. S., Jiang, Y., & Gao, B. (2019). Surface functional groups of carbon-based adsorbents and their roles in the removal of heavy metals from aqueous solutions: A critical review. *Chemical Engineering Journal, 366*, 608–621. doi:10.1016/j.cej.2019.02.119 PMID:34522159

Yelatontsev, D. (2023). Production of versatile biosorbent via eco-friendly utilization of non-wood biomass. *Chemical Engineering Journal, 451*, 138811. doi:10.1016/j.cej.2022.138811

Yilmaz, A. B., Mustafa, K. S., Deniz, Y., & Cemal, T. (2010). Metals (major, essential to non-essential) composition of different tissues of three demersal fish species from Iskenderun Bay. *Food Chemistry, 123*(2), 410–415. doi:10.1016/j.foodchem.2010.04.057

Yin, H., He, B., Peng, H., Ye, J., Yang, F., & Zhang, N. (2008). Removal of Cr (IV) and Ni (II) from aqueous solution by fused yeast: Study of cations release and biosorption mechanism. *Journal of Hazardous Materials, 158*(2-3), 568–576. doi:10.1016/j.jhazmat.2008.01.113 PMID:18346847

Yin, Q., Zhang, B., Wang, R., & Zhao, Z. (2017). Biochar as an adsorbent for inorganic nitrogen and phosphorus removal from water: A review. *Environmental Science and Pollution Research International, 24*(34), 26297–26309. doi:10.1007/s11356-017-0338-y PMID:29039039

Yoshizawa, K., Rimm, E. B., Morris, J. S., Spate, V. L., Hsieh, C. C., Spiegelman, D., Stampfer, M. J., & Willett, W. C. (2002). Mercury and the risk of coronary heart disease in men. *The New England Journal of Medicine, 347*(22), 1755–1760. doi:10.1056/NEJMoa021437 PMID:12456851

Young, T. R., & Xiao, Z. (2021). Principles and practice of determining metal–protein affinities. *The Biochemical Journal, 478*(5), 1085–1116. doi:10.1042/BCJ20200838 PMID:33710331

Yuan, J. H., Xu, R. K., & Zhang, H. (2011). The forms of alkalis in the biochar produced from crop residues at different temperatures. *Bioresource Technology, 102*(3), 3488–3497. doi:10.1016/j.biortech.2010.11.018 PMID:21112777

Yuan, W., Cheng, J., Huang, H., Xiong, S., Gao, J., Zhang, J., & Feng, S. (2019). Optimization of cadmium biosorption by *Shewanella putrefaciens* using a Box-Behnken design. *Ecotoxicology and Environmental Safety, 175*, 138–147. doi:10.1016/j.ecoenv.2019.03.057 PMID:30897412

Yu, L., Liou, I. W., Biggins, S. W., Yeh, M., Jalikis, F., Chan, L.-N., & Burkhead, J. (2019). Copper deficiency in liver diseases: A case series and pathophysiological considerations. *Hepatology Communications, 3*(8), 1159–1165. doi:10.1002/hep4.1393 PMID:31388635

Yu, O. Y., Raichle, B., & Sink, S. (2013). Impact of biochar on the water holding capacity of loamy sand soil. *International Journal of Energy and Environmental Engineering, 4*(1), 1–9. doi:10.1186/2251-6832-4-44

Yusoff, S. N. M., Kamari, A., Putra, W. P., Ishak, C. F., Mohamed, A., Hashim, N., & Isa, I. M. (2014). Removal of Cu(II), Pb(II) and Zn(II) Ions from Aqueous Solutions Using Selected Agricultural Wastes: Adsorption and Characterisation Studies. *Journal of Environmental Protection, 05*(04), 289–300. doi:10.4236/jep.2014.54032

Yusuf, A., Sodiq, A., Giwa, A., Eke, J., Pikuda, O., Eniola, J. O., Ajiwokewu, B., Sambudi, N. S., & Bilad, M. R. (2022). Updated review on microplastics in water, their occurrence, detection, measurement, environmental pollution, and the need for regulatory standards. *Environmental Pollution, 292*, 118421. doi:10.1016/j.envpol.2021.118421 PMID:34756874

Zahir, A., Rizwi, S. J., Haq, S. K., & Khan, R. H. (2005). Low dose mercury toxicity and human health. *Environmental Toxicology and Pharmacology, 20*(2), 351–360. doi:10.1016/j.etap.2005.03.007 PMID:21783611

Zeng, X., Xu, X., Boezen, H. M., & Huo, X. (2016). Children with health impairments by heavy metals in an e-waste recycling area. *Chemosphere, 148*, 408–415. doi:10.1016/j.chemosphere.2015.10.078 PMID:26829309

Zhang, G., Bai, J., Xiao, R., Zhao, Q., Jia, J., Cui, B., & Liu, X. (2017). Heavy metal fractions and ecological risk assessment in sediments from urban, rural and reclamation-affected rivers of the Pearl River Estuary, China. *Chemosphere, 184*, 278–288. doi:10.1016/j.chemosphere.2017.05.155 PMID:28601010

Zhang, H., Zhao, Y., Wang, Z., & Liu, Y. (2021). Distribution characteristics, bioaccumulation and trophic transfer of heavy metals in the food web of grassland ecosystems. *Chemosphere, 278*, 130407. doi:10.1016/j.chemosphere.2021.130407 PMID:33823346

Zhang, P., Yang, M., Lan, J., Huang, Y., Zhang, J., Huang, S., Yang, Y., & Ru, J. (2023). Water Quality Degradation Due to Heavy Metal Contamination: Health Impacts and Eco-Friendly Approaches for Heavy Metal Remediation. *Toxics, 11*(10), 828. doi:10.3390/toxics11100828 PMID:37888679

Zhang, T., Yang, Y., Li, X., Zhou, Z., & Wei, B. (2023). Ultrasonic–Thermal Regeneration of Spent Powdered Activated Carbon. *Sustainability (Basel), 15*(11), 9060. Advance online publication. doi:10.3390/su15119060

Zhang, X., Qin, Q., Sun, X., & Wang, W. (2022). Hydrothermal treatment: An efficient food waste disposal technology. *Frontiers in Nutrition, 9*, 986705. Advance online publication. doi:10.3389/fnut.2022.986705 PMID:36172524

Zhao, M., Xu, Y., Zhang, C., Rong, H., & Zeng, G. (2016). New trends in removing heavy metals from wastewater. *Applied Microbiology and Biotechnology, 100*(15), 6509–6518. doi:10.1007/s00253-016-7646-x PMID:27318819

Zhao, R., Jiang, D., Coles, N., & Wu, J. (2015). Effects of biochar on the acidity of a loamy clay soil under different incubation conditions. *Journal of Soils and Sediments, 15*(9), 1919–1926. doi:10.1007/s11368-015-1143-1

Zheng, N., Wang, Q. C., Liang, Z. Z., & Zheng, D. M. (2008). Characterization of heavy metal concentrations in the sediments of three freshwater rivers in Huludao City, Northeast China. *Environmental Pollution, 154*(1), 135–142. doi:10.1016/j.envpol.2008.01.001 PMID:18280624

Zhitkovich, A. (2011). Chromium in drinking water: Sources, metabolism, and cancer risks. *Chemical Research in Toxicology, 24*(10), 1617–1629. doi:10.1021/tx200251t PMID:21766833

Zhou, K., Yang, Z., Liu, Y., & Kong, X. (2015). Kinetics and equilibrium studies on biosorption of Pb(II) from aqueous solution by a novel biosorbent: *Cyclosorus interruptus. Journal of Environmental Chemical Engineering, 3*(3), 2219–2228. doi:10.1016/j.jece.2015.08.002

Zhou, Y., & Haynes, R. J. (2011). A comparison of inorganic solid wastes as adsorbents of heavy metal cations in aqueous solution and their capacity for desorption and regeneration. *Water, Air, and Soil Pollution*, *218*(1-4), 457–470. doi:10.1007/s11270-010-0659-7

Zhou, Y., & Haynes, R. J. (2012). A Comparison of Water Treatment Sludge and Red Mud as Adsorbents of As and Se in Aqueous Solution and Their Capacity for Desorption and Regeneration. *Water, Air, and Soil Pollution*, *223*(9), 5563–5573. doi:10.1007/s11270-012-1296-0

Zhu, N., Zhang, B., & Yu, Q. (2020). Genetic engineering-facilitated co-assembly of synthetic bacterial cells and magnetic nanoparticles for efficient heavy metal removal. *ACS Applied Materials & Interfaces*, *12*(20), 22948–22957. doi:10.1021/acsami.0c04512

Zinicovscaia, I., Yushin, N., Grozdov, D., Vergel, K., Popova, N., Artemiev, G., & Safonov, A. (2020). Metal Removal from Nickel-Containing Effluents Using Mineral–Organic Hybrid Adsorbent. *Materials (Basel)*, *13*(19), 4462. doi:10.3390/ma13194462 PMID:33050087

Zohrehvandi, S., & Bashiri, M. (2014). Evaluation of full experimental design method against fractional design method and Taguchi design method in machining operation. *Journal of Control Science and Engineering*, *2*, 43–51. doi:10.17265/2328-2231/2014.01.006

Zouboulis, A. I., Lazaridis, N. K., & Matis, K. A. (2002). Removal of toxic metal ions from aqueous systems by biosorptive flotation. *Journal of Chemical Technology and Biotechnology*, *77*(8), 958–964. doi:10.1002/jctb.663

Zyoud, A., Alkowni, R., Yousef, O., Salman, M., Hamdan, S., Helal, M., Jaber, S. F., & Hilal, H. S. (2019). Solar light-driven complete mineralization of aqueous gram-positive and gram-negative bacteria with ZnO photocatalyst. *Solar Energy*, *180*, 351–359. doi:10.1016/j.solener.2019.01.034

About the Contributors

Pinki Saini is presently working as Associate Professor at Centre of Food Technology, University of Allahabad. She completed degree of MSc and Ph.D. in Food Science and Technology from the G. B. Pant University of Agriculture and Technology, Pantnagar in 2005 and 2008 respectively. She has 15 years of teaching and 18 years of research experience. Published more than 75 research papers in journals and conferences. Completed 2 sponsored research projects, granted one patent and authored two books and more than 18 book chapters. Her research interests include bioprocessing of foods, fermentations studies, biosorption of heavy metals, biofilm formation and sustainable edible films from plant and microbial sources.

* * *

Lakshmi Kanthan Bharathi A. is a Research Scholar in the Department of Mechanical Engineering College of Engineering Guindy campus Anna University. He finished his B.E., in Mechanical Engineering and M.E., in Thermal Engineering with specialization in Refrigeration and Air-conditioning from Anna University Chennai. His research areas and interests include Desiccant Dehumidification, Heat transfer, Refrigeration and Air-conditioning, Phase change materials and Nano materials.

Hari Shankar Biswas received his B.Sc. degree in Chemistry Honours (2005) from R.K.M.V.C. College and his M.Sc. (2008) degree from Presidency College, Kolkata. He was awarded a Ph.D. (2015) in Material Science from the Saha Institute of Nuclear Physics, University of Calcutta. From 2008 to 2010 he was a School teacher at Keshub Academy, Kolkata. 2010 to now he is serving as an assistant professor in Chemistry at Surendranath College, Kolkata, India. His current research topic is Carbon thin Film, Graphene, Graphene Oxide-based nano Composite, Synthesis Characterization, and Application in the Environmental, Biological, and Technological fields. He is the author of 12 book chapters and published more than 26 papers in International Journals. He is a life member of the Indian Science Congress Association and the Indian Chemical Society. He is an Editorial board member of the American Journal of Chemical Engineering and a member of Rotary International..

Petronela Cozma is Lecturer within the Department of Environmental Engineering and Management, "Cristofor Simionescu" Faculty of Chemical Engineering and Environmental Protection - Gheorghe Asachi Technical University of Iasi, Romania. Research areas: environmental biotechnology (heavy metals biosorption, bioaccumulation, phytoremediation, phytomining), environmental risk, sustainable industrial production, waste treatment; Authorship: 7 books and chapters, 46 papers in WoS journals+proceedings,

19 papers in other journals included in international databases (others than WoS), over 150 presentations at scientific events, 2 national patents; Memberships in organizations and professional groups: Academic Organization of Environmental Engineering and Sustainable Development; Environmental Engineering and Impact Assessment Research Center; FEMS (Federation of European Microbiological Societies); EFB (European Federation of Biotechnology); Member of Editorial Production and Secretariat of: Environmental Engineering and Management Journal (JIF/2022=1.1); Bachelor/Master Thesis Supervisor: Environmental Engineering; Member in research grants: national grants 7; international 2; project director: 1 (Mobility projects for experienced researchers from the diaspora); Visibility: Over 309 citations (WoS), Hirsch index H=11 (WoS) (Scopus: over 355 citations, H=11; Google Academic: over 771 citations, H=16); Prizes: 14 international prizes; 7 awarding of research results; Web sites: , , , .

Megersa O. Dinka is a graduate with PhD from University of Natural Resources and Applied Life Science (Vienna) in 2010. He has MSc in Irrigation Engineering from ArbaMinch University (Ethiopia). He has about 18 years of experience as academician and 22 years of experience as a researcher. Currently, he is the head of Department of Civil Engineering Science at University of Johannesburg. He is specialized in the areas of water resource engineering specific to hydrology, hydraulics and water management aspects. He has taught various courses and modules at undergraduate and postgraduate levels successfully. Moreover, he also supervised a number of postgraduate students (31 M and 9 D) successfully. He has published more than 65 Journal articles, 1 book, 8 book chapters and 20 proceedings in international accredited publications.

Rakesh Dutta obtained his Ph.D. degree from Gauhati University, Guwahati, India in 2020. Presently he is working as an Assistant Professor in the Department of Chemistry in the Pandit Deendayal Upadhyaya Adarsha Mahavidyalaya, Behali, Assam. He has published 8 research articles in reputed journals. His research interests are Theoretical investigation on opto-electronic properties of organic molecules using Gaussian program package with the help of DFT and TD-DFT methods.

Bienvenu Gael Fouda Mbanga currently serving as a postdoctoral fellow within the Chemistry Department at Nelson Mandela University in South Africa, specializes in environmental analytical chemistry. His research focuses on forensic and wastewater treatment, particularly emphasizing the development of nanomaterials tailored for heavy metal removal and their subsequent application in latent fingerprint technologies.

Maria Gavrilescu is Full Professor within the Department of Environmental Engineering and Management, "Cristofor Simionescu" Faculty of Chemical Engineering and Environmental Protection - Gheorghe Asachi Technical University of Iasi, Romania. Research areas: chemical/biological process applied in environmental engineering and biotechnology, environmental biotechnology (heavy metals biosorption, bioaccumulation, phytoremediation, phytomining), environmental risk, sustainable industrial production and circular (bio)economy Authorship: 50 books and chapters, 11 student books, 224 papers in WoS journals+proceedings, 110 papers in other journals included in international databases (others than WoS), 39 papers in Conference Proceedings, 58 conferences/keynotes, over 347 presentations at scientific events, 51 research reports, 12 national patents Memberships: Academy of Romanian Scientists; various Scientific Societies; European Federation of Biotechnology – Environmental Biotechnology Division Editor-in-Chief Environmental Engineering and Management Journal (JIF/2022=1.1) PhD Supervisor:

Chemical Engineering, Environmental Engineering Project manager/member research grants 90: with industrial companies 21; national grants 49; international 9; original technologies 10 Over 6700 citations (WoS), Hirsch index H=40 (WoS) (Scopus: over 8100 citations, H=41; Google Academic: over 12700 citations, H=53) Prizes: Diplomas for Excellence in Research (Senate of the Technical University of Iasi); Best Paper Award, Springer; Prize awarded by The Romanian Academy. Included in the Ranking of the first 2% of the world's scientists chosen from 7 million people in 22 fields of activity (published by Stanford University) in the first quarter of the ranking (evaluated period: 1992-2022, 210 papers, Environmental Engineering) .

Bongibethu Msekeli Hlabano-Moyo is a research scientist and environmental/water resources lab manager at Iowa State University, USA. He has vast research expertise in water quality characterization through chromatographic and spectrometry techniques coupled with expertise in separation technologies inclusive of membrane processes and electro-deionization processes using adsorption media. Additionally, he has in the past been involved in process modelling and is currently leveraging that through data analytics.

Percy Hlangothi is an Associate Professor and Director of the Centre for Rubber Science and Technology at Nelson Mandela University, South Africa. His expertise is in materials chemistry, solid state physics, chemical kinetics and thermodynamics.

Yvonne Boitumelo Nthwane is a researcher at Tshwane University of Technology and University of Johannesburg in South Africa. Her research focus is on organometallic, pharmacology and medicinal chemistry.

Joan Nyika is a lecturer at the Technical University of Kenya and a postdoctoral research fellow at the University of Johannesburg. She holds a doctorate degree in Science, Engineering and Technology from the University of South Africa and MSc. in Land and Water Management from the University of Nairobi, Kenya. In her research work, she specializes in hydrobiogeochemistry and has published more than 40 journal articles and book chapters in the discipline. She is also involved in teaching, supervising and mentoring undergraduate and postgraduate students.

Bothwell Nyoni is a lecturer at the National University of Science and Technology in Zimbabwe and recently finished his PhD studies with Nelson Mandela University, South Africa. His research focus is on renewable energy and water treatment technologies.

Sandeep Poddar, Deputy Vice Chancellor (Research & Innovation), Member of Board of Studies, Lincoln University College, Malaysia, has been graduated from University of Calcutta in 1993 with Honours in Zoology, he has obtained Post Graduate Diploma in Dietetics from All India Institute of Hygiene and Public Health 1995, Master of Science in Zoology with specialization in Biochemical Genetics from Dayalbagh Educational Institute 1998 with distinction. He has completed PhD in Zoology from Vivekananda Institute of Medical Sciences on Cytotoxicity in 2004. After completing PhD he pursued Post Doctoral Research in different projects on Hemoglobinopathies and Oral Cancer mutation. He has published several research papers, filed patent and organized international conferences, and also edited books in Malaysia, Australia and India. Dr Sandeep is Fellow member of Indian Academic Research

Association, and World Researchers Associations and Life members of Different associations including Indian Science Congress Association. Dr Sandeep is founder Assistant Secretary of Dr Tarak Nath Podder Memorial Foundation, Kolkata, India.

Mihaela Rosca is a research assistant within the "Ion Ionescu de la Brad" Iasi University of Life Sciences, Faculty of Horticulture, Department of Horticultural Technology, Iasi, Romania and has the PhD Thesis in Environmental Engineering (2021), MSc in Environmental Management (2009) and BSc in Environmental Engineering in Industry (2012), all from "Gheorghe Asachi" Technical University of Iasi. Research areas: biosorption and bioaccumulation by microorganisms of persistent pollutants from contaminated environments, process modeling and optimization; assessing the impact of abiotic factors on plant growth and removal of environmental persistent pollutants using plants (phytoremediation); human health impacts and risks of pesticides in fruits and vegetables. Academic Editor: Journal of Applied Life Sciences and Environment (ALSE) indexed within EBSCO, FSTA, J-GATE, CABI, Dimensions and OpenAIRE. Authorship: 59 published papers (22 papers in ISI journals from which 5 papers are in Q1 Journals and 3 papers are in Q2 Journals after AIS, while the rest are in Q3 or Q4 Journals; 12 papers are published in journals indexex in international databases, 25 papers are included in ISI Conference Proceedings); 5 monographs published at national publishing houses, 2 book chapter published in Springer Science, 1 book chapter published at a national publishing house; over 120 presentations at international and national conferences with oral/poster communications. Member of 5 national research projects as research assistant. Over 390 citations (WoS), Hirsch index: H=11 (WoS), (Scopus: over 500 citations, H=12; Google Academic: over 685 citations, H=15) Prizes: Best paper award, Best paper of a young researcher .

Jitu Saikia obtained his PhD degree from CSIR-NEIST, Jorhat, India in 2019. Presently he is working as an Assistant Professor in the department of Chemistry in the Pandit Deendayal Upadhyaya Adarsha Mahavidyalaya, Behali, Assam. He has published 14 research articles in reputed journals. His research interests are Nanochemistry, Cementitious materials, Green Chemistry Technology, Porous materials, Surface Adsorption, Membrane Science and Waste Management.

Sudha Subburaj is a Research Scholar in the Department of Mechanical Engineering College of Engineering Guindy campus Anna University. She finished her B.Tech., in Food Technology and M.Tech., in Industrial Safety and hazards management from Anna University Chennai. Her research areas and interests include Risk Analysis, Process Safety Management, Resilience Analysis of Integrated Safety Systems, Fault Analysis and Recovery Study, Machine Learning Techniques.

Zikhona Tywabi-Ngeva is a lecturer at Nelson Mandela University in South Africa, whose research focuses on ultrafast laser-induced fabrication of nanomaterials and nanofluids for solar energy applications, isolation and characterization of nanocelluloses pineapple leaf fibres and their application in agricultural water-saving materials.

Bongani Yalala is a senior lecturer in the Department of Applied Chemistry at the National University of Science and Technology in Zimbabwe. His expertise is in environmental chemistry, chemical kinetics and thermodynamics with special focus on elemental speciation in polluted environments (water and soil systems); water quality assessment; remediation and mitigation aspects using low-cost adsorbents.

Index

C

D

E

F

G

H

I

K

M

O

P

R

S

T

W

www.ingramcontent.com/pod-product-compliance
Lightning Source LLC
LaVergne TN
LVHW062120300926
849161LV00010B/2499